Advanced Studies in the Mathematical Theory of Scattering

Dedicated to my wife

Waves and Scattering Set

coordinated by
Jean-Michel L. Bernard

Volume 3

Advanced Studies in the Mathematical Theory of Scattering

Jean-Michel L. Bernard

iSTE

WILEY

First published 2024 in Great Britain and the United States by ISTE Ltd and John Wiley & Sons, Inc.

ISTE Ltd
27-37 St George's Road
London SW19 4EU
UK

www.iste.co.uk

John Wiley & Sons, Inc.
111 River Street
Hoboken, NJ 07030
USA

www.wiley.com

Library of Congress Control Number: 2024940000

British Library Cataloguing-in-Publication Data
A CIP record for this book is available from the British Library
ISBN 978-1-78630-857-3

Contents

Introduction

This book is a collection of independent mathematical studies, describing the analytical reduction of complex generic problems in the theory of scattering and propagation of electromagnetic waves in the presence of imperfectly conducting objects. Their subjects are as follows:

– a global method for the scattering by a multimode plane;

– diffraction by an impedance curved wedge;

– scattering by impedance polygons;

– advanced properties of spectral functions in frequency and time domains;

– bianisotropic media and related coupling expressions;

– exact and asymptotic reductions of surface radiation integrals.

Each of our approaches can be qualified as analytical, when it leads to exact explicit expressions, or, as semi-analytical, when it drastically reduces the mathematical complexity of studied problems. Therefore, they can be used in mathematical physics and engineering, to analyze and model, as well as in applied mathematics, to calculate for a low computational cost, the scattered fields in electromagnetism.

All of these works derive from original methods initiated in our publications that we here detail, develop and extend.

The first chapter is devoted to original exact expressions of the diffraction by a multilayered plane that can be partly composed of metamaterials. In whole generality, we then determine the fields as depending on potentials attached to arbitrary passive or active modes whose combination will give the passivity of the complete system. Our expressions directly take account of primary sources composed of electric and magnetic dipoles with arbitrary orientations, and profit of a novel exact development of incomplete Bessel function as an exact series of error functions. This latter characteristic permits a complete uniform analysis for arbitrary complex parameters, contrary to previous known results with error functions that were only approximations. Exact and complete asymptotics (at any order) are described, allowing us to particularly analyze the contribution of guided waves (forward and backward) at any distance.

The second chapter concerns the diffraction of an impedance wedge with curved faces of arbitrary angle that supports distinct surface boundary conditions of impedance type. We then distinguish the domains above and below the tangent planes at the edge. Our method permits an asymptotic evaluation at arbitrary order of curvatures of both faces for arbitrary passive impedance parameters. The uniformity at the crossing of the tangent plane is a characteristic remaining at arbitrary order, permitting us to analyze reflected, guided waves, edge-diffracted waves, but also waves originating from the edge that creep along the faces (creeping waves when faces are convex).

The diffraction in free space of a imperfectly conducting polygons (finite or with semi-infinite faces) is a particularly delicate problem that we study in third chapter, using Sommerfeld–Maliuzhinets integral representation of fields in a novel manner to rigorously consider several discontinuities, without any approximations. Indeed, contrary to common methods which consider large facets to admit asymptotic coupling between edges, we consider a rigorous development valid for arbitrary dimensions of facets, by establishing novel spectral equations that we can solve exactly or asymptotically, from a novel analysis of properties of spectral functions. This is particularly permitted by using their single-face representations, which is perfectly adapted to directly consider boundary conditions on faces with piecewise smooth geometries, as in polygonal cases.

Chapter 4 explores spectral functions in Sommerfeld–Maliuzhinets integral representation, their properties in the complex plane and new developments of them and special functions, attached to the resolution of multiple problems concerning the diffraction by a wedge with impedance boundaries conditions (passive or active). By beginning the study in frequency domain, we also analyze the representation of fields in time domain, in particular for an efficient explicit expression of causality in the case of dispersive (not constant relatively to frequency) multimode faces.

The fifth chapter analyzes the coupling influence between two imperfectly conducting objects in presence of a third one, all constituted by bianisotropic media. After beginning with developments of integral equalities, in particular, a generalized reciprocity one, we derive different properties of fields that will permit a complete analysis of coupling influences. We then give an example of application for an efficient and simple numerical post-process suppression of the influence of one object, that is, its direct but also its coupling contributions, on a second object.

We conclude this book with the determination of explicit contour integral expressions for an efficient evaluation of surface radiation integrals at arbitrary distance. This reduction concerns radiation of plane or curved plates of arbitrary contours, when the fields, highly oscillatory or not, are analytically defined on them, which is particularly the case for physical optics radiation surface integrals, when the surface fields are defined in closed form from geometrical optics.

1

A Global Method for the Scattering by a Multimode Plane with Arbitrary Primary Sources and Complete Series with Error Functions

1.1. Introduction

In [1], we considered the field scattered by an arbitrary impedance plane in electromagnetism, and we here exploit this formalism to analyze the scattering by a structure composed of several homogeneous planar layers, with isotropy or uniaxial anisotropy, illuminated by arbitrary bounded sources. In this study, the plane is supposed to be either grounded, that is, a multilayer backed by an impedance plane, or not grounded, that is, a multilayer slab in free space; this will lead us to generalize our previous approach for a multilayer given in [2].

The field scattered of such structures is usually given by its plane wave expansion (Fourier representation) [3]-[6], which presents the particularity to have reflection coefficients that are meromorphic functions. Each one can be then modeled as a rational function with a set of N simple poles $\{-g_j\}_{j=1,..,N}$, which permits us to assume a multimode boundary condition of order N [2].

The Fourier expansion is well adapted in far field or for plane wave illuminations, but is not suitable for an analysis at any distance or for complex incident waves. Even when double Fourier integrals are reduced to single Fourier–Bessel integrals, calculation is lengthy and delicate because of functions in the integral that remain highly oscillating and, most often in literature [3]–[9], analytic expansions are not strictly convergent but asymptotic. Besides, an additional difficulty comes from that, and in multimode case, we have to take into account that the constants g_j can have real parts of any sign, which signifies that passive but also active modes are present, even if the complete system is strictly passive.

In this frame, after expanding potentials into a combination of Fourier–Bessel integrals depending on each g_j, we are led to transform them to derive a more efficient integral representation, which is able to take account of active modes. Among other specificities, the definition of a parameter ϵ, attached to each pole, is then particularly important to permit complete exact and asymptotic series with error functions. These series allow us to exhibit guiding waves terms near and far from the sources above the multilayer, generalizing [1] and refining [2].

Otherwise, our approach, as in [1], uses a new representation of potentials for the incident field, which possesses the originality to directly consider arbitrarily oriented electric and magnetic primary currents sources. Thus, we have no more to solve separately the problem for vertical or horizontal dipolar source as commonly done in the literature for passive impedance planes [7]–[14], isotropic or uniaxial slabs [15]–[17] or multilayers [3]–[6], [18]–[22]. In practice, the analytic method so developed can be applied in whole generality to various problems, in particular for the determination of coupling between antennas above an imperfectly reflective plane, or for the calculus of Green's functions for planar lines printed on a multilayer.

This chapter is organized as follows. In section 1.2, we give a discussion on the representation of the field with potentials, on the boundary conditions and on the positions of g_j in the complex plane when metamaterials can be present. Next, we give a global expression of potentials attached to the fields radiated by arbitrary bounded sources in free space in section 1.3, and above the multilayer in section 1.4, which we develop and expand for arbitrarily oriented dipoles in section 1.5. In sections 1.6 and 1.7, we then detail a compact expression of the special function involved in the potentials attached to each mode, intimately depending on a parameter ϵ that is necessary to correctly take account of active modes. The definition of ϵ will be useful for the development of exact (section 1.6) and asymptotic (section 1.7) expansions with error functions for arbitrary cases, allowing in particular a general analysis of guided waves in section 1.8, including backward waves, near and far from the sources.

1.2. Potentials, reflection coefficients and multimode boundary conditions

1.2.1. *Fields and potentials*

We consider the scattering by an imperfectly reflective plane when it is illuminated by the field radiated by a bounded primary source, which is composed of arbitrary electrical and magnetic currents J and M (see Figure 1.1). In the space of points r with Cartesian coordinates (x, y, z), this plane is defined by $z = 0$. A harmonic time dependence $e^{i\omega t}$, from now on assumed, is suppressed throughout. The constants ε_0 and μ_0 are, respectively, the permittivity and the permeability of the free space above the plane, and $k_0 = \omega(\mu_0\varepsilon_0)^{1/2}$ is its wavenumber. Each component of the scattered field is assumed to be regular in the domain $z > 0$, and $O(e^{-\gamma|r|})$ with $\gamma > 0$ as $|r| \to \infty$ when $|\arg(ik_0)| < \pi/2$ (note: no loss is a limit case).

The electric field E and the magnetic field H above the multilayer, following Harrington [23, p. 131] (see also Jones [3, p. 19]), can be written with two scalar potentials $\mathcal{E}$ and $\mathcal{H}$, as follows:

$$E = -ik_0\text{curl}(\mathcal{H}\,\widehat{z}) + (\text{grad}(\text{div}(.)) + k_0^2)(\mathcal{E}\widehat{z}),$$
$$Z_0H = ik_0\text{curl}(\mathcal{E}\,\widehat{z}) + (\text{grad}(\text{div}(.)) + k_0^2)(\mathcal{H}\,\widehat{z}), \qquad (1.1)$$

where the Helmhotz equations $(\Delta + k_0^2)\mathcal{E} = 0$ and $(\Delta + k_0^2)\mathcal{H} = 0$ are verified outside the sources, $Z_0 = (\mu_0/\varepsilon_0)^{1/2}$. Thereafter, we denote $(\mathcal{E}_i, \mathcal{H}_i)$ and $(\mathcal{E}_s,\mathcal{H}_s)$ the potentials corresponding, respectively, to the incident field (incoming wave) (E^i, H^i) and the scattered field (outgoing wave) (E^s, H^s), and we write (1.1) in the compact form:

$$(E{,}Z_0H) = (\mathcal{L}_1(\widehat{z}\mathcal{E}, \widehat{z}\mathcal{H}), \mathcal{L}_2(\widehat{z}\mathcal{E}, \widehat{z}\mathcal{H})) = \mathcal{L}(\widehat{z}\mathcal{E}, \widehat{z}\mathcal{H}),$$
$$\mathcal{L}_1(u{,}v) = ((\text{grad}(\text{div}(.)) + k_0^2)(u) - ik_0\text{curl}(v)),$$
$$\mathcal{L}_2(u{,}v) = (ik_0\text{curl}(u) + (\text{grad}(\text{div}(.)) + k_0^2)(v)). \qquad (1.2)$$

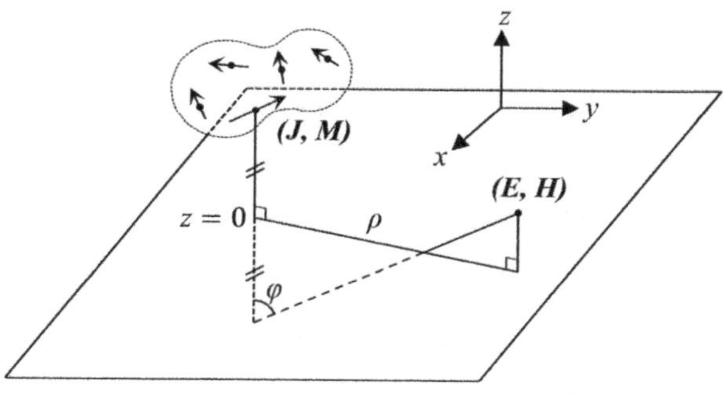

Figure 1.1. *Geometry: sources (J, M) and observation point above the plane $z = 0$*

1.2.2. *Multimode boundary conditions for a multilayer backed by an impedance plane*

Let us consider a multilayer plane composed of uniform isotropic (or z-axial anisotropic) layers. Any plane wave (E^i, H^i), incident at angle β with the normal $\hat{z}$, is then scattered as a reflected plane wave (E^s, H^s) that satisfies $E^s_z|_{z=0^+} = R_{e,e}E^i_z|_{z=0^-}$ (i.e. $H^s_\perp|_{z=0^+} = R_{e,e}H^i_\perp|_{z=0^-}$ in TM polarization), and $H^s_z|_{z=0^+} = R_{h,h}H^i_z|_{z=0^-}$ (i.e. $E^s_\perp|_{z=0^+} = R_{h,h}E^i_\perp|_{z=0^-}$ in TE polarization) [4], [5], [21] (see details in appendix A). If the multilayer is backed by a constant impedance plane, the reflection coefficients $R_{e,e}$ and $R_{h,h}$ are meromorphic functions of the variable $\cos\beta$, which we can model as rational functions [2] with simple poles, following:

$$R_{e,e}(\beta) = \prod_{j=1}^{N_e} \frac{\cos\beta - g^e_j}{\cos\beta + g^e_j},\ R_{h,h}(\beta) = \prod_{j=1}^{N_h} \frac{\cos\beta - g^h_j}{\cos\beta + g^h_j}, \tag{1.3}$$

for which we have the basic equalities (without superscripts e and h):

$$\prod_{j=1}^{N} \frac{\cos\beta - g_j}{\cos\beta + g_j} = (\pm1)^N + \sum_{j=1}^{N} a_j \frac{\left(\frac{-\cos\beta}{g_j}\right)^{\frac{1\mp1}{2}}}{\cos\beta + g_j},$$

$$\frac{a_j}{2g_j} = -\prod_{i\neq j}^{N} \frac{g_j + g_i}{g_j - g_i}, (-1)^N - 1 = \sum_{j=1}^{N} \frac{a_j}{g_j}, \tag{1.4}$$

where $N \geq 1$, and $\prod_{i\neq j}^{1} \frac{g_j+g_i}{g_j-g_i} \equiv 1$ for $N = 1$. The constants $(\pm1)^N$ refer to limit values when $|\cos\beta|^{\pm1} \to \infty$.

The $g^{e(,h)}_j$ are constants attached to complex modes with $\mathrm{Im}(g^{e(,h)}_j) \neq 0$, passive when $\mathrm{Re}(g^{e(,h)}_j) \geq 0$ or active when $\mathrm{Re}(g^{e(,h)}_j) < 0$, and ordered such that $|g^{e(,h)}_{j+1}| \geq |g^{e(,h)}_j|$, while, as considered in [26]–[28], we assume that:

when (1.3) applies, N_e and N_h and thus N are positive odd numbers, (1.5)

(note: this restriction on $N_{e(,h)}$ will be removed in the more general case of extended boundary conditions). Considering plane waves representation of fields (see appendix A), we can then write a multimode boundary conditions at $z = 0^+$ [2]:

$$\prod_{j=1}^{N_e}\left(\frac{\partial}{\partial z} - ik_0 g^e_j\right) E^s_z(z)|_{0^+} = \prod_{j=1}^{N_e}\left(\frac{\partial}{\partial z} + ik_0 g^e_j\right) E^i_z(-z)|_{0^+},$$

$$\prod_{j=1}^{N_h}\left(\frac{\partial}{\partial z} - ik_0 g^h_j\right) H^s_z(z)|_{0^+} = \prod_{j=1}^{N_h}\left(\frac{\partial}{\partial z} + ik_0 g^h_j\right) H^i_z(-z)|_{0^+}. \tag{1.6}$$

From the symmetry at normal incidence, the condition $R_{h,h}(0) = -R_{e,e}(0)$ must apply, which leads us to write:

$$\prod_{j=1}^{N_h} \frac{\pm 1 - g_j^h}{\pm 1 + g_j^h} = -\prod_{j=1}^{N_e} \frac{\pm 1 - g_j^e}{\pm 1 + g_j^e}, \tag{1.7}$$

and implies that $R_{h,h}(\pi) = -R_{e,e}(\pi)$. The condition (1.7) has crucial importance to avoid non-physical behaviors of fields derived from potentials, as examined further in this paper. Besides, the reader will notice that (1.7) implies $g_1^e = 1/g_1^h$ when $N_{e(,h)} = 1$, as well known for monomode (impedance) boundaries conditions [1]. The numbers $N_{e(,h)}$ correspond to truncated infinite products, where the less significant $g^{e(,h)}$ have been neglected, while some $g^{e(,h)}$ have to be modified so that $R_{h,h}(0) = -R_{e,e}(0)$ remains.

Considering (1.1), we can use:

$$E_z = \frac{\partial^2 \mathcal{E}}{\partial z^2} + k_0^2 \mathcal{E},\ Z_0 H_z = \frac{\partial^2 \mathcal{H}}{\partial z^2} + k_0^2 \mathcal{H}, \tag{1.8}$$

in (1.6), and we are led to search scattered potentials $\mathcal{E}_s$ and $\mathcal{H}_s$, satisfying the Helmholtz equation as $z > 0$, regular and exponentially vanishing as $z \to \infty$ when $|\arg(ik_0)| < \pi/2$, that verify as $z > 0$:

$$\prod_{i=1}^{N_e}\Big(\frac{\partial}{\partial z} - ik_0 g_j^e\Big)\mathcal{E}_s(z) = \prod_{j=1}^{N_e}\Big(\frac{\partial}{\partial z} + ik_0 g_j^e\Big)\mathcal{E}_i(-z),$$

$$\prod_{j=1}^{N_h}\Big(\frac{\partial}{\partial z} - ik_0 g_j^h\Big)\mathcal{H}_s(z) = \prod_{j=1}^{N_h}\Big(\frac{\partial}{\partial z} + ik_0 g_j^h\Big)\mathcal{H}_i(-z), \tag{1.9}$$

where $\mathcal{E}_i$ and $\mathcal{H}_i$ potentials are attached to radiation of arbitrary primary sources.

1.2.3. *Extended multimode boundary conditions*

More generally, we can consider an extended form when we want to include the case of a multilayer slab in free space, which is composed of isotropic [4]-[5] (or z-axial anisotropic [21]) layers. The reflection coefficients $R_{e,e}$ and $R_{h,h}$ remain meromorphic functions of $\cos\beta$, but we now model them in a more general form, following:

$$R_{e,e}(\beta) = R_0^e \frac{\prod_{j=1}^{N'_e}(\cos\beta - g_j^{e\prime})}{\prod_{j=1}^{N_e}(\cos\beta + g_j^e)},\ R_{h,h}(\beta) = R_0^h \frac{\prod_{j=1}^{N'_h}(\cos\beta - g_j^{h\prime})}{\prod_{j=1}^{N_h}(\cos\beta + g_j^h)}, \tag{1.10}$$

with simple complex poles $-g_j^{e(,h)}$, $N'_{e(,h)} \leq N_{e(,h)}$, constants R_0^e and R_0^h (this time for odd or even $N_{e(h)}$), for which we notice the basic equalities (without supercripts e and h):

$$R_0 \frac{\prod_{j=1}^{N'}(\cos\beta - g'_j)}{\prod_{j=1}^{N}(\cos\beta + g_j)} - a_{0\tau} = \sum_{j=1}^{N} a_j \frac{\left(\frac{-\cos\beta}{g_j}\right)^{\frac{1-\tau}{2}}}{\cos\beta + g_j} = \sum_{j=1}^{N} \frac{\tau \frac{a_j}{g_j}}{\left(\frac{\cos\beta}{g_j}\right)^{\tau} + 1},$$

$$a_{j\neq 0} = R_0 \frac{\prod_{i=1}^{N'}(-g'_i - g_j)}{\prod_{i\neq j}^{N}(g_i - g_j)}, \left|\begin{matrix} a_{0+} = \left|\begin{matrix} =R_0 \text{ if } N'=N \\ =0 \text{ if } N'<N \end{matrix}\right. \\ a_{0-} = R_0 (\prod_{i=1}^{N'} -g'_i)/(\prod_{i=1}^{N} g_i) \end{matrix}\right., \tag{1.11}$$

where the constants $a_{0\tau}$ refer to limit values when $|\cos\beta|^{\tau} \to \infty$ with $\tau = +1$ or -1. We have $N \geq N'$, $N \geq 1$, $N' \geq 0$, and we can let $\prod_{i\neq j}^{1}(g_i - g_j) \equiv 1$ for $N = 1$ and $\prod_{j=1}^{0}(\cos\beta - g'_j) \equiv 1$ for $N' = 0$.

As in the previous section, we consider that $R_{h,h}(0) = -R_{e,e}(0)$, and assume additionally that $R_{h,h}(\pi) = -R_{e,e}(\pi)$. This implies, after using (1.11) when $\cos\beta = \pm 1$:

$$\sum_{j=1}^{N_e} \frac{\tau_e \frac{a_j^e}{g_j^e}(g_j^e)^{\tau_e}}{(g_j^e)^{\tau_e} \pm 1} + \sum_{j=1}^{N_h} \frac{\tau_h \frac{a_j^h}{g_j^h}(g_j^h)^{\tau_h}}{(g_j^h)^{\tau_h} \pm 1} = -(a_{0\tau_e} + a_{0\tau_h}), \tag{1.12}$$

with $\tau_{e(,h)} = +1$ or -1. The condition (1.12), as previously noticed for (1.7), has a crucial importance to avoid non-physical behaviors of fields derived from potentials.

Considering (1.10) and plane waves representation of fields (see appendix A), we can write:

$$\prod_{j=1}^{N_e}\left(\frac{-\partial}{ik_0\partial z} + g_j^e\right) E_z^s(z)|_{0+} = R_0^e \prod_{j=1}^{N'_e}\left(\frac{-\partial}{ik_0\partial z} - g_j^{e'}\right) E_z^i(-z)|_{0+},$$

$$\prod_{j=1}^{N_h}\left(\frac{-\partial}{ik_0\partial z} + g_j^h\right) H_z^s(z)|_{0+} = R_0^h \prod_{j=1}^{N'_h}\left(\frac{-\partial}{ik_0\partial z} - g_j^{h'}\right) H_z^i(-z)|_{0+}, \tag{1.13}$$

which applies for any primary sources that illuminates the plane. Using (1.8), we can solve (1.13) with $(\mathcal{E}_s, \mathcal{H}_s)$ that satisfy as $z > 0$:

$$\prod_{i=1}^{N_e}\left(\frac{-\partial}{ik_0\partial z} + g_j^e\right) \mathcal{E}_s(z) = R_0^e \prod_{j=1}^{N'_e}\left(\frac{-\partial}{ik_0\partial z} - g_j^{e'}\right) \mathcal{E}_i(-z),$$

$$\prod_{j=1}^{N_h}\left(\frac{-\partial}{ik_0\partial z} + g_j^h\right) \mathcal{H}_s(z) = R_0^h \prod_{j=1}^{N'_h}\left(\frac{-\partial}{ik_0\partial z} - g_j^{h'}\right) \mathcal{H}_i(-z). \tag{1.14}$$

NOTE 1.1.– For z-uniaxial chiral layers [17], non-diagonal terms in (1.100) (see appendix A on plane waves representation of fields) do not vanish, and

$$\begin{pmatrix} \mathcal{P}_{N_e}^e(\frac{\partial}{\partial z})\mathcal{E}_s(z) \\ \mathcal{P}_{N_h}^h(\frac{\partial}{\partial z})\mathcal{H}_s(z) \end{pmatrix} = \begin{pmatrix} \mathcal{Q}_{N'}^{e,e}(\frac{\partial}{\partial z}) & \mathcal{Q}_{N'}^{e,h}(\frac{\partial}{\partial z}) \\ \mathcal{Q}_{N'}^{h,e}(\frac{\partial}{\partial z}) & \mathcal{Q}_{N'}^{h,h}(\frac{\partial}{\partial z}) \end{pmatrix} \begin{pmatrix} \mathcal{E}_i(-z) \\ \mathcal{H}_i(-z) \end{pmatrix}, \tag{1.15}$$

applies with polynomials $\mathcal{P}^{e(,h)}$ and $\mathcal{Q}^{e(,h)}$. By linearity, it can be solved by addition of terms similar to ones derived for the solution of equations (1.14) (note: in this case, adjoint characteristics have to be considered to apply the generalized reciprocity principle).

NOTE 1.2.– Letting $R = (\cos\beta - \mathcal{Z}(\cos\beta))/(\cos\beta + \mathcal{Z}(\cos\beta))$, $\mathcal{Z}$ is a rational function which is even as (1.3) applies [4], [5], [21] (see also Appendix A of [2]).

NOTE 1.3.– Considering (1.12), we can assume that, when a $g_j^{e(,h)} \to \mp 1$, it exists a $g_m^{h(,e)}$ such that $a_j^{e(,h)}(g_j^{e(,h)} \pm 1)^{-1} + a_m^{h(,e)}(g_m^{h(,e)} \pm 1)^{-1}$ is bounded.

1.2.4. *Properties of $R_{e,e}$ and $R_{h,h}$, and consequences on the $g_j^{e(,h)}$*

The multilayer is characterized by permittivity $\varepsilon(z)$ and permeability $\mu(z)$ as z varies from $z = 0$ to $-h_L$, which corresponds to a set of complex ε_j and μ_j specified in each layer $-h_j \leq z \leq -h_{j-1}$ with $1 \leq j \leq L$, $h_0 = 0$. Considering the hypotheses for (1.3) and (1.10), the last numbered layer shall be impenetrable or be bounded by the free space at $z = -h_L$. The plane wave spectrum of fields follows certain rules (see appendix A), and the reflection coefficients, considered for an harmonic time dependence $e^{i\omega t}$, are analytic functions, called thereafter R to simplify, which have some common elementary properties:

(a) R is an analytic function of the complex variables k_0, $\cos\beta$, $i\omega\varepsilon$, $i\omega\mu$ and layers depths. In the domain of passivity $\Omega_{\varepsilon,\mu}$ with $\mathrm{Re}(i\omega\varepsilon) > 0$ and $\mathrm{Re}(i\omega\mu) > 0$, R has no singularity when β varies from 0 to $i\infty + \arg(ik_0)$ as $\mathrm{Re}(ik_0 \sin\beta) = 0$, and at infinity. This regularity applies in the whole domain with $|\arg(i\omega)| \leq \pi/2$ from the causality principle. Its highest modulus is obtained for real ω from the maximum modulus principle, while, from (1.104), $|R| \leq 1$ when k_0 is real and $0 \leq \beta \leq \pi/2$.

(b) When $i\omega\varepsilon$ and $i\omega\mu$ are purely real positive (perfect lossy case), the multilayer is purely resistive, and, in these circumstances, R is real when k_0 and β are real.

(c) From (a) and (b), we can apply the Schwarz reflection principle (or edge-of-the-wedge theorem) [24, sect.5] (see also [25]), and deduce that R satisfies in $\Omega_{\varepsilon,\mu}$:

$$R((k_0)^*, (\cos\beta)^*, (i\omega\varepsilon)^*, (i\omega\mu)^*) = (R(k_0, \cos\beta, i\omega\varepsilon, i\omega\mu))^*. \tag{1.16}$$

(d) Using $(i\varepsilon)^* = i(-\varepsilon^*)$ and $(i\mu)^* = i(-\mu^*)$ in (1.16), we note, as we let change ε and μ for their "anticonjugate" $-\varepsilon^*$ and $-\mu^*$ (possible in practice with metamaterials, or plasmas), that, for real ω and k_0:

$$\begin{aligned}&\text{every pole } \cos\beta = (-g_j^{e(,h)}) \text{ of } R \text{ in (1.3) and (1.10) for } \varepsilon \text{ and } \mu,\\ &\text{gives us its conjugate } (-g_j^{e(,h)})^* \text{ as a pole of } R \text{ for } -\varepsilon^* \text{ and } -\mu^*.\end{aligned} \tag{1.17}$$

(e) The set $\{g_j^{e(,h)}\}$ has elements with positive and negative real parts [26]-[28], and for the class of problems including metamaterials, the total set to be considered, for real ω and k_0, is $\{g_j^{e(,h)}\} \cup \{(g_j^{e(,h)})^*\}$; this leaves no quarter-plane of the complex plane empty and explains why we have such a wide range of properties when multilayers include metamaterials.

Let us now consider the examples of two lossless systems for real positive ω, where, in the first case, ε and μ are with purely positive real values, and, in the second case, with negative real values. As defined in (d), both cases then belong, respectively, to anticonjugate classes. Pure imaginary $g_j^{e(,h)}$ are generally of finite number. From the analysis in [19] (respectively in [16]), the complex $g_j^{e(,h)}$ with non-null real part are of infinite number, and they are complex numbers with negative (respectively, positive) imaginary parts, associated with improper (respectively, proper) modes. Therefore, the domains of the $g_j^{e(,h)}$ for both cases of these examples are conjugated with each other, which plainly illustrates and confirms the property (d).

NOTE 1.4.– Considering appendix A, the conditions (a)–(e) also apply for anisotropic multilayers, when ε and μ are tensors.

NOTE 1.5.– $k_j = \omega(\mu_j\varepsilon_j)^{1/2}$, with $(\mu_j\varepsilon_j)^{1/2} = \sqrt{\mu_j}\sqrt{\varepsilon_j}$, can be $\neq \omega\sqrt{\mu_j\varepsilon_j}$, and thus, a term like $((\mu_j\varepsilon_j)^{1/2}/\varepsilon_j)\tan(\omega(\mu_j\varepsilon_j)^{1/2}d)$ can be $\neq \sqrt{\mu_j/\varepsilon_j}\tan(\omega\sqrt{\mu_j\varepsilon_j}d)$.

1.3. Incident potentials $(\mathcal{E}_i,\mathcal{H}_i)$ for arbitrary bounded primary sources

To solve problems with conditions (1.9) or (1.14), we need a correct explicit expression of $(\mathcal{E}_i, \mathcal{H}_i)$. We begin by considering the incident field (E^i, H^i) at point r of coordinates (x, y, z), radiated by arbitrary electric and magnetic bounded sources J and M in free space [3]:

$$E^i = \text{curl}(G * M) + \frac{i}{\omega\varepsilon_0}(\text{grad}(\text{div}(.)) + k_0^2)(G * J),$$

$$H^i = -\text{curl}(G * J) + \frac{i}{\omega\mu_0}(\text{grad}(\text{div}(.)) + k_0^2)(G * M), \qquad (1.18)$$

where $G(r) = -\frac{e^{-ik_0|r(x,y,z)|}}{4\pi|r(x,y,z)|}$ with $|r| = \sqrt{x^2+y^2+z^2}$ verifies $(\Delta + k_0^2)G = \delta$, $*$ is the convolution product, and J and M are generalized functions [29].

Considering arbitrary J and M in the domain $\pm z > 0$, the potentials $(\mathcal{E}_i, \mathcal{H}_i)$ in the representation of fields E^i and H^i with (1.1), which satisfy the Helmholtz equation as $\mp z > 0$ and exponentially vanish as $\mp z \to \infty$ when $|\arg(ik_0)| < \pi/2$ can be written when $\mp z > 0$ [1]:

$$(\mathcal{E}_i, \mathcal{H}_i)(x, y, z) = [\frac{\widehat{z}}{8\pi k_0^2}.(\mathcal{L}(Z_0 J, M) * \mathcal{W})](x, y, z), \qquad (1.19)$$

where $(A, B) * C \equiv (A * C, B * C)$, $\mathcal{L}$ is defined as in (1.2):

$$\mathcal{L}(Z_0 J, M) = (\mathcal{L}_1(Z_0 J, M), \mathcal{L}_2(Z_0 J, M)),$$
$$\mathcal{L}_1(u,v) = ((\text{grad}(\text{div}(.)) + k_0^2)(u) - ik_0\text{curl}(v)),$$
$$\mathcal{L}_2(u,v) = (ik_0\text{curl}(u) + (\text{grad}(\text{div}(.)) + k_0^2)(v)), \tag{1.20}$$

and $\mathcal{W}$ has the remarkable compact form [1]–[2]:

$$\mathcal{W}(r) = e^{ik_0|z|}E_1(ik_0(|r|+|z|)) + e^{-ik_0|z|}(E_1(ik_0(|r|-|z|)) + 2\ln\rho), \tag{1.21}$$

with $\rho = \sqrt{x^2 + y^2}$, E_1 being the exponential integral function [30], which verifies, as $\mp z > 0$:

$$\left(\frac{\partial^2}{\partial z^2} + k_0^2\right)\frac{\mathcal{W}(r)}{8\pi i k_0} = G(r),\ (\Delta + k_0^2)\mathcal{W}(r) = 0. \tag{1.22}$$

NOTE 1.6.– It is worth noticing that we have $\Delta_{xy}\ln(\rho) = 2\pi\delta(x)\delta(y)$, which implies that $\ln\rho$ in $\mathcal{W}$ has no influence on the expression of the field, except by its singularity at $\rho = 0$.

NOTE 1.7.– Let us divide the primary sources J and M into the sources above and below a plane $z = z_1$, $J_{z_1}^{\pm} = U_s(\pm(z - z_1))J$ and $M_{z_1}^{\pm} = U_s(\pm(z - z_1))M$, where U_s is the unit step function, with $U_s(z) = (\text{sign}(z) + 1)/2$, $U_s(z) = 1$ as $z > 0$, $U_s(z) = 1/2$ as $z = 0$ and $U_s(z) = 0$ as $z < 0$. We can then write, for arbitrary observation point at $z = z_1$:

$$(E^i,Z_0H^i)(x, y, z_1) = \sum_{\pm}(E^{\pm},Z_0H^{\pm})(x, y, z_1 \mp 0^{\pm}), \tag{1.23}$$

where $(E^{\pm},H^{\pm})$ are the field radiated by the sources $J_{z_1}^{\pm}$ and $M_{z_1}^{\pm}$, following $(E^{\pm},Z_0H^{\pm}) = \mathcal{L}(\hat{z}\mathcal{E}^{\pm},\hat{z}\mathcal{H}^{\pm})$ with $(\mathcal{E}^{\pm},\mathcal{H}^{\pm}) = \frac{\hat{z}}{8\pi k_0^2}.\mathcal{L}(Z_0J_{z_1}^{\pm}, M_{z_1}^{\pm}) * \mathcal{W}$, from (1.2) and (1.19).

1.4. Scattered potentials $(\mathcal{E}_s,\mathcal{H}_s)$ for arbitrary primary sources

1.4.1. *A global expression of* $(\mathcal{E}_s,\mathcal{H}_s)$ *for a multimode plane*

Considering the potentials $(\mathcal{E}_i, \mathcal{H}_i)$ attached to the radiation of arbitrary sources from (1.19), we can now express the scalar potentials $\mathcal{E}_s$ and $\mathcal{H}_s$, which satisfy the multimode conditions (1.9) or (1.14), as $z \geq 0$. From the method developed in [1]–[2], we have for conditions (1.9) (for odd numbers $N_{e(h)}$):

$$\mathcal{E}_s(x,y,z) = \pm\mathcal{E}_i(x,y,-z)$$
$$+\Big(\Big(\frac{\widehat{z}}{\omega\varepsilon_0}\frac{\text{grad}(\text{div}(J)) + k_0^2 J}{8\pi k_0} + \frac{\widehat{z}}{k}\frac{(-ik_0\,\text{curl}(M))}{8\pi k_0}\Big)$$
$$* \sum_{\epsilon'=-1,1}\Big(\Big(\prod_{j=1}^{N_e}\frac{\epsilon'+g_j^e}{\epsilon'-g_j^e} \mp 1\Big)\mathcal{V}_{\epsilon'} + \sum_{j=1}^{N_e}\frac{\epsilon' a_j^e \mathcal{K}_{g_j^e}}{(g_j^e-\epsilon')}\Big)\Big)(x,y,-z), \tag{1.24}$$

and

$$\mathcal{H}_s(x,y,z) = \mp\mathcal{H}_i(x,y,-z)$$
$$+\Big(\Big(\frac{\widehat{z}}{\omega\varepsilon_0}\frac{(ik_0\text{curl}(J))}{8\pi k_0} + \frac{\widehat{z}}{k}\frac{(\text{grad}(\text{div}(M)) + k_0^2 M)}{8\pi k_0}\Big)$$
$$* \sum_{\epsilon'=-1,1}\Big(\Big(\prod_{j=1}^{N_h}\frac{\epsilon'+g_j^h}{\epsilon'-g_j^h} \pm 1\Big)\mathcal{V}_{\epsilon'} + \sum_{j=1}^{N_h}\frac{\epsilon' a_j^h \mathcal{K}_{g_j^h}}{(g_j^h-\epsilon')}\Big)\Big)(x,y,-z), \tag{1.25}$$

while we have, for extended conditions (1.14) (for odd or even numbers $N_{e(h)}$):

$$\mathcal{E}_s(x,y,z) = a_{0T_e}^e\,\mathcal{E}_i(x,y,-z)$$
$$+\Big(\Big(\frac{\widehat{z}}{\omega\varepsilon_0}\frac{\text{grad}(\text{div}(J)) + k_0^2 J}{8\pi k_0} + \frac{\widehat{z}}{k}\frac{(-ik_0\,\text{curl}(M))}{8\pi k_0}\Big)$$
$$* \sum_{\epsilon'=-1,1}\Big(\big((R_{e,e}|_{\cos\beta=-\epsilon'}) - a_{0T_e}^e\big)\mathcal{V}_{\epsilon'} + \sum_{j=1}^{N_e}\frac{\epsilon' a_j^e \mathcal{K}_{g_j^e}}{(g_j^e-\epsilon')}\Big)\Big)(x,y,-z), \tag{1.26}$$

and

$$\mathcal{H}_s(x,y,z) = a_{0T_h}^h\,\mathcal{H}_i(x,y,-z)$$
$$+\Big(\Big(\frac{\widehat{z}}{\omega\varepsilon_0}\frac{(ik_0\,\text{curl}(J))}{8\pi k_0} + \frac{\widehat{z}}{k}\frac{(\text{grad}(\text{div}(M)) + k_0^2 M)}{8\pi k_0}\Big)$$
$$* \sum_{\epsilon'=-1,1}\Big(\big((R_{h,h}|_{\cos\beta=-\epsilon'}) - a_{0T_h}^h\big)\mathcal{V}_{\epsilon'} + \sum_{j=1}^{N_h}\frac{\epsilon' a_j^h \mathcal{K}_{g_j^h}}{(g_j^h-\epsilon')}\Big)\Big)(x,y,-z), \tag{1.27}$$

where $\mathcal{V}_{\epsilon'}$, $\mathcal{K}_g$ are given by,

$$\mathcal{V}_{\epsilon'}(x,y,-z) = e^{\epsilon' ik_0 z}(E_1(ik_0(|r|+\epsilon' z)) + (1-\epsilon')\ln\rho),$$
$$\mathcal{K}_g(x,y,-z) = e^{ik_0 gz}\mathcal{J}_g(\rho,-z), \tag{1.28}$$

with $\rho = \sqrt{x^2+y^2}$, $|r| = \sqrt{\rho^2+z^2}$, as $z \geq 0$. The function $\mathcal{J}_g(\rho, -z)$ is given by the integral:

$$\mathcal{J}_g(\rho,-z) = \frac{e^{-ik_0 gz}}{2}\int_{\mathcal{D}} \frac{H_0^{(2)}(k_0\rho\sin\beta)e^{-ik_0 z\cos\beta}}{\cos\beta+g}\sin\beta d\beta, \tag{1.29}$$

with $\mathrm{Re}(ik_0\sin\beta) = 0$ on $\mathcal{D}$ from $-i\infty - \arg(ik_0)$ to $i\infty + \arg(ik_0)$, $H_0^{(2)}$ being the Hankel function [30], which is a Fourier–Bessel integral also called a Sommerfeld-type integral [31]. Letting $g = \sin\theta_1$ with $|\mathrm{Re}(\theta_1)| \leq \pi/2$ for arbitrary g of the complex plane, we notice the cut $\mathrm{Re}(ik_0\cos\theta_1) = 0$ as g varies, due to poles of $(\cos\beta+g)^{-1}$ that can go through $\mathcal{D}$ when $\mathrm{Re}(g) < 0$. The function $\mathcal{J}_g$ will be rewritten and developed in an efficient way in section 1.6.

The choice of sign, $\pm$ in (1.24)–(1.25) and $\tau_{e(,h)} \equiv \pm$ in (1.26)–(1.27), is free (note: algebric modifications can apply, as detailed in appendix B).

NOTE 1.8.– In acoustics, the problem is no more vectorial. In this case, considering a multimode plane, described by the reflection coefficient R_a satisfying:

$$R_a = \prod_{j=1}^{N}\frac{\cos\beta-g_j}{\cos\beta+g_j} = 1+\sum_{j=1}^{N} a_j\frac{1}{\cos\beta+g_j},\ \frac{a_j}{2g_j} = -\prod_{i\neq j}^{N}\frac{g_j+g_i}{g_j-g_i}, \tag{1.30}$$

with $\prod_{i\neq j}^{1}\frac{g_j+g_i}{g_j-g_i} \equiv 1$ when $N = 1$, and illuminated by a point source at ($\rho = 0$, $z = h$) which radiates $u_{inc}(\rho,z) = e^{-ik_0R(\rho,z-h)}/k_0R(\rho,z-h)$, $R(\rho,z) = \sqrt{\rho^2+z^2}$, we can write the scattered field following:

$$u_s(\rho,z) = u_{inc}(\rho,-z) - i\sum_{j=1}^{N} a_j e^{ik_0 g_j(z+h)}\mathcal{J}_{g_j}(\rho,-z-h). \tag{1.31}$$

NOTE 1.9.– Let us note that we have [8], for $\mathrm{Re}(g) > 0$ and $\arg(ik) = \pi/2$, as $z > 0$:

$$e^{ik_0 gz}\mathcal{J}_g(\rho,-z) = \int_{-i\infty}^{0} e^{-ik_0 g z_1}\frac{e^{-ik_0R(-z-z_1)}}{k_0R(-z-z_1)}kdz_1, \tag{1.32}$$

where $R(z) = \sqrt{\rho^2+z^2}$ with $\mathrm{Im}(R(-z-z_1)) < 0$.

1.4.2. *Reciprocity principle from our expressions*

Let us consider the scattered fields (E_m^s,H_m^s) for $m = 1, 2$, excited by primary bounded sources (J_m, M_m) above the multilayer plane and corresponding to the potentials $(\mathcal{E}_{ms}, \mathcal{H}_{ms})$. Using (1.2) and integration by parts in sense of generalized functions [29], we have:

$$\int_{R^3} (Z_0 J_1(r).E_2^s(r) - M_1(r).Z_0 H_2^s(r))dV_r$$

$$= \int_{R^3} (\mathcal{L}_1(Z_0 J_1, M_1).\mathcal{E}_{2s}\,\widehat{z} - \mathcal{L}_2(Z_0 J_1, M_1).\mathcal{H}_{2s}\,\widehat{z})dV_r, \tag{1.33}$$

while, from (1.24)–(1.27), $(\mathcal{E}_{ms}, \mathcal{H}_{ms})$ expresses in the form:

$$(\mathcal{E}_{ms}, \mathcal{H}_{ms})(r) = \frac{\widehat{z}}{8\pi k_0^2}.((\mathcal{L}_1(Z_0 J_m, M_m),$$

$$\mathcal{L}_2(Z_0 J_m, M_m)) * \mathcal{R}_s)(x, y, -z), \tag{1.34}$$

with $\mathcal{R}_s(x - x', y - y', -z - z')$ unchanged as $r \leftrightarrow r'$. We can then write:

$$\int_{R^3} (Z_0 J_1(r).E_2^s(r) - M_1(r).Z_0 H_2^s(r))dV_r =$$

$$\frac{1}{8\pi k_0^2} \int_{R^3} \int_{R^3} dV_r dV_{r'} [\sum_{j=1,2} (-1)^{j-1} [(\mathcal{L}_j(Z_0 J_1, M_1)(r).\widehat{z})$$

$$\times (\mathcal{L}_j(Z_0 J_2, M_2)(r').\widehat{z})] \mathcal{R}_s(x - x', y - y', -z - z')$$

$$= \int_{R^3} (Z_0 J_2(r').E_1^s(r') - M_2(r').Z_0 H_1^s(r'))dV_{r'}. \tag{1.35}$$

Thus, we can exchange indices 1 and 2 without changing the result, and reciprocity theorem [3, p. 118] is satisfied.

1.5. When primary sources J and M are arbitrarily oriented dipoles

1.5.1. *Tensorial expressions of fields*

Considering (1.24)–(1.27), we now develop the expression of scattered field for the case with electric and magnetic point sources, $J = J_{r'}\delta(r - r')$ and $M = M_{r'}\delta(r - r')$, where $\delta(r)$ is the Dirac generalized function [29], while the expressions of scattered fields for volume and surface sources are analyzed in appendix C. From (1.1) with (1.24)–(1.27), we have for the scattered field (E^s, H^s):

$$E^s(r) = -ik_0\text{curl}(\mathcal{H}_s\,\widehat{z}) + (\text{grad}(\text{div}(.)) + k_0^2)(\mathcal{E}_s\widehat{z})$$

$$= \frac{1}{8\pi k_0^2}([\underline{\mathcal{E}}_{he}(r, r')].Z_0 J_{r'} - [\underline{\mathcal{D}}_{he}(r, r')].M_{r'}), \tag{1.36}$$

and

$$
\begin{aligned}
Z_0 H^s(r) &= ik_0 \mathrm{curl}(\mathcal{E}_s\,\widehat{z}) + (\mathrm{grad}(\mathrm{div}(.)) + k_0^2)(\mathcal{H}_s\,\widehat{z}) \\
&= \frac{1}{8\pi k_0^2}([\underline{\mathcal{D}}_{eh}(r,r')].Z_0 J_{r'} + [\underline{\mathcal{F}}_{eh}(r,r')].M_{r'}),
\end{aligned} \tag{1.37}
$$

where $[\underline{\mathcal{F}}_{he(,eh)}(r,r')]$ and $[\underline{\mathcal{D}}_{eh(,he)}(r,r')]$ are dyadic tensors, with $C.[\widehat{a}\,\widehat{b}] = (C.\widehat{a})\,\widehat{b}$ and $[\widehat{a}\,\widehat{b}].C = \widehat{a}\,(\widehat{b}.C)$, $r = x\widehat{x} + y\widehat{y} + z\widehat{z}$, $r' = x'\widehat{x}' + y'\widehat{y}' + z'\widehat{z}'$, $\widehat{x} \equiv \widehat{x}'$, $\widehat{y} \equiv \widehat{y}'$, $\widehat{z} \equiv \widehat{z}'$, which verify, according to the reciprocity principle stated in previous section:

$$
\begin{aligned}
D_r.[\underline{\mathcal{F}}_{he(,eh)}(r,r')].C_{r'} &= C_{r'}.[\underline{\mathcal{F}}_{he(,eh)}(r',r)].D_r, \\
D_r.[\underline{\mathcal{D}}_{he(,eh)}(r,r')].C_{r'} &= C_{r'}.[\underline{\mathcal{D}}_{eh(,he)}(r',r)].D_r,
\end{aligned} \tag{1.38}
$$

with $C_{r'} = c_x\widehat{x}' + c_y\widehat{y}' + c_z\widehat{z}'$, $D_r = d_x\widehat{x} + d_y\widehat{y} + d_z\widehat{z}$ being two constant vectors.

Noting that we have:

$$
\begin{aligned}
&\varsigma(\partial_z f * g)(x,y,\varsigma z) = \partial_z((f * g)(x,y,\varsigma z)), \\
&(\partial_x f * g)(x,y,\varsigma z) = \partial_x((f * g)(x,y,\varsigma z)), \\
&(\partial_y f * g)(x,y,\varsigma z) = \partial_y((f * g)(x,y,\varsigma z)), \\
&\partial_x^i \partial_y^j \partial_z^k \equiv \frac{\partial^{i+j+k}}{\partial x^i \partial y^j \partial z^k} \text{ and } \partial_{xy} \equiv \partial_x\partial_y,
\end{aligned} \tag{1.39}
$$

these tensors express the following:

$$
\begin{aligned}
&[\underline{\mathcal{D}}_{he(,eh)}(r,r')] = \\
&= ik_0[((\widehat{x}\partial_y - \widehat{y}\partial_x)(\widehat{x}'\partial_x + \widehat{y}'\partial_y)\varsigma\partial_z \mathcal{S}_{h(,e)}(r,r') \\
&+ (\widehat{x}\partial_x + \widehat{y}\partial_y)(\widehat{y}'\partial_x - \widehat{x}'\partial_y)(\partial_z \mathcal{S}_{e(,h)}(r,r')) \\
&+ (\widehat{x}\partial_y - \widehat{y}\partial_x)\widehat{z}'(\partial_z^2 + k_0^2)\mathcal{S}_{h(,e)}(r,r')) \\
&+ \widehat{z}(\widehat{y}'\partial_x - \widehat{x}'\partial_y)(\partial_z^2 + k_0^2)\mathcal{S}_{e(,h)}(r,r')] \\
&= ik_0[((\widehat{x}\widehat{x}' - \widehat{y}\widehat{y}')\partial_{xy} + \widehat{x}\widehat{y}'\partial_y^2 \\
&- \widehat{y}\widehat{x}'\partial_x^2)\partial_z(\varsigma\mathcal{S}_{h(,e)}(r,r') - \mathcal{S}_{e(,h)}(r,r'))] \\
&+ ik_0[-(\widehat{x}\widehat{y}' - \widehat{y}\widehat{x}')\partial_z(\partial_z^2 + k_0^2)\mathcal{S}_{e(,h)}(r,r') \\
&+ (\widehat{x}\partial_y - \widehat{y}\partial_x)\widehat{z}'(\partial_z^2 + k_0^2)\mathcal{S}_{h(,e)}(r,r')) \\
&+ \widehat{z}(\widehat{y}'\partial_x - \widehat{x}'\partial_y)(\partial_z^2 + k_0^2)\mathcal{S}_{e(,h)}(r,r')],
\end{aligned} \tag{1.40}
$$

and

$$
\begin{aligned}
&[\underline{\mathcal{F}}_{he(,eh)}(r,r')] = \\
&[(\widehat{x}\partial_x + \widehat{y}\partial_y)(\widehat{x}'\partial_x + \widehat{y}'\partial_y)\,(\varsigma\partial_z^2\mathcal{S}_{e(,h)}(r,r') + k_0^2\mathcal{S}_{h(,e)}(r,r'))] \\
&+ [k_0^2(\widehat{x}\widehat{x}' + \widehat{y}\widehat{y}')(\partial_z^2 + k_0^2))(\mathcal{S}_{h(,e)}(r,r')) \\
&+ (\widehat{x}\partial_x + \widehat{y}\partial_y)\widehat{z}'\partial_z(\partial_z^2 + k_0^2)\mathcal{S}_{e(,h)}(r,r') \\
&+ \widehat{z}(\widehat{x}'\partial_x + \widehat{y}'\partial_y)\varsigma\partial_z(\partial_z^2 + k_0^2)(\mathcal{S}_{e(,h)}(r,r')) \\
&+ \widehat{z}\widehat{z}'(\partial_z^2 + k_0^2)(\partial_z^2 + k_0^2)\mathcal{S}_{e(,h)}(r,r')],
\end{aligned} \tag{1.41}
$$

with $\varsigma = -1$. Using (1.24)–(1.27) with (1.107) of appendix B, the $\mathcal{S}_{e(h)}$ can be written as:

$$
\begin{aligned}
\mathcal{S}_e(r,r') &= c^e\mathcal{S}_i(r_{im} - r') + \sum_{j=1}^{N_e}\frac{\tau_e a_j^e}{g_j^e}\Big((g_j^e)^{\tau_e} \\
&\times \sum_{\epsilon'=-1,1}\frac{\mathcal{V}_{\epsilon'} + \epsilon'\mathcal{K}_{g_j^e}}{(g_j^e)^{\tau_e} - \epsilon'}\Big)(r_{im} - r'), \\
\mathcal{S}_h(r,r') &= c^h\mathcal{S}_i(r_{im} - r') + \sum_{j=1}^{N_h}\frac{\tau_h a_j^h}{g_j^h}\Big((g_j^h)^{\tau_h} \\
&\times \sum_{\epsilon'=-1,1}\frac{\mathcal{V}_{\epsilon'} + \epsilon'\mathcal{K}_{g_j^h}}{(g_j^h)^{\tau_h} - \epsilon'}\Big)(r_{im} - r'),
\end{aligned} \tag{1.42}
$$

where,

$$r_{im} \equiv (x, y, -z) \text{ and } r' \equiv (x', y', z'), \tag{1.43}$$

and, as (1.3)–(1.9) with (1.24)–1.25) applies (when $N_{e,(h)}$ are odd numbers),

$$c^e = -c^h = \pm 1 \text{ and } \tau_e = -\tau_h = \pm 1, \tag{1.44}$$

while, for the extended case, as (1.10)–(1.14) with (1.26)–(1.27) applies (when $N_{e,(h)}$ are odd or even numbers):

$$c^{e(,h)} = a_{0\tau_{e(,h)}}^{e(,h)} \text{ with } \tau_{e(,h)} = +1 \text{ or } -1. \tag{1.45}$$

The functions $\mathcal{V}_{\epsilon'}$, $\mathcal{K}_g$ and $\mathcal{S}_i$ satisfy the Helmholtz equation above the plane, and are defined, following:

$$
\begin{aligned}
&\mathcal{V}_{\epsilon'}(x,y,-z) = e^{\epsilon' ik_0|z|}(E_1(ik_0(|r| + \epsilon'|z|)) + (1-\epsilon')\ln\rho), \\
&\mathcal{K}_g(x,y,-z) = e^{ik_0 gz}\mathcal{J}_g(\rho,-z),\ \mathcal{S}_i \equiv \sum_{\epsilon'=-1,1}\mathcal{V}_{\epsilon'},
\end{aligned} \tag{1.46}
$$

for $z \geq 0$, $\rho = \sqrt{x^2 + y^2}$, $r^2 = \rho^2 + z^2$, and they verify:

$$\partial_z \mathcal{V}_{\epsilon'}(x, y, -z) = -\epsilon' \frac{e^{-ik_0|r|}}{|r|} + ik_0\epsilon' \mathcal{V}_{\epsilon'}(x, y, -z),$$

$$\partial_z \mathcal{K}_g(x, y, -z) = \frac{e^{-ik_0|r|}}{|r|} + ik_0 g\, \mathcal{K}_g(x, y, -z),$$

$$\partial_z((\mathcal{V}_{\epsilon'} + \epsilon' \mathcal{K}_g)(x, y, -z)) = ik_0\epsilon'(\mathcal{V}_{\epsilon'} + g\, \mathcal{K}_g)(x, y, -z),$$

$$(\partial_z^2 + k_0^2)\mathcal{V}_{\epsilon'} = (-\epsilon' \partial_z - ik_0) \frac{e^{-ik_0|r|}}{|r|},$$

$$(\partial_z^2 + k_0^2)\mathcal{K}_g = (\partial_z + ik_0 g) \frac{e^{-ik_0|r|}}{|r|} + (k_0^2 - (k_0 g)^2)\, \mathcal{K}_g,$$

$$\partial_z^2(\mathcal{V}_{\epsilon'} + \epsilon' \mathcal{K}_g) = -ik_0\epsilon'(\epsilon' - g) \frac{e^{-ik_0|r|}}{|r|} - k_0^2(\mathcal{V}_{\epsilon'} + g^2\epsilon'\, \mathcal{K}_g). \tag{1.47}$$

NOTE 1.10.– We notice that:

$$\begin{aligned}
&\mathrm{curl}_{r'}(C_r.[\underline{\mathcal{D}}_{eh(,he)}(r, r')]) = ik(C_r.[\underline{\mathcal{F}}_{eh(,he)}(r, r')]),\\
&\mathrm{curl}_{r}([\underline{\mathcal{D}}_{he(,eh)}(r, r')].C_{r'}) = ik([\underline{\mathcal{F}}_{eh(,he)}(r, r')].C_{r'}),\\
&\mathrm{curl}_{r'}(C_r.[\underline{\mathcal{F}}_{he(,eh)}(r, r')]) = -ik(C_r.[\underline{\mathcal{D}}_{he(,eh)}(r, r')]),\\
&\mathrm{curl}_{r}([\underline{\mathcal{F}}_{he(,eh)}(r, r')].C_{r'}) = -ik([\underline{\mathcal{D}}_{eh(,he)}(r, r')].C_{r'}).
\end{aligned} \tag{1.48}$$

NOTE 1.11.– We can use (1.36)–(1.37) for the incident field (E^i, H^i) in place of (E^s, H^s), if we take $\varsigma = 1$, instead of $\varsigma = -1$ in (1.40)–(1.41), and $\mathcal{S}_i(r - r')$ for $\mathcal{S}_{e(,h)}(r, r')$.

1.5.2. *Exact reduction of derivative orders after sums over ϵ' indices*

We can simplify in (1.40)–(1.42) the sums of terms over ϵ' indices, and then exactly reduce the derivative order of the following terms:

$$\begin{aligned}
\mathcal{U}_a &= (\partial_z^2 + k_0^2)\mathcal{S}_{e(,h)}(r, r'),\\
\mathcal{U}_b &= \partial_z(\mathcal{S}_{h(,e)}(r, r') + \mathcal{S}_{e(,h)}(r, r')),\\
\mathcal{U}_c &= -\partial_z^2\mathcal{S}_{e(,h)}(r, r') + k_0^2\mathcal{S}_{h(,e)}(r, r'),
\end{aligned} \tag{1.49}$$

for the field expressions (1.36)–(1.37).

For $\mathcal{U}_a = (\partial_z^2 + k_0^2)\mathcal{S}_{e(,h)}(r, r')$ term, we use (1.47), and noting that:

$$\sum_{\epsilon'=-1,1} (\partial_z^2 + k_0^2)\frac{(\mathcal{V}_{\epsilon'} + \epsilon'\mathcal{K}_g)}{(g - \epsilon')} = -2k_0^2\mathcal{K}_g,$$

$$\frac{1}{g}\sum_{\epsilon'=-1,1} (\partial_z^2 + k_0^2)\frac{(\mathcal{V}_{\epsilon'} + \epsilon'\mathcal{K}_g)}{(\frac{1}{g} - \epsilon')} = -2ik_0\frac{e^{-ik_0|r|}}{|r|} + 2k_0^2 g\mathcal{K}_g,$$

$$\sum_{\epsilon'=-1,1} (\partial_z^2 + k_0^2)\mathcal{V}_{\epsilon'} = -2ik_0\frac{e^{-ik_0|r|}}{|r|}. \tag{1.50}$$

we have the following reduction:

$$\mathcal{U}_a = -2k_0^2\Big(\Big(2c^{e(,h)} + (\tau^{e(,h)} - 1)\sum_{j=1}^{N_{e(,h)}} \frac{a_j^{e(,h)}}{g_j^{e(,h)}}\Big)\frac{ie^{-ik_0|r|}}{2k_0|r|}$$

$$+ \sum_{j=1}^{N_{e(,h)}} a_j^{e(,h)}\mathcal{K}_{g_j^{e(,h)}}\Big), \tag{1.51}$$

with $\sum_{j=1}^{N_{e(,h)}} \frac{a_j^{e(,h)}}{g_j^{e(,h)}} = -2$ (respectively, $= a_{0-} - a_{0+}$) when (1.3) (respectively, (1.10)) applies.

For $\mathcal{U}_b = \partial_z(\mathcal{S}_{h(,e)}(r, r') + \mathcal{S}_{e(,h)}(r, r'))$ and $\mathcal{U}_c = -\partial_z^2\mathcal{S}_{e(,h)}(r, r') + k_0^2\mathcal{S}_{h(,e)}(r, r')$, it will be done provided the condition (1.12) (or (1.7)) is used, which suppresses pure residual terms $\mathcal{V}_{\epsilon'}$ in the expressions of $\mathcal{U}_b$ and $\mathcal{U}_c$, in particular $\mathcal{V}_{-1}$ that diverges as $\rho \to \infty$. For that, we will consider (1.12), and note that:

$$\sum_{j=1}^{N_{e(,h)}} \frac{\frac{\tau_{e(,h)}a_j^{e(,h)}}{g_j^{e(,h)}}(g_j^{e(,h)})^{\tau_{e(,h)}}}{(g_j^{e(,h)})^{\tau_{e(,h)}} - \epsilon'} + \sum_{j=1}^{N_{h(,e)}} \frac{\frac{\tau_{h(,e)}a_j^{h(,e)}}{g_j^{h(,e)}}(g_j^{h(,e)})^{\tau_{h(,e)}}}{(g_j^{h(,e)})^{\tau_{h(,e)}} - \epsilon'}$$

$$+ c^{e(,h)} + c^{h(,e)} = 0, \text{ with } \sum_{\epsilon'=-1,1} \frac{\tau g^{\tau-1}}{g^\tau - \epsilon'} = \frac{2g^\tau}{g^2 - 1} \text{ as } \tau = \pm 1. \tag{1.52}$$

Therefore, using (1.47) and $\sum_{\epsilon'=-1,1} \frac{\epsilon'\tau g^{\tau-1}}{g^\tau - \epsilon'} = \frac{2}{g^2-1}$, the term $\mathcal{U}_b$ reduces to:

$$\mathcal{U}_b = 2ik_0\sum_{v=e,h}\Big[\frac{e^{-ik_0|r|}}{ik_0|r|}\sum_{j=1}^{N_v} \frac{a_j^v(1 - (g_j^v)^{\tau_v}\mathcal{B}')}{(g_j^v)^2 - 1}$$

$$+ \sum_{j=1}^{N_v} a_j^v\frac{(g_j^v\mathcal{K}_{g_j^v} - (g_j^v)^{\tau_v}\mathcal{B})}{(g_j^v)^2 - 1} - c^v\Big(\frac{e^{-ik_0|r|}}{2ik_0|r|}\mathcal{B}' + \frac{\mathcal{B}}{2}\Big)\Big]. \tag{1.53}$$

In this expression, $\mathcal{B}$ and $\mathcal{B}'$ are independent of v and their choice is free. Therefore, for example, when (1.12) applies and a $g_j^e = g$ is close to 1, or when (1.7) applies with $N_{e,h} = 1$ as $g_1^e = 1/g_1^h = g$ is close to 1, we can take, $\mathcal{B}' = 1$ and $\mathcal{B} = \mathcal{K}_g$ to ameliorate the calculus (note: when $g_j^e \to 1$, let us mention that a g_m^h also exists, such that $a_j^e(g_j^e - 1)^{-1} + a_m^h(g_m^h - 1)^{-1}$ is bounded).

In a similar manner, concerning $\mathcal{U}_c$, we begin by writing:

$$\mathcal{U}_c = \Big[-(\partial_z^2 + k_0^2) \sum_{\epsilon'=-1,1} \Big(\sum_{j=1}^{N_{e(,h)}} \frac{\frac{\tau_{e(,h)} a_j^{e(,h)}}{g_j^{e(,h)}} (g_j^{e(,h)})^{\tau_{e(,h)}} (\mathcal{V}_{\epsilon'} + \epsilon' \mathcal{K}_{g_j^{e(,h)}})}{(g_j^{e(,h)})^{\tau_{e(,h)}} - \epsilon'}$$

$$+ c^{e(,h)} \mathcal{V}_{\epsilon'} \Big)\Big] + \Big[k_0^2 \sum_{\epsilon'=-1,1} \Big(\sum_{j=1}^{N_{h,(e)}} \frac{\frac{\tau_{h(,e)} a_j^{h(,e)}}{g_j^{h(,e)}} (g_j^{h(,e)})^{\tau_{h(,e)}} (\mathcal{V}_{\epsilon'} + \epsilon' \mathcal{K}_{g_j^{h(,e)}})}{(g_j^{h(,e)})^{\tau_{h(,e)}} - \epsilon'}$$

$$+ c^{h(,e)} \mathcal{V}_{\epsilon'} + \sum_{j=1}^{N_{e(,h)}} \frac{\frac{\tau_{e(,h)} a_j^{e(,h)}}{g_j^{e(,h)}} (g_j^{e(,h)})^{\tau_{e(,h)}} (\mathcal{V}_{\epsilon'} + \epsilon' \mathcal{K}_{g_j^{e(,h)}})}{(g_j^{e(,h)})^{\tau_{e(,h)}} - \epsilon'} + c^{e(,h)} \mathcal{V}_{\epsilon'} \Big)\Big] \qquad (1.54)$$

and then obtain the following reduction:

$$\mathcal{U}_c = 2k_0^2 \Big(\Big[((\tau_{e(,h)} - 1) \sum_{j=1}^{N_{e(,h)}} \frac{a_j^{e(,h)}}{g_j^{h(,e)}} + 4c^{e(,h)}) \frac{ie^{-ik_0|r|}}{2k_0|r|}$$

$$+ \sum_{j=1}^{N_{e(,h)}} a_j^{e(,h)} \mathcal{K}_{g_j^{e(,h)}} \Big] + \Big[\sum_{j=1}^{N_{e(,h)}} a_j^{e(,h)} \frac{\mathcal{K}_{g_j^{e(,h)}} - (g_j^{e(,h)})^{\tau_{e(,h)}} \mathcal{B}}{(g_j^{e(,h)})^2 - 1}$$

$$+ \sum_{j=1}^{N_{h,(e)}} a_j^{h(,e)} \frac{\mathcal{K}_{g_j^{h(,e)}} - (g_j^{h(,e)})^{\tau_{h(,e)}} \mathcal{B}}{(g_j^{h(,e)})^2 - 1} - \frac{\mathcal{B}}{2}(c^{e(,h)} + c^{h(,e)}) \Big] \Big), \qquad (1.55)$$

where we have used (1.50)–(1.52) and $\sum_{\epsilon'=-1,1} \frac{\epsilon' \tau g^{\tau-1}}{g^\tau - \epsilon'} = \frac{2}{g^2-1}$ as $\tau = \pm 1$. Let us note that the choice of $\mathcal{B}$ is free, while $\sum_{j=1}^{N_{e(,h)}} \frac{a_j^{e(,h)}}{g_j^{e(,h)}} = -2$ (respectively, $= a_{0-} - a_{0+}$) as (1.3) (respectively, (1.10)) applies.

Therefore, after summations over ϵ' and use of analytical reductions (1.51), (1.53) and (1.55) in (1.40)–(1.41), the field expressions (1.36)–(1.37) are reduced to expansions with second-order partial derivatives, covering any cases of dipole sources with arbitrary directions over any multimode planes, with (1.12) (or (1.7)) satisfied.

NOTE 1.12.– $\mathcal{S}_{e(,h)}$ remains definite for $g^{e(,h)} = 1$, because $\mathcal{V}_{\epsilon'=1} + \mathcal{K}_g \to 0$ when $g \to 1$, while it is singular for $g^{e(,h)} = -1$. Moreover, we have $g\mathcal{K}_g \to -\frac{e^{-ik_0|r|}}{ik_0|r|}$ when $g \to \infty$.

1.6. Integral reduction of $\mathcal{J}_g$ and exact remarkable expansion with error functions

1.6.1. *A non-singular integral for* $\mathcal{J}_g$ *with* $g = \sin\theta_1$ *as* $|Re(\theta_1)| \leq \pi/2$

To avoid any restriction on the sign of the real part of $g_j^{e,h}$, we developed in [1]–[2] a novel expression of $\mathcal{J}_g$ for arbitrary $g = \sin\theta_1$ with $|\mathrm{Re}(\theta_1)| \leq \pi/2$ as $|\arg(ik_0)| \leq \pi/2$:

$$\mathcal{J}_g(\rho, -z) = i\int_b^{i\infty} e^{-a\cos\alpha}d\alpha, \tag{1.56}$$

where the parameters a and b, with $|\mathrm{Re}(b)| \leq \pi$, $\mathrm{Re}(a) > 0$, are defined following:

$$-ia\sin b = ik_0R(\cos\varphi + \sin\theta_1), a\cos b = ik_0R(1 + \sin\theta_1\cos\varphi),$$
$$e^{\mp ib} = \frac{ik_0R}{a}(1 \pm \sin\theta_1)(1 \pm \cos\varphi),\ a = \epsilon ik_0R\sin\varphi\cos\theta_1, \tag{1.57}$$

with $z = R\cos\varphi$, $\rho = R\sin\varphi$, $R = \sqrt{\rho^2 + z^2}$, ϵ =sign(Re($ik_0R\sin\varphi\cos\theta_1$)) (Re($a$) $= 0$ being a limit case), $0 \leq \varphi < \pi/2$. So described, we can derive from these equalities that:

$$e^{\mp ib} = \epsilon\cot\Big(\frac{\pi}{4} \mp \frac{\theta_1}{2}\Big)\cot\Big(\frac{\pi}{4} \mp (\frac{\pi}{4} - \frac{\varphi}{2})\Big),$$
$$b = i\ln(\epsilon\cot(\pi/4 - \theta_1/2)) - i\ln(\tan(\varphi/2)),$$
$$|\tan^2(\frac{b}{2})|^\epsilon = |\frac{a\cos b - \epsilon a}{a\cos b + \epsilon a}| = |\frac{1 + \sin(\theta_1 - \varphi)}{1 + \sin(\theta_1 + \varphi)}|$$
$$= |\frac{e^{\mathrm{Im}\theta_1}e^{i(\frac{\pi}{2} - \mathrm{Re}\theta_1)} + e^{-i\varphi}}{e^{\mathrm{Im}\theta_1}e^{i(\frac{\pi}{2} - \mathrm{Re}\theta_1)} + e^{i\varphi}}|^2 \leq 1, \tag{1.58}$$

and thus $\frac{1-\epsilon}{4}\pi \leq |\mathrm{Re}(b)| \leq \frac{3-\epsilon}{4}\pi$, sign(Re($b$)) $= -\epsilon$ sign(Im($\sin\theta_1$)), while we have sign(Im(a)) $= -$ sign(arg(ik_0)) when $\epsilon = -1$.

It is worth noticing that, as $g = \sin\theta_1$ varies in the complex plane, this expression has a correct cut as ϵ changes of sign for Re$g \leq 0$, is singular for $g = -1$, and is regular elsewhere (note: for Re$g > 0$, the change of sign of ϵ does not induce a cut as g varies [1]). After writing:

$$\mathcal{J}_g(\rho, -z) = i\Big(\int_0^{i\infty} e^{-a\cos\alpha} d\alpha - \int_0^{b} e^{-a\cos\alpha} d\alpha\Big)$$
$$= -(K_0(a) + \mathcal{M}(a,b)), \tag{1.59}$$

with K_0 a modified Bessel function [30], the aim of the calculus will be the reduction of $\mathcal{M}(a,b) = i\int_0^b e^{-a\cos\alpha} d\alpha$.

In this respect, we will consider it by regions, with a value of ϵ defined so that:

$$|\tan^2(\frac{b}{2})|^{\epsilon} \leq 1 \text{ with } \epsilon = +1 \text{ or } -1, \text{ as } |\mathrm{Re}(b)| \leq \pi, \tag{1.60}$$

as already noted in (1.58), for which we can exhibit a remarkable link between $\mathcal{M}$ in each region, which will lead us to exact and asymptotic expansions everywhere.

NOTE 1.13.– Considering the definitions of a and b given in (1.57)–(1.58), we have:

$$(-i\tan(\frac{b}{2}))^{\epsilon} = \frac{1+\sin(\theta_1-\varphi)}{\cos\varphi+\sin\theta_1} = \frac{\cos\varphi+\sin\theta_1}{1+\sin(\theta_1+\varphi)},$$
$$-2a\sin^2(\frac{b}{2}) = a\cos b - a = ik_0R|^{(1+\sin(\theta_1-\varphi)) \text{ when } \epsilon=1}_{(1+\sin(\theta_1+\varphi)) \text{ when } \epsilon=-1},$$
$$2a\cos^2(\frac{b}{2}) = a\cos b + a = ik_0R|^{(1+\sin(\theta_1+\varphi)) \text{ when } \epsilon=1}_{(1+\sin(\theta_1-\varphi)) \text{ when } \epsilon=-1}. \tag{1.61}$$

NOTE 1.14.– For derivatives of this expression, it is worth noticing that:

$$\frac{\partial\varphi(\rho,z)}{\partial z} = -\frac{\rho}{R^2}, \frac{\partial b}{\partial z} = \frac{i}{\sin\varphi}\frac{\rho}{R^2} = \frac{i}{R}, \frac{\partial a}{\partial z} = 0,$$
$$\frac{\partial\varphi(\rho,z)}{\partial \rho} = \frac{z}{R^2}, \frac{\partial b}{\partial \rho} = \frac{-i}{\sin\varphi}\frac{z}{R^2} = \frac{-i}{\rho}\frac{z}{R}, \frac{\partial a}{\partial \rho} = \epsilon i k\cos\theta_1. \tag{1.62}$$

NOTE 1.15.– $\mathcal{J}_g$ is also related to the incomplete cylindrical function [32] and to the leaky aquifer function [33]–[34] (see also appendix D).

1.6.2. *Exact remarkable expansion of* $\mathcal{M}(a,b)$ *with error functions*

1.6.2.1. *The link between the cases* $\epsilon = +1$ *and* $\epsilon = -1$

The function

$$\mathcal{M}(a,b) = i\int_0^b e^{-a\cos\alpha} d\alpha, \tag{1.63}$$

is an entire function (holomorphic for any complex a and b). It has particular properties in each subset defined in (1.60) with $\epsilon = \pm 1$ as $|\mathrm{Re} b| \leq \pi$, starting with a remarkable link between the cases:

$$\epsilon = +1 \text{ when } |\tan^2(\frac{b}{2})| \leq 1, \text{ and } \epsilon = -1 \text{ when } |\tan^2(\frac{b}{2})| \geq 1,$$

$$(\text{as previously noted, } \epsilon = +1 \text{ or } -1 \text{ when } |\tan^2(\frac{b}{2})|^{\epsilon} \leq 1). \tag{1.64}$$

Indeed, if $\epsilon = -1$ and thus $|\tan^2(\frac{b}{2})| \geq 1$, we have, as $|\mathrm{Re}(b)| \leq \pi$:

$$\begin{aligned}\mathcal{M}(a,b) &= i\int_{\nu\pi}^{b} e^{-a\cos\alpha}d\alpha + i\int_0^{\nu\pi} e^{-a\cos\alpha}d\alpha \\ &= \mathcal{M}(-a, b-\nu\pi) + i\pi\nu I_0(a),\end{aligned} \tag{1.65}$$

where ν =sign(Re(b)), $|\mathrm{Re}(b-\nu\pi)| \leq \pi$, and I_0 is a modified Bessel function [30]. Considering that $b - \nu\pi$ satisfies:

$$|\tan^2(\frac{b-\nu\pi}{2})| = 1/|\tan^2(\frac{b}{2})| \leq 1, \tag{1.66}$$

the regions, defined in (1.60) (and (1.64)) with $\epsilon = +1$ and $\epsilon = -1$, perfectly correspond to each other from (1.65).

NOTE 1.16.– We note that $i\pi I_0(a)$ =sign(Im(a))($K_0(-a) - K_0(a)$) for Im(a) $\neq 0$ [30].

1.6.2.2. *Exact series for $\mathcal{M}(a,b)$ with error functions as $|\mathrm{Re}(b)| \leq \pi$*

In the case $\epsilon = 1$, that is, as $|\tan^2(\frac{b}{2})| \leq 1$, we can use $\sin(\alpha/2) = \sin(b/2)$ $(1-t)^{1/2}$ in (1.63), and obtain, in first step:

$$\mathcal{M}(a,b) = -\int_0^1 \frac{e^{-a\cos b}e^{2a(-i\sin(b/2))^2 t}(-i\sin(b/2))}{(1-t)^{\frac{1}{2}}(1+(-i\sin(\frac{b}{2}))^2(1-t))^{\frac{1}{2}}}dt. \tag{1.67}$$

If we only expand it for small $\sin(b/2)$, we would obtain a series similar to the ones used by Chang and Fisher [13], but we want to go beyond and define an exact series for the case $\epsilon = 1$, that is for argitrary finite b with $|\tan^2(\frac{b}{2})| \leq 1$ as $|\mathrm{Re}(b)| \leq \pi$.

For that, noting that $\mathrm{Re}(1/\cos^2(b/2)) = \mathrm{Re}(1 + \tan^2(b/2)) \geq 0$, and thus $|\arg(\cos(b/2))| \leq \pi/4$, we can let:

$$(1 + (-i\sin(\frac{b}{2}))^2(1-t))^{1/2} = \cos(\frac{b}{2})(1 + \tan^2(\frac{b}{2})t)^{1/2}, \tag{1.68}$$

as $0 \leq t \leq 1$, and thus derive:

$$\mathcal{M}(a,b) = e^{-a\cos b}\int_0^1 \frac{i\tan(\frac{b}{2})e^{2a(-i\sin(b/2))^2 t}}{(1-t)^{1/2}(1+\tan^2(\frac{b}{2})t)^{1/2}}dt, \tag{1.69}$$

so that, using the binomial expansion of $(1+\tan^2(\frac{b}{2})t)^{-1/2}$, we have:

$$\mathcal{M}(a,b) = i\sum_{m\geq 0} c_m(\tan(\frac{b}{2}))^{2m+1}e^{-a\cos b}\int_0^1 \frac{e^{2a(-i\sin(b/2))^2 t}t^m}{(1-t)^{1/2}}dt$$

$$= ie^{-a\cos b}\sum_{m\geq 0} c_m(\tan(\frac{b}{2}))^{2m+1}\frac{\partial^m(\frac{\sqrt{\pi}e^s\operatorname{erf}(s^{1/2})}{s^{1/2}})}{\partial s^m}|_{s=-2a\sin^2(\frac{b}{2})}, \tag{1.70}$$

for $\tan^2(\frac{b}{2}) \neq -1$, that is, $|b| < \infty$. In these expressions, $c_m = (-1)^m(2m)!/((m!)^2 2^{2m})$ is the binomial coefficient for the function $(1+t)^{-1/2}$, and erf is the error function [30]. We remark that the function $W_m(s) = \frac{e^{-s}\partial^m}{\partial s^m}(\frac{e^s\operatorname{erf}(s^{1/2})}{s^{1/2}})$, detailed in appendix D, satisfies for large $m \gg |s|$:

$$W_m(s) = W_m(0) + O(s/m^{3/2}) = m^{-1/2} + O(m^{-3/2})$$

$$\text{with } W_m(0) = \Gamma(1+m)/\Gamma(3/2+m) = O(m^{-1/2}), \tag{1.71}$$

and $c_m = O(1/m^{1/2})$, while $(-1)^{m+1}c_m W_m(0)$ are real negative numbers, which confirms, from Abel's test theorem, the convergence of the series in (1.70) for finite b with $|\tan^2(\frac{b}{2})| \leq 1$, that is, $\tan^2(\frac{b}{2}) \neq -1$.

This convergence can be improved by letting:

$$W_m(s) = (W_m(s) - W_m(0)) + W_m(0), \tag{1.72}$$

and noticing that the sum with $W_m(0)$ is $e^{-a}\mathcal{M}(0,b) = ibe^{-a}$; it gives us:

$$\mathcal{M}(a,b) = ie^{-a}([\sqrt{\pi}\tan(\frac{b}{2})\sum_{m\geq 0} c_m(\tan(\frac{b}{2}))^{2m}$$

$$\times (W_m(-2a\sin^2(\frac{b}{2})) - W_m(0))] + b). \tag{1.73}$$

In the case $\epsilon = -1$, that is, as $|\tan^2(\frac{b}{2})| > 1$, we can then simply exploit the equations (1.65)–(1.66) to reduce the problem of calculus of $\mathcal{M}(a,b)$ to the one of $\mathcal{M}(-a, b-\nu\pi)$, and letting in equations (1.70)–(1.73), $a \to -a$ and $b \to b-\nu\pi$, with ν =sign(Re(b)), we obtain, as $\epsilon = -1$:

$$\mathcal{M}(a,b) = -ie^{-a\cos b}\cot(\frac{b}{2})\sum_{m\geq 0} c_m(\cot(\frac{b}{2}))^{2m}$$

$$\times \frac{\partial^m}{\partial s^m}\left(\frac{\sqrt{\pi}e^s}{s^{1/2}}\text{erf}(s^{1/2})\right)|_{s=2a\cos^2(\frac{b}{2})} + i\pi\nu I_0(a)$$

$$= ie^a([-\sqrt{\pi}\cot(\frac{b}{2})\sum_{m\geq 0} c_m(\cot(\frac{b}{2}))^{2m}$$

$$\times (W_m(2a\cos^2(\frac{b}{2})) - W_m(0))] + (b-\nu\pi)) + i\pi\nu I_0(a). \tag{1.74}$$

1.6.3. *Exact expansion of $\mathcal{J}_g(\rho,-z)$ for arbitrary parameters*

We can now use $\mathcal{M}(a,b)$ given in (1.70)–(1.74), and directly derive, from (1.59) and (1.61), an exact convergent expansion of $\mathcal{J}_g(\rho,-z)$ with error function, as $z = R\cos\varphi > 0$ and $\rho = R\sin\varphi > 0$, for arbitrary $g = \sin\theta_1$, when $|\arg(ik_0)| \leq \pi/2$ and $|\text{Re}(\theta_1)| \leq \pi/2$ (i.e. $|\arg(\cos\theta_1)| \leq \pi/2$), following:

$$\mathcal{J}_g(\rho,-z) = f_g(\rho,-z) \text{ if } \epsilon = 1,$$

$$\mathcal{J}_g(\rho,-z) = f_g(\rho,-z) \text{ if } \epsilon = -1, \text{Re}(g) > 0,$$

$$\mathcal{J}_g(\rho,-z) = f_g(\rho,-z) - C_g(\rho) \text{ if } \epsilon = -1, \text{Re}(g) < 0, \tag{1.75}$$

where $f_g(\rho,-z)$ verifies:

$$f_g(\rho,-z) = -ie^{-ik_0R(1+\sin\theta_1\cos\varphi)}\sum_{m\geq 0} c_m\left(i\frac{1+\sin(\theta_1-\varphi)}{\cos\varphi+\sin\theta_1}\right)^{2m+1}$$

$$\times \frac{\partial^m}{\partial s^m}\left(\frac{\sqrt{\pi}e^s\text{erf}(s^{1/2})}{s^{1/2}}\right)|_{s=ik_0R(1+\sin(\theta_1-\varphi))} - K_0(ik_0\rho\cos\theta_1)$$

$$= -ie^{-ik_0\rho\cos\theta_1}([\sqrt{\pi}\sum_{m\geq 0} c_m\left(i\frac{1+\sin(\theta_1-\varphi)}{\cos\varphi+\sin\theta_1}\right)^{2m+1}$$

$$\times (W_m(ik_0R(1+\sin(\theta_1-\varphi))) - W_m(0))] + b_g(\varphi)), \tag{1.76}$$

with

$$W_m(s) = \frac{e^{-s}\partial^m}{\partial s^m}(\frac{e^s\text{erf}(s^{1/2})}{s^{1/2}}),$$

$$W_m(0) = \Gamma(1+m)/\Gamma(3/2+m) = O(m^{-1/2}), \tag{1.77}$$

(see details on $W_m(s)$ in appendix D), while $C_g(\rho)$ and $b_g(\varphi)$ are given by:

$$C_g(\rho) = K_0(-ik_0\rho\cos\theta_1) - K_0(ik_0\rho\cos\theta_1),$$

$$b_g(\varphi) = i\ln(\cot(\frac{\pi}{4}-\frac{\theta_1}{2})\cot(\frac{\varphi}{2})) - ie^{ik_0\rho\cos\theta_1}K_0(ik_0\rho\cos\theta_1). \tag{1.78}$$

Since we have:

$$\cot(\frac{\pi}{4}-\frac{\theta_1}{2})\cot(\frac{\varphi}{2}) = \frac{4\cos^2(\frac{\pi}{4}-\frac{\theta_1}{2})\cos^2(\frac{\varphi}{2})}{\sin\varphi\cos\theta_1},$$
$$\rho = R\sin\varphi \text{ and } K_0(a) \sim -\ln(a) \text{ as } a \to 0, \tag{1.79}$$

we note that

$$b_g(\varphi), \text{ as } \mathcal{J}_g(\rho,-z), \text{ diverges as } R \to 0 \text{ or } \theta_1 \to -\pi/2,$$
$$\text{but is bounded elsewhere.} \tag{1.80}$$

In these expressions, $c_m = (-1)^m(2m)!/((m!)^2 2^{2m})$ is the binomial coefficient for the function $(1+t)^{-1/2}$, and erf is the error function [30] (note: other properties are given in appendix D).

NOTE 1.17.– We can write $C_g(\rho) = i\pi w I_0(ik_0\rho\cos\theta_1)$ with w = sign $(\arg(ik_0\cos\theta_1))$. Since $|\mathrm{Re}(ik_0)| \leq \pi/2$ and $|\mathrm{Re}(\theta_1)| \leq \pi/2$, $|\arg(ik_0) + \arg(\cos\theta_1)| > \pi/2$ only if $\arg(ik_0)$ and $\arg(\cos\theta_1)$ have the same sign, and thus, by definition, we note that w =sign$(\arg(ik_0))$ when $\epsilon = -1$.

NOTE 1.18.– Since $1+\sin(\theta_1-\varphi) = 1+\cos(\theta_1-(\varphi+\pi/2)) = 2\cos^2(\frac{\theta_1-(\varphi+\pi/2)}{2})$, we note the correspondence of first term of our series with the approximation obtained by Wait [36].

1.7. Asymptotic expansion of $\mathcal{J}_g$ with complementary error functions

1.7.1. *Asymptotics of* $\mathcal{M}(a,b)$

1.7.1.1. *For large* a *and arbitrary* b *with* $|Re(b)| \leq \pi$

We can let $\frac{\mathrm{erf}(u)}{u} = \frac{1-\mathrm{erfc}(\delta u)}{\delta u}$ with $\delta = +1$ or -1, in (1.70) and (1.74), with the goal of using the well-known asymptics of complementary error function erfc [30]. On this occasion, we remind that $\mathcal{M}(a,b)$ is an entire function with no cut, and that δ, which is initially a free parameter ($+1$ or -1), will be fixed only for a correct definition of asymptotics.

Subsequently, when $\epsilon = 1$, we write:

$$\mathcal{M}(a,b) = \lim_{M\to\infty}[ie^{-a}\tan(\frac{b}{2})\sum_{m=0}^{M} c_m(\tan(\frac{b}{2}))^{2m}(e^{-s}\frac{\partial^m(\frac{\sqrt{\pi}e^s}{s^{1/2}})}{\partial s^m})$$
$$-ie^{-a\cos b}\sum_{m=0}^{M} c_m(\tan(\frac{b}{2}))^{2m+1}\frac{\partial^m(\frac{\sqrt{\pi}e^s\mathrm{erfc}(s^{1/2})}{s^{1/2}})}{\partial s^m}]_{s^{1/2}=-i\delta\sqrt{2a}\sin(\frac{b}{2})}. \tag{1.81}$$

At this step, we then notice that the term with e^{-a} factor has the same asymptotics as $-\delta K_0(a)$. Indeed, using that:

$$\frac{\partial^m}{\partial s^m}\left(\frac{e^s}{s^{1/2}}\right) = \frac{e^s}{s^{1/2}} \sum_{p=0}^{m} \frac{(-1)^{p+1} m!(-1 \times 1 \times ... \times (2p-1))}{2^p s^p p!(m-p)!}, \tag{1.82}$$

with $s^{\frac{1}{2}} = \delta z^{\frac{1}{2}}$, we remark that, noting $z^{\frac{1}{2}} = -i\sqrt{2a}\sin(\frac{b}{2})$ and $v = \tan^2(\frac{b}{2})$, we can write, as $\epsilon = 1$ and thus $|v| < 1$, for large a and M:

$$\begin{aligned}
&-i\sqrt{\pi}e^{-a}\tan(\frac{b}{2})\sum_{m=0}^{M} c_m(\tan^2(\frac{b}{2}))^m (e^{-z}\frac{\partial^m}{\partial z^m}(\frac{e^z}{z^{\frac{1}{2}}}))\\
&= i\sqrt{\pi}e^{-a}\sum_{p=0}^{M}(\tan(\frac{b}{2}))^{2p+1}\frac{(-1)^p(-1\times ... \times (2p-1))}{2^p(-i\sqrt{2a}\sin(\frac{b}{2}))^{2p+1}}\\
&\times \frac{1}{p!}\sum_{m\geq p}^{M} c_m \frac{m!}{(m-p)!}v^{m-p} \sim i\sqrt{\pi}e^{-a}\tan(\frac{b}{2})\\
&\times \sum_{p\geq 0} v^p \frac{(-1)^p(-1\times ... \times (2p-1))}{2^p p!(-i\sqrt{2a}\sin(\frac{b}{2}))^{2p+1}} \frac{\partial^p(\frac{1}{(1+v)^{\frac{1}{2}}})}{\partial v^p}\\
&= \sqrt{\pi}e^{-a}\sum_{p\geq 0}\frac{(-1)^{p+1}(-1\times ... \times (2p-1))^2}{2^{2p}(2a)^{p+\frac{1}{2}}p!},
\end{aligned} \tag{1.83}$$

which is equal to the known asymptotics of $K_0(a)$ [30], and we then have:

$$\begin{aligned}
&\mathcal{M}(a,b) \sim -\delta K_0(a)\\
&- ie^{-a\cos b}\tan(\frac{b}{2})\sum_{m=0}^{M} c_m(\tan(\frac{b}{2}))^{2m}\frac{\partial^m}{\partial s^m}\left(\frac{\sqrt{\pi}e^s \mathrm{erfc}(s^{\frac{1}{2}})}{s^{\frac{1}{2}}}\right)\Big|_{s^{\frac{1}{2}}=-i\delta\sqrt{2a}\sin(\frac{b}{2})},
\end{aligned} \tag{1.84}$$

where $\delta = +\text{or}-1$ is chosen such that $\mathrm{Re}(s^{1/2}) > 0$, and $\frac{\partial^m}{\partial s^m}(\frac{e^s}{s^{1/2}}\mathrm{erfc}(s^{1/2})) = O(\frac{1}{s^{m+1}})$ for large s, with $\mathrm{erfc}(-u) = 2-\mathrm{erfc}(u)$ [30] so that this asymptotics can be considered as uniform as δ changes of sign.

In complements, when $\epsilon = -1$, we can use (1.65)–(1.66), so that, after letting $a \to -a$ and $b \to b - \nu\pi$ with $\nu = \mathrm{sign}(\mathrm{Re}(b))$ in (1.84) to express $\mathcal{M}(-a, b-\nu\pi)$, we derive that:

$$\mathcal{M}(a,b) \sim -\delta'' K_0(-a) + i\pi\nu I_0(a)$$
$$+ ie^{-a\cos b}\sum_{m=0}^{M} c_m(\cot(\frac{b}{2}))^{2m+1}\frac{\partial^m}{\partial s^m}\Big(\frac{\sqrt{\pi}e^s \text{erfc}(s^{\frac{1}{2}})}{s^{\frac{1}{2}}}\Big)\Big|_{s^{\frac{1}{2}}=-i\delta''\sqrt{-2a}\sin(\frac{b-\nu\pi}{2})}, \tag{1.85}$$

where $\delta'' = +\text{or}-1$ is chosen such that $\text{Re}(s^{1/2}) > 0$, with $\text{erfc}(-u) = 2-\text{erfc}(u)$. In the expression (1.85), we can use that, when $\text{Re}(a) > 0$:

$$-i\sqrt{-2a}\sin((b-\nu\pi)/2) = -\tau\sqrt{2a}\cos(b/2),$$
$$i\pi I_0(a) = \text{sign}(\text{Im}(a))(K_0(-a) - K_0(a)), \tag{1.86}$$

with $\tau = -\nu\text{sign}(\text{Im}(a))$. Letting $\delta = \delta''\tau$ and $s^{1/2} = -\delta\sqrt{2a}\cos(b/2)$, we can write for (1.85):

$$-\delta'' K_0(-a) + i\pi\nu I_0(a) = -\tau(\delta+1)K_0(-a) + \tau K_0(a),$$
$$\text{with } \delta = +\text{or} - 1 \text{ such that } \text{Re}(s^{1/2} = -\delta\sqrt{2a}\cos(b/2)) > 0. \tag{1.87}$$

(note: the term $\sqrt{(.)}$ has an always positive real part [30], in contrary of $(.)^{1/2}$ in sense of Riemann).

1.7.1.2. *Extension of domain of validity of asymptotics (1.84)*

In the previous section, we derived asymptotics with erfc functions from our exact series with erf functions, when a is large. We now proceed to consider an asymptotics in regions where $\text{Im}(b)$ is large without assuming a is too, when $\text{Re}(a) > 0$ and $|\text{Re}(b)| \leq \pi$.

Considering $s = -2a\sin^2(\frac{b}{2}) \neq 0$, $\text{Re}(s^{1/2} = -i\delta_\epsilon\sqrt{2a}\sin(\frac{b}{2})) > 0$ with $\delta_\epsilon = +1$ or -1 ($|\arg(s)| \neq \pi$), when $\text{Re}(a) > 0$ and $|\text{Re}(b)| \leq \pi$ ($\text{Re}(-i\delta_\epsilon b) > 0$), for any ϵ ($+1$ or -1), we can rearrange (1.67) and write:

$$\mathcal{M}(a,b) = i\delta_\epsilon\int_0^{\delta_\epsilon b} e^{-a\cos\alpha}d\alpha = -i\delta_\epsilon\int_{\delta_\epsilon b}^{i\infty} e^{-a\cos\alpha}d\alpha - \delta_\epsilon K_0(a)$$
$$= \delta_\epsilon\int_{-\chi\infty}^{0}\frac{-i\sin(\frac{\delta_\epsilon b}{2})e^{-a\cos(\delta_\epsilon b)}e^{st}}{(1-t)^{\frac{1}{2}}(1+(-i\sin(\frac{\delta_\epsilon b}{2}))^2(1-t))^{\frac{1}{2}}}dt - \delta_\epsilon K_0(a)$$
$$= \int_{-\chi\infty}^{0}\frac{-i\sin(\frac{b}{2})e^{-a\cos b}e^{st}}{(1-t)^{\frac{1}{2}}(1+(-i\sin(\frac{b}{2}))^2(1-t))^{\frac{1}{2}}}dt - \delta_\epsilon K_0(a)$$
$$= \int_{1}^{\chi\infty}\frac{-i\sin(\frac{b}{2})e^{-a\cos b}e^{s(1-t')}}{(t')^{\frac{1}{2}}(\cos^2(\frac{b}{2})(1+\tan^2(\frac{b}{2})(1-t')))^{\frac{1}{2}}}dt' - \delta_\epsilon K_0(a), \tag{1.88}$$

where we consider $st = s(1-t') \leq 0$ along the integration paths with $\chi = e^{-i\arg(s)}$, and note that $\mathrm{Re}(\cos(\frac{b}{2})) > 0$ as $|\mathrm{Re}(b)| < \pi$. For convergence, we are limited to not let the terms $(1-t)^{\frac{1}{2}}$ and $(1+(-i\sin(\frac{b}{2}))^2(1-t))^{\frac{1}{2}}$ vanish along this path, that is:

$$s - st \neq 0 \text{ and } 2a\cos^2(\frac{b}{2}) - st \neq 0 \text{ as } st \leq 0. \tag{1.89}$$

So defined, the expression (1.88) for $\mathcal{M}(a,b)$, as $\mathrm{Re}(a) > 0$ and $|\mathrm{Re}(b)| \leq \pi$:

$$\begin{aligned}&\text{is valid as } a \text{ and } b \text{ vary continuously, from region with } -2a\sin^2(\frac{b}{2}) < 0,\\ &\text{keeping } \delta_\epsilon \text{ constant, } +1 \text{ or } -1 \text{ with } \mathrm{Re}(s^{1/2} = -i\delta_\epsilon\sqrt{2a}\sin(\frac{b}{2})) > 0,\\ &\text{without crossing the domain with } 2a\cos^2(\frac{b}{2}) < 0,\end{aligned} \tag{1.90}$$

where we remind that $\mathcal{M}(a,b)$ is an entire function (no cut).

We can then expand the integral in (1.88), asymptotically for large $\sqrt{2a}\sin(\frac{b}{2})$. For that, we consider a Taylor expansion of $(1+\tan^2(\frac{b}{2})(1-t'))^{-1/2}$ in vicinity of $t' = 1$, then using $(t')^{1/2}$ as a new variable of integration, we derive, from complementary error function definition [30], an asymptotic series, which, as expected, perfectly corresponds to (1.84), initially defined for $\epsilon = 1$.

On this occasion, we have then enlarged the domain of validity of (1.84) with large $\sqrt{2a}\sin(\frac{b}{2})$ without considering large a, for arbitrary choice of $\epsilon = +1$ or -1, provided (1.90) is verified, which can be summarized as follows:

$$\begin{aligned}&\text{The asymptotics (1.84) of } \mathcal{M} \text{ for large } \sqrt{2a}\sin(\frac{b}{2}) \text{ with } \mathrm{Re}(-i\delta_\epsilon\sqrt{2a}\sin(\frac{b}{2})) > 0\\ &\text{was initially definite for } \epsilon = 1 \text{ as } a \text{ is large; Considering (1.90), it can also be used}\\ &\text{without assuming large } a, \text{ at fixed } \delta_\epsilon, \text{ for } \mathrm{Re}(a) > 0, |\mathrm{Re}(b)| \leq \pi, \text{ as } a \text{ and } b\\ &\text{continuously vary from region with } -2a\sin^2(\frac{b}{2}) < 0, \text{ without } 2a\cos^2(\frac{b}{2}) < 0.\end{aligned} \tag{1.91}$$

NOTE 1.19.– For $\mathrm{Re}(a) > 0$, $|\mathrm{Re}(b)| \leq \pi$, we note that $-2a\sin^2(\frac{b}{2}) < 0$ and $2a\cos^2(\frac{b}{2}) < 0$ are both satisfied only for $a > 0$, $|\mathrm{Re}(b)| = \pi$ with $b \neq \pi$. Except in vicinity of this case, this asymptotics is uniform when δ_ϵ changes of sign (i.e. as we cross the region $-2a\sin^2(\frac{b}{2}) < 0$), in particular for large a since $-2a\sin^2(\frac{b}{2}) = 2a\cos^2(\frac{b}{2}) - 2a$.

1.7.2. *Asymptotics of* $\mathcal{J}_g$ *for large* $|k_0\rho\cos\theta_1|$ *or* $|k_0R(1+\sin(\theta_1\mp\varphi))|$

Considering the expressions given in (1.57)–(1.58) and (1.61), we now use the previous asymptotic expansion (1.84)–(1.85) of $\mathcal{M}(a,b)$ when a or $2a\sin^2(\frac{b}{2})$ (and thus, in particular if k_0R or g) is large. On this occasion, we refine the asymptotics already given in [1] in a part of the domain with $\epsilon=-1$ and $\mathrm{Re}\theta_1<0$, considering that (1.84) is also the asymptotics of (1.88), provided (1.90)–(1.91) is satisfied.

Using, as $\epsilon=-1$, the equalities in (1.61), the condition (1.90)–(1.91) for a use of (1.84) is verified on the left of the path S' with $ik_0R(1+\sin(\theta_1+\varphi))\leq 0$ (Figure 1.2), and therefore, we will consider the asymptotics of $\mathcal{J}_g$, from (1.84), when $\epsilon=1$ and on the left of S' when $\epsilon=-1$, and, from (1.85), when $\epsilon=-1$ on the right of S' (see Figure 1.2).

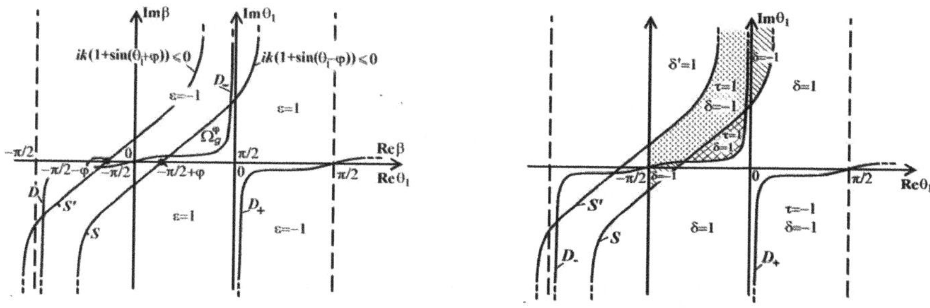

Figure 1.2. *For* $\arg(ik)>0$ *with* $k\equiv k_0$, $|Re\theta_1|<\pi/2$, *definitions of* S, S' *and* Ω_g^φ, *and values of* ϵ *(*$\epsilon=1$ *between* D_+ *and* D_-*, and* $\epsilon=-1$ *elsewhere),* τ, δ *and* δ'

1.7.2.1. *Asymptotics of* $\mathcal{J}_g$ *from (1.84) as* $\epsilon=1$, *and also on the left side of* S' *as* $\epsilon=-1$

Considering the complete domain of validity of (1.84), we can now write:

$$\mathcal{J}_g(\rho,-z)\sim-(1-\delta_\epsilon)K_0(a)+i\sqrt{\pi}e^{-a\cos b}\tan(\frac{b}{2})$$

$$\times\sum_{m\geq0}c_m(\tan(\frac{b}{2}))^{2m}\frac{\partial^m(\frac{e^s\mathrm{erfc}(s^{\frac{1}{2}})}{s^{\frac{1}{2}}})}{\partial s^m}\Big|_{s^{\frac{1}{2}}=-i\delta_\epsilon\sqrt{2a}\sin(\frac{b}{2})},\qquad(1.92)$$

when we have $\epsilon=1$, but also, in complements, when $\epsilon=-1$ on the left of S' (Figure 1.2) where (1.90) is verified, with the parameter δ_ϵ chosen $+$or-1 so that $\mathrm{Re}s^{1/2}>0$.

In this expression, δ_ϵ denotes $((1+\epsilon)\delta+(1-\epsilon)\delta')/2$ where the parameter δ (respectively, δ'), as $\epsilon=1$ (respectively, $\epsilon=-1$), changes of sign when we cross the line S with $ik_0R(1+\sin(\theta_1-\varphi))<0$ (respectively, the line S' with $ik_0R(1+\sin(\theta_1+\varphi))<0$) (Figure 1.2), that is, the line with $\text{Re}(\theta_1)=\varphi-\frac{\pi}{2}+\mathcal{G}(\text{Im}\theta_1)$ (respectively, with $\text{Re}(\theta_1)=-\varphi-\frac{\pi}{2}+\mathcal{G}(\text{Im}\theta_1)$), where $\mathcal{G}(x)=2\arctan(\tan(\frac{\arg(ik_0)}{2})\tanh(\frac{x}{2}))$.

Considering (1.57)–(1.58) and (1.61), we can use in (1.92):

$$-i\delta_\epsilon\sqrt{2a}\sin(\frac{b}{2})=\sqrt{ik_0R(1+\sin(\theta_1-\epsilon\varphi))},$$

$$\tan(\frac{b}{2})=i\frac{1+\sin(\theta_1-\epsilon\varphi)}{\cos\varphi+\sin\theta_1},\ a\cos b=ik_0R(1+\sin\theta_1\cos\varphi),$$

$$a=\epsilon ik_0R\sin\varphi\cos\theta_1 \text{ with } \epsilon=\text{sign}(\text{Re}(ik_0R\sin\varphi\cos\theta_1)),\tag{1.93}$$

where we exploit that $\sqrt{(.)}$ has an always positive real part [30] (in contrary of $(.)^{1/2}$ in sense of Riemann).

1.7.2.2. *Asymptotics of $\mathcal{J}_g$ from (1.85)–(1.87) on the right side of S' when $\epsilon=-1$*

On the right of the line S' (the line in Figure 1.2 with $ik_0R(1+\sin(\theta_1+\varphi))<0$), we can use (1.85)–(1.87) when $\epsilon=-1$, and write:

$$\mathcal{J}_g(\rho,-z)\sim-(1+\tau)K_0(a)+\tau(\delta+1)K_0(-a)$$

$$-i\sqrt{\pi}e^{-a\cos b}\sum_{m\geq 0}c_m(\cot(\frac{b}{2}))^{2m+1}\frac{\partial^m(\frac{e^s\text{erfc}(s^{\frac{1}{2}})}{s^{\frac{1}{2}}})}{\partial s^m}\Big|_{s^{\frac{1}{2}}=-\delta\sqrt{2a}\cos(\frac{b}{2})},\tag{1.94}$$

where δ is chosen $+$or-1 so that $\text{Re}(s^{1/2})>0$ and $\frac{\partial^m}{\partial s^m}(\frac{e^s}{s^{1/2}}\text{erfc}(s^{1/2}))$ is $O(\frac{1}{s^{m+1}})$ for large s. We have $\tau=-\nu\text{sign}(\text{Im}(a))$, $\nu=$sign$(\text{Re}(b))=-\epsilon\text{sign}(\text{Im}(\sin\theta_1))$, and since $\text{sign}(\text{Im}(a))=-\text{sign}(\arg(ik_0))$ as $\epsilon=-1$, $\tau=$sign$(\text{Im}(\sin\theta_1))\text{sign}(\arg(ik_0))$.

By definition, the parameter δ changes of sign when we cross the line S with $ik_0R(1+\sin(\theta_1-\varphi))<0$ (Figure 1.2), and considering (1.57)–(1.58) and (1.61), we can use in (1.94):

$$-\delta\sqrt{2a}\cos(\frac{b}{2})=\sqrt{ik_0R(1+\sin(\theta_1-\varphi))},$$

$$\cot(\frac{b}{2})=-i\frac{1+\sin(\theta_1-\varphi)}{\cos\varphi+\sin\theta_1},\ a\cos b=ik_0R(1+\sin\theta_1\cos\varphi),\tag{1.95}$$

where we exploit that $\text{Re}\sqrt{(.)}>0$ [30] (in contrary of $(.)^{1/2}$ in sense of Riemann).

1.7.2.3. *Concerning δ, δ' and δ_ϵ*

Considering (1.92) and (1.94), we note that, in the region with $\epsilon=1$ (Figure 1.2):

$$\delta_\epsilon=\delta=1 \text{ on the right side of } S,$$

$$\delta_\epsilon=\delta=-1 \text{ on the left side of } S,\tag{1.96}$$

while, in the zone with $\epsilon = -1$ (Figure 1.2):

$$\delta = \tau \text{ on the right side of } S,$$

$$\delta = -\tau \text{ on the left side of } S,$$

$$\delta_\epsilon = \delta' = 1 \text{ on the left side of } S', \tag{1.97}$$

with τ =sign(Im($\sin\theta_1$))sign(arg(ik_0)). From ϵ =sign(Re($ik_0\cos\theta_1$)), we have $\epsilon = 1$ between D_+ and D_-, and $\epsilon = -1$ elsewhere (Figure 1.2).

1.8. Forward and backward guided waves

1.8.1. *Guided waves from asymptotics* (*1.92*) *and* (*1.94*)

Excited guided waves refer to $K_0(\pm\epsilon ik_0\rho\cos\theta_1)e^{ik_0z\sin\theta_1}$ terms in the expression of $\mathcal{J}_g(\rho,-z)e^{ik_0z\sin\theta_1}$ with $g = \sin\theta_1$, for each mode attached to $g = g_j^{e(,h)}$. They behave like $\sqrt{\frac{\pi}{\pm 2\epsilon ik_0\rho\cos\theta_1}}e^{ik_0R\sin(\theta_1\mp\epsilon\varphi)}$ for large R.

Relatively to their asymptotic phase, these guided waves propagate like forward waves $e^{-ik_0(\rho\cos\theta_1 - z\sin\theta_1)}$ as $\pm\epsilon = 1$ and backward waves $e^{ik_0(\rho\cos\theta_1 + z\sin\theta_1)}$ as $\pm\epsilon = -1$. They are present in angular domains (depending on φ) where $e^{ik_0R\sin(\theta_1\mp\epsilon\varphi)}$ remain bounded at infinity, as already noticed by Tamir and Oliner [35] for the 2D case. We can then analyze the situation from (1.92)–(1.94) and Figure 1.2.

1.8.1.1. *In passive zone as Re*$(\theta_1) > 0$

In passive zone with Re$(\theta_1) > 0$, we have only $K_0(ik_0\rho\cos\theta_1)e^{ik_0z\sin\theta_1}$ term which is only present in the region (depending on φ) on the left of S where $\epsilon = 1$, that is clearly associated to the excitation of forward proper waves, vanishing as ρ or z increases (proper mode).

1.8.1.2. *In active zone as Re*$(\theta_1) < 0$

In the active zone with Re$(\theta_1) < 0$, the situation is more complex. Both $K_0(\pm ik_0\rho\cos\theta_1)e^{ik_0z\sin\theta_1}$ terms can be present, and excited waves can be attached to improper or backward modes:

a) In the region with $\epsilon = -1$ and Re$(\theta_1) < 0$, we have a mix of backward and forward waves in a limited domain of excitation Ω_g^φ, while, in the domain limited by S and S' also depending on φ, the forward waves disappear and backward term $K_0(-ik_0\rho\cos\theta_1)e^{ik_0z\sin\theta_1}$ remains alone. In these domains depending on φ, forward and backward waves are bounded, even if forward waves grows at fixed z as ρ increases.

b) In the region with $\epsilon = 1$ and $\text{Re}(\theta_1) < 0$, we have forward guided wave term $K_0(ik_0\rho\cos\theta_1)e^{ik_0 z\sin\theta_1}$ excited on the left of S, which grows at fixed ρ as z increases, but remains bounded because of its limited region of excitation depending on φ.

NOTE 1.20.– Let us note that, for meromorphic reflection coefficients, excited guided waves does not involve lateral waves (branch-cut waves) [5].

NOTE 1.21.– For positive $\text{Re}(\varepsilon)$ and $\text{Re}(\mu)$ with real k_0, quarter-plane with $\text{Re}(g_j^{e(,h)}) < 0$ and $\arg(ik)\text{Im}(g_j^{e(,h)}) > 0$ is generally empty, which bans the presence of backward waves. This explains why backward waves is generally attached to the presence of metamaterials or plasmas-like materials (see also (1.16) and properties (d) and (e)).

1.8.2. *Guided-like waves from the exact expansion* (*1.75*)–(*1.76*)

In addition to asymptotics given in (1.92)–(1.94), we have developed in (1.75)–(1.76) the exact expansions of $\mathcal{J}_g(\rho, -z)$ with error functions for any mode parameter $g = \sin\theta_1$. It allows us to consider in general, without any complex approximation, its exact behavior at any finite distance from the sources.

When the term $ik_0R(1 + \sin(\theta_1 - \varphi))$ is small, $K_0(\pm ik_0\rho\cos\theta_1)$ terms are predominant in (1.75)–(1.76). This guided-like waves behavior is important at small distance but also at large bounded distance for very small g when $\varphi \simeq \pi/2$ (closed to the plane). It has to be distinguished from actual guided waves behavior, attached to $K_0(\pm ik_0\rho\cos\theta_1)$ terms in asymptotics (1.92)–(1.94) for large unbounded R.

This behavior confirms, in a simple exact way, the comments of J.R. Wait given in [36]–[37], which expresses that surface waves like excitation is possible in limited region around the source, even if they are attached to modes not excited at very large distance; or said more precisely, every mode attached to a pole $g = \sin\theta_1$ of the reflection coefficients gives a guided-like waves contribution in vicinity of the source, provided $ik_0R(1 + \sin(\theta_1 - \varphi))$ is small, but only some of them develop excited guided waves in asymptotics at large unbounded distance.

1.9. References

[1] Bernard, J.M.L. (2018). Propagation over a constant impedance plane: Arbitrary primary sources and impedance, analysis of cut in active case, exact series, and complete asymptotics. *IEEE TAP*, 66, 12.

[2] Bernard, J.M.L. (2010). On the expression of the field scattered by a multimode plane. *The Quarterly Journal of Mechanics and Applied Mathematics*, 63(3), 237–266.

[3] Jones, D.S. (1964). *The Theory of Electromagnetism*. Pergamon Press, London.

[4] Wait, J.R. (1970). *Electromagnetic Waves in Stratified Media*. Pergamon Press, London.

[5] Brekhovskikh, L.M. (1980). *Waves in Layered Media*. Academic Press, London.

[6] Felsen, L.B. and Marcuvitz, N. (1973). *Radiation and Scattering of Waves*. Prentice Hall, Englewood Cliffs, NJ.

[7] Kawai, T., Hidaka, T., Nakajima, T. (1982). Sound propagation above an impedance boundary. *J. Sound Vibr.*, 83(1), 125–138.

[8] Maliuzhinets, G.D. (1948). About one generalisation of the Weyl formula for the wave field over an absorbing plane. *Dokl. Akad. Nauk.*, 60(3), 367–370.

[9] Thomasson, S.I. (1980). A powerful asymptotic solution for sound propagation above an impedance boundary. *Acustica*, 45(2), 122–125.

[10] Sarabandi, K., Casciato, M.D., Koh, I.S. (2002). Efficient calculation of the fields of a dipole radiating above an impedance surface. *IEEE Trans. AP*, 50(9), 1222–1234.

[11] Koh, I.S. and Yook, J.G. (2006). Exact closed-form expression of a Sommerfeld integral for the impedance plane problem. *IEEE Trans. AP*, 54(9), 2568–2576.

[12] Attenborough, K. (1980). Propagation of sound above a porous half space. *JASA*, 68(5), 1493–1501.

[13] Chang, D.C. and Fisher, R.J. (1974). A unified theory on radiation of a vertical electric dipole above dissipative earth. *Radio Science*, 9(12), 1129–1138.

[14] Lindell, I.V. (1995). *Methods for Electromagnetic Field Analysis*. IEEE Press, Piscataway.

[15] Holloway, C.L., Kuester, E.F., Haddab, A.H. (2019). Retrieval approach for determining surface susceptibilities and surface porosities of a symmetric metascreen from reflection and transmission coefficients. *PIER*, 166, 1–22.

[16] Shu, W. and Song, J.M. (2006). Mode spectrum of a grounded dielectric slab with double negative metamaterials. *PIER*, 65, 103–123.

[17] Dong, J.-F. and Li, J. (2012). The reflection and transmission of electromagnetic waves by a uniaxial chiral slab. *PIER*, 127, 389–404.

[18] Marin, M.A. and Pathak, P.H. (1992). An asymptotic closed-form representation for the grounded double-layer surface Green's function. *IEEE Trans. AP*, 40(11), 1357–1366.

[19] Valerio, G., Jackson, D.R., Galli, A. (2010). Fundamental properties of surface waves in lossless stratified structures. *Proc. R. Soc. A*, 466, 2447–2469.

[20] Knoesen, A., Moharam, M.G., Gaylord, T.K. (1985). Electromagnetic propagation at interfaces and in waveguides in uniaxial crystals. *Ann. Phys., B*, 38, 171–178.

[21] Loseth, L.O. and Ursin, B. (2007). Electromagnetic fields in planarly layered anisotropic media. *Geophys. J. Int.*, 170, 44–80.

[22] Qiu, C.W., Hu, L., Xu, X., Feng, Y. (2009). Spherical cloaking with homogeneous isotropic multilayered structures. *Phys. Rev. E,* 79, 047602.

[23] Harrington, R.F. (1961). *Time-Harmonic Electromagnetic Field.* McGraw-Hill, New York.

[24] Korevaar, J. and Wiengerinck, J. (1985). A representation of mixed derivatives with an application to the edge-of-the-wedge theorem. *Nedel. Akad. Wetensch. Indag. Math.*, 47(1), 77–86.

[25] Pascoe, J.E. (2019). The wedge of the edge theorem: Edge of the wedge type phenomenon within the common real boundary. *Canadian Math. Bulletin*, 62(2), 417–427.

[26] Bernard, J.M.L. (1987). Diffraction by a metallic wedge covered with a dielectric material. *Wave Motion*, 9, 543–561.

[27] Osipov, A.V. (1994). Diffraction by a wedge with higher-order boundary conditions. *Radio Science*, 31(6), 1705–1720.

[28] Tuzhilin, A.A. (1973). Diffraction of plane sound wave in an angular domain with abslutely hard and slippery face bounded by thin elastic plates. *Different Urav.*, 9(10), 1875–1888.

[29] Vladimirov, V.S. (1979). *Generalized Functions in Mathematical Physics*. MIR Publishers, Moscow.

[30] Abramowitz, M. and Stegun, I. (1972). *Handbook of Mathematical Functions*. Dover Publications Inc., New York.

[31] Sommerfeld, A. (1909). Uber die ausbreitung der wellen in der drahtlosen telegraphie. *Analen Physik*, 28, 665–736.

[32] Agrest, M.M. and Maksimov, M.S. (1971). *Theory of Incomplete Cylindrical Functions and their Applications*. Springer, New York.

[33] Harris, F.E. (2008). Incomplete Bessel, generalized incomplete gamma or leaky aquifer functions. *J. Comp. Appl. Math.*, 215, 260–269.

[34] Temme, N.M. (2009). The leaky aquifer function revisited. *Int. J. Quantum Chem.*, 109, 2826–2830.

[35] Tamir, T. and Oliner, A.A. (1963). Guided complex waves. Part 1: Fields at an interface. *Proc. IEEE*, 110(2), 310–324.

[36] Wait, J.R. (1957). Excitation of surface waves on conducting, stratified, dielectric clad, corrugated surface. *Journal of Research National Bureau of Standards*, 59(6), 365–377.

[37] Wait, J.R. (1998). The ancient and modern history of EM ground-wave propagation. *IEEE Antennas and Propagation Magazine*, 40(5), 7–24.

[38] Jones, D.S. (1994). *Methods in Electromagnetism Wave Propagation*. IEEE Press, Piscataway.

[39] Bernard, J.M.L. (2013). On the scattering by a cavity in an impedance plane in three dimensions: Boundary integral equations and novel Green's function. *Quarterly Journal of Mechanics and Applied Mathematics*, 66(4), 609–639.

1.A. Appendix A: complex spectral properties

1.A.1. *Wave spectrum and reflection coefficients*

We consider the plane wave spectrum (F_E, F_H) attached to the representation [3] of an electromagnetic field (E, H), in a half-space free of sources $\pm z \geq 0$, except possibly at positions of some sources on boundary $z = 0$:

$$E(x, y, z) = \frac{i}{4\pi} \int_{R^2} F_E(p, q) \frac{e^{-2\pi i(px+qy\pm sz)}}{s} dpdq,$$

$$H(x, y, z) = \frac{i}{4\pi} \int_{R^2} F_H(p, q) \frac{e^{-2\pi i(px+qy\pm sz)}}{s} dpdq,$$

$$\widehat{u} = \frac{2\pi}{k_0}(p, q, \pm s)^t,\ Z_0 F_H = \widehat{u} \wedge F_E,\ \widehat{u}.F_E = 0, \tag{1.98}$$

with $\text{Re}(is) \geq 0$, where $2\pi|_q^p = k_0 \sin\beta|_{\sin\phi}^{\cos\phi}$ and $2\pi s = k_0 \cos\beta$, $0 < \phi \leq 2\pi$, β varies from 0 to $i\infty + \arg(ik_0)$ as $\text{Re}(ik_0 \sin\beta) = 0$. From regularity of electromagnetic field as $\pm z \geq 0$, except possibly at $z = 0$, $F_{E,H}(p, q)$ is analytic on R^2, except possibly at infinity. We note that $(\frac{2\pi}{k_0})^2 dpdq = \sin\beta\cos\beta d\beta d\phi$ (see note 1.23), $\widehat{u}$ is the unit vector of propagation of each plane wave composing the spectrum, with $\widehat{u}.\widehat{u} = 1$ ($\neq \widehat{u}.\widehat{u}^*$ when $\widehat{u}$ is complex) and $Z_0 = (\mu_0/\varepsilon_0)^{1/2}$.

The representation (1.98) applies for the field scattered by plane $z = 0$ with (F_E^s, F_H^s) spectrum, on and above it, and the field incident on this plane with (F_E^i, F_H^i) spectrum, on and below it. Denoting $\widehat{u}_\perp = (\widehat{z} \wedge \widehat{u})/\sin\beta$ and $\widehat{u}_\parallel = (\widehat{u} \wedge (\widehat{z} \wedge \widehat{u}))/\sin\beta$, $F_{Ez} = \widehat{z}.F_E$ and $F_{Hz} = \widehat{z}.F_H$, we note in general that:

$$F_H = \frac{-\widehat{u}_\perp F_{Ez}/Z_0 + \widehat{u}_\parallel F_{Hz}}{\sin\beta},\ F_E = \frac{\widehat{u}_\perp Z_0 F_{Hz} + \widehat{u}_\parallel F_{Ez}}{\sin\beta}, \tag{1.99}$$

whether $\widehat{u}$ is complex or not. Considering a multilayer invariant by translation at fixed z, each element of the incident spectrum, at $\widehat{u} = \frac{2\pi}{k_0}(p, q, -s)^t$, induces a plane wave at $\widehat{u} = \frac{2\pi}{k_0}(p, q, s)^t$ in the scattering spectrum. In these circumstances, we generally have:

$$F_{Ez}^s = R_{e,e} F_{Ez}^i + R_{e,h}(Z_0 F_{Hz}^i),\ Z_0 F_{Hz}^s = R_{h,h}(Z_0 F_{Hz}^i) + R_{h,e} F_{Ez}^i, \tag{1.100}$$

where the reflection coefficients $R_{e,e}$, $R_{h,h}$, $R_{e,h}$ and $R_{h,e}$ depend on frequency and on multilayer parameters, while varying with p, q, and s in general, or more simply with s when layers have z-axial bi-anisotropy [17]. When meromorphic, these coefficients are generally with simple poles. For isotropic [3]–[6] or z-axial anisotropic layers [20]–[22], we have TE and TM polarizations decoupling at interfaces, and incident TE (respectively, TM) waves give reflected TE (respectively, TM) ones, which leads to $R_{e,h} = R_{h,e} = 0$.

NOTE 1.22.– Due to the regularity of the scatterer at $z = 0$, the scattered fields cannot be more singular than primary sources at $z = 0$ (if any), and thus, the reflection coefficients are bounded as $|p|$ or $|q| \to \infty$.

NOTE 1.23.– For a change of variables in double integrals, the more simple is to use the formalism of exterior differential forms. So, using $d\beta \wedge d\phi = -d\phi \wedge d\beta$, we can write:

$$
\begin{aligned}
d(\sin\beta\cos\phi) &= \cos\beta\cos\phi d\beta - \sin\beta\sin\phi d\phi,\\
d(\sin\beta\sin\phi) &= \cos\beta\sin\phi d\beta + \sin\beta\cos\phi d\phi,\\
d(\sin\beta\cos\phi)\wedge d(\sin\beta\sin\phi) &= \sin\beta\cos\beta(d\beta\wedge d\phi).
\end{aligned}
\quad (1.101)
$$

1.A.2. *Complex Poynting vector and reflection coefficients*

The expression of the energy W outgoing in the region without sources $\pm z > z_0$ is given by an integral of complex Poynting vector $\frac{1}{2}(E \wedge H^*)$ [3] along the plane $z = z_0$ that can be transformed with (1.98) from Parseval equality, following:

$$
\begin{aligned}
W = W' + iW'' &= \frac{1}{2}\int_{z=z_0} \pm\widehat{z}.((E\wedge H^*))\, dxdy\\
&= \frac{\pi k_0}{Z_0(4\pi)^2|k_0|^2}\int_{R^2} e^{-2\pi i(s-s^*)z_0}\Big(\frac{(s^*+s)(F_E F_E^*)}{2ss^*}\\
&+ \frac{(s^*-s)(F_E F_E^* - 2|F_{Ez}|^2)}{2ss^*}\Big)dpdq,
\end{aligned}
\quad (1.102)
$$

where we have used $k_0^*\widehat{z}.(F_E\wedge(u^*\wedge F_E^*)) = 2\pi(s^*(F_E F_E^*) - (s^*-s)F_{Ez}^* F_{Ez})$ (note: if we change F_E for $Z_0 F_H$ in the integral, we obtain W^*). The real and imaginary parts of integral term appears clearly, and for real k_0, we notice that:

$$
\begin{aligned}
W' &= \int_0^{2\pi}\int_0^{\frac{\pi}{2}} \frac{(F_E F_E^*)}{2Z_0(4\pi)^2}\sin\beta d\beta d\phi,\\
W'' &= \int_0^{2\pi}\int_{\frac{\pi}{2}}^{\frac{\pi}{2}+i\infty} e^{-2ik_0 z_0\cos\beta}\frac{(F_E F_E^* - 2|F_{Ez}|^2)}{2iZ_0(4\pi)^2}\sin\beta d\beta d\phi.
\end{aligned}
\quad (1.103)
$$

By definition, if the scatterer is passive, the term W' refering to scattered wave is smaller than the one refering to incident field. Therefore, we can use in (1.103) that, $\sin\beta(F_E F_E^*) = (|Z_0 F_{Hz}|^2 + |F_{Ez}|^2)/\sin\beta$ as $0 \le \beta \le \pi/2$, then (1.100), and thus obtain, for real k_0, as $0 \le \beta \le \pi/2$:

$$
|R_{e,e}|^2 + |R_{h,e}|^2 \le 1,\ |R_{h,h}|^2 + |R_{e,h}|^2 \le 1. \quad (1.104)
$$

1.B. Appendix B: some equalities derived from (1.4) and (1.11)

When we let, in (1.4) and (1.11), $\cos\beta = -\epsilon'$ with $\epsilon' = +1$ or -1, we have respectively:

$$\prod_{j=1}^{N} \frac{\epsilon' + g_j}{\epsilon' - g_j} - (\pm 1)^N = \pm \sum_{j=1}^{N} \frac{a_j}{g_j} \frac{(g_j)^{\pm 1}}{(g_j)^{\pm 1} - \epsilon'},$$

$$R_0 \frac{\prod_{j=1}^{N'}(-\epsilon' - g'_j)}{\prod_{j=1}^{N}(-\epsilon' + g_j)} - a_{0\pm} = \pm \sum_{j=1}^{N} \frac{a_j}{g_j} \frac{(g_j)^{\pm 1}}{(g_j)^{\pm 1} - \epsilon'}, \tag{1.105}$$

where we note that, as (1.4) applies, $(\pm 1)^N = \pm 1$ since N is odd. Considering that:

$$\sum_{\epsilon'=-1,1} \frac{g_j}{1 - \epsilon' g_j} = \sum_{\epsilon'=-1,1} \frac{\epsilon'}{1 - \epsilon' g_j} = \frac{2g_j}{1 - g_j^2}, \tag{1.106}$$

we can then write, for a use in (1.24)–(1.25) and in (1.26)–(1.27):

$$\sum_{\epsilon'=-1,1} \Big(\Big(\prod_{j=1}^{N} \frac{\epsilon' + g_j}{\epsilon' - g_j} \mp 1 \Big) \mathcal{V}_{\epsilon'} + \sum_{j=1}^{N} \frac{\epsilon' a_j \mathcal{K}_{g_j}}{(g_j - \epsilon')} \Big) =$$

$$= \sum_{j=1}^{N} \frac{\pm a_j}{g_j} \Big((g_j)^{\pm 1} \sum_{\epsilon'=-1,1} \frac{\mathcal{V}_{\epsilon'} + \epsilon' \mathcal{K}_{g_j}}{(g_j)^{\pm 1} - \epsilon'} \Big),$$

$$\sum_{\epsilon'=-1,1} \Big(\Big(R_0 \frac{\prod_{j=1}^{N'}(-\epsilon' - g'_j)}{\prod_{j=1}^{N}(-\epsilon' + g_j)} - a_{0\pm} \Big) \mathcal{V}_{\epsilon'} + \sum_{j=1}^{N} \frac{\epsilon' a_j \mathcal{K}_{g_j}}{(g_j - \epsilon')} \Big)$$

$$= \sum_{j=1}^{N} \frac{\pm a_j}{g_j} \Big((g_j)^{\pm 1} \sum_{\epsilon'=-1,1} \frac{\mathcal{V}_{\epsilon'} + \epsilon' \mathcal{K}_{g_j}}{(g_j)^{\pm 1} - \epsilon'} \Big). \tag{1.107}$$

1.C. Appendix C: other types of bounded sources

Our expressions (1.24)–(1.27) can be used indifferently for dipole, volume and surface primary sources (in the sense of generalized functions) illuminating a multilayer, and this appendix is devoted to particular simplifications or expansions in the two latter cases.

1.C.1. *Bounded volume sources*

Let us consider the case of volume sources, $J = 1_\Omega J_0$ and $M = 1_\Omega M_0$ when J_0 and M_0 are analytic, where $1_\Omega(r') = \frac{1}{4\pi} \int_{\partial\Omega} \widehat{n}.\text{grad}(\frac{1}{|r-r'|}) dS$ is the indicator function for Ω, such that $1_\Omega(r') = 1$ in Ω, $= 0$ in $R^3\backslash\overline{\Omega}$, and fractional on $\partial\Omega$ ($\widehat{n}$ is the unit normal, directed inside Ω).

Since $\mathcal{V}_{\epsilon'}(x, y, -z)$ and $\mathcal{K}_g(x, y, -z)$ satisfy the Helmhotz equation for $z > 0$, we can write curl(curl(.)), which is equal to grad(div(.)) – Δ(.), in place of grad(div(.)) $+ k_0^2$ in our expressions of potentials above the plane.

Letting $A = 1_\Omega(r')A_0(r')$ and $S = \partial\Omega$, we notice the following identities for $\text{curl}(A)$ and $\text{curl}(\text{curl}(A))$ terms:

$$\begin{aligned}\text{curl}(A) &= 1_\Omega(r')\text{curl}(A_0) + \widehat{n}\wedge A_0\delta_S,\\ \text{curl}(\text{curl}(A)) &= 1_\Omega(r')\text{curl}(\text{curl}(A_0)) +\\ &+ \widehat{n}\wedge\text{curl}(A_0)\delta_S + \text{curl}(\widehat{n}\wedge A_0\delta_S),\end{aligned} \tag{1.108}$$

and thus simplify our expression of potentials (note: integrals reduce to surface ones if J_0 and M_0 are constants). Since J and M are distributions, we have $\int_{R^3} C.\partial_i f = -\int_{R^3}\partial_i C.f$, where ∂_i denotes a derivative along a Cartesian coordinates, and then note that $\int_{R^3}\text{curl}(C).f = \int_{R^3} C.\text{curl}(f)$, $\int_{R^3}\text{curl}(\text{curl}(C)).f = \int_{R^3} C.\text{curl}(\text{curl}(f))$, while $\int_{R^3}\text{grad}(\text{div}(C)).f = -\int_{R^3}\text{div}(C).\text{div}(f) = \int_{R^3} C.\text{grad}(\text{div}(f))$.

1.C.2. *Bounded surface sources*

The equivalent (or Huygens) surface sources, $M = -\widehat{n}\wedge[E']\,\delta_S$ and $J = \widehat{n}\wedge[H']\,\delta_S$, where $[E']$ and $[H']$ denote fields discontinuities in free space through S along normal $\widehat{n}$ are commonly used in boundary integral equation (exact method) or in physical optics (approximation) [23], [38]. In this case, we can use:

$$\begin{aligned}Z_0\text{div}(J) &= Z_0\text{div}_S(\widehat{n}\wedge[H'])\,\delta_S - Z_0(\widehat{n}\wedge[H']).\widehat{\nu}\delta_{\partial S},\\ \text{div}(M) &= -\text{div}_S(\widehat{n}\wedge[E'])\,\delta_S + (\widehat{n}\wedge[E']).\widehat{\nu}\delta_{\partial S},\end{aligned} \tag{1.109}$$

where Div$\equiv$div$_S$ is the surface divergence [38, p.330], $\widehat{\nu}$ is the geodesic normal to ∂S directed outside S and δ_S (respectively, $\delta_{\partial S}$) is the Dirac surface (respectively, line) function, while $\text{div}_S(\widehat{n}\wedge[H']) = -i\omega\varepsilon_0\widehat{n}.[E']$ and $\text{div}_S(\widehat{n}\wedge[E']) = i\omega\mu_0\widehat{n}.[H']$ from Maxwell equations, which directly simplify our expression of potentials.

1.D. Appendix D: some miscellaneous results on $\mathcal{J}_g(\rho, -z)$ and $W_m(s)$

Using (1.56), we have, for arbitrary $g = \sin\theta_1$, $|\text{Re}\theta_1| \le \pi/2$, and $|\arg(ik_0)| \le \pi/2$:

$$\begin{aligned}\mathcal{J}_g(\rho, -z) &= \pm i\int_{\pm b}^{i\infty} e^{-a\cos\alpha}d\alpha - (1\mp 1)K_0(a),\\ \mathcal{J}_g(\rho, -z) &= -\mathcal{J}_{-g}(\rho, z) - 2K_0(a) \text{ as } \rho\neq 0,\end{aligned} \tag{1.110}$$

with $z = R\cos\varphi$ and $\rho = R\sin\varphi$, the terms a and b being defined in (1.57)–(1.58) with $|\mathrm{Re}(b)| \leq \pi$, $\mathrm{Re}(a) > 0$ ($\mathrm{Re}(a) = 0$ being a limit case). The expression (1.56) of $\mathcal{J}_g(\rho, -z)$, regular as $z > 0$ with $0 \leq \varphi < \pi/2$, gives an analytical continuation below $z = 0$, by letting $z \leq 0$ with $\varphi \geq \pi/2$ when $\rho \neq 0$ (impossible with a direct use of (1.29)), which presents a logarithmic singularity $\sim -K_0(\epsilon ik\rho\cos\theta_1)$ at $\rho = 0$ as $z \leq 0$. Letting $\mathcal{K}_g(x, y, -z) = e^{ik_0gz}\mathcal{J}_g(\rho, -z)$, we have:

$$
\begin{aligned}
&(\partial_z - ik_0g)\,\mathcal{K}_g(x, y, -z) = \frac{e^{-ik_0|r|}}{|r|},\\
&(\Delta + k_0^2)(\mathcal{K}_g(x, y, -z)) = 0 \text{ as } z > 0,\\
&(\Delta + k_0^2)(\mathcal{K}_g(x, y, -z)) = 4\pi e^{ikgz}\delta(x)\delta(y) \text{ as } z < 0, \qquad (1.111)
\end{aligned}
$$

where $R = |r| = \sqrt{\rho^2 + z^2}$ and δ is the Dirac function [29] (see also our works in [39]). Using $\frac{\partial\varphi}{\partial z} = -\frac{\rho}{R^2}$ and $\frac{\partial\varphi}{\partial\rho} = \frac{z}{R^2}$, we have $\frac{\partial b}{\partial z} = \frac{i}{\sin\varphi}\frac{\rho}{R^2} = \frac{i}{R}$ and $\frac{\partial a}{\partial z} = 0$, while $\frac{\partial b}{\partial\rho} = \frac{-i}{\sin\varphi}\frac{z}{R^2} = \frac{-i}{\rho}\frac{z}{R}$ and $\frac{\partial a}{\partial\rho} = \epsilon ik\cos\theta_1$. We can then write, for arbitrary g and $|\arg(ik_0)| \leq \pi/2$:

$$
\begin{aligned}
&\frac{\partial\mathcal{J}_g(\rho, -z)}{\partial z} = \frac{e^{-ik(R+gz)}}{R},\\
&\frac{\partial\mathcal{J}_g(\rho, -z)}{\partial\rho} = -\frac{z}{\rho}\frac{e^{-ik(R+gz)}}{R} + \epsilon ik\cos\theta_1\Big(\frac{\partial\mathcal{J}_g(\rho, -z)}{\partial a}\Big|_{b \text{ is fixed}}\Big),\\
&(\Delta + k_0^2)(\mathcal{K}_g(x, y, -z)) = 4\pi e^{ikgz}U_s(-z)\delta(x)\delta(y), \qquad (1.112)
\end{aligned}
$$

where $U_s(z) = (1{+}\mathrm{sign}(z))/2$ is the unit step function.

Besides, we note that, for the particular case with $\mathrm{Re}g > 0$ and $\arg(ik_0) = \pi/2$, that is, $g = \sin\theta_1$ as $0 \leq \mathrm{Re}\theta_1 \leq \pi/2$ [8]:

$$
\begin{aligned}
\mathcal{J}_g(\rho, -z) &= -\int_{\nu_z = ik(R+gz)}^{\infty} \frac{e^{-\nu}}{\sqrt{\nu^2 - (ik\rho\cos\theta_1)^2}}\,d\nu\\
&= e^{-ik_0gz}\int_{-i\infty}^{0} e^{-ik_0gz_1}\frac{e^{-ik_0R(-z_1-z)}}{R(-z_1 - z)}\,dz_1, \qquad (1.113)
\end{aligned}
$$

with $R(-z) = \sqrt{\rho^2 + z^2}$, $\mathrm{Im}(R(-z - z_1)) < 0$.

In other respects, for the calculus of our exact series with the error functions in the expressions (1.75)–(1.78) for $\mathcal{J}_g(\rho, -z)$, it is worth noticing [30] that:

$$
\begin{aligned}
W_m(s) &= \frac{e^{-s}\partial^m}{\partial s^m}\Big(\frac{e^s \mathrm{erf}(s^{1/2})}{s^{1/2}}\Big) = \int_0^1 \frac{e^{-st}(1-t)^m}{\sqrt{\pi}t^{1/2}} dt \\
&= \frac{m! M(\frac{1}{2}, m+\frac{3}{2}, -s)}{\Gamma(m+\frac{3}{2})}, \\
\frac{\partial^m}{\partial s^m}\Big(\frac{e^s \mathrm{erfc}(s^{1/2})}{m! s^{1/2}}\Big) &= \frac{(-1)^m}{\sqrt{\pi}} U(m+1, m+\frac{3}{2}, s) \\
&= \frac{(-1)^m}{\sqrt{\pi} s^{m+1}}(1+O(s^{-1})), \qquad (1.114)
\end{aligned}
$$

where M and U are confluent hypergeometric functions [30]. These special functions are well-tabulated, and we remark in particular that, for large $m \gg |s|$, $M(1/2, m+3/2, -s) = 1 + O(s/m)$ and $W_m(0) = \frac{\Gamma(m+1)}{\Gamma(m+3/2)} = m^{-1/2} + O(m^{-3/2})$.

NOTE 1.24.– Letting $V_m^v = W_m(s) - vW_m(0)$, with $v = 0$ or 1, we derive, from [30]:

$$
\begin{aligned}
&2sV_{m+1}^v(s) + V_m^v(s)(1 - 2s + 2m) - 2mV_{m-1}^v(s) \\
&= v\frac{s\Gamma(m+1)}{\Gamma(m+\frac{5}{2})}, \qquad (1.115)
\end{aligned}
$$

NOTE 1.25.– We can write [30], as $\mathrm{Re}(u) > 0$ and $M \geq 0$:

$$
\begin{aligned}
\mathrm{erfc}(u) &= \frac{e^{-u^2}}{\sqrt{\pi}u}\sum_{m=0}^{M-1}\frac{(-1)^m(2m)!}{2^m m!(2u^2)^m} + \mathcal{R}_M(u), \\
\mathcal{R}_M(u) &= \frac{(-1)^M(2M)!}{2^M M!(2u^2)^M u}\int_0^\infty \frac{e^{-(t+u^2)}/\sqrt{\pi}}{(\frac{t}{u^2}+1)^{M+\frac{1}{2}}} dt \qquad (1.116)
\end{aligned}
$$

NOTE 1.26.– We have [3]:

$$
\begin{aligned}
\frac{e^{-ik_0|r|}}{k_0|r|} &= \frac{-i}{2}\int_{\mathcal{D}} H_0^{(2)}(k_0\rho\sin\beta)e^{-ik_0|z|\cos\beta}\sin\beta d\beta, \\
(\Delta + k_0^2)(\frac{-e^{-ik_0|r|}}{4\pi|r|}) &= \delta(x)\delta(y)\delta(z). \qquad (1.117)
\end{aligned}
$$

2

Diffraction by an Impedance Curved Wedge with Arbitrary Angle and Uniform Higher Order Asymptotics

2.1. Introduction

The methods generally considered for the diffraction by a wedge with curved faces are most often different as you consider a wedge of narrow or large angle [4],[6]–[11]. An original approach, initiated in [1, 2] for faces of constant curvatures, and whose results have been used by Borovikov in [3] for his developments on nonconstant curvature case, permits a new global asymptotic study for an impedance curved wedge of any angle, with arbitrary distinct faces impedances, and for points far or close to the curved surface. Contrary to [4], we can then consider simultaneously diffracted phenomena appearing in different orders. It then becomes possible to study the case where two orders have comparable contributions, for example, the case of a curved plane with a discontinuity of curvature and materials with low contrasts.

In this approach, we reduce the boundary value problem on the field to an asymptotic one on the spectral function, attached to the representation of the field in the form of a Sommerfeld–Maliuzhinets integral. The functional equations then obtained are solved, and the asymptotics of the spectral function is developed. Besides, an expression, allowing the transition to an asymptotic expansion with fractional powers, is also given for directions close to the tangents at the edge, where, in particular, creeping waves due to curvature can appear. A new presentation of our method is given here, more developed and detailed than in [1, 2]–[5], and a section is specifically devoted to fringe waves. In noting that our expressions apply for any wedge angle, for arbitrary distinct faces impedances, and for curvatures at several orders, they allow us to recover all known results on diffraction coefficients, from the case of a discontinuity of curvature to the case of a curved half plane [6]–[13].

2.2. Higher order asymptotics in the region without creeping waves terms

2.2.1. *Spectral function and asymptotic boundary conditions on both faces*

We consider a wedge with tangent planes at the edge defined by $\varphi = \pm\Phi$, in cylindrical coordinates (ρ, φ), and curved circular faces of radius $a^{\pm}$, with $\Phi > \pi/2$ (note: $\Phi = \pi/2$ is obtained by letting $\Phi \to \pi/2$), illuminated by an unit plane wave, normal to the edge:

$$u^{inc}(\rho, \varphi) = e^{ik\rho\cos(\varphi-\varphi_\circ)}, \tag{2.1}$$

with $\varphi_\circ$ the angle of incidence. The total field u verifies outside the wedge $(\Delta + k^2)u = 0$, with $k = \frac{\omega}{c}$ the wave number of the free space, $|\arg(ik)| < \pi/2$ and $e^{i\omega t}$ time dependence. It can be then represented as an integral of Sommerfeld–Maliuzhinets [1, 2], [5], [14]–[17], for $|\varphi| < \Phi$:

$$u(\rho, \varphi) = \frac{1}{2\pi i}\int_{\gamma} f(\alpha + \varphi)e^{ik\rho\cos\alpha}d\alpha, \tag{2.2}$$

with f to be defined. The contour γ (Figure 2.1) is composed of two symmetric infinite loops of the complex plane, γ_+ and γ_-, respectively, above and below the real axis, and within which f is regular. Considering, from symmetry of γ, that we can add any constant to f without changing u, and that $f(\pm i\infty)$ exists when u is definite at the edge [5], [14]–[17], we will let $f(+i\infty) = -f(-i\infty)$.

After deformation of γ to the steepest descent path SDP, composed of two paths $SDP_{\pm}$ centered on the stationary phase points $\alpha = \pm\pi$, we can write:

$$\begin{aligned} u(\rho,\varphi) &= u_i + \sum u_r^{\pm} + \sum u_s^{\pm} \\ &+ \frac{e^{-ik\rho}}{2\pi i}\sum_{\pm}\int_{SDP_{\pm}} f(\alpha + \varphi)e^{ik\rho(\cos\alpha+1)}d\alpha, \end{aligned} \tag{2.3}$$

where

– the terms u_i and $u_r^{\pm}$ correspond to real poles of $f(\alpha+\varphi)$ (if any), captured during deformation of γ to SDP, deriving from an asymptotics of geometrical optics fields when $ka^{\pm}$ is large. The first term gives the incident field in the illuminated zone, while the second term expresses the reflection by each face (note: if present, some secular terms are in the higher order asymptotic development of this term);

– the terms $u_s^{\pm}$ correspond to complex poles of $f(\alpha + \varphi)$ (if any), captured during deformation of γ to SDP, for terms of complex plane guided waves excited by the edge (note: these terms are different from creeping guided waves due to faces curvature);

– the last term, named u_d, is principally radiated cylindrically when $\rho \to \infty$. Approximating $f(\alpha+\varphi)$ on $SDP_{\pm}$ by its value at the stationary phase points $\alpha = \pm\pi$, we obtain:

$$u_d \sim \frac{-e^{-i\pi/4-ik\rho}}{\sqrt{2\pi k\rho}}[f(\pi+\varphi) - f(-\pi+\varphi)], \tag{2.4}$$

where $F(\varphi) = f(\pi+\varphi) - f(-\pi+\varphi)$ is denoted the diffraction (or far field) coefficient.

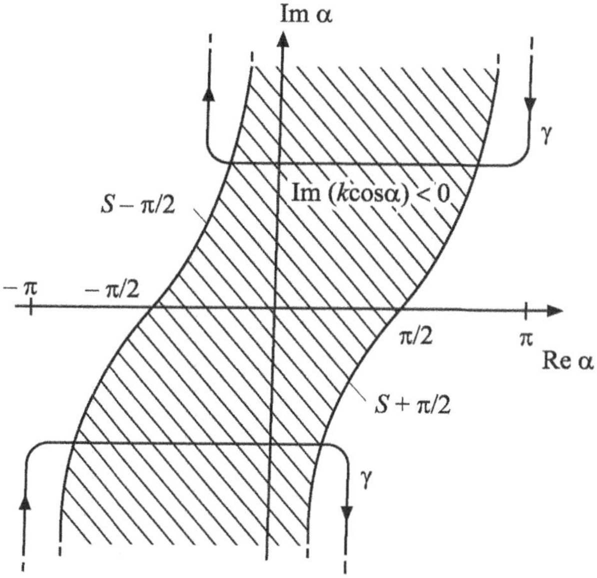

Figure 2.1. *Complex path of integration*

We assume impedance boundary conditions on the curved surface of the wedge:

$$(\frac{\partial}{\partial n} - ik\sin\theta^{\pm})u|_{\text{curved face}\pm} = 0, \tag{2.5}$$

where $\frac{\partial(.)}{\partial n} = \widehat{n}.\text{grad}(.)$, $\widehat{n}$ is the normal outward the scatterer surface, and $\sin\theta^{\pm}$ is some complex constants with $\text{Re}(\sin\theta^{\pm}) > 0$, attached to the choice of materials composing the scatterer, with the case $\text{Re}(\sin\theta^{\pm}) = 0$ being taken in the sense of the limit.

We approximate (2.5) for each semi-circular face $S_\pm$ in entire powers of the curvature $1/a^\pm$ for large $a^\pm$ [1, 2, 5], and then consider asymptotic conditions $\mathcal{B}^\pm u = 0$ on semi-plane $\varphi = \pm\Phi$, tangent to $S_\pm$ at the edge. For that, we begin by using Cartesian coordinates (x, y), respectively, attached to each face, where the plane $\varphi = \pm\Phi$ refers to the semi-infinite straight line with $\mp x \geq 0$ at $y = 0$, while the curved faces $S_\pm$ refer to the curved semi-infinite line with $y = a^\pm\sqrt{1-(x/a^\pm)^2} - a^\pm \sim -x^2/2a^\pm$ as $\mp x \geq 0$. Using Taylor expansions, we have:

$$u \sim 1 + y(\partial_y u|_{y=0}) + \frac{y^2}{2}(\partial_{y^2} u|_{y=0}),\ \frac{\partial}{\partial n}|_{s_\pm} = \frac{\frac{x}{(a+y)}\frac{\partial}{\partial x} + \frac{\partial}{\partial y}}{\sqrt{1+(\frac{x}{(a+y)})^2}} \sim \frac{\partial}{\partial y} + \frac{x}{a^\pm}\frac{\partial}{\partial x},$$

$$u|_{s_\pm} \sim (1 - \frac{x^2}{2a^\pm}\frac{\partial u}{\partial y})|_{y=0},\ \frac{\partial u}{\partial n}|_{s_\pm} \sim (\frac{\partial u}{\partial y} - \frac{x^2}{2a^\pm}(\frac{\partial^2 u}{\partial y^2}) + \frac{x}{a^\pm}\frac{\partial u}{\partial x})|_{y=0}, \tag{2.6}$$

and we can then write (2.5) for large $a^\pm$ as follows:

$$[(\frac{\partial}{\partial y} - ik\sin\theta^\pm - \frac{x^2}{2a^\pm}\frac{\partial}{\partial y^2} + \frac{x}{a^\pm}\frac{\partial}{\partial x}$$
$$+ ik\sin\theta^\pm\frac{x^2}{2a^\pm}\frac{\partial}{\partial y} + O(x/(a^\pm)^2))u]|_{\varphi=\pm\Phi} = 0. \tag{2.7}$$

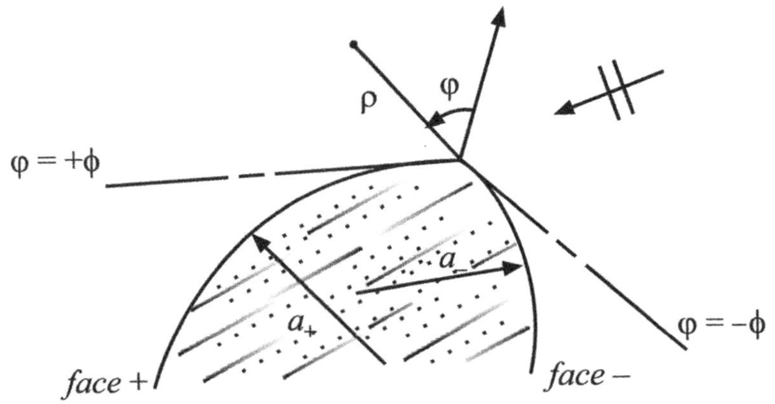

Figure 2.2. *Geometry of the curved impedance wedge*

Considering the angle $\varphi_e = \varphi \pm (\pi/2 - \Phi)$ referring to each face, we have:

$$x = -\rho\sin\varphi_e,\ y = \rho\cos\varphi_e,$$

$$\frac{\partial}{\partial x} = \frac{\partial\rho}{\partial x}\frac{\partial}{\partial\rho} + \frac{\partial\varphi_e}{\partial x}\frac{\partial}{\partial\varphi_e} = -\sin\varphi_e\frac{\partial}{\partial\rho} - \frac{\cos\varphi_e}{\rho}\frac{\partial}{\partial\varphi_e},$$

$$\frac{\partial}{\partial y} = \frac{\partial \rho}{\partial y}\frac{\partial}{\partial \rho} + \frac{\partial \varphi_e}{\partial y}\frac{\partial}{\partial \varphi_e} = \cos\varphi_e \frac{\partial}{\partial \rho} - \frac{\sin\varphi_e}{\rho}\frac{\partial}{\partial \varphi_e},$$
$$\frac{\partial^2}{\partial y^2}|_{\cos\varphi_e=0} = \frac{1}{\rho}\frac{\partial}{\partial \rho} + \frac{1}{\rho^2}\frac{\partial^2}{\partial \varphi^2}, \tag{2.8}$$

which leads us to write:

$$\frac{\partial}{\partial y}|_{\varphi=\pm\Phi} = \mp\frac{\partial}{\rho\partial\varphi}|_{\varphi=\pm\Phi}, \quad \frac{x}{a^\pm}\frac{\partial}{\partial x}|_{\varphi=\pm\Phi} = \frac{\rho}{a^\pm}\frac{\partial}{\partial \rho}|_{\varphi=\pm\Phi},$$
$$\frac{x^2}{2a^\pm}\frac{\partial^2}{\partial y^2}|_{\varphi=\pm\Phi} = \frac{\rho^2}{2a^\pm}(\frac{1}{\rho}\frac{\partial}{\partial \rho} + \frac{1}{\rho^2}\frac{\partial^2}{\partial \varphi^2})|_{\varphi=\pm\Phi}, \tag{2.9}$$

and thus, to express (2.7) the following:

$$[(\mp\frac{\partial}{\rho\partial\varphi} - ik\sin\theta^\pm - \frac{1}{2a^\pm}(\frac{\partial^2}{\partial\varphi^2} - \frac{\rho\partial}{\partial\rho}$$
$$\pm ik\sin\theta^\pm\frac{\rho\partial}{\partial\varphi}) + O(\frac{\rho}{(a^\pm)^2}))u]|_{\varphi=\pm\Phi} = 0. \tag{2.10}$$

Applying (2.10) to the expression (2.2) of u, with a correct use of integration by parts (see note 2.3), then Maliuzhinets inversion theorem [14], we then derive the asymptotic functional equation for f:

$$B^\pm_{\alpha\mp\frac{\pi}{2}}f(\alpha\pm\Phi) - B^\pm_{-\alpha\mp\frac{\pi}{2}}f(-\alpha\pm\Phi) = 0, \tag{2.11}$$

with,

$$B^\pm_{\alpha\mp\frac{\pi}{2}} = (\sin\alpha \pm \sin\theta^\pm \pm \frac{D^\pm_\alpha(.)}{ika^\pm} + O(\frac{1}{(ka^\pm)^2})),$$
$$D^\pm_\alpha(.) = \frac{1}{2}[\frac{\partial^2(.)}{\partial\alpha^2} - \frac{\partial}{\partial\alpha}(\cot\alpha(.)) \pm \sin\theta^\pm\frac{\partial}{\partial\alpha}(\frac{1}{\sin\alpha}\frac{\partial}{\partial\alpha}(.))]. \tag{2.12}$$

Besides, noticing that $B^\pm_{\alpha+2\pi} = B^\pm_\alpha$, we note in complements that:

$$B^\pm_{\alpha\pm\frac{\pi}{2}}f(\pi+\alpha\pm\Phi) - B^\pm_{-\alpha\pm\frac{\pi}{2}}f(-\pi-\alpha\pm\Phi) = 0,$$
$$B^\pm_{-\alpha\pm\frac{\pi}{2}}f(\pi-\alpha\pm\Phi) - B^\pm_{\alpha\pm\frac{\pi}{2}}f(-\pi+\alpha\pm\Phi) = 0,$$
$$B^\pm_{\alpha\pm\frac{\pi}{2}}F(\alpha\pm\Phi) + B^\pm_{-\alpha\pm\frac{\pi}{2}}F(-\alpha\pm\Phi) = 0. \tag{2.13}$$

NOTE 2.1.– For geometries with concave face(s), that is, with negative a^+ or a^-, we can take in consideration the analytical continuation of $B^\pm_\alpha$ with respects to $a^\pm$.

NOTE 2.2.– The circle of convergence of the Taylor expansion of an analytic function is limited by first encountered singularity. In our case, the field derivatives can be singular only at the edge point.

NOTE 2.3.– For a correct use of the inversion theorem [14] on partial differential equations, we have to note that letting n, m, p, l some positive integers, and $q = 0$ or 1, we can write:

$$(ik\rho)^n(\frac{\partial}{ik\partial\rho})^m(ik\rho)^k(\frac{\partial}{ik\rho\partial\varphi})^p(\frac{\partial}{ik\partial\rho})^l(\frac{\partial}{ik\rho\partial\varphi})^q(u(\rho,\varphi)) =$$

$$= \frac{1}{2\pi i}\int_{\gamma_+}\Big[(\frac{\partial}{\partial\alpha}(\frac{(.)}{\sin\alpha}))^n\Big((\cos\alpha)^m(\frac{\partial}{\partial\alpha}(\frac{(.)}{\sin\alpha}))^{k-p}$$

$$[(\cos\alpha)^l(\sin\alpha)^q(\frac{\partial}{\partial\alpha})^p f(\alpha+\varphi,\varphi_o)]\Big)$$

$$-(\frac{\partial}{\partial\alpha}(\frac{(.)}{\sin\alpha}))^n\Big((\cos\alpha)^m(\frac{\partial}{\partial\alpha}(\frac{(.)}{\sin\alpha}))^{k-p}$$

$$[(\cos\alpha)^l(-\sin\alpha)^q(\frac{-\partial}{\partial\alpha})^p f(-\alpha+\varphi,\varphi_o)]\Big)\Big]e^{ik\rho\cos\alpha}d\alpha. \quad (2.14)$$

2.2.2. Detailed calculation of first term influenced by the curvatures

We can search the solution of the functional equations (2.11), derived asymptotically in vicinity of the edge (relatively to large $ka^{\pm}$), in the form:

$$f = \sum_{n\geq 0} f_n/k^n, \quad (2.15)$$

We can then consider the terms of same powers in k in (2.11) and obtain a succession of functional equations of order m on the $f_{n\leq m}$ that we can solve, with f_0 the spectral function for a wedge with plane faces [15]. In practice, this will allow us to express in particular the far field function $F(\varphi) = f(\pi+\varphi) - f(-\pi+\varphi)$ (also named the diffraction coefficient) in the region $|\varphi| < \Phi$, except for $\varphi \sim \pm\Phi$ where the influence of curvature induces creeping waves terms requiring fractional powers of k.

Let us detail the derivation of expressions given initially in [1, 2] for f_1, which is the first term influenced by the curvatures. In this process, the term $f_0(\alpha)$ will be the only term with a pole at $\alpha = \varphi_o$ in the band $|\mathrm{Re}\alpha| \leq \Phi$, inducing the incident term u_i. It is given by $f_0(\alpha) = \sigma(\alpha)\Psi(\alpha)/\Psi(\varphi_0)$ for passive wedge with $\mathrm{Re}(\sin\theta^{\pm}) > 0$ [15], where $\sigma(\alpha) = \mu\cos(\mu\varphi_o)/(\sin(\mu\alpha) - \sin(\mu\varphi_o))$, $\mu = \pi/2\Phi$, φ_o is the angle of incidence and Ψ is a meromorphic function without poles or zeros in the strip $|\mathrm{Re}\alpha| \leq \Phi$, verifying:

$$(\sin\alpha \pm \sin\theta^{\pm})\Psi(\alpha \pm \Phi) - (-\sin\alpha \pm \sin\theta^{\pm})\Psi(-\alpha \pm \Phi) = 0. \quad (2.16)$$

This function expresses [16], as $\mathrm{Re}(\sin\theta^{\pm})>0$, following:

$$\Psi(\alpha)=B_N\prod_{\pm}\Big(\prod_{m=0}^{N}[\Gamma(\frac{1}{2}+\frac{1}{4\Phi}(\alpha\pm\Phi+\theta^{\pm}+m\pi))$$
$$\Gamma(\frac{1}{2}-\frac{1}{4\Phi}(\alpha\pm\Phi-(\theta^{\pm}+m\pi)))\Gamma(\frac{1}{2}+\frac{1}{4\Phi}(\alpha\pm\Phi+\pi-\theta^{\pm}+m\pi))$$
$$\Gamma(\frac{1}{2}-\frac{1}{4\Phi}(\alpha\pm\Phi-(\pi-\theta^{\pm}+m\pi)))]^{(-1)^{m+1}}$$
$$\times\exp\Big(\int_0^{\infty}(-e^{-\nu\pi})^{N+1}\frac{e^{-\nu\theta^{\pm}}+e^{-\nu(\pi-\theta^{\pm})}}{(1+e^{-\nu\pi})}\frac{(1-\cosh(\nu(\alpha\pm\Phi)))}{\nu\sinh(2\nu\Phi)}d\nu\Big)\Big),\tag{2.17}$$

for $|\mathrm{Re}\alpha|\le\Phi+(N+1)\pi$, with N an arbitrary positive integer, B_N a constant and Γ the Gamma function [18].

By letting $f_1(\alpha)/k=\Psi(\alpha)\chi(\alpha)/k$, we then have from (2.11):

$$\chi(\alpha\pm\Phi)-\chi(-\alpha\pm\Phi)=K^{\pm}(\alpha),$$
$$K^{\pm}(\alpha)=\mp\frac{D^{\pm}_{\alpha}(f_0(\alpha\pm\Phi))-D^{\pm}_{-\alpha}(f_0(-\alpha\pm\Phi))}{ia^{\pm}\Psi(\alpha\pm\Phi)(\sin\alpha\pm\sin\theta^{\pm})},\tag{2.18}$$

where f_1 and χ is regular in the band $|\mathrm{Re}\alpha|<\Phi$, and $\chi(\pm i\infty)=0$. We can then write [1], for $|\mathrm{Re}(\alpha')|<\Phi$:

$$\chi(\alpha')=\frac{-i}{8\Phi}\sum_{\pm}(\int_{-i\infty}^{i\infty}\pm K^{\pm}(\alpha)\,W^{\pm}(\alpha',\alpha)\,d\alpha),\tag{2.19}$$

where $W^{\pm}(\alpha',\alpha)=\tan(\frac{\pi}{4\Phi}(\alpha'\pm\Phi-\alpha))$. To develop this expression, we remark first that:

$$D^{\pm}_{\alpha}(.)=\frac{1}{2}\frac{\partial}{\partial\alpha}\Big(\frac{(\sin\alpha\pm\sin\theta^{\pm})}{\sin\alpha}[\partial_{\alpha}(.)\ -(\partial_{\alpha}\ln(\sin\alpha\pm\sin\theta^{\pm}))(.)]\Big),$$
$$D^{\pm}_{\alpha}(f_0(\alpha\pm\Phi))=\frac{\partial}{2\partial\alpha}\Big(f_0(\alpha\pm\Phi)\frac{\sin\alpha\pm\sin\theta^{\pm}}{\sin\alpha}\Big[\partial_{\alpha}\ln\Big(\frac{f_0(\alpha\pm\Phi)}{(\sin\alpha\pm\sin\theta^{\pm})}\Big)\Big]\Big),\tag{2.20}$$

and thus,

$$D^{\pm}_{\alpha}(f_0(\alpha\pm\Phi))=\frac{\partial}{2\partial\alpha}\Big(\frac{\Psi(\alpha\pm\Phi)(\sin\alpha\pm\sin\theta^{\pm})}{\Psi(\varphi_0)\sin\alpha}$$
$$\times\Big[\sigma(\alpha\pm\Phi)\partial_{\alpha}\ln\Big(\frac{\Psi(\alpha\pm\Phi)}{-\sin\alpha\pm\sin\theta^{\pm}}\frac{-\sin\alpha\pm\sin\theta^{\pm}}{\sin\alpha\pm\sin\theta^{\pm}}\Big)+\partial_{\alpha}\sigma(\alpha\pm\Phi)\Big]\Big),$$

$$D^{\pm}_{\alpha}(f_0(\alpha \pm \Phi)) - D^{\pm}_{-\alpha}(f_0(-\alpha \pm \Phi)) = \frac{\partial}{\partial \alpha}\left(\frac{\Psi(\alpha \pm \Phi)(\sin\alpha \pm \sin\theta^{\pm})}{\Psi(\varphi_0)\sin\alpha}\right.$$

$$\left.\times\left[\sigma(\alpha \pm \Phi)\partial_{\alpha}\ln\left(\frac{\Psi(\alpha \pm \Phi)}{-\sin\alpha \pm \sin\theta^{\pm}}\right) + \partial_{\alpha}\sigma(\alpha \pm \Phi)\right]\right), \tag{2.21}$$

where we have used that $\sigma(\alpha \pm \Phi)$, $\frac{\Psi(\alpha\pm\Phi)}{(-\sin\alpha\pm\sin\theta^{\pm})}$ and $\partial_{\alpha}\ln\left(\frac{-\sin\alpha\pm\sin\theta^{\pm}}{\sin\alpha\pm\sin\theta^{\pm}}\right)$ are even functions. We can now write:

$$K^{\pm}(\alpha)W^{\pm}(\alpha',\alpha) = \mp\frac{W^{\pm}(\alpha',\alpha)}{ia^{\pm}\Psi(\alpha \pm \Phi)(\sin\alpha \pm \sin\theta^{\pm})}$$

$$\times\frac{\partial}{\partial\alpha}\left[\frac{\Psi(\alpha \pm \Phi)(\sin\alpha \pm \sin\theta^{\pm})}{\Psi(\varphi_0)\sin\alpha}\right.$$

$$\left.(\sigma(\alpha \pm \Phi)\partial_{\alpha}\ln\left(\frac{\Psi(\alpha \pm \Phi)}{-\sin\alpha \pm \sin\theta^{\pm}}\right) + \partial_{\alpha}\sigma(\alpha \pm \Phi))\right]. \tag{2.22}$$

It is worth noticing that $K^{\pm}(\alpha)$ is not singular at $\alpha = 0$, and we can choose to use integration by parts for the expression of χ based on $u(\partial_{\alpha}v) = -(\partial_{\alpha}u)v + \partial_{\alpha}(uv)$. For this, we note that:

$$\partial_{\alpha}\left(\frac{W^{\pm}(\alpha',\alpha)}{\Psi(\alpha \pm \Phi)(\sin\alpha \pm \sin\theta^{\pm})}\right) =$$

$$= \frac{-W^{\pm}(\alpha',\alpha)\partial_{\alpha}\ln(\Psi(\alpha \pm \Phi)(\sin\alpha \pm \sin\theta^{\pm})) + \partial_{\alpha}W^{\pm}(\alpha',\alpha)}{\Psi(\alpha \pm \Phi)(\sin\alpha \pm \sin\theta^{\pm})}, \tag{2.23}$$

and $\Psi(\alpha) \sim K_0\exp(\mp i\mu\alpha)$ as $\mathrm{Im}\alpha \to \pm\infty$ (in passive case), and we can then write:

$$\chi(\alpha') = \frac{i}{8\Phi\Psi(\varphi_0)}\sum_{\pm}\frac{1}{ia^{\pm}}\int_{-i\infty}^{i\infty} d\alpha$$

$$\left[(-W^{\pm}(\alpha',\alpha)\partial_{\alpha}\ln(\Psi(\alpha \pm \Phi)(\sin\alpha \pm \sin\theta^{\pm}))\right.$$

$$+ \partial_{\alpha}W^{\pm}(\alpha',\alpha))(-\sigma(\alpha \pm \Phi)\partial_{\alpha}\ln\left(\frac{\Psi(\alpha \pm \Phi)}{(-\sin\alpha \pm \sin\theta^{\pm})}\right)$$

$$\left.- \partial_{\alpha}\sigma(\alpha \pm \Phi))\frac{1}{\sin\alpha}\right], \tag{2.24}$$

for $|\mathrm{Re}(\alpha')| < \Phi$, whose regularity in this band implies that, as expected, $f_1(\alpha')$ has no poles of incident field (note: $f_1(\alpha')$ can have poles of GO reflected waves depending on φ_o outside the band $|\mathrm{Re}(\alpha')| < \Phi$). In this expression, we note that $\Psi(\alpha \pm \Phi)(\sin\alpha \pm \sin\theta^{\pm})$ is an even function, while the term within brackets,

$o(1)$ as $|\text{Im}(\alpha)| \to \infty$, remains regular on the imaginary axis as $\text{Re}(\sin\theta^{\pm}) > 0$ (note: the case $\text{Re}(\sin\theta^{\pm}) = 0$ is taken in the sense of the limit).

We can then extend the domain of analyticity [2]. For that, we shift the integration path with the real quantities $\pm d^{\pm}$ along the real axis, for each term concerning the faces $\pm$, adding the contribution of poles captured during the shift, and we obtain the following expression:

$$\chi(\alpha') = \frac{1}{8\Phi\Psi(\varphi_\circ)} \sum_{\pm} \frac{1}{a^{\pm}} \left(\int_{-i\infty \pm d^{\pm}}^{i\infty \pm d^{\pm}} d\alpha \right.$$

$$\left[(-W^{\pm}(\alpha',\alpha)\partial_\alpha \ln((-\sin\alpha \pm \sin\theta^{\pm})\Psi(-\alpha \pm \Phi)) \right.$$

$$-\, \partial_{\alpha'} W^{\pm}(\alpha',\alpha))(-\sigma(\alpha \pm \Phi)\partial_\alpha \ln\left(\frac{\Psi(-\alpha \pm \Phi)}{(\sin\alpha \pm \sin\theta^{\pm})} \right)$$

$$\left. -\, \partial_\alpha \sigma(\alpha \pm \Phi)) \frac{1}{\sin\alpha} \right]$$

$$\left. \mp 2\pi i \,\text{sign}(d^{\pm}) \sum_{0 < \pm \text{sign}(d^{\pm}) Re\alpha'_s < |d^{\pm}|} \text{Residue}[...]|_{\alpha = \alpha'_s} \right), \tag{2.25}$$

with $W^{\pm}(\alpha',\alpha) = \tan(\frac{\pi}{4\Phi}(\alpha' \pm \Phi - \alpha))$ and arbitrary real constants $d^{\pm}$, for $\text{Re}(\sin\theta^{\pm}) > 0$. We have used $\partial_\alpha W^{\pm}(\alpha',\alpha) = -\partial_{\alpha'} W^{\pm}(\alpha',\alpha)$ in this expression, and we can note, from parity on the initial path of integration, that we can take $\frac{1}{2}(W^{\pm}(\alpha',\alpha) - W^{\pm}(\alpha',-\alpha))$ in place of $W^{\pm}(\alpha',\alpha)$ in (2.25). This expression is an analytical continuation of (2.24) for $\text{Re}(\alpha') \in]-3\Phi + d^{+}, \Phi + d^{+}[\cap] - \Phi - d^{-}, 3\Phi - d^{-}[$, which allows, completed by (2.18), a definition for any α'.

NOTE 2.4.– For more general wedge impedances, $s^{\pm} = \text{sign}(\text{Re}(\sin\theta^{\pm}))$ can be negative (active impedance case), and χ then derives from a more general definition of f_0:

$$f_0(\alpha) = P_n(\sin(\mu\alpha))\Psi(\alpha, \sin\theta^{+}, \sin\theta^{-})\sigma(\alpha)/\Psi(\varphi_\circ, \sin\theta^{+}, \sin\theta^{-}) - A_0,$$

$$\sigma(\alpha) = \mu\cos(\mu\varphi_\circ)/(\sin(\mu\alpha) - \sin(\mu\varphi_\circ)),$$

$$\Psi(\alpha, \sin\theta^{+}, \sin\theta^{-}) = \Psi^{+}(\alpha, \sin\theta^{+})\Psi^{-}(\alpha, \sin\theta^{-}), \tag{2.26}$$

where the expression of Ψ for general case derives from $\Psi^{\pm}(\alpha, -\sin\theta^{\pm}) = 1/\Psi^{\pm}(\alpha, \sin\theta^{\pm})$ as $\theta^{\pm} \neq 0$. The term P_n is a polynomial of order n that satisfies several conditions. Since $f_0(\alpha) \sim 1/(\alpha - \varphi_\circ)$ in vicinity of $\alpha = \varphi_\circ$, we have $P_n(\sin(\mu\varphi_\circ)) = 1$. From continuity of u at $\rho = 0$, we have $f_0(\alpha) = O(1)$ as $|\text{Im}\alpha| \to \infty$, and thus $n = \sum_{\pm}(1 - s^{\pm})/2$, while A_0 is such $f_0(i\infty) = -f_0(-i\infty)$ (note: in passive case, $n = 0$ and $A_0 = 0$). Later, the n zeros of P_n are defined to avoid any nonphysical poles of Ψ and any divergent wave at infinity. So, for a curved wedge with not strictly passive impedances, we shall take care to consider the general definition of f_0 for the derivation of f_n.

2.3. Several orders asymptotic expressions in a region with creeping waves terms

2.3.1. *Plane wave illumination: observation at infinity*

We note, from (2.18) and residues terms in (2.25), that $f_1(\pm(\pi+\Phi))$, and thus, the asymptotic expression (3) of u_d are singular for $\varphi = \pm\Phi$. This mathematical (not physical) singularity is the sign of a limit of validity. It physically corresponds to the necessity to take account, near tangent planes $\varphi = \pm\Phi$, of creeping waves, guided by curved convex surface and excited by multipoles edge sources. Therefore, near the tangents of the faces $\pm\Phi$, we will consider the mutipoles sources deduced from the asymptotics of far field function F in entire powers of $1/k$ for $|\varphi| < \Phi$, thereafter named $F_<$, then the radiation of them in presence of a cylinder of radius $a^\pm$.

So developed [1, 2, 5] (see appendix A for details), the transition expression for convex faces is given, when $|\pm\varphi-\Phi| < \pi/2$ and $\pm\varphi_\circ - \Phi < -\pi$, as follows:

$$F(\varphi) \equiv \frac{e^{ika^\pm \sin(\pm\varphi-\Phi)}}{i\pi ka^\pm} \int_{\mathcal{E}} \frac{e^{-i\nu(\pm\varphi-\Phi)}\mathcal{M}^\pm(\pm\nu)}{H_\nu^{(2)\prime}(ka^\pm) + Q^\pm\, H_\nu^{(2)}(ka^\pm)}\, d\nu, \qquad (2.27)$$

with $Q^\pm = -i\sin\theta^\pm$, $H_\nu^{(2)}$ the Hankel function [18], and $\mathcal{E}$ a path from $-\infty - i\epsilon$ to $+\infty - i\epsilon$ with $\mathrm{Im}\nu < 0$. The path $\mathcal{E}$ is such that the singularities of the integrand, due to $\mathcal{M}^\pm$ and to zeros of $H_\nu^{(2)\prime}(ka^\pm) + Q^\pm H_\nu^{(2)}(ka^\pm)$, are, respectively, above and below $\mathcal{E}$, which corresponds, when we let $\nu = ka^\pm \sin\alpha'$ with $\mathrm{Im}\nu < 0$ and $|\arg(ik)| < \pi/2$, to take α' from $-i\infty - \arg(ik) + 0^+$ to $i\infty + \arg(ik) + 0^+$ with $|\mathrm{Re}\alpha'| \leq \pi/2$. From our analysis (see details in appendices A and B), the asymptotic expansion of $\mathcal{M}^\pm$ is given by:

$$\mathcal{M}^\pm(\pm ka^\pm \sin\alpha') = \frac{i\pi ka^\pm}{2\cos\alpha'} \mathcal{O}^\pm_{\pm\alpha'}(F_<(\pm(\alpha' - \pi/2 + \Phi))),$$

$$\mathcal{O}^\pm_\alpha = \frac{\pm 2}{i\pi ka^\pm}((1 + \sum_{p\geq 0} \frac{\mathcal{P}^\pm_p(.)}{(a^\pm)^{p+2}})B^\pm_\alpha), \qquad (2.28)$$

that we can consider at arbitray order of development of $B^\pm_\alpha$ verifying (2.11), while the $\mathcal{P}^\pm_p$ are explicit differential operators of $\sin\alpha$ independent of $a^\pm$. By letting

$\alpha = \alpha' - \pi/2$ in (2.28), and using (2.12), we then have:

$$\begin{aligned}&\mathcal{M}^{\pm}(\pm ka^{\pm}\cos\alpha)\\ &= \frac{1}{\sin\alpha}[\frac{-i\pi ka^{\pm}}{2}\mathcal{O}^{\pm}_{\pm(\alpha+\pi/2)}(.)]F_{<}(\pm(\alpha+\Phi))\\ &= \frac{1}{\sin\alpha}[(\sin\alpha-\sin\theta^{\pm})(.)-D^{\pm}_{\pm\alpha+\pi}(.)/ika^{\pm}\\ &+O(\frac{1}{(ka^{\pm})^{2}})]\ F_{<}(\pm(\alpha+\Phi)),\end{aligned} \tag{2.29}$$

where it is worth noticing that, at each order of its asymptotic expression, $\mathcal{M}^{\pm}(\pm ka^{\pm}\cos\alpha)$ has no singularity at $\alpha = 0$ when $\pm\varphi_{\circ} - \Phi \neq -\pi$, with $\mathcal{O}^{\pm}_{\alpha\pm\frac{\pi}{2}}F_{<}(\alpha\pm\Phi)$ regular at $\alpha = 0$, $\mathcal{O}^{\pm}_{\alpha\pm\frac{\pi}{2}}F_{<}(\alpha\pm\Phi)|_{\alpha=0} = 0$, and $\mathcal{O}^{\pm}_{\alpha\pm\frac{\pi}{2}}F_{<}(\alpha\pm\Phi)$ $+\mathcal{O}^{\pm}_{-\alpha\pm\frac{\pi}{2}}F_{<}(-\alpha\pm\Phi) = 0$.

2.3.2. *Plane wave illumination: observation at finite distance*

To consider an observation point (ρ,φ) at finite distance, we can use that:

$$u(\rho,\varphi) = -\frac{1}{2\pi i}\int_{\mathcal{S}_{\circ}} F(\alpha\pm(\Phi-\pi/2))e^{-ik\rho\cos(\alpha\pm(\Phi-\pi/2)-\varphi)}d\alpha, \tag{2.30}$$

is valid for $0 < \pm\varphi < \Phi$ when no source in the zone $\Phi-\pi < \pm\varphi < \Phi$, with $\mathcal{S}_{\circ}$ correctly chosen from $-\arg(ik) - i\infty$ to $\arg(ik) + i\infty$. We then use (2.27) in (2.30) with $\mathcal{L}^{\pm}_{<} \equiv \frac{\mathcal{M}^{\pm}}{i\pi ka^{\pm}}$, and write:

$$\begin{aligned}u(\rho,\varphi) &= \frac{-1}{2\pi i}\int_{\mathcal{E}}\frac{\mathcal{L}^{\pm}_{<}(\pm\nu)}{H^{(2)'}_{\nu}(ka^{\pm})+Q^{\pm}H^{(2)}_{\nu}(ka^{\pm})}\\ &\int_{\mathcal{S}_{\circ}} e^{-ik\rho\cos(\alpha\pm(\Phi-\pi/2)-\varphi)-i\nu(\pm\alpha-\pi/2)-ika^{\pm}\cos\alpha}d\alpha d\nu\\ &= -\int_{\mathcal{E}}\frac{\mathcal{L}^{\pm}_{<}(\pm\nu)}{H^{(2)'}_{\nu}(ka^{\pm})+Q^{\pm}H^{(2)}_{\nu}(ka^{\pm})}\\ &\frac{e^{\mp i\nu\Omega}}{2i}(\frac{e^{i\nu\pi/2}}{\pi}\int_{\mathcal{S}_{\circ}} e^{-ikR\cos(\alpha-\Omega)-i\nu(\pm(\alpha-\Omega))}d\alpha)d\nu,\end{aligned} \tag{2.31}$$

where $R\cos\Omega = a^{\pm} + \rho\cos(\varphi\pm(\pi/2-\Phi))$ and $R\sin\Omega = \rho\sin(\varphi\pm(\pi/2-\Phi))$, that is, relatively to the curvature center of each face, for:

$$R = \sqrt{\rho^{2}+(a^{\pm})^{2}+2\rho a^{\pm}\cos(\Phi-\pi/2\mp\varphi)} \tag{2.32}$$

the distance to the point (ρ, φ), and Ω the angle from the edge to (ρ, φ). On last line of (2.31), the integral is equal to the function $\pi e^{-i\nu\pi/2} H_\nu^{(2)}(kR)$, and we obtain:

$$u(\rho,\varphi) = \frac{-1}{2i} \int_{\mathcal{E}} \frac{\mathcal{L}_<^\pm(\pm\nu)\, H_\nu^{(2)}(kR)\, e^{\mp i\nu\Omega}}{H_\nu^{(2)'}(ka^\pm) + Q^\pm H_\nu^{(2)}(ka^\pm)} d\nu \text{ with } \mathcal{L}_<^\pm \equiv \frac{\mathcal{M}^\pm}{i\pi ka^\pm}, \qquad (2.33)$$

for $|\pm\varphi - \Phi| < \pi/2$ when $\pm\varphi_\circ - \Phi < -\pi$. We note that, for large R, we recover (2.27), since $H_\nu^{(2)}(kR)e^{\mp i\nu\Omega} \sim \frac{-e^{-ikR-i\pi/4}}{\sqrt{2\pi kR}} \frac{2e^{-i\nu(\pm\Omega-\pi/2)}}{i}$, and for $\rho \gg |a^\pm|$, $R \sim (\rho - a^\pm \sin(\pm\varphi - \Phi))$ and $\Omega \sim \varphi \pm (\pi/2 - \Phi)$.

When we go beyond the frontier $\varphi = \pm\Phi$ of the convex face $\pm$, there are no more stationnary phase points for this integral. Considering the residues terms at zeros $\nu_p^\pm = ka^\pm \cos\alpha_p^\pm$ of $[H_\nu^{(2)'}(ka^\pm) + Q^\pm H_\nu^{(2)}(ka^\pm)]$ in the zone $\text{Im}\nu < 0$, then closing the contour $\mathcal{E}$ at infinity, we obtain, when $\pm\varphi > \Phi$:

$$u(\rho,\varphi) = \sum_p \frac{\pi\mathcal{L}_<^\pm(\pm\nu_p^\pm)\, H_{\nu_p^\pm}^{(2)}(kR)e^{\mp i\nu_p^\pm\Omega}}{\frac{\partial}{\partial\nu}(H_\nu^{(2)'}(ka^\pm) + Q^\pm H_\nu^{(2)}(ka^\pm))|_{\nu_p^\pm}}. \qquad (2.34)$$

The expressions (2.33) for u, for diffracted field at large distance and for associated far field function F, are then uniform for $\varphi \sim \pm\Phi$. These expressions permit the continuous transformation of the field into creeping waves depending on $\nu_p^\pm$ (a development in entire powers then changes continuously to one in fractional powers as φ varies), for any wedge angle.

Besides, we remark that, contrary to expressions of Albertsen and Christiansen, the operators $\mathcal{L}_<^\pm \equiv \frac{\mathcal{M}^\pm}{i\pi ka^\pm}$ go beyond first order. We can then consider a second-order development of f without making diverge the expression (2.33), which permits us to take account, in whole generality, of the influence of both orders, and for example to study at low contrasts, the case of a curved plane with a discontinuity of materials and curvature.

NOTE 2.5.– For an extension to the case with concave (negative $a^\pm$) face $\pm$, we shall, after determining $\mathcal{M}^\pm$, replace $H_\nu^{(2)}(kR)$, $H_\nu^{(2)}(ka^\pm)$, and $H_\nu^{(2)'}(ka^\pm)$, by $2J_\nu(k|R|)$, $2J_\nu(k|a^\pm|)$, and $2J_\nu'(k|a^\pm|)$, to take conveniently account of whispering gallery modes [9] . On this occasion, let us note that:

$$J_\nu(z) = \frac{H_\nu^{(1)}(z) + H_\nu^{(2)}(z)}{2}, \ \frac{J_\nu(z)}{(z)^\nu} \text{ is an even function of } z,$$

$$J_\nu'(z) = \frac{J_{\nu-1}(z) - J_{\nu+1}(z)}{2} = J_{\nu-1}(z) - \frac{\nu}{z}J_\nu(z). \qquad (2.35)$$

NOTE 2.6.– In noticing that the diffraction coefficient $F_<$ for a cylinder $\mathcal{C}_{a^\pm,\infty}$ of large radius $a^\pm$ without discontinuity is null, we can continue the solution for $\Phi - \pi < \pm\varphi_0 < \Phi$, either by analytical continuation or by taking account of additional terms coming from contribution of $\mathcal{C}_{a^\pm,\infty}$ in (2.34).

NOTE 2.7.– For $\nu \approx ka^\pm$, $H_\nu^{(2)'}(ka^\pm) + Q^\pm H_\nu^{(2)}(ka^\pm) \sim \frac{-i}{(m^\pm)^2\sqrt{\pi}}(w_2'(\tau) - q^\pm w_2(\tau))$ [18], where $m^\pm = (ka^\pm/2)^{1/3}, q^\pm = -im^\pm \sin\theta^\pm, \nu = ka^\pm + m^\pm\tau$, $w_2 = \sqrt{\pi}(Bi - iAi)$, and thus, when $\pm\varphi \sim \Phi$, an approximation of (2.27) is given by:

$$F(\varphi) \simeq \frac{e^{ika^\pm \sin(\pm\varphi-\Phi)}}{i\pi ka^\pm}\int_{\mathcal{E}'} \frac{e^{-i\nu(\pm\varphi-\Phi)}\mathcal{M}^\pm(\pm\nu)}{\frac{-i}{(m^\pm)^2\sqrt{\pi}}(w_2'(\tau) - q^\pm w_2(\tau))} m^\pm d\tau$$

$$\simeq \frac{e^{ika^\pm(\sin(\pm\varphi-\Phi)-(\pm\varphi-\Phi))}}{2\sqrt{\pi}}\int_{\mathcal{E}'} \frac{e^{-i\tau(m^\pm(\pm\varphi-\Phi))}\mathcal{M}^\pm(\pm\nu)}{(w_2'(\tau) - q^\pm w_2(\tau))} d\tau,$$

$$\text{with } ka^\pm(\sin(\pm\varphi - \Phi) - (\pm\varphi - \Phi)) \sim -(m^\pm(\pm\varphi - \Phi))^3/3, \tag{2.36}$$

where we note that $ka^\pm \cos\alpha = ka^\pm + m^\pm(-(m^\pm)^2(2\sin(\alpha/2))^2)$ [1].

2.3.3. *Source and observation points at finite distance*

We can then derive an expression at finite distance for observation point (ρ', φ'), when primary point source at (ρ, φ) radiates:

$$u^i = \frac{i}{2}H_0^{(2)}(k\sqrt{\rho^2 + (\rho')^2 - 2\rho\rho'\cos(\varphi - \varphi')})$$

$$\sim -\frac{e^{-ik\rho' - i\pi/4}}{\sqrt{2\pi k\rho'}}e^{ik\rho\cos(\varphi-\varphi')}, \tag{2.37}$$

where $e^{ik\rho\cos(\varphi-\varphi')}$ is the far field function of u^i for large ρ'. We can then use the expression for the wedge illuminated by the plane wave $e^{ik\rho\cos(\varphi-\varphi')}$, considering the reciprocity, by exchanging the directions of observation and incidence in (2.33). Therefore, letting:

$$F_<(\pm(\alpha' + \Phi)) = \frac{1}{2\pi i}\int_\gamma f(\alpha + \varphi, \pm(\alpha' + \Phi))e^{ik\rho\cos\alpha}d\alpha, \tag{2.38}$$

we obtain, for convex faces, when $\pm\varphi - \Phi < -\pi$ and $\varphi' \sim \pm\Phi$:

$$u(\rho', \varphi', \rho, \varphi) = \frac{-1}{2i}\int_{\mathcal{E}} \frac{\mathcal{L}^\pm(\pm\nu')\, H_{\nu'}^{(2)}(kR')\, e^{\mp i\nu'\Omega'}}{H_{\nu'}^{(2)'}(ka^\pm) + Q^\pm H_{\nu'}^{(2)}(ka^\pm)} d\nu', \tag{2.39}$$

where,

$$\mathcal{L}^{\pm}(\pm ka^{\pm}\cos\alpha') = \mathcal{A}^{\pm}_{\alpha'}(\frac{1}{2\pi i}\int_{\gamma} f(\alpha+\varphi, \pm(\alpha'+\Phi))e^{ik\rho\cos\alpha}d\alpha),$$

$$R'\cos\Omega' = a + \rho'\cos(\varphi' \pm (\pi/2-\Phi)), R'\sin\Omega' = \rho'\sin(\varphi' \pm (\pi/2-\Phi)),$$

$$\mathcal{A}^{\pm}_{\alpha} = \frac{-\mathcal{O}^{\pm}_{\pm(\alpha+\pi/2)}}{2\sin\alpha}, R' = \sqrt{\rho'^2 + (a^{\pm})^2 + 2\rho' a^{\pm}\cos(\Phi - \pi/2 \mp \varphi')}, \quad (2.40)$$

with Ω' the angle, relatively to the cylinder center, from the edge to the observation point (ρ', φ'). Let us notice that this expressions remains valid, by reciprocity, when the roles of (ρ, φ) and (ρ', φ'), as source and observation points, are exchanged, and by continuation for $|\varphi| < \Phi$.

When we go beyond the frontier $\varphi' = \pm\Phi$, there are no more stationary phase points for this integral, and considering the residues terms at zeros $\nu^{\pm}_{p}$ of $[H^{(2)'}_{\nu}(ka^{\pm}) + Q^{\pm}H^{(2)}_{\nu}(ka^{\pm})]$ in the zone $\text{Im}\nu < 0$, we obtain, when $\pm\varphi' > \Phi$:

$$u(\rho', \varphi', \rho, \varphi) =$$

$$= \sum_{p} \frac{\pi\mathcal{L}^{\pm}(\pm\nu^{\pm}_{p})\, H^{(2)}_{\nu^{\pm}_{p}}(kR')e^{\mp i\nu^{\pm}_{p}\Omega'}}{\frac{\partial}{\partial\nu}(H^{(2)'}_{\nu}(ka^{\pm}) + Q^{\pm}H^{(2)}_{\nu}(ka^{\pm}))|_{\nu^{\pm}_{p}}}, \quad (2.41)$$

where

$$\mathcal{L}^{\pm}(\pm ka^{\pm}\cos\alpha_p) =$$

$$= \mathcal{A}^{\pm}_{\alpha_p}(\frac{1}{2\pi i}\int_{\gamma} f(\alpha+\varphi, \pm(\alpha_p+\Phi))e^{ik\rho\cos\alpha}d\alpha)$$

$$= \mathcal{A}^{\pm}_{\alpha_p}(u_{i,p} + \sum u^{\pm}_{r,p} + \sum u^{\pm}_{s,p}$$

$$+ \frac{e^{-ik\rho}}{2\pi i}\int_{SDP} f(\alpha+\varphi, \pm(\alpha_p+\Phi))\, e^{ik\rho(\cos\alpha+1)}d\alpha), \quad (2.42)$$

with $\mathcal{A}^{\pm}_{\alpha} = \frac{-\mathcal{O}^{\pm}_{\pm(\alpha+\pi/2)}}{2\sin\alpha}$, as $|\varphi| < \Phi$. This expression can be used for arbitrary ρ'. However, except for its edge diffracted (SDP) part, its use for large ρ, because of the presence of secular terms in $\sum u^{\pm}_{r}$, is limited to the region of φ where $u^{\pm}_{r,p} = 0$ (see note below and further section on fringe waves).

For $l\varphi > \Phi$ and $l'\varphi' > \Phi$ (l,l' are equal to+or−), the problem is more complex and we need to express $\mathcal{L}^{\pm}(\pm\nu')$ as an integral of previous form. We then obtain for the diffracted part of the field:

$$u_{da} = \pi^2 \sum_{p,q} \mathcal{A}^l_{\alpha^l_p} \mathcal{A}^{l'}_{\alpha^{l'}_q} (f(\pi + l(\alpha^l_p + \Phi), l'(\alpha^{l'}_q + \Phi))$$

$$- f(-\pi + l(\alpha^l_p + \Phi), l'(\alpha^{l'}_q + \Phi)))$$

$$\times \frac{H^{(2)}_{\nu^l_p}(kR)\, e^{-li\nu^l_p \Omega}}{\frac{\partial}{\partial \nu}(H^{(2)\prime}_{\nu}(ka^l) + Q^l H^{(2)}_{\nu}(ka^l))|_{\nu=\nu^l_p}}$$

$$\times \frac{H^{(2)}_{\nu^{l'}_q}(kR')\, e^{-l' i\nu^{l'}_q \Omega'}}{\frac{\partial}{\partial \nu}(H^{(2)\prime}_{\nu}(ka^{l'}) + Q^{l'} H^{(2)}_{\nu}(ka^{l'}))|_{\nu=\nu^{l'}_q}}. \qquad (2.43)$$

This expression has a form similar to the one defined by Thompson for the particular case of a discontinuity of materials on a circular cylinder.

NOTE 2.8.– If we denote $f_{a^{\pm},\infty}$ the spectral function attached to radiation of asymptotic currents induced on a cylinder $\mathcal{C}_{a^{\pm},\infty}$ of radius $a^{\pm}$ without discontinuity (respectively, for the PO contribution of face $\pm$), we can analytically continue (2.39)–(2.41) for $\pm\varphi - \Phi > -\pi$, and avoid the problem attached to secular terms in $u^{\pm}_{r,p}$ terms, by letting $f(\alpha) - f_{a^{\pm},\infty}(\alpha \mp (\Phi - \frac{\pi}{2}))$ in place of $f(\alpha)$, and add the exact contribution of $\mathcal{C}_{a^{\pm},\infty}$ in (2.39)–(2.41) (respectively, of the PO currents of face $\pm$). A further section on fringe waves refers in generality to this problem.

NOTE 2.9.– It is worth noticing that $\mathcal{A}^{\pm}_{\alpha'}(\sum u^{\pm}_r + \sum u^{\pm}_s)$ with $\varphi_o = \pm(\alpha' + \Phi)$ has additional secular terms (polynomial terms) in powers of ρ for the $\mathcal{A}^{\pm}_{\alpha'}(\sum u^{\pm}_r)$ part because exponential terms in $\sum u^{\pm}_r$ (reflected geometrical optics waves) are dependent on φ_o, but no additional secular terms for $\mathcal{A}^{\pm}_{\alpha'}(\sum u^{\pm}_s)$ because exponential terms in $u^{\pm}_s$ (guided complex plane waves excited by the edge) are not dependent on φ_o.

2.4. Some developed expressions concerning $f = \sum_{0 \leq n \leq m} f_n/k^n$ for arbitrary wedge angle

2.4.1. *A developed expression of* χ *for* $f_1(\alpha) = \Psi(\alpha)\chi(\alpha)$

We give here a developed expression of (2.25) for $\chi(\alpha')$ when $\mathrm{Re}(\sin\theta^{\pm}) > 0$, $\mathrm{Re}\alpha' \neq 0$, in order to express second-order term of the far field radiation function $f(\pi + \varphi) - f(-\pi + \varphi)$ with $f_1(\alpha') = \Psi(\alpha')\chi(\alpha')$.

In letting ϵ =sign(Reα'), and considering $a^\delta = a^+$ (respectively, $a^\delta = a^-$) when $\delta = +1$ (respectively, $\delta = -1$), we choose to take $d^+ = (1+\epsilon)\Phi$ and $d^- = (1-\epsilon)\Phi$ in (2.25). We then have, for Re$(\alpha') \in]-3\Phi, 0[\cup]0, 3\Phi[$:

$$\begin{aligned}
\chi(\alpha') &= \frac{1}{8\Phi\Psi(\varphi_o)} \sum_{\pm} \frac{1}{a^{\pm\epsilon}} \\
&\frac{1}{2}\left(\int_{-i\infty-0^+}^{+i\infty-0^+} + \int_{-i\infty+0^+}^{+i\infty+0^+}\right)\Bigg[\Big(\tan(\frac{\pi}{4\Phi}(\alpha' - \epsilon\Phi - \alpha)) \\
&\times \partial_\alpha \ln((-\sin(\alpha + (1\pm 1)\epsilon\Phi) \pm \epsilon \sin\theta^{\pm\epsilon})\Psi(-\alpha - \epsilon\Phi)) \\
&+ \partial_{\alpha'} \tan(\frac{\pi}{4\Phi}(\alpha' - \epsilon\Phi - \alpha)\Big)\Big(\sigma(\alpha - \epsilon\Phi) \\
&\times \partial_\alpha \ln\left(\frac{\Psi(-\alpha - \epsilon\Phi)}{(\sin(\alpha + (1\pm 1)\epsilon\Phi) \pm \epsilon\sin\theta^{\pm\epsilon})}\right) + \partial_\alpha \sigma(\alpha - \epsilon\Phi)\Big) \\
&\frac{1}{\sin(\alpha + (1\pm 1)\epsilon\Phi)}\Bigg] d\alpha \\
&+ \frac{-2\pi i\epsilon}{8\Phi\Psi(\varphi_0)} \frac{1}{a^\epsilon} \sum_{0<\epsilon Re\alpha'_s<2\Phi} \text{Residue}\Big[G(\alpha)\Big]|_{\alpha=\text{pole }\alpha'_s},
\end{aligned} \tag{2.44}$$

where,

$$\begin{aligned}
G(\alpha) &= (W^\epsilon(\alpha', \alpha)\partial_\alpha \ln(\Psi(-\alpha + \epsilon\Phi)(\sin\alpha - \epsilon\sin\theta^\epsilon)) \\
&+ \partial_{\alpha'} W^\epsilon(\alpha', \alpha))(\sigma(\alpha + \epsilon\Phi)\partial_\alpha \ln\left(\frac{\Psi(-\alpha + \epsilon\Phi)}{(\sin\alpha + \epsilon\sin\theta^\epsilon)}\right) \\
&+ \partial_\alpha \sigma(\alpha + \epsilon\Phi))\frac{1}{\sin\alpha},
\end{aligned} \tag{2.45}$$

with,

$$\begin{aligned}
W^\epsilon(\alpha', \alpha) &= \tan(\frac{\pi}{4\Phi}(\alpha' + \epsilon\Phi - \alpha)),\ \partial_\alpha W^\epsilon(\alpha', \alpha) = -\partial_{\alpha'} W^\epsilon(\alpha', \alpha), \\
\sigma(\alpha \pm \epsilon\Phi) &= \frac{\frac{\pi}{2\Phi}\cos\frac{\pi}{2\Phi}\varphi_o}{(\pm\epsilon\cos\frac{\pi}{2\Phi}\alpha - \sin\frac{\pi}{2\Phi}\varphi_o)}, \\
\partial_\alpha \sigma(\alpha \pm \epsilon\Phi) &= \partial_{\varphi_o} \frac{\frac{\pi}{2\Phi}\sin\frac{\pi}{2\Phi}\alpha}{(\cos\frac{\pi}{2\Phi}\alpha \mp \epsilon\sin\frac{\pi}{2\Phi}\varphi_o)}.
\end{aligned} \tag{2.46}$$

Developing the residue terms, we have:

$$\frac{-2\pi i\epsilon}{8\Phi\Psi(\varphi_0)}\frac{1}{a^\epsilon}\sum_{0<\pm\text{sign}(d^\pm)Re\alpha'_s<|d^\pm|}\text{Residue}\Big[G(\alpha)\Big]|_{\alpha=\text{pole }\alpha'_s}$$

$$= \frac{-2\pi i\epsilon}{8\Phi\Psi(\varphi_0)}\frac{1}{a^\epsilon}\Bigg[\Bigg(\partial_{\varphi_o}\Big(\,((W^\epsilon(\alpha',\alpha)\partial_\alpha \ln(\Psi(-\alpha+\epsilon\Phi)(\sin\alpha-\epsilon\sin\theta^\epsilon))$$

$$+\,\partial_{\alpha'}W^\epsilon(\alpha',\alpha))\frac{1}{\sin\alpha})|_{\alpha=\epsilon\Phi-\varphi_o}\Big)\times -1$$

$$+\Big((W^\epsilon(\alpha',\alpha)\partial_\alpha \ln(\Psi(-\alpha+\epsilon\Phi)(\sin\alpha-\epsilon\sin\theta^\epsilon))+\partial_{\alpha'}W^\epsilon(\alpha',\alpha))$$

$$\times\frac{\partial_\alpha \ln\left(\frac{\Psi(-\alpha+\epsilon\Phi)}{(\sin\alpha+\epsilon\sin\theta^\epsilon)}\right)}{\sin\alpha})|_{\alpha=\epsilon\Phi-\varphi_o}\times -1\Big)$$

$$+\Big((\sigma(\alpha+\epsilon\Phi)\partial_\alpha \ln\Big(\frac{\Psi(-\alpha+\epsilon\Phi)}{(\sin\alpha+\epsilon\sin\theta^\epsilon)}\Big)+\partial_\alpha\sigma(\alpha+\epsilon\Phi))\frac{W^\epsilon(\alpha',\alpha)}{\sin\alpha}\Big)|_{\alpha=\epsilon\theta^\epsilon}$$

$$+\Big(G_1+G_2\times\frac{1}{2}(1+\text{sign}(2\Phi-\text{Re}(\pi+\theta^\epsilon)))+G_3\Big)\Bigg], \tag{2.47}$$

with $\text{sign}(x)=\pm1$ as $\pm x>0$, $\text{sign}(0)=0$, and:

$$G_1=\lim_{\alpha\to\alpha_1}G(\alpha)(\alpha-\alpha_1)=\Big((W^\epsilon(\alpha',\alpha)\partial_\alpha \ln(\Psi(-\alpha+\epsilon\Phi)(\sin\alpha-\epsilon\sin\theta^\epsilon))$$

$$+\,\partial_{\alpha'}W^\epsilon(\alpha',\alpha))(\sigma(\alpha+\epsilon\Phi)\partial_\alpha \ln\Big(\frac{\Psi(-\alpha+\epsilon\Phi)}{(\sin\alpha+\epsilon\sin\theta^\epsilon)}\Big)$$

$$+\,\partial_{\varphi_o}\frac{\frac{\pi}{2\Phi}\sin(\frac{\pi}{2\Phi}\alpha)}{(\cos\frac{\pi}{2\Phi}\alpha-\epsilon\sin\frac{\pi}{2\Phi}\varphi_o)})\Big)|_{\alpha=\alpha_1=\epsilon\pi}\times -1,$$

$$G_2=\lim_{\alpha\to\alpha_2}G(\alpha)(\alpha-\alpha_2)=\Big((W^\epsilon(\alpha',\alpha)\partial_\alpha \ln(\Psi(-\alpha+\epsilon\Phi)(\sin\alpha-\epsilon\sin\theta^\epsilon))$$

$$+\,\partial_{\alpha'}W^\epsilon(\alpha',\alpha))\frac{\sigma(\alpha+\epsilon\Phi)}{\sin\alpha}\Big)|_{\alpha=\alpha_2=\epsilon(\pi+\theta^\epsilon)}\times -1,$$

$$G_3=\lim_{\alpha\to\alpha_3}G(\alpha)(\alpha-\alpha_3)=\Big((\sigma(\alpha+\epsilon\Phi)\partial_\alpha \ln\Big(\frac{\Psi(-\alpha+\epsilon\Phi)}{(\sin\alpha+\epsilon\sin\theta^\epsilon)}\Big)$$

$$+\,\partial_\alpha\sigma(\alpha+\epsilon\Phi))\frac{W^\epsilon(\alpha',\alpha)}{\sin\alpha}\Big)|_{\alpha=\alpha_3=\epsilon(\pi-\theta^\epsilon)}, \tag{2.48}$$

as $\alpha_1=\epsilon\pi$, $\alpha_2=\epsilon(\pi+\theta^\epsilon)$, $\alpha_3=\epsilon(\pi-\theta^\epsilon)$. When $\theta^\epsilon=0$, $G(\alpha)$ has a pole of third order at $\alpha=\epsilon\pi$, but it is worth noticing that $G_1+G_2+G_3$ remains finite when $\theta^\epsilon\to 0$ (see appendix C). We now compare the results obtained from this expression (or directly from functional equations) with the expressions known for some particular wedge angles.

2.4.2. *Perfectly conducting case: curved half plane and discontinuity of curvature*

In the limit case $\sin\theta^{\pm} \rightarrow \infty$, we have $\Psi \equiv 1$ from the functional equation satisfied by Ψ. We can then write:

$$f_0(\alpha) = \frac{\pi}{2\Phi} \frac{\cos(\frac{\pi}{2\Phi}\varphi_o)}{(\sin(\frac{\pi}{2\Phi}\alpha) - \sin(\frac{\pi}{2\Phi}\varphi_o))} \text{ and } f_1(\alpha) = \chi(\alpha), \tag{2.49}$$

for which, from (2.44) to (2.48), we have:

$$\chi(\alpha') = \Big(\frac{1}{8\Phi}\partial_{\alpha'}\partial_{\varphi_o}\int_{-i\infty}^{i\infty} d\alpha \\ \Big[\frac{\mathrm{tg}(\frac{\pi}{4\Phi}(\alpha' - \epsilon\Phi - \alpha)\frac{\pi}{2\Phi}\sin\frac{\pi}{2\Phi}\alpha}{\cos(\frac{\pi}{2\Phi}\alpha) + \epsilon\sin(\frac{\pi}{2\Phi}\varphi_o)}(\frac{1}{a^{-\epsilon}\sin\alpha} + \frac{1}{a^{\epsilon}\sin(\alpha + 2\epsilon\Phi)})\Big]\Big) \\ + i\frac{\epsilon\pi}{4\Phi a^{\epsilon}}\partial_{\alpha'}\partial_{\varphi_o}\Big[\frac{\mathrm{tg}(\frac{\pi}{4\Phi}(\alpha' + \varphi_o))}{\sin(\epsilon\Phi - \varphi_o)} + \frac{\frac{\epsilon\pi}{2\Phi}\sin(\frac{\pi}{2\Phi}\pi)\mathrm{tg}(\frac{\pi}{4\Phi}(\alpha' + \epsilon(\Phi - \pi)))}{\cos(\frac{\pi}{2\Phi}\pi) - \epsilon\sin(\frac{\pi}{2\Phi}\varphi_o)}\Big], \tag{2.50}$$

for $\mathrm{Re}(\alpha') \in] - 3\Phi, 0[\cup]0, 3\Phi[$ with $\epsilon = \mathrm{sign}(\mathrm{Re}\alpha')$.

In the case of a curved half-plane, we have $\Phi = \pi$, and letting $\epsilon a^{\epsilon} = a$, we can write:

$$\chi(\alpha') = (0) + i\frac{1}{4a}\partial_{\alpha'}\partial_{\varphi_o}[\frac{1}{\sin(\varphi_o)}\mathrm{tg}(\frac{1}{4}(\alpha' + \varphi_o)) \\ - \frac{1}{2}\mathrm{tg}(\frac{1}{4}\alpha')\frac{1}{\sin(\frac{1}{2}\varphi_o)}]. \tag{2.51}$$

This gives us the second-order term, that is, the first term depending on curvatures of faces, for the far field radiation function, following:

$$(f_1(\pi + \varphi) - f_1(-\pi + \varphi))/k = \frac{i}{4ka}\partial_{\varphi}\partial_{\varphi_o}(\frac{\mathrm{tg}((\varphi + \varphi_o)/2)}{\cos(\varphi/2)\cos(\varphi_o/2)}), \tag{2.52}$$

which is conform to the result given in the paper of Borovikov [7], first developed by V.B. Filippov [8].

For the discontinuity of curvature, $\Phi = \frac{\pi}{2}$ and (2.50) gives us:

$$\chi(\alpha') = \frac{1}{4\pi}\partial_{\alpha'}\partial_{\varphi_o}\int_{-i\infty}^{i\infty} d\alpha\,[\frac{\cos\alpha'}{(\cos\alpha + \epsilon\sin\alpha')}\frac{1}{(\cos\alpha + \epsilon\sin\varphi_o)}(\frac{1}{a^+} - \frac{1}{a^-})]$$

$$
\begin{aligned}
&+\frac{i}{2a^{\epsilon}}\partial_{\alpha'}\partial_{\varphi_\circ}\left(\frac{\mathrm{tg}(\frac{1}{2}(\alpha'+\varphi_\circ))}{\cos(\varphi_\circ)}\right)\\
&=\frac{i}{2\pi}\,\partial_{\alpha'}\partial_{\varphi_\circ}\left(\frac{\epsilon\cos\alpha'}{(\sin\varphi_\circ-\sin\alpha')}(\frac{\pi/2-\epsilon\alpha'}{\cos\alpha'}-\frac{\pi/2-\epsilon\varphi_\circ}{\cos\varphi_\circ})(\frac{1}{a^+}-\frac{1}{a^-})\right.\\
&\left.+\frac{\pi}{a^{\epsilon}}\frac{\mathrm{tg}(\frac{1}{2}(\alpha'+\varphi_\circ))}{\cos(\varphi_\circ)}\right)\\
&=\frac{i}{2\pi}\,\partial_{\alpha'}\partial_{\varphi_\circ}\left(\epsilon\frac{1}{2}(\mathrm{tg}(\frac{1}{2}(\alpha'+\varphi_\circ))+\mathrm{cotg}(\frac{1}{2}(-\alpha'+\varphi_\circ))\right.\\
&(\frac{\pi/2-\epsilon\alpha'}{\cos\alpha'}-\frac{\pi/2-\epsilon\varphi_\circ}{\cos\varphi_\circ})(\frac{1}{a^+}-\frac{1}{a^-})\\
&\left.+\frac{\pi}{2}(\epsilon(\frac{1}{a^+}-\frac{1}{a^-})+(\frac{1}{a^+}+\frac{1}{a^-}))\frac{\mathrm{tg}(\frac{1}{2}(\alpha'+\varphi_\circ))}{\cos(\varphi_\circ)}\right)\\
&=\frac{i}{2\pi}\,\partial_{\alpha'}\partial_{\varphi_\circ}\left(\frac{-\cos\alpha'}{(\sin\varphi_\circ-\sin\alpha')}(\frac{\alpha'}{\cos\alpha'}-\frac{\varphi_\circ}{\cos\varphi_\circ})(\frac{1}{a^+}-\frac{1}{a^-})\right.\\
&\left.+\frac{\pi}{2}\frac{\mathrm{tg}(\frac{1}{2}(\alpha'+\varphi_\circ))}{\cos(\varphi_\circ)}(\frac{1}{a^+}+\frac{1}{a^-})\right),
\end{aligned}
\tag{2.53}
$$

which leads us to the following expression for the far field radiation function:

$$
\begin{aligned}
&(f_1(\pi+\varphi)-f_1(-\pi+\varphi))/k\ =\\
&=-i\,\partial_{\varphi}\partial_{\varphi_\circ}\left(\frac{1}{(\sin\varphi+\sin\varphi_\circ)}\right)(\frac{1}{ka^+}-\frac{1}{ka^-})\\
&=-2i\frac{\cos\varphi\cos\varphi_\circ}{(\sin\varphi+\sin\varphi_\circ)^3}(\frac{1}{ka^+}-\frac{1}{ka^-}),
\end{aligned}
\tag{2.54}
$$

that is conform to the diffraction coefficient of Kaminetzky and Keller [6].

2.4.3. *Discontinuity of curvature in an impedance surface*

In this case, we have $\Phi=\pi/2$ and $\sin\theta^{\pm}=\sin\theta$, and note that the far field function verifies $F(\varphi)=(f_1(\pi+\varphi)-f_1(-\pi+\varphi))/k$. Rewriting the functional equation in f_1 from (2.11), by letting $\alpha=\pm\pi/2+\varphi$, we have:

$$
\begin{aligned}
&f_1(\pm\pi+\varphi)\frac{\cos\varphi+\sin\theta}{-\cos\varphi+\sin\theta}-f_1(-\varphi)=\\
&=\frac{D^{\pm}_{\pm\pi/2+\varphi}(f_0(\pm\pi+\varphi))-D^{\pm}_{\mp\pi/2-\varphi}(f_0(-\varphi))}{ia^{\pm}(\cos\varphi-\sin\theta)}.
\end{aligned}
\tag{2.55}
$$

From $\sin\theta^{\pm} = \sin\theta$ and $f_0(\pi+\varphi) = f_0(-\pi+\varphi)$, we have:

$$D^{+}_{\pi/2+\varphi}(f_0(\pi+\varphi)) - D^{+}_{-\pi/2-\varphi}(f_0(-\varphi))$$
$$= D^{-}_{-\pi/2+\varphi}(f_0(\pi+\varphi)) - D^{-}_{\pi/2-\varphi}(f_0(-\varphi)), \tag{2.56}$$

and thus,

$$f_1(\pi+\varphi) - f_1(-\pi+\varphi) = i(\frac{1}{a^+} - \frac{1}{a^-})\times$$
$$\frac{D^{+}_{\pi/2+\varphi}(f_0(\pi+\varphi)) - D^{+}_{-\pi/2-\varphi}(f_0(-\varphi))}{(\cos\varphi + \sin\theta)}. \tag{2.57}$$

Using $f_0(\alpha) = \frac{\cos\alpha+\sin\theta}{\cos\varphi_o+\sin\theta}\,\frac{\cos\varphi_o}{\sin\alpha-\sin\varphi_o}$, and expanding D^+, we obtain:

$$f_1(\pi+\varphi) - f_1(-\pi+\varphi) = (\frac{1}{a^+} - \frac{1}{a^-})\times$$
$$\frac{2i\,\cos\varphi\,\cos\varphi_o(1+\sin\varphi\,\sin\varphi_o - \sin^2\theta)}{(\cos\varphi_o+\sin\theta)(\cos\varphi+\sin\theta)(\sin\varphi+\sin\varphi_o)^3}, \tag{2.58}$$

which is perfectly in agreement with the expression given by Kaminetzky and Keller [6].

2.5. Rewriting the field with fringe waves of second order

2.5.1. *Definition of fringe waves* u_{fr} *when* $|\varphi| < \Phi$

We consider, as previously, the integral representation of Sommerfeld–Maliuzhinets:

$$u(\rho,\varphi,\varphi_o) = \frac{1}{2\pi i}\int_\gamma f(\alpha+\varphi,\varphi_o)e^{ik\rho\cos\alpha}d\alpha, \tag{2.59}$$

for the field radiated by the wedge in the exterior sector $|\varphi| < \Phi$, when it is illuminated by a plane wave $u^{inc}(\rho,\varphi)$ upcoming from the direction φ_o. By application of asymptotic decomposition of the spectral function f in inverse powers of wavenumber k, the expression of field u has GO (geometrical optics) term u_{OGa}, which corresponds to a Taylor expansion, in vicinity of the edge, of the exact term u_{OGe} that derives from the field for a circular cylinder without discontinuity. The term u_{OGa} has secular terms in powers of $k\rho$ describing efficiently the GO field at short distance of the edge, but not appropriate for the field at large distance.

If we know the analytical expression of u_{OGe} on the curved faces, we can improve the solution and correct the influence of secular terms. For that, we will distinguish the field term due to radiation of currents perturbated by the edge, named fringe wave u_{fr}, from the terms due to currents ignoring it (currents due to GO). Later, we will replace, in the expression of the total field, the field $u^{\pm}_{opa}$ due to surface currents derived from u_{OGa} at $\varphi = \pm\Phi$, by the field $u^{\pm}_{op}$ due to surface currents derived from u_{OGe} on the curved surfaces, with relative impedances $\sin\theta^{\pm}$. This process corresponds to:

$$u(\rho,\varphi,\varphi_o) = [u_{fr}] + [\sum_{\pm} U(\pi - |\varphi_o \mp \Phi|)u^{\pm}_{op} + u^{inc}], \tag{2.60}$$

with,

$$u_{fr}(\rho,\varphi) = u(\rho,\varphi,\varphi_o) - u^{inc}(\rho,\varphi) - \sum_{\pm} U(\pi - |\varphi_o \mp \Phi|)u^{\pm}_{opa},$$

$$u^{\pm}_{opa} = -\frac{i}{4}\int_{S'^{\pm}} (u_{OGa}\frac{\partial H_0^{(2)}(kR)}{\partial n} - \frac{\partial u_{OGa}}{\partial n}H_0^{(2)}(kR))d\rho',$$

$$u^{\pm}_{op} = -\frac{i}{4}\int_{S^{\pm}} (u_{OGe}(\frac{\partial H_0^{(2)}(kR)}{\partial n} + kQH_0^{(2)}(kR))a^{\pm}d\varphi_c, \tag{2.61}$$

where $S^{\pm}$ and $S'^{\pm}$ are, respectively, the circular curved faces $\pm$ of the wedge with $0 \leq \pm\varphi_c \leq \pi/2$ (i.e. asymptotically for large $a^{\pm}$ as $0 \leq \pm a^{\pm}\varphi_c \leq \infty$), and the tangent surfaces at the edge $\varphi' = \pm\Phi$ from $\rho' = 0$ to ∞. The term U is the Heaviside function, while $u^{inc}(\rho,\varphi) = e^{ik\rho\cos(\varphi-\varphi_o)}$, $Q = -i\sin\theta^{\pm}$ and $R = \sqrt{\rho^2 + \rho'^2 - 2\rho\rho'\cos(\varphi - \varphi')}$ is the distance of integration point to observation point (ρ,φ). At large distance, we will be able to use [18]:

$$H_0^{(2)}(kR) = \sqrt{\frac{2}{\pi kR}}\, e^{-ikR+i\pi/4}(1 + O(\frac{1}{kR})),$$

$$\partial_n H_0^{(2)}(kR) = \sqrt{\frac{2}{\pi kR}}\, e^{-ikR-i\pi/4}\partial_n(kR)(1 + O(\frac{1}{kR})), \tag{2.62}$$

where we note that

$$\partial_n(kR)|_{\varphi'=\pm\Phi} = \mp\frac{\partial(kR)}{\rho'\partial\varphi'}|_{\varphi'=\pm\Phi} = \pm\frac{k\rho\sin(\varphi\mp\Phi)}{R},$$

$$R = \rho - \rho'\cos(\varphi - \varphi') + \frac{(\rho'\sin(\varphi-\varphi'))^2}{R + \rho - \rho'\cos(\varphi-\varphi')},$$

$$R = \rho' - \rho\cos(\varphi' - \varphi) + \frac{(\rho\sin(\varphi'-\varphi))^2}{R + \rho' - \rho\cos(\varphi'-\varphi)}. \tag{2.63}$$

NOTE 2.10.– The far field function $(\frac{-e^{-i\pi/4}}{\sqrt{2\pi k\rho'}}e^{-ik\rho'})^{-1}u(\rho',\varphi',\rho,\varphi)|_{\rho'\to\infty}$ for a wedge illuminated by a point source radiating $u^i = \frac{i}{2}H_0^{(2)}(kR)$ in free space, with $R = \sqrt{\rho^2 + \rho'^2 - 2\rho\rho'\cos(\varphi - \varphi')}$, is, when reciprocity principle applies, the field $u(\rho,\varphi,\varphi')$ in presence of the wedge illuminated by a plane wave radiating $u^{inc} = e^{ik\rho\cos(\varphi-\varphi')}$.

2.5.2. The field term $u^{\pm}_{opa}$ and the expression of u_{fr} at large distance

We can determine u_{OGa} from the behavior at poles of GO (for incident and reflected waves) of the spectral function $f \equiv f_0 + f_1/k$. By use of functional equations (see (2.143) in appendix D), we can write at $\varphi = \pm\Phi$:

$$|^{1}_{\frac{\partial}{\partial n}} u_{OGa}(\rho, \varphi = \pm\Phi, \varphi_\circ = \pm(-\alpha + \Phi)) = \Big(\Big[\Big|^{1}_{ik\sin\theta^{\pm}}\big(\frac{2e^{ik\rho\cos\alpha}\sin\alpha}{\sin\alpha + \sin\theta^{\pm}}\big)\Big]$$

$$-\frac{1}{2ika^{\pm}}\Big[\big(\frac{2\sin\theta^{\pm}}{\sin\alpha+\sin\theta^{\pm}}\big)\big(\frac{\sin\alpha}{\sin\theta^{\pm}}\big(\frac{\partial}{\partial\alpha} + \mathrm{cotg}\alpha\big)\frac{\partial}{\partial\alpha}$$

$$-\sin\theta^{\pm}\frac{\partial}{\partial\alpha}\frac{1}{\sin\alpha}\frac{\partial}{\partial\alpha}\big)\big(\Big|^{1}_{-ik\sin\alpha}\frac{e^{ik\rho\cos\alpha}}{\sin\alpha+\sin\theta^{\pm}}\big)\Big]\Big)\Big|_{\alpha=\mp\varphi_\circ+\Phi}. \tag{2.64}$$

We can then use the field $u^{\pm}_{opa}$, expressed following:

$$u^{\pm}_{opa} = \frac{1}{2\pi i}\int_\gamma f^{\pm}_{opa}(\alpha+\varphi,\varphi_\circ))e^{ik\rho\cos\alpha}d\alpha, \tag{2.65}$$

and write,

$$\Big[\big(\frac{-e^{-i\pi/4}}{\sqrt{2\pi k\rho}}e^{-ik\rho}\big)^{-1}u^{\pm}_{opa}(\rho,\varphi)\Big]_{\rho\to\infty}$$
$$= f^{\pm}_{opa}(\pi+\varphi) - f^{\pm}_{opa}(-\pi+\varphi)$$
$$= \frac{1}{2}\int_{S'^{\pm}}(iku_{OGa}(\rho', \pm\Phi)(\pm\sin(\varphi\mp\Phi))$$
$$+\frac{\partial u_{OGa}}{\partial n}(\rho', \pm\Phi))e^{ik\rho'\cos(\varphi\mp\Phi)}d\rho', \tag{2.66}$$

as $\mathrm{Re}(ik(\cos(\mp\varphi_\circ+\Phi)+\cos(\varphi\mp\Phi))) < 0$, where we note that, developing (2.66) with (2.64), we have:

$$f^{\pm}_{opa}(\pi+\alpha') - f^{\pm}_{opa}(-\pi+\alpha')$$
$$= -\Big(\big[\big(\frac{\sin\alpha}{\sin\alpha+\sin\theta^{\pm}}\big)\big]\frac{(\pm\sin(\alpha'\mp\Phi)+\sin\theta^{\pm})}{\cos\alpha+\cos(\alpha'\mp\Phi)}$$
$$-\frac{1}{2ika^{\pm}}\big[\big(\frac{\sin\theta^{\pm}}{\sin\alpha+\sin\theta^{\pm}}\big)\big(\frac{\sin\alpha}{\sin\theta^{\pm}}\big(\frac{\partial}{\partial\alpha}+\mathrm{cotg}\alpha\big)\frac{\partial}{\partial\alpha}$$
$$-\sin\theta^{\pm}\frac{\partial}{\partial\alpha}\frac{1}{\sin\alpha}\frac{\partial}{\partial\alpha}\big)\big]$$
$$\frac{(\pm\sin(\alpha'\mp\Phi)-\sin\alpha)}{(\sin\alpha+\sin\theta^{\pm})(\cos\alpha+\cos(\alpha'\mp\Phi))}\Big)\Big|_{\alpha=\mp\varphi_\circ+\Phi}. \tag{2.67}$$

Considering that we have not GO fields terms in u_{fr}, and neglecting the exponentially vanishing contributions of complex poles, we then obtain at large distance, for $|\mathrm{Re}\varphi_\circ| < \Phi$ and $|\mathrm{Re}\varphi| < \Phi$:

$$u_{fr}(\rho,\varphi,\varphi_\circ) \sim -\frac{e^{-i\pi/4}}{\sqrt{2\pi k\rho}}e^{-ik\rho}(f(\pi+\varphi,\varphi_\circ) - f(-\pi+\varphi,\varphi_\circ)$$

$$-\sum_{\pm} U(\pi - |\varphi_\circ \mp \Phi|)(f^{\pm}_{opa}(\pi+\varphi) - f^{\pm}_{opa}(-\pi+\varphi))). \tag{2.68}$$

NOTE 2.11.– $f^{\pm}_{opa}(\pi+\alpha') - f^{\pm}_{opa}(-\pi+\alpha') \to \pm A$ as $\mathrm{Im}(\alpha') \to \pm\infty$, with generally $A \neq 0$, in contrary to $f(\pi+\alpha') - f(-\pi+\alpha')$ that vanishes at infinity (note: $f^{\pm}_{opa}(\alpha)$ is generally not bounded for large $\mathrm{Im}\alpha$ in contrary to $f(\alpha)$).

2.5.3. *The expressions of u_{fr} for $|\varphi_\circ| < \Phi$ and $|\varphi| < \Phi$*

Considering (2.59)–(2.61) and (2.65) with (2.67), we can develop an analytical expression of u_{fr} at any distance for $|\varphi| < \Phi$ and $|\varphi_\circ| < \Phi$, in the form of a Sommerfeld–Maliuzhinets integral, which we will be able to continue by use of transition function.

By definition in (2.61) and use of (2.59) and (2.65), we have u_{fr}, for $|\varphi| < \Phi$, following:

$$u_{fr}(\rho,\varphi,\varphi_\circ) = \frac{1}{2\pi i}\int_\gamma (f(\alpha+\varphi,\varphi_\circ) - \frac{1}{2}\mathrm{cotg}(\frac{1}{2}(\alpha+\varphi-\varphi_\circ))$$

$$-\sum_{\pm} U(\pi - \mid \varphi_\circ \mp \Phi \mid) f^{\pm}_{opa}(\alpha+\varphi,\varphi_\circ))e^{ik\rho\cos\alpha}d\alpha, \tag{2.69}$$

where $f^{\pm}_{opa}$ refers to $u^{\pm}_{opa}$ from (2.65) to(2.67). The spectral function $f^{\pm}_{opa}$ in (2.69) can be expressed from the integral expression of $u^{\pm}_{opa}$ along $S'^{\pm}$ given in (2.61), when we replace $H^{(2)}_0(kR)$ by its spectral function $f^{H^{(2)}_\circ}$ [17]:

$$f^{H^{(2)}_\circ}(\alpha) = \frac{-1}{2\pi}\int_S e^{ik\rho'\cos(\alpha'-\varphi')}(\tan(\frac{1}{2}(\alpha-\alpha')) - g_\circ(\alpha'))d\alpha', \tag{2.70}$$

for $R = \sqrt{\rho^2+\rho'^2-2\rho\rho'\cos(\varphi-\varphi')}$, $|\varphi'| > \frac{\pi}{2}$, $|\varphi| < |\varphi'|$, $|\arg(ik)| < \pi/2$, when $\alpha \in]\mathcal{S}-\pi, \mathcal{S}+\pi[$ (i.e. the domain limited by $\mathcal{S}-\pi$ and $\mathcal{S}+\pi$), with $\mathcal{S}$ the path from $-i\infty - \arg(ik)$ to $i\infty + \arg(ik)$ with $\mathrm{Im}(k\sin\alpha) = 0$. Because of oddness of the path γ, the choice of $g_\circ$ is arbitrary and we can take $g_\circ(\alpha') = -\tan(\frac{\alpha'}{2})$ to ameliorate convergence. This gives us, after comparison with (2.66):

$$f^{\pm}_{opa}(\alpha'') = \frac{1}{4\pi i}\int_{-i\infty-\epsilon}^{+i\infty+\epsilon} (f^{\pm}_{opa}(\pi+\alpha') - f^{\pm}_{opa}(-\pi+\alpha'))$$

$$(\tan(\frac{1}{2}(\alpha''-\alpha')) + \tan(\frac{\alpha'}{2}))d\alpha', \tag{2.71}$$

as $|\text{Re}(\alpha'')| < \pi - \epsilon$, where it is worth noticing that, from (2.67), the integral term can be expressed analytically. To express (2.71) in closed form, we shall use in (2.67):

$$\begin{aligned}
\frac{\sin(\alpha' \mp \Phi)}{\cos\alpha + \cos(\alpha' \mp \Phi)} &= \frac{1}{2}(\tan(\frac{1}{2}((\alpha' \mp \Phi) - \alpha)) + \tan(\frac{1}{2}(\alpha' \mp \Phi + \alpha))),\\
\frac{\sin\alpha}{\cos\alpha + \cos(\alpha' \mp \Phi)} &= \frac{1}{2}(-\tan(\frac{1}{2}(\alpha' \mp \Phi - \alpha)) + \tan(\frac{1}{2}(\alpha' \mp \Phi + \alpha))),\\
\frac{\sin(\alpha' \mp \Phi) \mp \sin\alpha}{\cos\alpha + \cos(\alpha' \mp \Phi)} &= \tan(\frac{1}{2}((\alpha' \mp \Phi) \mp \alpha)) = \tan(\frac{1}{2}(\alpha' \mp (\alpha + \Phi))),\\
\frac{1}{\sin\alpha + \sin\theta^\pm} &= \frac{1}{2\cos\theta^\pm}(\cot(\frac{1}{2}(\theta^\pm + \alpha)) - \tan(\frac{1}{2}(\theta^\pm - \alpha))),\\
\tan(\frac{1}{2}(\alpha'' - \alpha'))\tan(\frac{1}{2}(\alpha' + w)) &= 1 - \frac{\tan(\frac{1}{2}(\alpha'' - \alpha')) + \tan(\frac{1}{2}(\alpha' + w))}{\tan(\frac{1}{2}(\alpha'' + w))},
\end{aligned} \tag{2.72}$$

with $\tan(\frac{1}{2}(\alpha' + v)) = -2\partial_{\alpha'}(\ln(\cos(\frac{1}{2}(\alpha' + v))))$, and thus:

$$\text{pv}\int_{-i\infty-\epsilon}^{+i\infty+\epsilon} \tan(\frac{1}{2}(\alpha' + v))d\alpha' = 2i(v - 2n\pi) \text{ when } |\text{Re}(v - 2n\pi)| < \pi. \tag{2.73}$$

NOTE 2.12.– The reader can verify by inspection that we have, for $\Phi = \pi/2$ and $\sin\theta^\pm = \sin\theta$, $f^\pm_{opa}(\pi + \alpha') - f^\pm_{opa}(-\pi + \alpha') = \pm f(\pm\pi + \alpha')$ in this particular case.

2.5.4. *The term of radiation $u^\pm_{op}$ derived from u_{OGe}*

2.5.4.1. *The term $u_{OGe}|_{S^\pm}$ and $\frac{\partial u_{OGe}}{\partial n}|_{S^\pm}$*

We consider the integral expression of $u^\pm_{op}$ in (2.61) where the term u_{OGe} is obtained from the expression of the field on an impedance circular cylinder of radius $a^\pm$ illuminated at angle φ_o, for large $ka^\pm$. Neglecting waves that make more than one turn on each curved face $S^\pm$ of the wedge, we have:

$$u_{OGe}|_{S^\pm} = \frac{2i}{\pi k a^\pm}\int_{\mathcal{E}} \frac{e^{-i\nu(|(\varphi_o \pm (\pi/2 - \Phi)) - \varphi_c| - \pi/2) + ika^\pm \sin(\pm\varphi_o - \Phi)}}{H^{(2)\prime}_\nu(ka^\pm) + Q^\pm H^{(2)}_\nu(ka^\pm)} d\nu, \tag{2.74}$$

with $Q^\pm = -i\sin\theta^\pm$, and $\frac{\partial u_{OGe}}{\partial n}|_{S^\pm} = ik\sin\theta^\pm u_{OGe}|_{S^\pm}$. In these expressions, $\mathcal{E}$ is a path from $-\infty - i\epsilon$ to $+\infty - i\epsilon$ with $\text{Im}\nu < 0$, above the poles of integrand. Letting $\nu = ka^\pm \sin\alpha'$ with $|\text{Re}\alpha'| \le \pi/2$ and $|\arg(ik)| < \pi/2$, α' then goes from $-i\infty - \arg(ik) + 0^+$ to $i\infty + \arg(ik) + 0^+$ on the integration path.

Considering the expression of $|(\varphi_o \pm (\pi/2 - \Phi)) - \varphi_c|$, we will be led to distinguish the case where a part of the curved face $S^\pm$ is in shadow zone from the case where the whole face $S^\pm$ is illuminated.

2.5.4.2. $u_{op}^{\pm}$ as $\pi/2 \leq |\varphi_\circ \mp \Phi| \leq \pi$

In the case with $\pi/2 \leq |\varphi_\circ \mp \Phi| \leq \pi$, and thus, $0 \leq \mp\varphi_\circ - (\pi/2 - \Phi) = -\varphi^{\pm} \leq \pi/2$, the curved face $S^{\pm}$ is partially in shadow region. We have $\pm\varphi_c > 0$ on $S^{\pm}$, and when $0 \leq -\varphi^{\pm} \leq \pi/2$, we can then write in (2.74):

$$|(\varphi_\circ \pm (\pi/2 - \Phi)) - \varphi_c| = \pm\varphi_c - \varphi^{\pm}. \tag{2.75}$$

Considering $u_{op}^{\pm}$ in (2.61), we have, letting $\varphi_r = \pm\varphi_c$:

$$\begin{aligned}
u_{op}^{\pm}(\rho,\varphi) &= -\frac{i}{4}\int_{S^{\pm}} u_{OGe}\Big(\frac{\partial H_\circ^{(2)}(kR)}{\partial n} + kQH_\circ^{(2)}(kR)\Big)a^{\pm}d\varphi_c \\
&= \frac{e^{ika^{\pm}\sin(\pm\varphi_\circ - \Phi)}}{2\pi k}\int_{\mathcal{E}} \frac{e^{-i\nu(\mp\varphi_\circ + (\Phi - \pi))}}{H_\nu^{(2)\prime}(ka^{\pm}) + Q^{\pm}H_\nu^{(2)}(ka^{\pm})} \\
&\times \Big(\Big(-\sin\varphi_c\frac{\partial .}{\partial x_c} + \cos\varphi_c\frac{\partial .}{\partial y_c} + kQ\Big)\int_0^{\pi/2} e^{-i\nu\varphi_r}H_\circ^{(2)}(kR)d\varphi_r\Big)d\nu \\
&= \frac{e^{ika^{\pm}\sin(\pm\varphi_\circ - \Phi)}}{2\pi k}\int_{\mathcal{E}} \frac{e^{-i\nu(\mp\varphi_\circ + (\Phi - \pi))}}{H_\nu^{(2)\prime}(ka^{\pm}) + Q^{\pm}H_\nu^{(2)}(ka^{\pm})}\Big(\int_0^{\pi/2} e^{-i\nu\varphi_r} \\
&\Big(\Big(\frac{-\sin\varphi_c(x_c - x') + \cos\varphi_c(y_c - y')}{R}\frac{\partial .}{\partial R} + kQ\Big)H_\circ^{(2)}(kR)\Big)d\varphi_r\Big)d\nu,
\end{aligned} \tag{2.76}$$

with R the distance of observation point with the integration point, that is, $R = \sqrt{(x' - x_c)^2 + (y' - y_c)^2}$ where $x_c = -a^{\pm}\sin\varphi_c$ and $y_c = a^{\pm}\cos\varphi_c$, $x' = -\rho\sin\varphi_1$ and $y' = a^{\pm} + \rho\cos\varphi_1$, with $\varphi_r = \pm\varphi_c$ and $\varphi_1 = \varphi \pm (\pi/2 - \Phi)$, and the edge is at $\varphi_c = 0$ (or $\rho = 0$).

2.5.4.3. $u_{op}^{\pm}$ as $0 \leq |\varphi_\circ \mp \Phi| \leq \pi/2$

We consider the case with $0 \leq |\varphi_\circ \mp \Phi| \leq \pi/2$, and thus, $-\pi/2 \leq \mp\varphi_\circ - (\pi/2 - \Phi) = -\varphi^{\pm} \leq 0$, when the curved face $S^{\pm}$ is completely in light region. We have $\pm\varphi_c > 0$ on $S^{\pm}$, and when $-\pi/2 \leq -\varphi^{\pm} \leq 0$, we can write in (2.74):

$$\begin{aligned}
&|(\varphi_\circ \pm (\pi/2 - \Phi)) - \varphi_c| = \mp\varphi_c + \varphi^{\pm} \text{ for } 0 \leq \pm\varphi_c \leq \varphi^{\pm}\ (I), \\
&|(\varphi_\circ \pm (\pi/2 - \Phi)) - \varphi_c| = \pm\varphi_c - \varphi^{\pm} \text{ for } \pm\varphi_c \geq \varphi^{\pm}\ (II).
\end{aligned} \tag{2.77}$$

Letting $\varphi_r = \pm\varphi_c$ as we consider faces $\pm$, we have:

$$u_{op}^{\pm} = u_{op}^{(I)} + u_{op}^{(II)}, \tag{2.78}$$

where,

$$u_{op}^{(I)}(\rho,\varphi) = -\frac{i}{4}\int_{(I)} (u_{OGe}(\frac{\partial H_\circ^{(2)}(kR)}{\partial n} + kQH_\circ^{(2)}(kR))a^\pm d\varphi_c$$

$$= \frac{e^{ika^\pm \sin(\pm\varphi_\circ - \Phi)}}{2\pi k}\int_{\mathcal{E}} d\nu \frac{e^{i\nu(\mp\varphi_\circ+\Phi)}}{H_\nu^{(2)\prime}(ka^\pm) + Q^\pm H_\nu^{(2)}(ka^\pm)}\Big(\int_0^{\varphi^\pm} d(\varphi_r)e^{i\nu\varphi_r}$$

$$(\frac{-\sin\varphi_c(x_c - x') + \cos\varphi_c(y_c - y')}{R}\frac{\partial.}{\partial R} + kQ)H_\circ^{(2)}(kR)\Big), \tag{2.79}$$

and,

$$u_{op}^{(II)}(\rho,\varphi) = -\frac{i}{4}\int_{(II)} (u_{OGe}(\frac{\partial H_\circ^{(2)}(kR)}{\partial n} + kQH_\circ^{(2)}(kR))a^\pm d\varphi_c$$

$$= \frac{e^{ika^\pm \sin(\pm\varphi_\circ - \Phi)}}{2\pi k}\int_{\mathcal{E}} d\nu \frac{e^{-i\nu(\mp\varphi_\circ+(\Phi-\pi))}}{H_\nu^{(2)\prime}(ka^\pm) + Q^\pm H_\nu^{(2)}(ka^\pm)}\Big(\int_{\varphi^\pm}^{\pi/2} d\varphi_r e^{-i\nu\varphi_r}$$

$$(\frac{-\sin\varphi_c(x_c - x') + \cos\varphi_c(y_c - y')}{R}\frac{\partial.}{\partial R} + kQ)H_\circ^{(2)}(kR)\Big), \tag{2.80}$$

where $R = \sqrt{(x' - x_c)^2 + (y' - y_c)^2}$ is the distance of the observation point with the integration point, while:

$$x_c = -a^\pm \sin\varphi_c, y_c = a^\pm \cos\varphi_c, \varphi_r = \pm\varphi_c,$$
$$x' = -\rho\sin\varphi_1,\ y' = a^\pm + \rho\cos\varphi_1,$$
$$\varphi^\pm = \pm\varphi_\circ + (\pi/2 - \Phi),\ \varphi_1 = \varphi \pm (\pi/2 - \Phi). \tag{2.81}$$

2.5.4.4. *About $u_{op}^\pm$ when observation point is on $S^\pm$*

When the observation point (x', y') is on the curved face $S^\pm$, we have $x' = -a^\pm \sin\varphi_{c,obs}$ and $y' = a^\pm \cos\varphi_{c,obs}$ with $\varphi_{c,obs} = 2\varphi_1 \mp \pi$. We can then write in (2.76)–(2.80):

$$-\sin\varphi_c(x_c - x') + \cos\varphi_c(y_c - y') = 2a^\pm(\sin((\varphi_c - \varphi_{c,obs})/2))^2,$$
$$R = |a^\pm|\sqrt{(2 - 2\cos(\varphi_c - \varphi_{c,obs})} = 2|a^\pm||\sin((\varphi_c - \varphi_{c,obs})/2)|, \tag{2.82}$$

and $R = 2|a^\pm|\cos(\varphi_1 - \varphi_c/2)$ with $\varphi_c = \pm\varphi_r$. Letting $\varphi_c = \pm\varphi_r$, we will then write asymptotically for large $ka^\pm$ in (2.76)–(2.80):

$$H_0^{(2)}(kR) \sim \sqrt{\frac{2}{\pi a^\pm}}e^{i\pi/4}e^{-ik2|a^\pm|\cos((\varphi_r \mp 2\varphi_1)/2)}. \tag{2.83}$$

2.5.5. Continuation of $u^{\pm}_{fr}(\rho', \varphi', \rho, \varphi)$ in shadow zone when $\pm\varphi' > \Phi$

When we go beyond the frontier $\varphi' = \pm\Phi$, we can refer to the integral expression (2.39) and then (2.41) to deduce a continuation of (2.60). We have from (2.41), when $\pm\varphi' > \Phi$:

$$u(\rho', \varphi', \rho, \varphi) =$$

$$= \sum_p \frac{\pi \mathcal{L}^{\pm}(\pm\nu_p^{\pm})\, H^{(2)}_{\nu_p^{\pm}}(kR') e^{\mp i\nu_p^{\pm}\Omega'}}{\frac{\partial}{\partial\nu}(H^{(2)\prime}_{\nu}(ka^{\pm}) + Q^{\pm} H^{(2)}_{\nu}(ka^{\pm}))|_{\nu_p^{\pm}}}, \tag{2.84}$$

where

$$\mathcal{L}^{\pm}(\pm ka^{\pm}\cos\alpha_p) =$$

$$= \mathcal{A}^{\pm}_{\alpha_p}(\frac{1}{2\pi i}\int_{\gamma} f(\alpha + \varphi, \pm(\alpha_p + \Phi)) e^{ik\rho\cos\alpha} d\alpha)$$

$$= \mathcal{A}^{\pm}_{\alpha_p}(u_{i,p} + \sum u^{\pm}_{r,p} + \sum u^{\pm}_{s,p}$$

$$+ \frac{e^{-ik\rho}}{2\pi i}\int_{SDP} f(\alpha + \varphi, \pm(\alpha_p + \Phi)) e^{ik\rho(\cos\alpha + 1)} d\alpha), \tag{2.85}$$

with $\mathcal{A}^{\pm}_{\alpha} = \frac{-\mathcal{O}^{\pm}_{\pm(\alpha+\pi/2)}}{2\sin\alpha}$. We then deduce, from this expression, a continuation of $u^{\pm}_{fr}$ for $\pm\varphi' > \Phi$, following:

$$u^{\pm}_{fr}(\rho', \varphi', \rho, \varphi) =$$

$$= \sum_p \frac{\pi \mathcal{L}^{\pm}_{fr}(\pm\nu_p^{\pm})\, H^{(2)}_{\nu_p^{\pm}}(kR') e^{\mp i\nu_p^{\pm}\Omega'}}{\frac{\partial}{\partial\nu}(H^{(2)\prime}_{\nu}(ka^{\pm}) + Q^{\pm} H^{(2)}_{\nu}(ka^{\pm}))|_{\nu_p^{\pm}}},$$

$$\text{with } \mathcal{L}^{\pm}_{fr}(\pm ka^{\pm}\cos\alpha_p) = \mathcal{A}^{\pm}_{\alpha_p}(u_{fr}(\rho, \varphi, \pm(\alpha_p + \Phi))) \text{ as } |\varphi| < \Phi, \tag{2.86}$$

where,

$$\mathcal{A}^{\pm}_{\alpha} = \frac{1}{i\pi ka^{\pm}\sin\alpha}[(\sin\alpha - \sin\theta^{\pm}) - \frac{D^{\pm}_{\pm\alpha+\pi}(.)}{ika^{\pm}} + 0(1/(ka^{\pm})^2)],$$

$$D^{\pm}_{\alpha}(.) = \frac{1}{2}[\frac{\partial^2(.)}{\partial\alpha^2} - \frac{\partial}{\partial\alpha}(\cot\alpha(.)) \pm \sin\theta^{\pm}\frac{\partial}{\partial\alpha}(\frac{1}{\sin\alpha}\frac{\partial}{\partial\alpha}(.))],$$

$$R'\cos\Omega' = a^{\pm} + \rho'\cos(\varphi' \pm (\frac{\pi}{2} - \Phi)),\ R'\sin\Omega' = \rho'\sin(\varphi' \pm (\frac{\pi}{2} - \Phi)),$$

$$R' \sim (\rho' - a^{\pm}\sin(\pm\varphi' - \Phi)) \text{ and } \Omega' \sim \varphi' \pm (\pi/2 - \Phi) \text{ for } \rho' \gg |a^{\pm}|. \tag{2.87}$$

This expression permits the continuation of (2.60) for $\pm\varphi' > \Phi$, with an expression correcting the problem due to secular terms in the asymptotic expression of $u^{\pm}_{r}$ for large $a^{\pm}$, which appears when we use (2.85) for $\pm\varphi' > \Phi$ as $\pm\varphi - \Phi > -\pi$ for large ρ.

2.6. References

[1] Bernard, J.M.L. (1991). Diffraction par un dièdre à faces courbes non parfaitement conducteur. *Rev. Tech. Th.*, 23(2), 321–330.

[2] Bernard, J.M.L. (2002). The diffraction by a curved impedance wedge: Diffracted and creeping waves. In *Proceedings of the URSI General Assembly*, Maastricht, 18–23 August, hal-03654088v2.

[3] Borovikov, V.A. (1998). Diffraction by an impedance wedge with curved sides. *J. Comm. Techn. Elect.*, 43(12), 1337–1346.

[4] Albertsen, N.C. and Christiansen, P.L. (1978). Hybrid diffraction coefficients for first and second order discontinuities of two dimensional scatterers. *SIAM J. Appl. Math.*, 34(2), 398–414.

[5] Bernard, J.M.L. (2020). New insight in integral representation theory for the solution of complex canonical diffraction problems. In *Advances in Mathematical Methods for Electromagnetics*, Kobayashi, K. and Smith, P.D. (eds). Institution of Engineering and Technology, London.

[6] Kaminetzky, L. and Keller, J.B. (1972). Diffraction coefficients for high order edges and vertices. *SIAM J. Appl. Math.*, 22(1), 109–134.

[7] Borovikov, V.A. (1979). Diffraction by a wedge with curved faces. *Akust. Zh.*, 25(6), 825–835.

[8] Filippov, V.B. (1974). Diffraction by a curved half-plane. *Zap. Nauchn. Sem. LOMI (Leningrad)*, 42, 244–249.

[9] Buyukaksoy, A. and Uzgoren, G. (1985). High frequency scattering by a cylindrically curved impedance surface. *AEU*, 39, 351–358.

[10] Montarnal, P. and Bouche, D. (1996). Fringe wave diffraction coefficients for higher order discontinuities. *Ann. des Telecom.*, 51, 3–4.

[11] Michaeli, A. (1989). Transition functions for high-frequency diffraction by a curved perfectly conducting wedge. Part I: Canonical solution for a curved sheet. *IEEE Trans. AP*, 37(9), 1073–1079.

[12] Lafitte, O. (2002). The wave diffracted by a wedge with mixed boundary conditions. Report, Mémoires de la Société mathématique de France, Paris.

[13] Gerard, P. and Lebeau, G. (1993). Diffusion d'une onde par une coin. *Journal of the AMS*, 6(2), 341–423.

[14] Maliuzhinets, G.D. (1958). Inversion formula for the Sommerfeld integral. *Sov. Phys. Dokl.,* 3, 52–56.

[15] Maliuzhinets, G.D. (1958). Excitation, reflection and emission of surface waves from a wedge with given face impedances. *Sov. Phys. Dokl.,* 3, 752–755.

[16] Bernard, J.M.L. (1987). Diffraction by a metallic wedge covered with a dielectric material. *Wave Motion*, 9, 543–561.

[17] Bernard, J.M.L. (2006). A spectral approach for scattering by impedance polygons. *Quart. J. Mech. Appl. Math.*, 59(4), 517–550.

[18] Abramowitz, M. and Stegun, I. (1972). *Handbook of Mathematical Functions*. Dover Publications Inc., New York.

2.A. Appendix A: on an integral expression with asymptotic kernel for the field near tangence of faces at $\varphi = \pm\Phi$, permitting an explicit transition at arbitrary order

An asymptotic development rarely covers the whole angular space. We consider a wedge composed of two semi-circular faces of radius $a^{\pm}$, with tangent plane at the edge $\varphi = \pm\Phi$ and relative impedance $\sin\theta = \sin\theta^{\pm}$. Let us assume that we know the asymptotic development of field as $|\varphi| < \Phi$ for large $a^{\pm}$ at arbitrary order. We present here an integral transform of it for a global transition near and beyond the tangence $\varphi = \pm\Phi$.

2.A.1. *Expression when the far field function F has no singularity*

Considering the transition attached to each face $\pm$ separately, we begin to assume first that the field u is composed of waves outgoing from a cylindrical surface of radius $a = a^{\pm}$. This circular surface is a continuation of the semi-circular face $\pm$, and the impedance boundary condition with relative impedance $\sin\theta = \sin\theta^{\pm}$ only applies on the semi-circular face $\pm$.

The field u then satisfies the wave equation in 2D:

$$\Delta u + k^2 u = 0, \tag{2.88}$$

in a domain Ω^a exterior to a circular cylinder of radius a, with k the wavenumber and time convention $e^{i\omega t}$. We define the Cartesian coordinates x and y with origin at the edge of the curved wedge, such that the surface $\partial\Omega^a = \mathcal{C}$ satisfies $x^2 + (y+a)^2 = a^2$, and consider the term S following:

$$(\frac{\partial.}{\partial n} - ik\sin\theta)u|_{\mathcal{C}} = S, \tag{2.89}$$

where $\frac{\partial(.)}{\partial n} = \widehat{n}.\text{grad}(.)$ with $\widehat{n}$ the outward normal, and θ is supposed to be a constant.

At this step, the field has been assumed to be cylindrically radiated, and the presence of plane waves in the field can only be considered as a limit case in the final expression. We then develop u in cylindrical functions $\mathrm{e}^{-im\varphi_c} H_m^{(2)}(k\rho_c)$, with $H_m^{(2)}$ the Hankel function of order m [18]. The (ρ_c,φ_c) polar coordinates is defined such that the cylinder C has its center at $\rho_c = 0$ and its surface at $\rho_c = a$, while $\varphi_c = 0$ corresponds to the y-axis. Considering that $\frac{\partial.}{\partial n}|_C \equiv \frac{\partial.}{\partial \rho_c}$, we obtain:

$$u = \frac{1}{2\pi k} \sum_m e^{-im\varphi_c} \frac{\int_C e^{im\varphi_c} S \, d\varphi_c}{H_m^{(2)'}(ka) + Q\, H_m^{(2)}(ka)} H_m^{(2)}(k\rho_c), \tag{2.90}$$

where $Q = -i \sin\theta$.

The angle φ was initially used in Sommerfeld–Maliuzhinets representation such that each face is given by $\varphi = \pm\Phi$, and we now consider a new definition of angle, φ_e, so that:

$$\begin{aligned} &\varphi_e = \varphi \pm (\pi/2 - \Phi), \\ &\{\varphi = \pm\Phi\} \equiv \{\varphi_e = \pm\pi/2\}, \\ &x = -\rho_c \sin\varphi_c = -\rho \sin\varphi_e, \\ &y = \rho_c \cos\varphi_c - a = \rho\cos\varphi_e, \end{aligned} \tag{2.91}$$

where it is worth noticing that φ, φ_e and φ_c grows with $-x$.

We can expand the field u in plane waves, following:

$$u(\rho, \varphi_e) = \frac{-1}{2\pi i} \int_S F^{\pm}(\alpha) e^{-ik\rho\cos(\alpha - \varphi_e)} d\alpha, \tag{2.92}$$

as $|\varphi_e| \leq \pi/2$, that is, within the half space free of sources $y > 0$, where $F^{\pm}(\alpha) = F(\alpha \mp (\pi/2 - \Phi))$ is the far field function referring to face $\varphi = \pm\Phi$, and the path S goes from $-\arg(ik) - i\infty$ to $\arg(ik) + i\infty$ with $\mathrm{Im}(k \sin\alpha) = 0$. At this step, as for (2.90), the field u is assumed to be radiated cylindrically at large distance, and we have:

$$u = \frac{-e^{-i\pi/4}}{\sqrt{2\pi k\rho}} e^{-ik\rho} F^{\pm}(\varphi_e) + O((k\rho)^{-3/2}), \tag{2.93}$$

with $F^{\pm}(\varphi_e) = F(\varphi)$. In other respects, we can consider the development of u at (ρ'_c, φ'_c), in Helmhotz representation, by the radiation of simple and double layers currents on the cylinder of radius $\rho_c = a$:

$$\begin{aligned} &u(\rho'_c, \varphi'_c) = \frac{-i}{4} \int_C (u \frac{\partial H_0^{(2)}(kR)}{\partial \rho_c} - \frac{\partial u}{\partial \rho_c} H_0^{(2)}(kR))|_{\rho_c = a} a d\varphi_c, \\ &R = \sqrt{\rho_c^2 + \rho_c'^2 - 2a\rho'_c \cos((\varphi_c - \varphi'_c)}, \end{aligned} \tag{2.94}$$

and using that, for ρ and $\rho'_c = \sqrt{\rho^2 + a^2 + 2\rho a \cos\varphi_e} \to \infty$:

$$\begin{aligned} &H_0^{(2)}(kR) \sim \sqrt{\frac{2}{\pi k R}}\, e^{-ikR + i\pi/4}(1 + O(\frac{1}{kR})), \\ &R \sim \rho'_c - a\cos(\varphi_c - \varphi'_c),\ \rho'_c \sim \rho + a\cos\varphi_e,\ \varphi'_c \sim \varphi_e, \end{aligned} \tag{2.95}$$

we derive, after comparison with (2.93):

$$F^{\pm}(\varphi_e) = \frac{-e^{-ika\cos\varphi_e}}{2}\int_C (u\frac{\partial e^{ik\rho_c\cos(\varphi_c-\varphi_e)}}{\partial \rho_c} - \frac{\partial u}{\partial \rho_c}e^{ik\rho_c\cos(\varphi_c-\varphi_e)})ad\varphi_c. \quad (2.96)$$

Then, using S to express the normal derivative of field, we can write:

$$F^{\pm}(\varphi_e) = \frac{e^{-ika\cos\varphi_e}}{2}\int_C (iu(-k\ \cos(\varphi_c-\varphi_e)+k\sin\theta) + S)\ e^{ika\cos(\varphi_c-\varphi_e)}ad\varphi_c. \quad (2.97)$$

From $e^{ika\cos(\varphi_c-\varphi_e)} = \sum_m i^m J_m(ka)e^{-im(\varphi_e-\varphi_c)}$ and (2.90), we then have:

$$F^{\pm}(\varphi_e) = \frac{e^{-ika\cos\varphi_e}}{2}(\sum_m i^m\ e^{-im\varphi_e}\int_C S\ e^{im\varphi_c}ad\varphi_c \times [J_m(ka) - \frac{J_m'(ka)+Q\ J_m(ka)}{H_m^{(2)'}(ka)+Q\ H_m^{(2)}(ka)}H_m^{(2)}(ka)]), \quad (2.98)$$

which, from $\mathrm{W}(J_m, H_m^{(2)}) = -2i/\pi ka$, gives us:

$$F^{\pm}(\varphi_e) = \frac{e^{-ika\cos\varphi_e}}{i\pi ka}\sum_m e^{-im\varphi_e}\frac{i^m\mathcal{M}(m)}{H_m^{(2)'}(ka)+Q\ H_m^{(2)}(ka)}, = \sum_m e^{\mp im\varphi_e}d_m\mathcal{M}(\pm m), \quad (2.99)$$

where

$$\mathcal{M}(\nu) = \int_C S\ e^{i\nu\varphi_c}a\,d\varphi_c,$$
$$d_\nu = d_{-\nu} = \frac{\frac{e^{-ika\cos\varphi_e}}{i\pi ka}i^\nu}{H_\nu^{(2)'}(ka)+Q\ H_\nu^{(2)}(ka)}. \quad (2.100)$$

Considering the previous series as a sum of residues for an integral expression, as G.N. Watson classically suggests, we can write:

$$\sum_m e^{\mp im\varphi_e}d_m\mathcal{M}(\pm m) = \Big(\int_{-\infty-i\epsilon}^{+\infty-i\epsilon} + \int_{+\infty+i\epsilon}^{-\infty+i\epsilon}\Big)d_\nu\mathcal{M}(\pm\nu)\frac{e^{-i\nu(\pm\varphi_e-\pi)}}{2i\sin\nu\pi}d\nu, \quad (2.101)$$

with $\epsilon > 0$. The function $(\frac{e^{i\nu\pi}}{2i\sin\nu\pi})$ is then developed as $(\sum_{p\geq 0} e^{-i\nu 2p\pi})$ if Im$\nu < 0$, and $(-\sum_{p<0} e^{-i\nu 2p\pi})$ if Im$\nu > 0$. For our problem, we choose to only take the term $p = 0$ by considering $|\pm\varphi_e - \frac{\pi}{2}| < \pi/2$, omitting the waves corresponding to $\varphi_e \pm 2p\pi$ with $p \neq 0$ which virtually make more than a wheel in φ_e. We then obtain:

$$F^{\pm}(\varphi_e) \equiv \frac{e^{-ika\cos\varphi_e}}{i\pi ka} \int_{-\infty-i\epsilon}^{+\infty-i\epsilon} e^{-i\nu(\pm\varphi_e - \pi/2)} \frac{\mathcal{M}(\pm\nu)}{H_\nu^{(2)\prime}(ka) + Q\, H_\nu^{(2)}(ka)} d\nu, \tag{2.102}$$

with $\epsilon > 0$. Considering the far field function F referring to φ, with $F^{\pm}(\alpha) = F(\alpha \mp (\pi/2 - \Phi))$ and $\varphi_e = \varphi \pm (\pi/2 - \Phi)$, we obtain:

$$F(\varphi) \equiv \frac{e^{ika\sin(\pm\varphi - \Phi)}}{i\pi ka} \int_{-\infty-i\epsilon}^{+\infty-i\epsilon} e^{-i\nu(\pm\varphi - \Phi)} \frac{\mathcal{M}(\pm\nu)}{H_\nu^{(2)\prime}(ka) + Q\, H_\nu^{(2)}(ka)} d\nu. \tag{2.103}$$

It remains to develop an asymptotic expression of $\mathcal{M}$. For that, we begin to give an asymptotic development of x_c for the points (x_c, y_c) of $\mathcal{C}$:

$$x_c = -a\sin\varphi_c \sim -a \sum_{0\leq p<N} \frac{(-1)^p}{(2p+1)!} (\widetilde{s}/a)^{2p+1}, \tag{2.104}$$

where the curvilinear abscissa $\widetilde{s}$ (growing as φ, φ_c and $-x$) can go asymptotically from $-\infty$ to $+\infty$. Thereafter, we can consider a correspondence between points of $\mathcal{C}$ at $\widetilde{s}$ and points of the plane $y = 0$ following $x = x_1(\widetilde{s})$, in vicinity of the origin $x = 0$. We can then develop S in (2.89), from an differential operator $\mathcal{B}$ applied to the field on the plane $y = 0$ when expressed with (2.92), following:

$$S \sim [\mathcal{B}\, u]|_{x=x_1(\widetilde{s}), y=0} = \frac{-1}{2\pi i} \int_S B'_\alpha(F^{\pm}(\alpha)) e^{ikx_1(\widetilde{s})\sin\alpha} d\alpha,$$

$$u(\rho, \varphi_e) = \frac{-1}{2\pi i} \int_S F^{\pm}(\alpha)\, \mathrm{e}^{-ik(-x\sin\alpha + y\cos\alpha)} d\alpha, \tag{2.105}$$

where $\mathcal{B}$ derives from Taylor expansion of u, with $x = -\rho\sin\varphi_e, y = \rho\cos\varphi_e$. We remark that $x_1(\widetilde{s})$ is not yet specified, and thus, that $\mathcal{B}$ and B'_α are not yet completely defined. Considering the asymptotic development $\widetilde{S}$ of S given in (2.105), we can write for $\mathcal{M}(\nu)$:

$$\mathcal{M} \sim \widetilde{\mathcal{M}} = \int_{-\infty}^{+\infty} \widetilde{S} e^{i\nu(\widetilde{s}/a)}\, d\widetilde{s}, \tag{2.106}$$

and thus,

$$\widetilde{\mathcal{M}}(\nu) = \frac{-1}{2\pi i} \int_S B'_\alpha(F^{\pm}(\alpha)) \int_{-\infty}^{+\infty} e^{i(\nu(\widetilde{s}/a) + kx_1(\widetilde{s})\sin\alpha)}\, d\widetilde{s} d\alpha. \tag{2.107}$$

Letting $\nu = ka\sin\Lambda$, and assuming:

$$x_1(\tilde{s}) \sim -\tilde{s} + a\sum_{p\geq 0} g_p(\tilde{s}/a)^{p+3}, \tag{2.108}$$

we can write:

$$e^{i(\nu(\tilde{s}/a)+kx_1(\tilde{s})\sin\alpha)} \sim e^{i2\pi\tilde{s}(\frac{k}{2\pi}(\sin\Lambda-\sin\alpha))}(1+\sum_{n\geq 3}\frac{b_n(\tilde{s})^n}{n!}),$$

$$\sum_{n\geq 3}\frac{b_n(\tilde{s})^n}{n!}e^{i2\pi\tilde{s}(\frac{k}{2\pi}(\sin\Lambda-\sin\alpha))} = \sum_{n\geq 3}\mathcal{P}_{\Lambda,n}(e^{i2\pi\tilde{s}(\frac{k}{2\pi}(\sin\Lambda-\sin\alpha))}),$$

$$b_n = \frac{\partial^n}{\partial(\tilde{s})^n}e^{ika\sin\alpha\sum_{p\geq 0}g_p(\tilde{s}/a)^{p+3}}|_{(\tilde{s})=0},\ \mathcal{P}_{\Lambda,n} = \frac{b_n}{n!}\frac{(ik)^{-n}\partial^n}{\partial(\sin\Lambda)^n}. \tag{2.109}$$

Using (2.109) for (2.107), we then obtain:

$$\tilde{\mathcal{M}} \equiv \frac{-1}{2\pi i}(1+\sum_{p\geq 3}\mathcal{P}_{\Lambda,p})\int_S B'_\alpha(F^{\pm}(\alpha)).\delta(\frac{k}{2\pi}(\sin\alpha-\sin\Lambda))d\alpha,$$

$$\tilde{\mathcal{M}} \equiv \frac{-1}{2\pi i}(1+\sum_{p\geq 3}\mathcal{P}_{\Lambda,p})\int_{-\infty}^{+\infty}\frac{2\pi}{k}\delta(\frac{k}{2\pi}(\sin\alpha-\sin\Lambda))\frac{B'_\alpha(F^{\pm}(\alpha))}{\cos\alpha}d(\frac{k\sin\alpha}{2\pi}),$$

$$\tilde{\mathcal{M}}(ka\sin\Lambda) = \frac{1}{-ik}(1+\sum_{p\geq 3}\mathcal{P}_{\Lambda,p})\frac{B'_\Lambda(F^{\pm}(\Lambda))}{\cos\Lambda}, \tag{2.110}$$

where $\sum_{p\geq 3}\mathcal{P}_{\Lambda,p} = O(1/a^2)$. Writing $F^{\pm}(\alpha) = F(\alpha\mp(\pi/2-\Phi))$, and

$$\sum_{n\geq 3}\mathcal{P}_{\Lambda,n} \equiv \sum_{p\geq 0}\mathcal{P}^a_{\Lambda,p}/\,a^{p+2}, \tag{2.111}$$

we can write:

$$F(\varphi) \equiv \frac{e^{ika\sin(\pm\varphi-\Phi)}}{i\pi ka}\int_{-\infty-i\epsilon}^{+\infty-i\epsilon}e^{-i\nu(\pm\varphi-\Phi)}\frac{\mathcal{M}(\pm\nu)}{H_\nu^{(2)\prime}(ka)+Q\,H_\nu^{(2)}(ka)}d\nu,$$

$$\mathcal{M}(ka\sin\alpha) = \frac{i\pi ka}{2\cos\alpha}\mathcal{O}_\alpha(F(\alpha\mp(\pi/2-\Phi))),$$

$$\mathcal{O}_\alpha = \frac{2\cos\alpha}{\pi k^2 a}\,[(1+\sum_{p\geq 0}\mathcal{P}^a_{\alpha,p}/a^{p+2})\frac{B'_\alpha}{\cos\alpha}], \tag{2.112}$$

that we can consider at arbitrary order. The $\mathcal{P}_{\alpha,p}$ and $\mathcal{P}^a_{\alpha,p}$ are linear differential operators of the variable $\sin\alpha$, with constant coefficients and the $\mathcal{P}^a_{\alpha,p}$ are independent of the radius a.

We remark that $\mathcal{M}$, denoted as $\widetilde{\mathcal{M}}$, does not depend (as $\mathcal{O}_\alpha$) on the choice of the function x_1 while $\mathcal{P}_{\alpha,p}$ and B'_α depend on it. Therefore, we can decide to use (2.6)–(2.12), that is, $(x_c, y_c)|_C \to (x = x_c, y)|_{y=0}$ with $x_1(\widetilde{s}) = x_c = -a\sin(\widetilde{s}/a)$, which gives us, for transition function (2.112) at $\pm\varphi \sim$ and $> \Phi$, referring to each face of radius $a = a^\pm$:

$$B'_\alpha = -ik(\pm B^\pm_\alpha),\ \mathcal{M} \equiv \mathcal{M}^\pm,\ \mathcal{O}_\alpha \equiv \mathcal{O}^\pm_\alpha,\ \mathcal{P}^a_{\alpha,p} \equiv \mathcal{P}^\pm_p,$$

$$\mathcal{M}^\pm(\pm ka^\pm \sin\alpha') = \frac{i\pi ka^\pm}{2\cos\alpha'}\mathcal{O}^\pm_{\pm\alpha'}(F(\pm(\alpha' - \pi/2 + \Phi))),$$

$$\mathcal{M}^\pm(\pm ka^\pm \cos\alpha) = \frac{-i\pi ka^\pm}{2\sin\alpha}\ \mathcal{O}^\pm_{\pm(\alpha+\pi/2)}(F(\pm(\alpha + \Phi))),$$

$$\mathcal{O}^\pm_\alpha = \frac{\pm 2}{i\pi ka^\pm}((1 + \sum_{p\geq 0}\frac{\mathcal{P}^\pm_p(.)}{(a^\pm)^{p+2}})B^\pm_\alpha), \tag{2.113}$$

that we can consider at arbitray order of development of $B^\pm_\alpha$.

The expression (2.112) with (2.113) is an integral expression of F valid for $\pm\varphi \sim$ and $> \Phi$ (above and below the tangent plane) that only necessitates a higher order asymptotic expression $F_<(\varphi)$ of $F(\varphi)$ for $|\varphi| < \Phi$ (note: let us notice that we can also obtain this result by a different method, using multipolar sources at the edge, previously given in appendix of [1]).

NOTE 2.13.– Let us notice that, for the choice $x_1(s) = -\widetilde{s}$, the coefficients g_p, and thus the functions $\mathcal{P}_{\alpha,p}$ and $\mathcal{P}^a_{\alpha,p}$, vanish.

2.A.2. *Expression when F has a singularity*

The integration path $\mathcal{E}$ in ν is from $-\infty - i\epsilon$ to $+\infty - i\epsilon$ with $\text{Im}\nu < 0$, and letting $\nu = ka^\pm \sin\alpha'$ with $|\text{Re}\alpha'| \leq \pi/2$ and $|\arg(ik)| < \pi/2$, the integration path in α' goes from $-i\infty - \arg(ik) + 0^+$ to $i\infty + \arg(ik) + 0^+$. Let us consider a singularity of $\mathcal{M}(\pm\nu)$ due to a GO pole of $F_<(\pm(\alpha' - \pi/2 + \Phi))$ at $\pm(\alpha' - \pi/2 + \Phi) = \pm\pi + \varphi_{1o} = \varphi^\pm_s$, that is, $\alpha' = \pm\varphi^\pm_s + \pi/2 - \Phi$, and stationary phase point at $\pm(\alpha' - \pi/2 + \Phi) = \varphi$, that is, $\alpha' = \pm\varphi + \pi/2 - \Phi$.

Studying the problem by shifting and deformation of path at stationary phase points, we capture a real singularity as $\pm\varphi > \pm\varphi^\pm_s$, that is, when $\pm(\varphi^\pm_s - \varphi) = \pi \pm (\varphi_{1o} - \varphi) > 0$, and we can write:

$$\frac{e^{ika\sin(\pm\varphi - \Phi)}}{i\pi ka}\int_{-\infty - i\epsilon}^{+\infty - i\epsilon} e^{-i\nu(\pm\varphi - \Phi)}\frac{\mathcal{M}(\pm\nu)}{H^{(2)'}_\nu(ka) + Q\, H^{(2)}_\nu(ka)}d\nu$$

$$= fp(F(\varphi)) + \sum_{\varphi_s} C_s\mathcal{D}_a(\varphi^\pm_s, \varphi)/2, \tag{2.114}$$

where $\mathcal{D}_a(\varphi_s^{\pm},\varphi) \to \delta(\varphi_s^{\pm}-\varphi)$ when $ka \to \infty$, C_s derives from residues of F at GO poles, and fp denotes the finite part (see appendix for details).

2.A.3. *Why we need to consider* $\sin\theta = \sin\theta^{\pm}$ *for the transition function at* $\varphi = \pm\Phi$

We note that the use of transition integrals for F is greatly influenced by the regularity of kernels (meromorphicity, no cut) of $\widetilde{\mathcal{M}}(\nu)$ in vicinity of $\nu = \pm ka$. This makes possible the deformation of the path of integration without having to take account of artificial singularity at $\nu = \pm ka$, and permits an uniform transition, between the asymptotics of F in entire and fractional powers of $(ka^{\pm})^{-1}$, respectively, above and below the tangent faces $\varphi = \pm\Phi$.

In this respect, it can be shown that the choice of $\sin\theta = \sin\theta^{\pm}$, where the $\sin\theta^{\pm}$ refer to definitions of boundary conditions (2.5) and asymptotic functional equations (2.13) following:

$$
\begin{aligned}
&(\frac{\partial}{\partial n} - ik\sin\theta^{\pm})u|_{\text{curved face}\pm} = 0,\\
&B^{\pm}_{\alpha\pm\frac{\pi}{2}}F(\alpha\pm\Phi) + B^{\pm}_{-\alpha\pm\frac{\pi}{2}}F(-\alpha\pm\Phi) = 0,
\end{aligned}
\tag{2.115}
$$

allows this necessary regularity in vicinity of $\nu = \pm ka^{\pm}$.

2.A.3.1. *Regularity from Laplace transform when* $\sin\theta = \sin\theta^{\pm}$

We can write, by definition:

$$
S \sim \mathcal{B}\,u|_{x,y=0} = \frac{-1}{ik}\int_{-\infty}^{+\infty}\frac{B'_{\alpha}(F^{\pm}(\alpha))}{\cos\alpha}e^{i2\pi x(\frac{k}{2\pi}\sin\alpha)}d(\frac{k}{2\pi}\sin\alpha). \tag{2.116}
$$

In other respects, when $\sin\theta = \sin\theta^{\pm}$, $a = a^{\pm}$ and $\mp x > 0$ as $\varphi = \pm\Phi$, we can write, as $|\arg(ik)| < \pi/2$:

$$
\mathcal{B}\,u|_{\mp x>0,y=0} = 0, \text{ and } \mathcal{B}u|_{\pm x>0,y=0}e^{\mp ikx} < e^{\mp\epsilon x} \to 0 \text{ as } \pm x \to \infty, \tag{2.117}
$$

with $\epsilon > 0$, and we then have by usual properties of Laplace transform:

$$
\frac{B'_{\alpha}(F^{\pm}(\alpha))}{\cos\alpha} = -ik\int_{-\infty}^{+\infty}\mathcal{B}\,u|_{y=0}e^{-ikx\sin\alpha}dx < \infty, \tag{2.118}
$$

for $\text{Re}(\pm ik\sin\alpha - ik) > -\epsilon$. So, following (2.117)–(2.118), this function is a regular function of $\sin\alpha$ in the vicinity $\alpha = \pm\pi/2$. Therefore, from (2.110) and definition of $\mathcal{P}^a_{\alpha,n}$, $\widetilde{\mathcal{M}}(ka\sin\alpha)$ is then also regular in this vicinity.

2.A.3.2. *Regularity from functional equations when* $\sin\theta = \sin\theta^{\pm}$

We note that, using $B'_{\alpha} = -ik(\pm B^{\pm}_{\alpha})^{\pm}$ with $\sin\theta = \sin\theta^{\pm}$ and (2.13), we have:

$$B'_{\alpha\pm\pi/2}F(\alpha\pm\Phi) + B'_{-\alpha\pm\pi/2}F(-\alpha\pm\Phi) = 0. \tag{2.119}$$

Therefore, considering that $\mathcal{P}^{a}_{-\alpha\pm\pi/2,p} \equiv \mathcal{P}^{a}_{\alpha\pm\pi/2,p}$, we obtain:

$$\mathcal{O}_{\alpha\pm\pi/2}(F(\alpha\pm\Phi)) + \mathcal{O}_{-\alpha\pm\pi/2}(F(-\alpha\pm\Phi)) = 0, \tag{2.120}$$

with,

$$\mathcal{O}_{\alpha\pm\pi/2} \equiv \frac{2\sin\alpha}{\pi k^2 a}[(1+\sum_{p\geq 0}\mathcal{P}^{a}_{\alpha\pm\pi/2,p})\frac{B'_{\alpha\pm\pi/2}}{\sin\alpha}],$$

$$\tilde{\mathcal{M}}(\pm ka\sin\alpha) = \frac{i\pi ka}{2\cos\alpha}\mathcal{O}_{\pm\alpha}(F(\pm(\alpha-\pi/2+\Phi)),$$

$$\tilde{\mathcal{M}}(\pm ka\cos\alpha) = \frac{-i\pi ka}{2\sin\alpha}\mathcal{O}_{\pm(\alpha+\pi/2)}(F(\pm(\alpha+\Phi)), \tag{2.121}$$

At each order of its asymptotic expression, the function $\tilde{\mathcal{M}}(\pm ka\cos\alpha)$ is then regular at $\alpha = 0$ when $\pm\varphi_{\circ} - \Phi \neq -\pi$, with $\mathcal{O}^{\pm}_{\alpha\pm\frac{\pi}{2}}F_{<}(\alpha\pm\Phi)$ regular at $\alpha = 0$ which is null at $\alpha = 0$ from (2.120). It is even, and, relatively to ν, it has no cut at $\nu = \pm ka$.

2.B. Appendix B: first-order asymptotic of transition integral above tangent faces $\varphi = \pm\Phi$

Let us consider the transition integral:

$$G(\varphi) = \frac{e^{ika^{\pm}\sin(\pm\varphi-\Phi)}}{i\pi ka^{\pm}}\int_{\mathcal{E}}\frac{e^{-i\nu(\pm\varphi-\Phi)}\mathcal{M}^{\pm}(\pm\nu)}{H_{\nu}^{(2)\prime}(ka^{\pm}) + Q^{\pm}H_{\nu}^{(2)}(ka^{\pm})}d\nu, \tag{2.122}$$

and the first-order term of its asymptotics as $\mathcal{M}^{\pm}(\pm\nu)$ is slowly oscillating for large ka, when a stationary phase point exists, that is, as $-\pi/2 < \pm\varphi - \Phi < 0$. The path $\mathcal{E}$ is just below the real axis from $-\infty - i0^{+}$ to $+\infty - i0^{+}$, which corresponds, as we let $\nu = ka\sin\alpha$, to α that belongs to the path $\mathcal{S}$, from $-i\infty - \arg(ik) + 0^{+}$ to $i\infty + \arg(ik) + 0^{+}$, as $|\arg(ik)| \leq \pi/2$.

2.B.1. *When* F *has no singularity*

We first consider the case where F and thus $\mathcal{M}^{\pm}(\pm\nu)$ has no singularities. From Debye first-order asymptotics of $H_{\nu}^{(2)}$ and its derivative, as $\nu = ka\sin\alpha$ with $|\text{Re}\alpha| < \pi/2$:

$$H_\nu^{(2)}(ka) \sim \sqrt{2/\pi ka\cos\alpha}\, e^{-ika\cos\alpha+ika\sin\alpha(\pi/2-\alpha)} e^{i\pi/4},$$

$$H_\nu^{(2)'}(ka) \sim -i\cos\alpha\sqrt{2/\pi ka\cos\alpha}\, e^{-ika\cos\alpha+ika\sin\alpha(\pi/2-\alpha)} e^{i\pi/4}, \qquad (2.123)$$

and $\mathcal{M}^\pm(\pm ka^\pm \sin\alpha) = \frac{i\pi ka^\pm}{2\cos\alpha}\mathcal{O}^\pm_{\pm\alpha}(F(\pm(\alpha-\pi/2+\Phi)))$ with $\mathcal{O}^\pm_\alpha \sim \frac{2}{i\pi ka}(\cos\alpha + \sin\theta^\pm)$ at first order, we have the first-order term of $G(\varphi)$ following:

$$G(\varphi) \sim \frac{e^{ika^\pm \sin(\pm\varphi-\Phi)}}{e^{i\pi/4}} \int_{\mathcal{E}} \frac{e^{-ika(\pm\varphi-\Phi)\sin\alpha}\frac{F(\pm\alpha\pm(-\pi/2+\Phi))}{\pi ka}}{\cos\alpha\sqrt{2/\pi ka\cos\alpha}e^{-ika\cos\alpha+ika\sin\alpha(\pi/2-\alpha)}} d\nu, \qquad (2.124)$$

with $\nu = ka\sin\alpha$. The phase term in the integral $-ikag(\alpha)$ verifies $(g(\alpha))' = 0$ at $\alpha = \varphi_1 = \pm\varphi + \pi/2 - \Phi$, and:

$$\begin{aligned} g(\alpha) &= (\pm\varphi - \Phi)\sin\alpha - \cos\alpha + (\pi/2-\alpha)\sin\alpha,\\ (g(\alpha))' &= (\pm\varphi - \Phi)\cos\alpha + (\pi/2-\alpha)\cos\alpha,\\ (g(\alpha))'' &= -(\pm\varphi - \Phi)\sin\alpha - \cos\alpha - (\pi/2-\alpha)\sin\alpha,\\ g(\alpha) &\sim \sin(\pm\varphi - \Phi) - \cos(\varphi_1)(\alpha-\varphi_1)^2/2. \end{aligned} \qquad (2.125)$$

We then have, when $-\pi/2 < \pm\varphi - \Phi < 0$:

$$\begin{aligned} G(\varphi) &\sim \frac{e^{ika^\pm \sin(\pm\varphi-\Phi)}}{e^{i\pi/4}} \int_S \frac{e^{-ika(\pm\varphi-\Phi)\sin\alpha}\frac{F(\pm\alpha\pm(-\pi/2+\Phi))}{\pi}}{\sqrt{2/\pi ka\cos\alpha}e^{-ika\cos\alpha+ika\sin\alpha(\pi/2-\alpha)}} d\alpha\\ &\sim \frac{F(\varphi)}{e^{i\pi/4}\sqrt{2\pi/(ka\cos\varphi_1)}} \int_S e^{i(ka\cos\varphi_1)(\alpha-\varphi_1)^2/2} d\alpha = F(\varphi), \end{aligned} \qquad (2.126)$$

where we have used that:

$$\int_S e^{iu\alpha^2/2} d\alpha = \sqrt{\frac{2\pi}{u}} e^{i\pi/4}. \qquad (2.127)$$

2.B.2. *When F has a simple real pole*

Let us consider $F(\pm\alpha \pm (-\pi/2+\Phi))$ with a simple pole of incident field type:

$$F(\alpha') \sim \mp 1/(\alpha' - \varphi_s^\pm), \qquad (2.128)$$

with $\varphi_s^\pm = \pm\pi + \varphi_{o1}$, where the angle φ_{o1} refers to a plane wave direction with $\mp\varphi_{o1} > \pi - \Phi$, that is, $\pm\varphi_s^\pm - \Phi < 0$. For $\pm\alpha \pm (-\pi/2+\Phi) = \varphi_s^\pm$, we note that $\alpha = \pm\varphi_s^\pm + (\pi/2 - \Phi)$. By a shift and a deformation of initial path to steepest descent path (SDP), centered on stationary phase points $\alpha = \pm\varphi + (\pi/2 - \Phi)$, we can capture a singularity on real axis as $\pm\varphi_s^\pm > \pm\varphi$, that is, when $\pm(\varphi_s^\pm - \varphi) = \pi \pm (\varphi_{o1} - \varphi) > 0$, and we then have:

$$G(\varphi) \sim fp(F(\varphi)) +$$
$$+ A\, U_c(\pm(\varphi_s^\pm - \varphi))\sqrt{ka}\, e^{ika(\sin(\pm\varphi - \Phi) + (\pm\varphi - \Phi)\cos(\pm\varphi_{o1} - \Phi))},$$
$$A = \frac{2\pi i Res(F(\varphi_s^\pm))e^{-i\pi/4}}{\sqrt{2\pi/\cos\alpha}\, e^{-ika\cos\alpha + ika\sin\alpha(\pi/2 - \alpha)}}|_{\alpha = \pm\varphi_s^\pm + \pi/2 - \Phi}, \tag{2.129}$$

where $U_c(x)$ is the Heaviside function for x real, fp denotes the finite part. The term in addition of $F(\varphi)$ has a contribution that has a Dirac-like contribution as $ka \to \infty$. Indeed, with a stationary phase point for $\cos(\pm\varphi_{o1} - \Phi) = -\cos(\pm\varphi - \Phi)$, that is, for $\varphi = \pm\pi + \varphi_{o1} = \varphi_s^\pm$ with $-\pi/2 < \pm\varphi_s^\pm - \Phi < 0$, it contributes as a Dirac function for $ka \to \infty$ (see (2.132)), following:

$$G(\varphi) \sim fp(F(\varphi)) + U_c(\pm(\varphi_s^\pm - \varphi))(2\pi i Res(F(\varphi_s^\pm))$$
$$\times [e^{-i\pi/4}\sqrt{\frac{ka\sin(\Phi \mp \varphi_s^\pm)}{2\pi}}e^{ika\sin(\Phi\mp\varphi_s^\pm)(\varphi - \varphi_s^\pm)^2/2}])$$
$$\sim fp(F(\varphi)) + 2\pi i U_c(\pm(\varphi_s^\pm - \varphi))Res(F(\varphi_s^\pm))[\delta(\varphi - \varphi_s^\pm)]$$
$$\sim fp(F(\varphi)) + i\pi Res(F(\varphi_s^\pm))[\delta(\varphi - \varphi_s^\pm)], \tag{2.130}$$

that we can extend to φ complex. On this occasion, we note the connection with the distributions:

$$\frac{1}{x + i0^\mp} = fp(\frac{1}{x}) \pm i\pi\delta. \tag{2.131}$$

NOTE 2.14.– For (2.130), we use a definition of approximated (nascent) Dirac function:

$$\frac{1}{b}(\frac{e^{-x^2}}{\sqrt{\pi}})|_{x/b} \to \delta(x) \text{ as } b \to 0,\ |\arg(b)| \leq \pi/4, |c|\delta(cx) = \delta(x),$$
$$\frac{1}{b}(\frac{e^{-x^2}}{\sqrt{\pi}})|_{\alpha/b} = e^{\mp i\pi/4}\sqrt{\frac{u}{2\pi}}e^{\pm iu\alpha^2/2} \to \delta(\alpha), \text{ as } b = \frac{e^{\pm i\pi/4}}{\sqrt{u/2}} \to 0. \tag{2.132}$$

NOTE 2.15.– In case of pole in vicinity of the stationary phase point, we note that we can write, considering $\alpha_1 > 0$:

$$g^{\pm}(\alpha_1) = -\alpha_1 \int_{-e^{\pm i\pi/4}\infty}^{+e^{\pm i\pi/4}\infty} \frac{e^{\pm iu\alpha^2/2}}{\alpha - \alpha_1} d\alpha = -\alpha_1 \int_{-\infty}^{+\infty} \frac{e^{-ut^2/2}}{t - e^{\mp i\pi/4}\alpha_1} dt$$

$$= \frac{-\alpha_1}{2} \int_{-\infty}^{+\infty} e^{-ut^2/2} \left(\frac{1}{t - e^{\mp i\pi/4}\alpha_1} + \frac{1}{-t - e^{\mp i\pi/4}\alpha_1}\right) dt$$

$$= -e^{\mp i\frac{\pi}{4}} \int_{-\infty}^{+\infty} \frac{\alpha_1^2 e^{-u\frac{t^2}{2}}}{t^2 \pm i\alpha_1^2} dt = -\alpha_1^2 e^{\mp i\frac{\pi}{4}} \int_0^{e^{\mp i\pi/4}\infty} e^{\mp pi\alpha_1^2} \int_{-\infty}^{+\infty} e^{-\frac{(u+2p)t^2}{2}} dt dp$$

$$= -\alpha_1^2 e^{\mp i\frac{\pi}{4}} \int_0^{e^{\mp i\pi/4}\infty} e^{\mp ip\alpha_1^2} \sqrt{\frac{\pi}{\frac{u}{2} + p}} dp$$

$$= \frac{\pm 2\pi i}{\sqrt{\frac{u}{2}}} \left[\frac{x e^{\pm ix^2} e^{\pm i\frac{\pi}{4}}}{\sqrt{\pi}} \int_x^{e^{\mp i\pi/4}\infty} e^{\mp it^2} dt\right]_{x=\alpha_1\sqrt{\frac{u}{2}}} = 2\pi \frac{e^{\pm i\pi/4}}{\sqrt{2\pi u}} F_*^{\pm}\left(\alpha_1 \sqrt{\frac{u}{2}}\right), \tag{2.133}$$

with $F_*^{\pm}(x) = \pm 2ix e^{\pm ix^2} \int_x^{\infty} e^{\mp it^2} dt$. We remark that:

$$\int_{-x}^{\infty} e^{\mp it^2} dt = \int_{-\infty}^{x} e^{\mp it^2} dt, \int_{-\infty}^{+\infty} e^{\mp it^2} dt = \sqrt{\pi} e^{\mp i\pi/4},$$

$$F_*^{\pm}(x) \sim 1 \text{ as } x \to \infty, g^{\pm}(\alpha_1) \sim \sqrt{\frac{2\pi}{u}} e^{\pm i\pi/4} \text{ as } \alpha_1 \to \infty, \tag{2.134}$$

and that the expression can be analytically continued for $\alpha_1 < 0$, from:

$$F_*^{\pm}(x) - F_*^{\pm}(-x) = 2x\sqrt{\pi} e^{\pm i\pi/4} e^{\pm ix^2},$$

$$g^{\pm}(\alpha_1)|_{\alpha_1<0} = g^{\pm}(-\alpha_1) \pm 2\pi i\alpha_1 e^{\pm iu\alpha_1^2/2}. \tag{2.135}$$

2.C. Appendix C: regularity of a sum of residues at coalescent poles

Let us consider:

$$G(\alpha) = \frac{F(\alpha)}{(\alpha - \alpha_1)(\alpha - \alpha_2)(\alpha - \alpha_3)},$$

$$\text{with } F(\alpha) = F(\alpha_j) + (\partial_\alpha F)(\alpha_j)(\alpha - \alpha_j)$$

$$+ (\partial_{\alpha^2} F)(\alpha_j)(\alpha - \alpha_j)^2/2 + (\alpha - \alpha_j)^3 \mathcal{F}_j(\alpha), \tag{2.136}$$

for $1 \leq j \leq 3$, with $F(\alpha)$ regular at any $\alpha = \alpha_j$. We then have:

$$\begin{aligned}
&G_1 + G_2 + G_3 = \\
&= \frac{F(\alpha_1)}{(\alpha_1 - \alpha_2)(\alpha_1 - \alpha_3)} + \frac{F(\alpha_2)}{(\alpha_2 - \alpha_1)(\alpha_2 - \alpha_3)} + \frac{F(\alpha_3)}{(\alpha_3 - \alpha_1)(\alpha_3 - \alpha_2)} \\
&= \frac{-(\alpha_2 - \alpha_1)^2(\alpha_1 - \alpha_3)(\partial_{\alpha^2} F(\alpha_1)/2 + (\alpha_2 - \alpha_1)\mathcal{F}_1(\alpha_2))}{(\alpha_1 - \alpha_2)(\alpha_1 - \alpha_3)(\alpha_2 - \alpha_3)} \\
&+ \frac{(\alpha_3 - \alpha_1)^2(\alpha_1 - \alpha_2)(\partial_{\alpha^2} F(\alpha_1)/2 + (\alpha_3 - \alpha_1)\mathcal{F}_1(\alpha_3))}{(\alpha_1 - \alpha_2)(\alpha_1 - \alpha_3)(\alpha_2 - \alpha_3)} \\
&= \frac{\partial_{\alpha^2} F(\alpha_1)(\frac{(\alpha_2-\alpha_1)-(\alpha_3-\alpha_1)}{(\alpha_2-\alpha_3)})}{2} + \frac{(\alpha_2 - \alpha_1)^2\mathcal{F}_1(\alpha_2) - (\alpha_3 - \alpha_1)^2\mathcal{F}_1(\alpha_3)}{(\alpha_2 - \alpha_3)} \\
&= \frac{\partial_{\alpha^2} F(\alpha_1)}{2} + (\alpha_2 + \alpha_3 - 2\alpha_1)\mathcal{F}_1(\alpha_2) + \frac{(\alpha_3 - \alpha_1)^2(\mathcal{F}_1(\alpha_2) - \mathcal{F}_1(\alpha_3))}{(\alpha_2 - \alpha_3)},
\end{aligned} \tag{2.137}$$

or, more generally, letting α_j, α_k and α_l with j, k, and l distinct, we have:

$$\begin{aligned}
&G_1 + G_2 + G_3 = \\
&= \frac{\partial_{\alpha^2} F(\alpha_j)}{2} + (\sum_{i \neq j} \alpha_i - 2\alpha_j)\mathcal{F}_j(\alpha_k) + \frac{(\alpha_l - \alpha_j)^2(\mathcal{F}_j(\alpha_k) - \mathcal{F}_j(\alpha_l))}{(\alpha_k - \alpha_l)},
\end{aligned} \tag{2.138}$$

which is regular even when $\alpha_k = \alpha_l$.

2.D. Appendix D: analytical expression of u_{OGa} from behavior at reflection poles of f

The incident pole of f, corresponding to shadow frontier, is only present in the term f_0. The function $f_1(\alpha \pm \Phi)$ is then regular for $0 < \mp\text{Re}\alpha < 2\Phi$. On the contrary, it has a pole, corresponding to reflected wave on each face $\varphi = \pm\Phi$, when $\alpha \pm \Phi = \pm 2\Phi - \varphi_o$ in the zone $0 < \pm\text{Re}\alpha < 2\Phi$, as $|\text{Re}\varphi_o| < \Phi$. This pole, at $\alpha = -\varphi_o \pm \Phi = \pm\alpha'$, contributes as $0 < \pm\text{Re}\alpha = \text{Re}\alpha' \leq \pi$. In vicinity of it, we have, from functional equations on f_1:

$$\begin{aligned}
f_1(\alpha \pm \Phi) &= \mp \frac{1}{ia^{\pm}(\sin\alpha \pm \sin\theta^{\pm})}(D_\alpha f_0(\alpha \pm \Phi) - D_{-\alpha} f_0(-\alpha \pm \Phi)) \\
&+ \text{term regular at reflection poles},
\end{aligned} \tag{2.139}$$

where $f_0(\alpha) = \frac{\Psi(\alpha)}{\Psi(\varphi_o)} \frac{\mu \cos \mu\varphi_o}{\sin \mu\alpha - \sin \mu\varphi_o}$ with $\frac{\Psi(\alpha \pm \Phi)}{\Psi(-\alpha \pm \Phi)} = \frac{-\sin\alpha \pm \sin\theta^{\pm}}{\sin\alpha \pm \sin\theta^{\pm}}$, and $\Phi > \pi/2$. Considering an integral on a closed contour around the singularity and an arbitrary regular function h, we can derive the residue of $h(\alpha)f_1(\alpha \pm \Phi)$ at $\alpha = \pm\alpha'$, after a use of integration by parts, following:

$$\mp Res(h(\alpha)\frac{(D^{\pm}_{\alpha}(f_0(\alpha \pm \Phi)) - D^{\pm}_{-\alpha}(f_0(-\alpha \pm \Phi))}{ia^{\pm}(\sin\alpha \pm \sin\theta^{\pm})})|_{\alpha\pm\Phi=-\varphi_o\pm 2\Phi} =$$

$$= -V_1^{\pm}(\frac{h(\alpha)}{\pm\sin\alpha + \sin\theta^{\pm}})|_{\alpha=-\varphi_o\pm\Phi} Res(f_0(\alpha \pm \Phi) - f_0(-\alpha \pm \Phi))$$

$$- V_2^{\pm}(\frac{h(\alpha)}{\pm\sin\alpha + \sin\theta^{\pm}})|_{\alpha=-\varphi_o\pm\Phi} Res(f_0(\alpha \pm \Phi) + f_0(-\alpha \pm \Phi)), \quad (2.140)$$

where,

$$V_1^{\pm}(.) = \frac{1}{2ia^{\pm}}[\frac{\partial^2(.)}{\partial\alpha^2} + \text{cotg}\alpha\frac{\partial}{\partial\alpha}(.)],$$

$$V_1^{\pm}(\frac{h(\alpha)}{\pm\sin\alpha + \sin\theta^{\pm}})|_{\alpha=-\varphi_o\pm\Phi} = V_1^{\pm}(\frac{h(\pm\alpha)}{\sin\alpha + \sin\theta^{\pm}})|_{\alpha=\mp\varphi_o+\Phi},$$

$$V_2^{\pm}(.) = \frac{1}{2ia^{\pm}}[\pm\sin\theta^{\pm}\frac{\partial}{\partial\alpha}(\frac{1}{\sin\alpha}\frac{\partial}{\partial\alpha}(.))],$$

$$V_2^{\pm}(\frac{h(\alpha)}{\pm\sin\alpha + \sin\theta^{\pm}})|_{\alpha=-\varphi_o\pm\Phi} = V_2^{\pm}(\frac{\pm h(\pm\alpha)}{\sin\alpha + \sin\theta^{\pm}})|_{\alpha=\mp\varphi_o+\Phi}, \quad (2.141)$$

and

$$Res(f_0(\alpha \pm \Phi) - f_0(-\alpha \pm \Phi))|_{\alpha=-\varphi_o\pm\Phi}$$

$$= (\frac{-\sin\alpha \pm \sin\theta^{\pm}}{\sin\alpha \pm \sin\theta^{\pm}} - 1)Res\frac{\mu\cos\mu\varphi_o}{\sin\mu(-\alpha \pm \Phi) - \sin\mu\varphi_o}|_{\alpha=-\varphi_o\pm\Phi}$$

$$= (\frac{-\sin\alpha \pm \sin\theta^{\pm}}{\sin\alpha \pm \sin\theta^{\pm}} - 1)(-1)|_{\alpha=-\varphi_o\pm\Phi} = \frac{2\sin\alpha}{\sin\alpha + \sin\theta^{\pm}}|_{\alpha=\mp\varphi_o+\Phi},$$

$$Res(f_0(\alpha \pm \Phi) + f_0(-\alpha \pm \Phi))|_{\alpha=-\varphi_o\pm\Phi}$$

$$= (\frac{-\sin\alpha \pm \sin\theta^{\pm}}{\sin\alpha \pm \sin\theta^{\pm}} + 1)Res\frac{\mu\cos\mu\varphi_o}{\sin\mu(-\alpha \pm \Phi) - \sin\mu\varphi_o}|_{\alpha=-\varphi_o\pm\Phi}$$

$$= (\frac{-\sin\alpha \pm \sin\theta^{\pm}}{\sin\alpha \pm \sin\theta^{\pm}} + 1)(-1)|_{\alpha=-\varphi_o\pm\Phi} = \frac{-2\sin\theta^{\pm}}{\sin\alpha + \sin\theta^{\pm}}|_{\alpha=\mp\varphi_o+\Phi}. \quad (2.142)$$

where we note that $(\sin\alpha \pm \sin\theta^{\pm})f_0(\alpha \pm \Phi) - (-\sin\alpha \pm \sin\theta^{\pm})f_0(-\alpha \pm \Phi) = 0$.

Following the Sommerfeld–Maliuzhinets expression of fields, the contribution of f_0 for $|^1_{\frac{\partial}{\partial n}} u_{OGa}$ at $\varphi = \pm\Phi$ corresponds to the residue of $h(\alpha)(f_0(\alpha \pm \Phi) - f_0(-\alpha \pm \Phi))$ at $\alpha = \pm\alpha' = -\varphi_o \pm \Phi$ with $h(\alpha) = |^1_{ik\sin\theta^{\pm}} e^{ik\rho\cos\alpha}$, while the contribution of f_1 for it, with $\frac{\partial}{\partial n} \equiv \mp\frac{\partial}{\rho\partial\varphi}$ at $\varphi = \pm\Phi$, corresponds to the residue of

$h(\alpha)f_1(\alpha \pm \Phi)/k$ at $\alpha = \pm\alpha'$ with $h(\alpha) = |^{1}_{\mp ik \sin\alpha}\, e^{ik\rho\cos\alpha}$, so that we have:

$$|^{1}_{\frac{\partial}{\partial n}}\, u_{OGa}(\rho, \varphi = \pm\Phi) = A_0 + A_1, \tag{2.143}$$

with

$$A_0 = |^{1}_{ik\sin\theta^{\pm}} \left(\frac{2e^{ik\rho\cos\alpha}\sin\alpha}{\sin\alpha + \sin\theta^{\pm}}\right)|_{\alpha=\mp\varphi_\circ+\Phi},$$

$$\begin{aligned} A_1 &= \frac{-1}{2ika^{\pm}}\Big(\frac{2\sin\alpha}{\sin\alpha+\sin\theta^{\pm}}\Big[\frac{\partial^2(.)}{\partial\alpha^2} + \mathrm{cotg}\alpha\frac{\partial}{\partial\alpha}(.)\Big] \\ &+ \frac{-2\sin\theta^{\pm}}{\sin\alpha+\sin\theta^{\pm}}\Big[\sin\theta^{\pm}\frac{\partial}{\partial\alpha}\Big(\frac{1}{\sin\alpha}\frac{\partial}{\partial\alpha}(.)\Big)\Big]\Big)\Big(\frac{h(\pm\alpha)}{\sin\alpha+\sin\theta^{\pm}}\Big)|_{\alpha=\mp\varphi_\circ+\Phi} \\ &= \frac{-1}{2ika^{\pm}}\frac{2\sin\theta^{\pm}}{\sin\alpha+\sin\theta^{\pm}}\Big(\frac{\sin\alpha}{\sin\theta^{\pm}}\Big[\frac{\partial^2(.)}{\partial\alpha^2} + \mathrm{cotg}\alpha\frac{\partial}{\partial\alpha}(.)\Big] \\ &- \Big[\sin\theta^{\pm}\frac{\partial}{\partial\alpha}\Big(\frac{1}{\sin\alpha}\frac{\partial}{\partial\alpha}(.)\Big)\Big]\Big)\Big(|^{1}_{-ik\sin\alpha}\frac{e^{ik\rho\cos\alpha}}{\sin\alpha+\sin\theta^{\pm}}\Big)|_{\alpha=\mp\varphi_\circ+\Phi}. \end{aligned} \tag{2.144}$$

NOTE 2.16.– Considering that the GO part attached to each face $\varphi = \pm\Phi$ is only dependent on $\mp\varphi_\circ + \Phi$ and $\sin\theta^{\pm}$, we can also determine it by taking $f_0(\alpha)$ with $\Phi = \pi/2$ and $\theta^+ = \theta^- = \theta$, that is, $f_0(\alpha) = \frac{\cos\alpha+\sin\theta}{\cos\varphi_\circ+\sin\theta}\frac{\cos\varphi_\circ}{\sin\alpha-\sin\varphi_\circ}$.

3

Spectral Equations for Scattering by Impedance Polygons: Properties and Solutions

3.1. Introduction

The Sommerfeld–Maliuzhinets spectral representation of fields is not limited to the study of the diffraction by isolated impedance wedges, and we developed in [1]–[2] an analytical method concerning the determination of spectral function for the scattering by an impedance polygonal object (convex or not). This method is now quoted and used by other authors [3], but alternative ways exist for cavities, perfectly conducting object, and some specific geometries; this is by asymptotic or exact reductions [4]–[13], or by iterative ones, as in [14]. Our approach is exact and concerns general impedance polygons, either finite, or infinite when the facets composing the surface are terminated by two semi-infinite planes. For that, we consider special features of single-face expression of spectral function, that we defined in [1]–[2], to develop exact functional difference equations, then Fredholm integral equations, that can be solved exactly, but also asymptotically, considering small or large facets. We detail novel properties here. Existence and uniqueness of solutions are, in particular, analyzed in a new, original manner, while some refined asymptotics are presented.

3.2. Generalities

Let us consider the case of diffraction in free space of a plane wave:

$$u^i(\rho,\varphi) = e^{ik\rho\cos(\varphi-\varphi_o)}, \tag{3.1}$$

by a scatterer enclosed in a wedge-shaped region, defined in cylindrical coordinates (ρ,φ) as the domain complementary to the free space sector of radiation, $-\Phi_r \leq$

$\varphi \leq \Phi_l$ (Figure 3.1), with $-\Phi_r \leq \varphi_\circ \leq \Phi_l$. The characteristics of the scatterer are supposed to be independent of z coordinate. An implicit harmonic dependence on time $e^{i\omega t}$ is understood and henceforth suppressed. In expression (3.1), k denotes the wave number of the exterior medium with $|\arg(ik)| < \frac{\pi}{2}$, while $|\arg(ik)| = \frac{\pi}{2}$ is considered as a limit case. We assume that the total field in the free space region, $u = u_s + u^i$, satisfies the Helmholtz equation:

$$(\Delta + k^2)u(\rho, \varphi) = 0, \tag{3.2}$$

with analyticity with respect to ρ and φ as $-\Phi_r < \varphi < \Phi_l$, except possibly at the origin O. Some elementary properties are assumed to hold for the field:

– (a') the only incoming plane wave, from the free space sector with origin O, $-\Phi_r \leq \varphi \leq \Phi_l$, is the incident field;

– (b') the field u tends to a finite value independent of φ as $\rho \to 0$, and the derivatives $\partial_\rho u$ and $\partial_\varphi u/\rho$ are locally summable with respect to ρ in the vicinity of the origin. This property applies for an origin taken at any point out of or upon the scatterer;

– (c') the field, except possibly its geometrical optics part when $\text{Im}(k) \neq 0$, does not grow at infinity. In addition, some bounds on the far field are assumed. We consider here that the field is $O(e^{ik\rho\cos(\varphi-\varphi_\circ)})$ for large ρ, $|\arg(ik)| < \pi/2$, which is a standard assumption [1]–[2] when the scattering object has passive impedance boundary conditions.

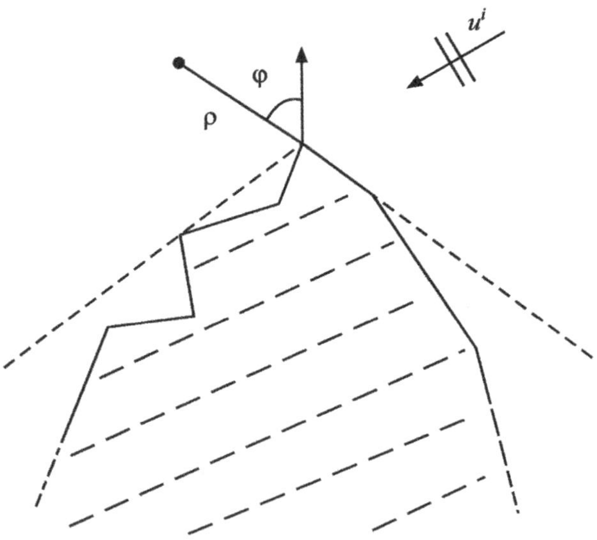

Figure 3.1. *A polygonal surface in a wedge-shaped sector*

The total field u for $-\Phi_r \leq \varphi \leq \Phi_l$ can be then represented as a Sommerfeld–Maliuzhinets integral [1]–[2], [17]–[20]:

$$u(\rho,\varphi) = \frac{1}{2\pi i}\int_\gamma f(\alpha+\varphi)e^{ik\rho\cos\alpha}d\alpha, \tag{3.3}$$

which satisfies the Helmholtz equation. In this representation, f is an analytical function and the path γ consists of two branches: one, named γ_+, goes from $(i\infty+\arg(ik) + (a_1 + \frac{\pi}{2}))$ to $(i\infty+\arg(ik) - (a_2 + \frac{\pi}{2}))$ with $0 < a_{1,2} < \pi$, as $\mathrm{Im}\alpha \geq d$, above all the singularities of the integrand; and the other, named γ_-, is obtained by inversion of γ_+ with respect to $\alpha = 0$.

We can develop the expression, considering a deformation of γ to the steepest descent path SDP for large $k\rho$, composed of two paths $SDP_\pm$ centered on the stationary phase points $\alpha = \pm\pi$. Assuming that, between γ and SDP, $f(\alpha+\varphi,\varphi_o)e^{ik\rho\cos\alpha}$ is meromorphic and vanishes at infinity, we have:

$$\begin{aligned} u(\rho,\varphi) &= u_i + \sum u_r^\pm + \sum u_{tr}^\pm + \sum u_s^\pm \\ &+ \frac{e^{-ik\rho}}{2\pi i}\int_{SDP} f(\alpha+\varphi)e^{ik\rho(\cos\alpha+1)}d\alpha, \end{aligned} \tag{3.4}$$

where:

– the terms u_i, $u_r^\pm$ and $u_{tr}^\pm$, for incident, reflected and transmitted geometrical optics plane waves (if any), correspond to real poles of $f(\alpha+\varphi)$ captured during deformation of γ to SDP;

– the terms $u_s^\pm$ correspond to complex poles of $f(\alpha+\varphi)$ captured during deformation of γ to SDP, for terms of guided complex plane waves (if any) excited by the edge;

– the last term, named u_d, is principally radiated cylindrically when $\rho \to \infty$. Approximating $f(\alpha+\varphi)$ on $SDP_\pm$ by its value at the stationary phase points $\alpha = \pm\pi$, we obtain:

$$u_d \sim \frac{-e^{-i\pi/4-ik\rho}}{\sqrt{2\pi k\rho}}[f(\pi+\varphi) - f(-\pi+\varphi)], \tag{3.5}$$

with $F(\varphi) = f(\pi+\varphi) - f(-\pi+\varphi)$ the diffraction (or far field) coefficient.

This representation is commonly applied for a wedge. We investigate its use for a scatterer with several discontinuities, as in [1]–[2], detailing properties and ameliorating the method. This study requires us to express shifts of the origin, and for this, to use particular relations of f with radiation by one face [1]–[2]. By choice, we take in what follows $\Phi_r = \Phi_l = \Phi$.

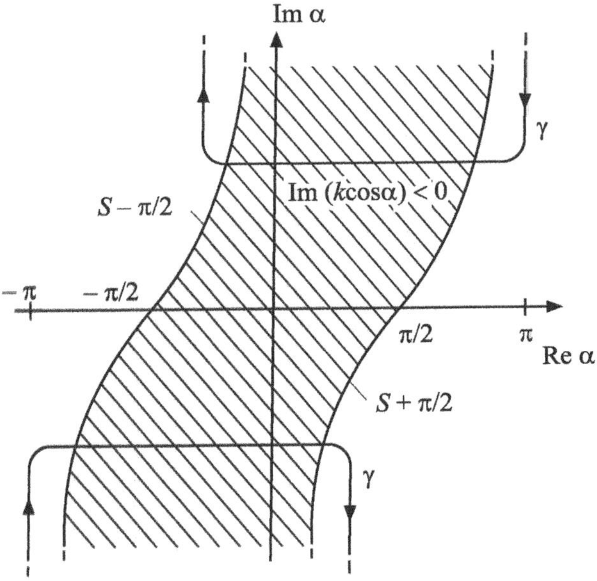

Figure 3.2. *Complex path of integration*

3.3. Single-face expression of f for a scatterer enclosed by a surface with two semi-infinite polygonal faces

Let us consider a polygonal surface, located inside the domain $|\varphi| > \Phi$, and enclosing the scatterer. This surface is composed of two joined semi-infinite polygonal faces $L^{\pm}_{0,\infty}$, denoted $+$ and $-$, respectively, defined with $m^{\pm}$ straight segments of lengths $d^{\pm}_{j}$ with tangent angles $\pm\Phi^{\pm}_{j}$, $j = 1, 2, ...,$ ($m^{\pm} = 0$ when no finite segments), and a semi-infinite plane with tangent angles $\pm\Phi^{\pm}_{e}$, $\Phi \leq \Phi^{\pm}_{e}$.

The spectral function f satisfies a single-face expression [1], as $\frac{\pi}{2} < \Phi^{\pm}_{e} \mp \varphi_{o} < \frac{3\pi}{2}$ and $\frac{\pi}{2} < \Phi^{\pm}_{e} \mp \varphi < \frac{3\pi}{2}$:

$$f(\pm\pi + \varphi) = \frac{1}{2}\int_{L^{\pm}_{0,\infty}} \mp(u(M)\nabla_{M}(e^{ik\hat{r}.OM}) - \nabla_{M}(u(M))e^{ik\hat{r}.OM}).\hat{n}^{\pm}ds(M), \tag{3.6}$$

where $s(M)$ is the length abscissa on each face, at points M of $L^{\pm}_{0,\infty}$ from origin to ∞, the term $\hat{r}$ is the unit vector of the direction of observation with angle φ, and $\hat{n}^{\pm}$ is the outward unit normal to face at M.

Considering the geometry of $L^{\pm}_{0,\infty}$, and the notation $\widehat{n}^{\pm}\nabla_M(.) = \partial(.)/\partial n = \mp\partial(.)/\rho_j\partial\varphi_j$ on straight segments $\rho_j \in [0, d^{\pm}_j]$ at $\varphi_j = \pm\Phi^{\pm}_j$, we can write the single-face expression of f [1] in the form:

$$f(\alpha) = \sum_{1\leq j\leq m^{\pm}} \frac{e^{-ik\sum_{1\leq i<j} d^{\pm}_i \cos(\alpha\mp\Phi^{\pm}_i)}}{2} \int_0^{d^{\pm}_j} (-iku(\rho'_j, \pm\Phi^{\pm}_j)\sin(\alpha\mp\Phi^{\pm}_j)$$

$$\pm \frac{\partial u}{\partial n}(\rho'_j, \pm\Phi^{\pm}_j))e^{-ik\rho'_j\cos(\alpha\mp\Phi^{\pm}_j)}d\rho'_j + e^{-ik\sum_{1\leq i\leq m^{\pm}} d^{\pm}_i\cos(\alpha\mp\Phi^{\pm}_i)} f^{\pm}_{e,m^{\pm}}(\alpha), \tag{3.7}$$

with $m^{\pm} \geq 0$ ($m^{\pm} = 0$ when no finite segments on face $\pm$), $\sum_{1\leq i\leq 0} \equiv 0$, $\sum_{1\leq i<1} \equiv 0$, where $f^{\pm}_{e,m^{\pm}}(\alpha)$ is the analytic continuation of the integral expression:

$$f^{\pm}_{e,m^{\pm}}(\alpha' \pm \Phi^{\pm}_e) = \frac{1}{2}\int_0^{\infty} (-iku(\rho'_e, \pm\Phi^{\pm}_e)\sin\alpha'$$

$$\pm \frac{\partial u}{\partial n}(\rho'_e, \pm\Phi^{\pm}_e))e^{-ik\rho'_e\cos\alpha'}d\rho'_e, \tag{3.8}$$

valid as $\text{Re}(ik(\cos\alpha' - \cos(\Phi^{\pm}_e \mp \varphi_o))) > 0$, $|\text{Re}\alpha'| < \pi$, $|\arg(ik)| < \frac{\pi}{2}$.

The function $f^{\pm}_{e,m^{\pm}}$ is the spectral function corresponding to the Sommerfeld–Maliuzhinets representation of the field in cylindrical coordinates $(\rho^{\pm}_e, \varphi^{\pm}_e)$, with origin $Q^{\pm}_{e,m^{\pm}}$ at the edge of the semi-plane $\varphi^{\pm}_e = \pm\Phi^{\pm}_e$, radiating in the free space sector $\Phi^{\pm}_e - \delta^{\pm}_e \leq \pm\varphi^{\pm}_e \leq \Phi^{\pm}_e$ where $\delta^{\pm}_e$ is a strictly positive constant determined by the geometry.

In accordance with the conditions (a'), (b') and (c') on the field u, the spectral functions f and $f^{\pm}_{e,m^{\pm}}$ have the following properties:

(a) $f(\alpha) - 1/(\alpha - \varphi_o)$ has no singularity in the band $-\Phi \leq \alpha \leq \Phi$, even at infinity, while $f^{\pm}_{e,m^{\pm}}(\alpha) - u^i(Q^{\pm}_{e,m^{\pm}})/(\alpha - \varphi_o)$ has no singularity in the band $\Phi^{\pm}_e - \delta^{\pm}_e \leq \pm\text{Re}\alpha \leq \Phi^{\pm}_e$, where $u^i(Q^{\pm}_{e,m^{\pm}}) = \exp(ik\sum_{1\leq i\leq m^{\pm}} d^{\pm}_i\cos(\varphi_o \mp \Phi^{\pm}_i))$.

(b) $f(\alpha)$ is regular in the band $-\Phi^{-}_e \leq \text{Re}(\alpha) \leq \Phi^{+}_e$, except at $\alpha = \varphi_o$ and possibly as $|\text{Im}\alpha| \to \infty$, and thus $f^{\pm}_{e,m^{\pm}}(\alpha)$ too (note: $f^{\pm}_{e,m^{\pm}} = e^{A^{\pm}}f + B^{\pm}$, where $A^{\pm}$ and $B^{\pm}$ are entire functions). The function $f(\alpha)$ has no singularities, except the poles of geometrical optics, in the region where $e^{ik\rho\cos(\alpha-\varphi)}$ diverges as $\rho \to \infty$, $|\text{Re}\alpha| \leq \pi$, $|\varphi| < \Phi$, and verifies for large $\text{Im}\alpha$ in this domain, $f(\alpha) = \frac{\mp i}{2}u(\rho = 0) + O(e^{\pm ic_{\pm}\alpha}\alpha^{d_{\pm}})$ as $\pm\text{Im}\alpha > 0$, $c_{\pm} > 0$, $d_{\pm} \geq 0$. Thus, $f(\alpha)$ remains finite when $-\Phi - \pi/2 + \arg(ik) \leq \pm\text{Re}(\alpha) \leq \pi/2 + \arg(ik) + \Phi$ as $\text{Im}\alpha \to \pm\infty$. Therefore, the spectral functions $f^{\pm}_{e,m^{\pm}}$, referring to geometry at origin $Q^{\pm}_{e,m^{\pm}}$, have behaviors of the same type.

The spectral functions $f_{a,p}^{\pm}$ and $f_{b,p}^{\pm}$, attached to shifts of the origin at opposite ends of the segment p on the face $\pm$, with $f_{b,p}^{\pm}(\alpha) = f_{a,p+1}^{\pm}(\alpha)$ for $1 \leq p < m^{\pm}$ and $f_{b,m^{\pm}}^{\pm}(\alpha) = f_{e,m^{\pm}}^{\pm}(\alpha)$, are the spectral functions associated with the representations of the field in coordinates $(\rho_{a,p}^{\pm}, \varphi_{a,p}^{\pm})$ and $(\rho_{b,p}^{\pm}, \varphi_{b,p}^{\pm})$ with origins at opposite ends $Q_{a,p}^{\pm}$ and $Q_{b,p}^{\pm}$ of segment p. From single-face representation of spectral function (3.7), we have:

$$f_{a,p}^{\pm}(\alpha) = \frac{1}{2}\int_{0}^{d_p^{\pm}} (-iku(\rho'_{a,p}, \varphi)\sin(\alpha - \varphi) \pm \frac{\partial u}{\partial n}(\rho'_{a,p}, \varphi))$$
$$\times\, e^{-ik\rho'_{a,p}\cos(\alpha-\varphi)} d\rho'_{a,p}|_{\varphi=\pm\Phi_p^{\pm}} + e^{-ik\, d_p^{\pm}\cos(\alpha\mp\Phi_p^{\pm})} f_{b,p}^{\pm}(\alpha),$$

$$f_{b,p}^{\pm}(\alpha) = \frac{1}{2}\int_{0}^{d_p^{\pm}} (-iku(\rho'_{b,p}, \varphi)\sin(\alpha - \varphi) \mp \frac{\partial u}{\partial n}(\rho'_{b,p}, \varphi))$$
$$\times\, e^{-ik\rho'_{b,p}\cos(\alpha-\varphi)} d\rho'_{b,p}|_{\varphi=\pm(\Phi_p^{\pm}-\pi)} + e^{-ik\, d_p^{\pm}\cos(\alpha\mp(\Phi_p^{\pm}-\pi))} f_{a,p}^{\pm}(\alpha), \quad (3.9)$$

and we note that the functions $f_{1,p}^{\pm}$ and $f_{2,p}^{\pm}$, combinations of functions $f_{a,p}^{\pm}$ and $f_{b,p}^{\pm}$ following:

$$f_{1,p}^{\pm}(\alpha) = f_{a,p}^{\pm}(\alpha) - e^{-ik\, d_p^{\pm}\cos(\alpha\mp\Phi_p^{\pm})} f_{b,p}^{\pm}(\alpha)$$
$$f_{2,p}^{\pm}(\alpha) = f_{b,p}^{\pm}(\alpha) - e^{ik\, d_p^{\pm}\cos(\alpha\mp\Phi_p^{\pm})} f_{a,p}^{\pm}(\alpha) = -e^{ik\, d_p^{\pm}\cos(\alpha\mp\Phi_p^{\pm})} f_{1,p}^{\pm}(\alpha), \quad (3.10)$$

are directly related to the radiation by the finite segment p.

NOTE 3.1.– Let us define $v^{\pm}$ so that $v^{\pm}(M) = 0$ at $M \in L_{0,\infty}^{\pm}$, with non-null finite $\nabla v^{\pm}(M)$ on regular parts of $L_{0,\infty}^{\pm}$. We have $dv^{\pm}(M) = d(OM).\nabla v^{\pm}(M) = 0$ on regular parts of each face. Therefore, with a constant a and the normal $\widehat{z}$ to 2D plane, the outward unit normal to face $L_{0,\infty}^{\pm}$ is written $\widehat{n}^{\pm} = \frac{a\nabla v^{\pm}(M)}{|a\nabla v^{\pm}(M)|} = \frac{\mp(\widehat{z}\wedge d(\overline{OM}))}{|d(\overline{OM})|}$, while noting that:

$$ds = |d(\overline{OM})| = \mp\widehat{n}^{\pm}.(\widehat{z}\wedge d(\overline{OM})),\ \widehat{n}^{\pm}ds = \mp(\widehat{z}\wedge d(\overline{OM})). \quad (3.11)$$

3.4. Polygonal surface with impedance boundary conditions

3.4.1. *Functional equations on $f_{e,m^{\pm}}^{\pm}$ due to conditions on semi-infinite planes*

In the case of an infinite polygonal scatterer with impedance boundary conditions along both semi-infinite planes $\varphi_e^{\pm} = \pm\Phi_e^{\pm}$, we have:

$$\frac{\partial u}{\partial n^{\pm}}(\rho_e^{\pm}, \pm\Phi_e^{\pm}) - ik\sin\theta_e^{\pm} u(\rho_e^{\pm}, \pm\Phi_e^{\pm}) = 0, \quad (3.12)$$

where $\partial(.)/\partial n^{\pm} = \hat{n}^{\pm}\nabla(.) = \mp\partial(.)/\rho_e^{\pm}\partial\varphi_e^{\pm}$, $0 < |\mathrm{Re}(\theta_e^{\pm})| \leq \pi/2$, $\hat{n}^{\pm}$ is the outward normal to the face $\varphi_e^{\pm} = \pm\Phi_e^{\pm}$, and from (3.8), the functions $f_{e,m^{\pm}}^{\pm}$ satisfy the functional difference equations:

$$(\sin\alpha \pm \sin\theta_e^{\pm})f_{e,m^{\pm}}^{\pm}(\alpha \pm \Phi_e^{\pm}) - (-\sin\alpha \pm \sin\theta_e^{\pm})f_{e,m^{\pm}}^{\pm}(-\alpha \pm \Phi_e^{\pm}) = 0. \tag{3.13}$$

If the scatterer is finite, the segments of both faces form a closed surface, and we can take $Q_{e,m^+}^{+} = Q_{e,m^-}^{-}$ with $\Phi_e^{+} + \Phi_e^{-} = 2\pi$. The fields on both semi-infinite planes $\varphi_e^{\pm} = \pm\Phi_e^{\pm}$ are then equal and their normal derivatives are opposite:

$$u(\rho_e^{+},\Phi_e^{+}) = u(\rho_e^{-},-\Phi_e^{-}),\ \frac{\partial u}{\partial n^{+}}(\rho_e^{+},\Phi_e^{+}) = -\frac{\partial u}{\partial n^{-}}(\rho_e^{-},-\Phi_e^{-}), \tag{3.14}$$

and we derive from (3.8) that:

$$f_{e,m^+}^{+}(\alpha+\pi) = f_{e,m^-}^{-}(\alpha-\pi). \tag{3.15}$$

Considering the definition of $f(\alpha)$ and $f_{e,m^{\pm}}^{\pm}$ given in (3.7) with properties (a) and (b), the functional equations (3.13) or (3.15) imply [1] that these functions are meromorphic with simple poles, regular in the band $-\Phi_e^{-} \leq \mathrm{Re}\alpha \leq \Phi_e^{+}$ except for the pole at $\alpha = \varphi_o$, and possibly at infinity when $\Phi \leq \pm\mathrm{Re}\alpha \leq \Phi_e^{\pm}$ (note: for finite scatterers, $f(\alpha) - \frac{1}{2}\cot(\frac{1}{2}(\alpha - \varphi_o))$ have no poles, and this function is an entire function).

3.4.2. *Functional equations due to boundary conditions on finite segments*

Considering impedance boundary conditions on the segments in the form:

$$\begin{aligned}&[\frac{\partial u}{\partial n}(\rho_{a,p},\pm\Phi_p^{\pm}) - ik\sin\theta_p^{\pm}u(\rho_{a,p},\pm\Phi_p^{\pm})]_{0\leq\rho_{a,p}\leq d_p^{\pm}} = 0,\\ &[\frac{\partial u}{\partial n}(\rho_{b,p},\pm(\Phi_p^{\pm}-\pi)) - ik\sin\theta_p^{\pm}u(\rho_{b,p},\pm(\Phi_p^{\pm}-\pi))]_{0\leq\rho_{b,p}\leq d_p^{\pm}} = 0,\end{aligned} \tag{3.16}$$

with $0 < |\mathrm{Re}(\theta_p^{\pm})| \leq \pi/2$, we can use the definition of $f_{1,p}^{\pm}$ and $f_{2,p}^{\pm}$ in (3.10) with the equalities given in (3.9), and deduce:

$$\begin{aligned}&[(\sin\alpha \pm \sin\theta_p^{\pm})f_{1,p}^{\pm}(\alpha+\varphi) - (-\sin\alpha \pm \sin\theta_p^{\pm})f_{1,p}^{\pm}(-\alpha+\varphi)]|_{\varphi=\pm\Phi_p^{\pm}} = 0,\\ &[(\sin\alpha \mp \sin\theta_p^{\pm})f_{2,p}^{\pm}(\alpha+\varphi) - (-\sin\alpha \mp \sin\theta_p^{\pm})f_{2,p}^{\pm}(-\alpha+\varphi)]|_{\varphi=\pm(\Phi_p^{\pm}-\pi)} = 0.\end{aligned} \tag{3.17}$$

3.4.3. *A remarkable relation between* $f^{\pm}_{a,p}$ *and* $f^{\pm}_{b,p}$ *deriving from* (3.17) *when* (3.10) *applies*

It is known from (3.10) with (3.9) that $f^{\pm}_{1,p}$ has a 2π period, but it is remarkable that this property also derives, in an independent manner, from the difference equations (3.17). Considering (3.17) with (3.10), this 2π periodicity will then imply the equivalence between the fields represented with $f^{\pm}_{a,p}$ and $f^{\pm}_{b,p}$.

Indeed, $f^{\pm}_{2,p}(\alpha \pm \Phi^{\pm}_{p}) = -e^{ik\,d^{\pm}_{p}\cos(\alpha)} f^{\pm}_{1,p}(\alpha \pm \Phi^{\pm}_{p})$ from (3.10), and we derive from (3.17):

$$[(\sin\alpha \pm \sin\theta^{\pm}_{p}) f^{\pm}_{1,p}(\alpha+\varphi) + (\sin\alpha \mp \sin\theta^{\pm}_{p}) f^{\pm}_{1,p}(-\alpha+\varphi)]|_{\varphi=\pm\Phi^{\pm}_{p}} = 0,$$

$$[(\sin\alpha \mp \sin\theta^{\pm}_{p}) f^{\pm}_{1,p}(\alpha+\varphi) + (\sin\alpha \pm \sin\theta^{\pm}_{p}) f^{\pm}_{1,p}(-\alpha+\varphi)]|_{\varphi=\pm(\Phi^{\pm}_{p}-\pi)} = 0, \tag{3.18}$$

then,

$$f^{\pm}_{1,p}(\alpha \pm 2\Phi^{\pm}_{p}) + \frac{(\sin(\alpha \pm \Phi^{\pm}_{p}) \mp \sin\theta^{\pm}_{p})}{(\sin(\alpha \pm \Phi^{\pm}_{p}) \pm \sin\theta^{\pm}_{p})} f^{\pm}_{1,p}(-\alpha) = 0,$$

$$f^{\pm}_{1,p}(\alpha \pm 2(\Phi^{\pm}_{p}-\pi)) + \frac{(\sin(\alpha \pm \Phi^{\pm}_{p}) \mp \sin\theta^{\pm}_{p})}{(\sin(\alpha \pm \Phi^{\pm}_{p}) \pm \sin\theta^{\pm}_{p})} f^{\pm}_{1,p}(-\alpha) = 0. \tag{3.19}$$

This allows, after substraction, to deduce the 2π periodicity of $f^{\pm}_{1,p}$:

$$f^{\pm}_{1,p}(\alpha) = f^{\pm}_{1,p}(\alpha+2\pi), \tag{3.20}$$

which gives us, after writing $f^{\pm}_{1,p}(\alpha) = f^{\pm}_{a,p}(\alpha) - e^{-ik\,d^{\pm}_{p}\cos(\alpha\mp\Phi^{\pm}_{p})} f^{\pm}_{b,p}(\alpha)$ from (3.10):

$$f^{\pm}_{a,p}(\pi+\alpha) - f^{\pm}_{a,p}(-\pi+\alpha) = e^{ik\,d^{\pm}_{p}\cos(\alpha\mp\Phi^{\pm}_{p})}(f^{\pm}_{b,p}(\pi+\alpha) - f^{\pm}_{b,p}(-\pi+\alpha)). \tag{3.21}$$

Consequently, the equations (3.17), when (3.10) applies, implies that the far field given by $f^{\pm}_{a,p}$ is the far field given by $f^{\pm}_{b,p}$ with phase center shifted by Δ, so that the field given by $f^{\pm}_{a,p}$ is the one given by $f^{\pm}_{b,p}$, relatively to two distinct origins that correspond to ends of the segment p.

In whole generality, it is remarkable that functional difference equations for boundary conditions on each segment, when expressed with $f^{\pm}_{a,n}$ and $f^{\pm}_{b,n}$, has for additional consequence, the equivalence of fields represented with any $f^{\pm}_{a,n}$ (or $f^{\pm}_{b,n}$), referring to any origins on the impedance polygons.

3.5. Formulation of the three-part polygonal problem: spectral functions in Sommerfeld–Maliuzhinets representation and functional difference equations in complex plane

3.5.1. *Definition*

We consider the diffraction of an incident plane wave by a semi-infinite impedance polygon divided into three parts, each one characterized by relative surface impedances $\sin\theta_-$, $\sin\theta_1$ and $\sin\theta_+$, with $0 < \mathrm{Re}(\theta_\pm) \leq \pi/2$ and $0 < \mathrm{Re}(\theta_1) \leq \pi/2$ (strict passivity). This means that $m^+ = 1$, $m^- = 0$, $\Phi_1^+ = \frac{\pi}{2}$, $\Phi_e^+ = \frac{\pi}{2} + \Phi_b$, $\Phi_e^- = -\frac{\pi}{2} - \Phi_a$, $f \equiv f_a = f_{a,1}^+ = f_e^-$, $f_b = f_{b,1}^+ = f_e^+$.

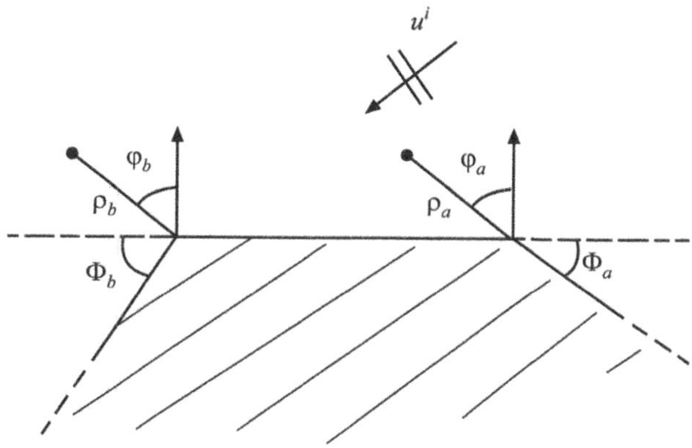

Figure 3.3. *Geometry of the three-part polygon*

The functions f_a and f_b are the spectral functions associated with the Sommerfeld–Maliuzhinets representation of the field, in coordinate systems (ρ_a, φ_a) and (ρ_b, φ_b), with origins at opposite ends of the finite segment of length Δ (see geometry in Figure 3.3). We have, in (ρ_a, φ_a) coordinates:

$$\begin{aligned}&(\rho_a \in]0,\infty[, \varphi_a = -\frac{\pi}{2} - \Phi_a) \text{ with } \partial u/\partial n - ik\sin\theta_- u = 0,\\ &(\rho_a \in [0,\Delta], \varphi_a = \frac{\pi}{2}) \text{ with } \partial u/\partial n - ik\sin\theta_1 u = 0,\end{aligned} \tag{3.22}$$

with the incident field $u^i = e^{ik\rho_a \cos(\varphi_a - \varphi_o)}$, and, in coordinates (ρ_b, φ_b):

$$\begin{aligned}&(\rho_b \in [0,\Delta], \varphi_b = -\frac{\pi}{2}) \text{ with } \partial u/\partial n - ik\sin\theta_1 u = 0,\\ &(\rho_b \in]0,\infty[, \varphi_b = \frac{\pi}{2} + \Phi_b) \text{ with } \partial u/\partial n - ik\sin\theta_+ u = 0,\end{aligned} \tag{3.23}$$

with the incident field $u^i = e^{ik(\rho_b \cos(\varphi_b - \varphi_o) + \Delta \sin\varphi_o)}$.

The properties (a) and (b) give us that:

$$(f_a(\alpha) - 1/(\alpha - \varphi_\circ)) \text{ and } (f_b(\alpha) - e^{ik\Delta \sin\varphi_\circ}/(\alpha - \varphi_\circ)), \tag{3.24}$$

are regular in the respective strips $-\frac{\pi}{2} - \Phi_a \leq \text{Re}\alpha \leq \frac{\pi}{2}$ and $-\frac{\pi}{2} \leq \text{Re}\alpha \leq \frac{\pi}{2} + \Phi_b$, with $-\frac{\pi}{2} - \Phi_a < \varphi_\circ < \frac{\pi}{2} + \Phi_b$, and verify $f_{a(b)}(\alpha) = \frac{\mp i}{2}u(\rho_{a(b)} = 0) + O(e^{\pm ic_{a(,b)\pm}\alpha}\alpha^{d_{a(,b)\pm}})$ in these regions as $\text{Im}\alpha \to \pm\infty$, $c_{a(,b)\pm} > 0$.

Considering the definitions, we have $f_e^- = f_a$, $f_e^+ = f_b$, and we obtain, relatively to integration on piecewise straight faces:

$$f_a(\alpha) = e^{-ik\Delta\sin\alpha} f_b(\alpha) - \int_0^\Delta \frac{(-iku(\rho_a, \frac{\pi}{2})\cos\alpha - \frac{\partial u}{\partial n}(\rho_a, \frac{\pi}{2}))e^{-ik\rho_a \sin\alpha}}{2} d\rho_a,$$

$$f_b(\alpha) = e^{ik\Delta\sin\alpha} f_a(\alpha) - \int_0^\Delta \frac{(iku(\rho_b, -\frac{\pi}{2})\cos\alpha + \frac{\partial u}{\partial n}(\rho_b, -\frac{\pi}{2}))e^{+ik\rho_b \sin\alpha}}{2} d\rho_b, \tag{3.25}$$

for any complex angle α, while, relatively to integration on straight semi-infinite faces:

$$f_a(\alpha - \varphi_-) = \frac{1}{2}\int_0^\infty (-iku(\rho_a, -\varphi_-)\sin\alpha - \frac{\partial u}{\partial n}(\rho_a, -\varphi_-))e^{-ik\rho_a\cos\alpha} d\rho_a, \tag{3.26}$$

as $\text{Re}(ik(\cos(\varphi_\circ + \varphi_-) - \cos\alpha)) < 0$, $|\text{Re}\alpha| < \pi$, with $\varphi_- = \frac{\pi}{2} + \Phi_a$, and

$$f_b(\alpha + \varphi_+) = \frac{1}{2}\int_0^\infty (-iku(\rho_b, \varphi_+)\sin\alpha + \frac{\partial u}{\partial n}(\rho_b, \varphi_+))e^{-ik\rho_b\cos\alpha} d\rho_b, \tag{3.27}$$

as $\text{Re}(ik(\cos(\varphi_\circ - \varphi_+) - \cos\alpha)) < 0$, $|\text{Re}\alpha| < \pi$, with $\varphi_+ = \frac{\pi}{2} + \Phi_b$.

3.5.2. *About functional difference equations in complex plane*

3.5.2.1. *Functional equations for the central strip and semi-infinite planes*

We now derive functional equations on the spectral functions from the boundary conditions on the central strip (of finite size) and on the semi-infinite planes of the three-part impedance polygon (Figure 3.3). Using the boundary conditions on the central strip (3.22), and letting:

$$f_1(\alpha + \frac{\pi}{2}) = f_a(\alpha + \frac{\pi}{2}) - e^{-ik\Delta\cos\alpha} f_b(\alpha + \frac{\pi}{2}),$$

$$f_2(\alpha - \frac{\pi}{2}) = f_b(\alpha - \frac{\pi}{2}) - e^{-ik\Delta\cos\alpha} f_a(\alpha - \frac{\pi}{2}) = -e^{-ik\Delta\cos\alpha} f_1(\alpha - \frac{\pi}{2}), \tag{3.28}$$

we obtain, as described in (3.17):

$$(\sin\alpha + \sin\theta_1) f_1(\alpha + \frac{\pi}{2}) - (-\sin\alpha + \sin\theta_1) f_1(-\alpha + \frac{\pi}{2}) = 0,$$
$$(\sin\alpha - \sin\theta_1) f_2(\alpha - \frac{\pi}{2}) - (-\sin\alpha - \sin\theta_1) f_2(-\alpha - \frac{\pi}{2}) = 0, \quad (3.29)$$

while, using the boundary conditions on both semi-infinite planes (3.23), we obtain, as described in (3.13):

$$(\sin\alpha + \sin\theta_+) f_b(\alpha + \frac{\pi}{2} + \Phi_b) - (-\sin\alpha + \sin\theta_+) f_b(-\alpha + \frac{\pi}{2} + \Phi_b) = 0,$$
$$(\sin\alpha - \sin\theta_-) f_a(\alpha - \frac{\pi}{2} - \Phi_a) - (-\sin\alpha - \sin\theta_-) f_a(-\alpha - \frac{\pi}{2} - \Phi_a) = 0. \quad (3.30)$$

3.5.2.2. *A remarkable relation between f_a and f_b deriving from (3.29) when (3.28) applies*

We follow a development similar to the one used to obtain (3.19)–(3.21). Considering the relation between f_1 and f_2 in (3.28), we have from (3.29):

$$(\sin\alpha + \sin\theta_1) f_1(\alpha + \frac{\pi}{2}) - (-\sin\alpha + \sin\theta_1) f_1(-\alpha + \frac{\pi}{2}) = 0,$$
$$(\sin\alpha - \sin\theta_1) f_1(\alpha - \frac{\pi}{2}) - (-\sin\alpha - \sin\theta_1) f_1(-\alpha - \frac{\pi}{2}) = 0, \quad (3.31)$$

and thus,

$$(\cos\alpha + \sin\theta_1) f_1(\pi + \alpha) + (\cos\alpha - \sin\theta_1) f_1(-\alpha) = 0,$$
$$-(\cos\alpha + \sin\theta_1) f_1(-\pi + \alpha) - (\cos\alpha - \sin\theta_1) f_1(-\alpha) = 0. \quad (3.32)$$

This implies, after addition,

$$f_1(\pi + \alpha) = f_1(-\pi + \alpha), \quad (3.33)$$

and, from the expression of f_1 with f_a and f_b given in (3.28):

$$f_a(\pi + \alpha) - f_a(-\pi + \alpha) = e^{ik\Delta \sin\alpha}(f_b(\pi + \alpha) - f_b(-\pi + \alpha)). \quad (3.34)$$

Consequently, equations (3.29), which initially concern the impedance boundary condition on the segment, has for additional remarkable consequence (3.34) when they are expressed with f_a and f_b; this signifies that the far field given by f_a is the one given by f_b with phase center shifted by Δ, so that the field given by f_a is the one given by f_b relatively to the two origins at both ends of the segment.

NOTE 3.2.– Let us consider the poles of the meromorphic function f_1. They are 2π-periodic from (3.33), and are, by definition of f_1, also poles of f_a, or f_b, or both. The presence of these 2π-periodic poles is in contradiction with the functional equations due to impedance conditions on the semi-infinite faces (as (a) and (b) are satisfied), and thus the integral equations derived from them (note: it is different for finite scatterers, since $f(\alpha) - \frac{1}{2}\cot(\frac{1}{2}(\alpha - \varphi_\circ))$ has no pole). It finally implies that f_1 has no pole, and is then an entire function. This property is in agreement with the integral expression of f_1 with u on the finite segment, derived from (3.9)–(3.10).

3.5.3. *Functional difference equations for* $f_{br}(\alpha) = f_b(\alpha + \frac{\Phi_b}{2})$, $f_{ar}(\alpha) = f_a(\alpha - \frac{\Phi_a}{2})$

We can now derive, from (3.29) to (3.30), difference equations on $f_{br}(\alpha) = f_b(\alpha + \frac{\Phi_b}{2})$, $f_{ar}(\alpha) = f_a(\alpha - \frac{\Phi_a}{2})$, corresponding to both discontinuities when influencing each other. If we let $f_{br}(\alpha - \frac{\Phi_b}{2}) = f_b(\alpha)$ with $\Phi_+ = \frac{\pi}{2} + \frac{\Phi_b}{2}$, we have:

$$(\sin\alpha + \sin\theta_+)f_{br}(\alpha + \Phi_+) - (-\sin\alpha + \sin\theta_+)f_{br}(-\alpha + \Phi_+) = 0, \qquad (3.35)$$

and

$$\begin{aligned}&(\sin\alpha - \sin\theta_1)f_{br}(\alpha - \Phi_+) - (-\sin\alpha - \sin\theta_1)f_{br}(-\alpha - \Phi_+)\\ &= e^{-ik\Delta\cos\alpha}((\sin\alpha - \sin\theta_1)f_a(\alpha - \frac{\pi}{2}) - (-\sin\alpha - \sin\theta_1)f_a(-\alpha - \frac{\pi}{2}))\\ &= 2e^{-ik\Delta\cos\alpha}S_b^-(\alpha).\end{aligned} \qquad (3.36)$$

If we let $f_{ar}(\alpha + \frac{\Phi_a}{2}) = f_a(\alpha)$ with $\Phi_- = \frac{\pi}{2} + \frac{\Phi_a}{2}$, we have:

$$\begin{aligned}&(\sin\alpha + \sin\theta_1)f_{ar}(\alpha + \Phi_-) - (-\sin\alpha + \sin\theta_1)f_{ar}(-\alpha + \Phi_-)\\ &= e^{-ik\Delta\cos\alpha}((\sin\alpha + \sin\theta_1)f_b(\alpha + \frac{\pi}{2}) - (-\sin\alpha + \sin\theta_1)f_b(-\alpha + \frac{\pi}{2}))\\ &= 2e^{-ik\Delta\cos\alpha}S_a^+(\alpha),\end{aligned} \qquad (3.37)$$

and

$$(\sin\alpha - \sin\theta_-)f_{ar}(\alpha - \Phi_-) - (-\sin\alpha - \sin\theta_-)f_{ar}(-\alpha - \Phi_-) = 0. \qquad (3.38)$$

From the properties of f_a and f_b, the functions:

$$(f_{ar}(\alpha) - 1/(\alpha - \varphi_{\circ,a}) \text{ and } (f_{br}(\alpha) - e^{ik\Delta\sin\varphi_\circ}/(\alpha - \varphi_{\circ,b})), \qquad (3.39)$$

are, respectively, regular in the bands $|\text{Re}\alpha| \le \Phi_-$ and $|\text{Re}\alpha| \le \Phi_+$, with $\varphi_{\circ,a} = \varphi_\circ + \frac{\Phi_a}{2}$ and $\varphi_{\circ,b} = \varphi_\circ - \frac{\Phi_b}{2}$, $-\frac{\pi}{2} - \Phi_a < \varphi_\circ < \frac{\pi}{2} + \Phi_b$, and, in these regions,

$$\begin{aligned}&f_{ar}(\alpha) = \mp\frac{i}{2}u(\rho_a = 0) + O(e^{\pm ic_{a,\pm}\alpha}\alpha^{d_{a,\pm}}),\\ &\text{and } f_{br}(\alpha) = \mp\frac{i}{2}u(\rho_b = 0) + O(e^{\pm ic_{b,\pm}\alpha}\alpha^{d_{b,\pm}}),\end{aligned} \qquad (3.40)$$

as $\text{Im}\alpha \to \pm\infty$, $c_{a(,b),\pm} > 0$, $d_{a(,b),\pm} \ge 0$.

Besides, we remark that, for $\Delta = 0$, the three-part polygon is reduced to a wedge with exterior angle $2\Phi_d = (2(\Phi_+ + \Phi_-) - \pi)$ and face impedances $\sin\theta_\pm$, where $\Phi_+ = \frac{\pi}{2} + \frac{\Phi_b}{2}$, $\Phi_- = \frac{\pi}{2} + \frac{\Phi_a}{2}$. In this case, we have $f_{br}(\alpha - \Phi_- + \frac{\pi}{2}) = f_{ar}(\alpha + \Phi_+ - \frac{\pi}{2}) = f_0^{ab}(\alpha)$, where f_0^{ab}, referring to the solution for an impedance wedge (see appendix A), is given by:

$$f_0^{ab}(\alpha, \varphi_o) = \frac{\pi}{2\Phi_d} \frac{\Psi_{+-}(\alpha)\cos(\frac{\pi\varphi_{o,d}}{2\Phi_d})}{\Psi_{+-}(\varphi_{o,d})(\sin(\frac{\pi\alpha}{2\Phi_d}) - \sin(\frac{\pi\varphi_{o,d}}{2\Phi_d}))}, \tag{3.41}$$

where $\varphi_{o,d} = \varphi_o + (\Phi_a - \Phi_b)/2$, $-\Phi_d < \varphi_{o,d} < \Phi_d$, with $\Phi_d = (\Phi_a + \Phi_b + \pi)/2$.

3.6. The integral expressions and integral equations for the three-part impedance polygon

3.6.1. *Elementary integral solutions for difference equations*

Considering the theory of functional equations, the analytic function $\chi(\alpha)$ satisfying:

$$\chi(\alpha \pm \Phi) - \chi(-\alpha \pm \Phi) = \vartheta^\pm(\alpha), \tag{3.42}$$

and regular as $|\text{Re}\alpha| \leq \Phi$, is given in the strip $|\text{Re}\alpha| < \Phi$ [20, 22] by:

$$\chi(\alpha) = \frac{\chi(i\infty) + \chi(-i\infty)}{2} + \frac{-i}{8\Phi}\int_{-i\infty}^{+i\infty} d\alpha' \Big(\vartheta^+(\alpha')\tan\Big(\frac{\pi}{4\Phi}(\alpha + \Phi - \alpha')\Big)$$

$$- \vartheta^-(\alpha')\tan\Big(\frac{\pi}{4\Phi}(\alpha - \Phi - \alpha')\Big)\Big), \tag{3.43}$$

when the functions $\vartheta^\pm(\alpha)$ are regular and summable on the imaginary axis, with $\chi(i\infty) - \chi(-i\infty) = 0$. The functional equations (3.35)–(3.38) can be reduced into the form (3.42), by using the solutions $\Psi_{+1}(\alpha, \Phi_+)$ (respectively, $\Psi_{1-}(\alpha, \Phi_-)$), without pole or zero and $O(\cos(\pi\alpha/2\Phi_+))$ (respectively, $O(\cos(\pi\alpha/2\Phi_-))$) in the band $|\text{Re}\alpha| \leq \Phi_+ = \frac{\pi}{2} + \frac{\Phi_b}{2}$ (respectively, $|\text{Re}\alpha| \leq \Phi_- = \frac{\pi}{2} + \frac{\Phi_a}{2}$), of the equations without second members of the type:

$$(\sin\alpha + \sin\theta_l)\Psi_{lr}(\alpha + \Phi, \Phi) - (-\sin\alpha + \sin\theta_l)\Psi_{lr}(-\alpha + \Phi, \Phi) = 0,$$

$$(\sin\alpha - \sin\theta_r)\Psi_{lr}(\alpha - \Phi, \Phi) - (-\sin\alpha - \sin\theta_r)\Psi_{lr}(-\alpha - \Phi, \Phi) = 0, \tag{3.44}$$

with $l \equiv +$, $r \equiv 1$, $\Phi = \Phi_+ = \frac{\pi}{2} + \frac{\Phi_b}{2}$ for $\Psi_{+1}(\alpha, \Phi_+)$, and $l \equiv 1$, $r \equiv -$, $\Phi = \Phi_- = \frac{\pi}{2} + \frac{\Phi_a}{2}$ for $\Psi_{1-}(\alpha, \Phi_-)$ (see details on Ψ_{lr}, named $\Psi_{+-}(\alpha, \Phi)$, in appendix A).

3.6.2. *Coupled integral expressions for* $f_{br}(\alpha)$ *and* $f_{ar}(\alpha)$

Considering (3.35)–3.38), we first consider the solution for $S_b^- \equiv 0$ (respectively, $S_a^+ \equiv 0$) with a unique pole due to incident plane wave as $|\text{Re}\alpha| < \Phi_+$ (respectively, $|\text{Re}\alpha| < \Phi_-$), which is directly related to impedance wedge case (see appendix A). Substracting it from the general solution, we obtain unknowns regular as $|\text{Re}\alpha| < \Phi_+$ (respectively, $|\text{Re}\alpha| < \Phi_-$). As previously noted, we can then use Ψ_{lr} of (3.44), then (3.42), and thus obtain integral expressions of the type (3.43).

Therefore, we can write [1], $f_{br}(\alpha)$ for $-\Phi_+ < \text{Re}\alpha < 3\Phi_+$, following:

$$\frac{f_{br}(\alpha)}{\Psi_{+1}(\alpha)} = \frac{i}{4\Phi_+}\int_{-i\infty}^{+i\infty}\frac{S_b^-(\alpha')e^{-ik\Delta\cos\alpha'}\tan(\frac{\pi}{4\Phi_+}(\alpha-\Phi_+-\alpha'))}{(\sin\alpha'-\sin\theta_1)\Psi_{+1}(\alpha'-\Phi_+)}d\alpha' + \chi_b^i(\alpha)$$

$$= \frac{-i}{4\Phi_+}\int_{-i\infty}^{+i\infty}\frac{\frac{S_b^-(\alpha')}{(\sin\alpha'-\sin\theta_1)}}{\Psi_{+1}(\alpha'-\Phi_+)}\frac{e^{-ik\Delta\cos\alpha'}\sin(\frac{\pi\alpha'}{2\Phi_+})}{\cos(\frac{\pi}{2\Phi_+}(\alpha-\Phi_+))+\cos(\frac{\pi\alpha'}{2\Phi_+})}d\alpha' + \chi_b^i(\alpha)$$

$$= \frac{-i}{4\Phi_+}\int_{-i\infty}^{+i\infty}\frac{f_{ar}(\alpha'-\frac{\pi}{2}+\frac{\Phi_a}{2})}{\Psi_{+1}(\alpha'-\Phi_+)}\frac{e^{-ik\Delta\cos\alpha'}\sin(\frac{\pi\alpha'}{2\Phi_+})}{\cos(\frac{\pi}{2\Phi_+}(\alpha-\Phi_+))+\cos(\frac{\pi\alpha'}{2\Phi_+})}d\alpha' + \chi_b^i(\alpha), \tag{3.45}$$

and, $f_{ar}(\alpha)$ for $-3\Phi_- < \text{Re}\alpha < \Phi_-$,

$$\frac{f_{ar}(\alpha)}{\Psi_{1-}(\alpha)} = \frac{-i}{4\Phi_-}\int_{-i\infty}^{+i\infty}\frac{S_a^+(\alpha')e^{-ik\Delta\cos\alpha'}\tan(\frac{\pi}{4\Phi_-}(\alpha+\Phi_--\alpha'))}{(\sin\alpha'+\sin\theta_1)\Psi_{1-}(\alpha'+\Phi_-)}d\alpha' + \chi_a^i(\alpha)$$

$$= \frac{i}{4\Phi_-}\int_{-i\infty}^{+i\infty}\frac{\frac{S_a^+(\alpha')}{(\sin\alpha'+\sin\theta_1)}}{\Psi_{1-}(\alpha'+\Phi_-)}\frac{e^{-ik\Delta\cos\alpha'}\sin(\frac{\pi\alpha'}{2\Phi_-})}{\cos(\frac{\pi}{2\Phi_-}(\alpha+\Phi_-))+\cos(\frac{\pi\alpha'}{2\Phi_-})}d\alpha' + \chi_a^i(\alpha)$$

$$= \frac{i}{4\Phi_-}\int_{-i\infty}^{+i\infty}\frac{f_{br}(\alpha'+\frac{\pi}{2}-\frac{\Phi_b}{2})}{\Psi_{1-}(\alpha'+\Phi_-)}\frac{e^{-ik\Delta\cos\alpha'}\sin(\frac{\pi\alpha'}{2\Phi_-})}{\cos(\frac{\pi}{2\Phi_-}(\alpha+\Phi_-))+\cos(\frac{\pi\alpha'}{2\Phi_-})}d\alpha' + \chi_a^i(\alpha), \tag{3.46}$$

with $\varphi_{o,b} = \varphi_o - \frac{\Phi_b}{2}$, as $-\frac{\pi}{2} < \varphi_o < \frac{\pi}{2} + \Phi_b$, and $\varphi_{o,a} = \varphi_o + \frac{\Phi_a}{2}$, as $-\frac{\pi}{2} - \Phi_a < \varphi_o < \frac{\pi}{2}$, while the source terms are given by:

$$\chi_b^i(\alpha) = e^{ik\Delta\sin\varphi_o}\chi_{+1}^i(\alpha) = \frac{\pi}{2\Phi_+}\Big(\frac{e^{ik\Delta\sin\varphi_o}\cos(\frac{\pi\varphi_{o,b}}{2\Phi_+})}{\Psi_{+1}(\varphi_{o,b})(\sin(\frac{\pi\alpha}{2\Phi_+})-\sin(\frac{\pi\varphi_{o,b}}{2\Phi_+}))}\Big),$$

$$\chi_a^i(\alpha) = \frac{\pi}{2\Phi_-}\Big(\frac{\cos(\frac{\pi\varphi_{o,a}}{2\Phi_-})}{\Psi_{1-}(\varphi_{o,a})(\sin(\frac{\pi\alpha}{2\Phi_-})-\sin(\frac{\pi\varphi_{o,a}}{2\Phi_-}))}\Big), \tag{3.47}$$

with $\text{Re}(\theta_\pm) \neq 0$, $\text{Re}(\theta_1) \neq 0$, $\Phi_+ = \frac{\pi}{2} + \frac{\Phi_b}{2}$ and $\Phi_- = \frac{\pi}{2} + \frac{\Phi_a}{2}$.

NOTE 3.3.– We use, in (3.45), that $(\sin\alpha' - \sin\theta_1)\Psi_{+1}(\alpha' - \Phi_+)$ is even, and thus:

$$\frac{2S_b^-(\alpha')}{(\sin\alpha' - \sin\theta_1)\Psi_{+1}(\alpha' - \Phi_+)} = \frac{f_{ar}(\alpha' - \frac{\pi}{2} + \frac{\Phi_a}{2})}{\Psi_{+1}(\alpha' - \Phi_+)} - \frac{f_{ar}(-\alpha' - \frac{\pi}{2} + \frac{\Phi_a}{2})}{\Psi_{+1}(-\alpha' - \Phi_+)}, \tag{3.48}$$

and in (3.46), that $(\sin\alpha' + \sin\theta_1)\Psi_{1-}(\alpha' + \Phi_-)$ is even, and thus:

$$\frac{2S_a^+(\alpha')}{(\sin\alpha' + \sin\theta_1)\Psi_{1-}(\alpha' + \Phi_-)} = \frac{f_{br}(\alpha' + \frac{\pi}{2} - \frac{\Phi_b}{2})}{\Psi_{1-}(\alpha' + \Phi_-)} - \frac{f_{br}(-\alpha' + \frac{\pi}{2} - \frac{\Phi_b}{2})}{\Psi_{1-}(-\alpha' + \Phi_-)}, \tag{3.49}$$

with $f_{br}(\alpha - \frac{\Phi_b}{2}) = f_b(\alpha)$, $f_{ar}(\alpha + \frac{\Phi_a}{2}) = f_a(\alpha)$.

3.6.3. *Integral equations when* $\Phi_{a,b} > -\frac{\pi}{2}$, *as* $|\arg(ik)| < \pi/2$

3.6.3.1. *Definition of integral equations*

From previous integral expressions (3.45)–(3.46), we can write:

$$\frac{f_{br}(\alpha + \frac{\pi}{2} - \frac{\Phi_b}{2})}{\Psi_{+1}(\alpha + \frac{\pi}{2} - \frac{\Phi_b}{2})} = \frac{-i}{4\Phi_+}\int_{-i\infty}^{+i\infty} \frac{S_b^-(\alpha')}{(\sin\alpha' - \sin\theta_1)}\frac{1}{\Psi_{+1}(\alpha' - \Phi_+)}$$
$$\times \frac{e^{-ik\Delta\cos\alpha'}\sin(\frac{\pi\alpha'}{2\Phi_+})}{\cos(\frac{\pi}{2\Phi_+}(\alpha - \Phi_b)) + \cos(\frac{\pi\alpha'}{2\Phi_+})}d\alpha' + \chi_b^i(\alpha + \frac{\pi}{2} - \frac{\Phi_b}{2}), \tag{3.50}$$

as $-\Phi_+ <\text{Re}(\alpha + \frac{\pi}{2} - \frac{\Phi_b}{2}) < 3\Phi_+$, and,

$$\frac{f_{ar}(\alpha - \frac{\pi}{2} + \frac{\Phi_a}{2})}{\Psi_{1-}(\alpha - \frac{\pi}{2} + \frac{\Phi_a}{2})} = \frac{i}{4\Phi_-}\int_{-i\infty}^{+i\infty} \frac{S_a^+(\alpha')}{(\sin\alpha' + \sin\theta_1)}\frac{1}{\Psi_{1-}(\alpha' + \Phi_-)}$$
$$\times \frac{e^{-ik\Delta\cos\alpha'}\sin(\frac{\pi\alpha'}{2\Phi_-})}{\cos(\frac{\pi}{2\Phi_-}(\alpha + \Phi_a)) + \cos(\frac{\pi\alpha'}{2\Phi_-})}d\alpha' + \chi_a^i(\alpha - \frac{\pi}{2} + \frac{\Phi_a}{2}), \tag{3.51}$$

as $-3\Phi_- <\text{Re}(\alpha - \frac{\pi}{2} + \frac{\Phi_a}{2}) < \Phi_-$, when $-\min(\frac{\pi}{2}, \frac{\pi}{2} + \Phi_a) < \varphi_\circ < \min(\frac{\pi}{2}, \frac{\pi}{2} + \Phi_b)$.

When $\Phi_{a,b} > -\frac{\pi}{2}$, we can take α purely imaginary in this system of equations, while noticing that the functions:

$$\frac{S_b^-(\alpha)}{(\sin\alpha - \sin\theta_1)} = \frac{1}{2}\Big(f_{ar}(\alpha - \frac{\pi}{2} + \frac{\Phi_a}{2}) + \frac{(\sin\alpha + \sin\theta_1)f_{ar}(-\alpha - \frac{\pi}{2} + \frac{\Phi_a}{2})}{(\sin\alpha - \sin\theta_1)}\Big),$$
$$\frac{S_a^+(\alpha)}{(\sin\alpha + \sin\theta_1)} = \frac{1}{2}\Big(f_{br}(\alpha + \frac{\pi}{2} - \frac{\Phi_b}{2}) + \frac{(\sin\alpha - \sin\theta_1)f_{br}(-\alpha + \frac{\pi}{2} - \frac{\Phi_b}{2})}{(\sin\alpha + \sin\theta_1)}\Big), \tag{3.52}$$

vanish at infinity (in contrary to $f_{ar}(\alpha - \frac{\pi}{2} + \frac{\Phi_a}{2})$ and $f_{br}(\alpha + \frac{\pi}{2} - \frac{\Phi_b}{2})$ in general).

Choosing these functions as unknowns, we then derive the following system of integral equations for imaginary α:

$$\frac{S_a^+(\alpha)}{(\sin\alpha+\sin\theta_1)} = \frac{-i}{8\Phi_+}\int_{-i\infty}^{+i\infty}\frac{S_b^-(\alpha')}{(\sin\alpha'-\sin\theta_1)}\frac{e^{-ik\Delta\cos\alpha'}\sin(\frac{\pi\alpha'}{2\Phi_+})}{\Psi_{+1}(\alpha'-\Phi_+)}$$
$$\times\left[\frac{\Psi_{+1}(\alpha+\frac{\pi}{2}-\frac{\Phi_b}{2})}{\cos(\frac{\pi}{2\Phi_+}(\alpha-\Phi_b))+\cos(\frac{\pi\alpha'}{2\Phi_+})}+\frac{\frac{(\sin\alpha-\sin\theta_1)}{(\sin\alpha+\sin\theta_1)}\Psi_{+1}(-\alpha+\frac{\pi}{2}-\frac{\Phi_b}{2})}{\cos(\frac{\pi}{2\Phi_+}(\alpha+\Phi_b))+\cos(\frac{\pi\alpha'}{2\Phi_+})}\right]d\alpha'$$
$$+\frac{1}{2}\left[\Psi_{+1}(\alpha+\frac{\pi}{2}-\frac{\Phi_b}{2})\chi_b^i(\alpha+\frac{\pi}{2}-\frac{\Phi_b}{2})\right.$$
$$\left.+\frac{(\sin\alpha-\sin\theta_1)\Psi_{+1}(-\alpha+\frac{\pi}{2}-\frac{\Phi_b}{2})}{(\sin\alpha+\sin\theta_1)}\chi_b^i(-\alpha+\frac{\pi}{2}-\frac{\Phi_b}{2})\right], \tag{3.53}$$

and,

$$\frac{S_b^-(\alpha)}{(\sin\alpha-\sin\theta_1)} = \frac{i}{8\Phi_-}\int_{-i\infty}^{+i\infty}\frac{S_a^+(\alpha')}{(\sin\alpha'+\sin\theta_1)}\frac{e^{-ik\Delta\cos\alpha'}\sin(\frac{\pi\alpha'}{2\Phi_-})}{\Psi_{1-}(\alpha'+\Phi_-)}$$
$$\times\left[\frac{\Psi_{1-}(\alpha-\frac{\pi}{2}+\frac{\Phi_a}{2})}{\cos(\frac{\pi}{2\Phi_-}(\alpha+\Phi_a))+\cos(\frac{\pi\alpha'}{2\Phi_-})}+\frac{\frac{(\sin\alpha+\sin\theta_1)}{(\sin\alpha-\sin\theta_1)}\Psi_{1-}(-\alpha-\frac{\pi}{2}+\frac{\Phi_a}{2})}{\cos(\frac{\pi}{2\Phi_-}(\alpha-\Phi_a))+\cos(\frac{\pi\alpha'}{2\Phi_-})}\right]d\alpha'$$
$$+\frac{1}{2}\left[\Psi_{1-}(\alpha-\frac{\pi}{2}+\frac{\Phi_a}{2})\chi_a^i(\alpha-\frac{\pi}{2}+\frac{\Phi_a}{2})\right.$$
$$\left.+\frac{(\sin\alpha+\sin\theta_1)\Psi_{1-}(-\alpha-\frac{\pi}{2}+\frac{\Phi_a}{2})}{(\sin\alpha-\sin\theta_1)}\chi_a^i(-\alpha-\frac{\pi}{2}+\frac{\Phi_a}{2})\right], \tag{3.54}$$

where we note that S_b^- and S_a^+ are, from the characteristics of kernels, odd meromorphic functions of the complex plane.

3.6.3.2. *Equations in* $L^2(iR)\times L^2(iR)$ *as* $|\arg(ik)|<\pi/2$

Let us analyze the asymptotic behavior of kernels in previous integral equations. For that, we begin by noting that:

$$\frac{\pi}{2}-\frac{\Phi_{a(,b)}}{2}=\frac{\pi}{2}+\frac{\Phi_{a(,b)}}{2}-\Phi_{a(,b)}=\Phi_{-(,+)}-\Phi_{a(,b)},$$
$$\frac{(\sin\alpha\mp\sin\theta_1)}{(\sin\alpha\pm\sin\theta_1)}=1-\frac{\pm2\sin\theta_1}{(\sin\alpha\pm\sin\theta_1)}, \tag{3.55}$$

and,

$$\Psi_{+1,1-}(\alpha) = A_0 \cos(\frac{\pi\alpha}{2\Phi_{+,-}})(1 + O(\alpha^\nu e^{-|\mu \mathrm{Im}\alpha|})),$$

$$\Psi_{+1}(-\alpha + \frac{\pi}{2} - \frac{\Phi_b}{2}) = -\Psi_{+1}(\alpha + \frac{\pi}{2} - \frac{\Phi_b}{2})e^{\frac{\mp i\pi\Phi_b}{\Phi_+}}(1 + O(\alpha^\nu e^{-|\mu \mathrm{Im}\alpha|})),$$

$$\Psi_{1-}(-\alpha - \frac{\pi}{2} + \frac{\Phi_a}{2}) = -\Psi_{1-}(\alpha - \frac{\pi}{2} + \frac{\Phi_a}{2})e^{\frac{\pm i\pi\Phi_a}{\Phi_-}}(1 + O(\alpha^\nu e^{-|\mu \mathrm{Im}\alpha|})), \quad (3.56)$$

when $\mathrm{Im}\alpha \to \pm\infty$, with $\Phi_+ = \frac{\pi}{2} + \frac{\Phi_b}{2}$, $\Phi_- = \frac{\pi}{2} + \frac{\Phi_a}{2}$. Letting the kernels in the form:

$$K_{a,b}(\alpha', \alpha) = k_{a,b}(\alpha', \alpha)e^{-ik\Delta\cos\alpha'}, \quad (3.57)$$

then using (3.55) and (3.56), we note that the $k_{a,b}(\alpha', \alpha)$, independent of $k\Delta$, are bounded functions on $iR \times iR$ that verify:

$$k_{a,b}(\alpha', \alpha) = O\Bigg(\frac{\left(\frac{\cos(a_1\alpha)\cos(a_1\alpha')}{\cos(a_1\alpha)+c\cos(a_1\alpha')} + \frac{d\cos(a_1\alpha)}{1+b_1\cos(\alpha)}\right)}{\cos(a_1\alpha) + c\cos(a_1\alpha')}\Bigg), \quad (3.58)$$

as $|\mathrm{Im}\alpha'| > A$ and $|\mathrm{Im}\alpha| > B$, with positive real constants A, B, a_1, $b_1 > 0$, $c > 0$ and d, while the source term is $O(1/\cos(a_1\alpha))$ as $|\mathrm{Im}\alpha| \to \infty$.

The kernels in (3.53)–(3.54) are then square integrable as $\mathrm{Im}(k\Delta) < 0$, that is, $|\arg(ik)| < \pi/2$ when $k\Delta \neq 0$, and we can consider the coupled equations (3.53)–(3.54) with:

$$\int_{-\infty}^{+\infty} |\frac{S_b^-(ix)}{(\sin(ix) - \sin\theta_1)}|^2 dx < \infty, \int_{-\infty}^{+\infty} |\frac{S_a^+(ix)}{(\sin(ix) + \sin\theta_1)}|^2 dx < \infty, \quad (3.59)$$

as a system of Fredholm equations of second kind in $L^2(iR) \times L^2(iR)$. Besides, we note that the solution of the functional equations (3.35)–(3.38) when $k\Delta = 0$ is the one for an impedance wedge, uniquely defined from [18] with [24].

We now explain how to extend, by a slight deformation of the integration path, the domain where our equations can be qualified of L_2 Fredholm type, in order to enclose the case $|\arg(ik)| = \pi/2$.

3.6.4. *Modification of integration path, and extension including k real*

3.6.4.1. *The new path of integration C_ϵ*

We note that the exponential term $e^{-ik\Delta\cos\alpha'}$ in the kernels of (3.45)–(3.46), and of (3.53)–(3.54), vanish at infinity when $\alpha' \to \pm i\infty$ as $|\arg(ik)| < \pi/2$, $k\Delta \neq 0$.

However, it is possible to deform the domain of integration (possible by analyticity) so that $e^{-ik\Delta\cos\alpha'} \to 0$ at infinity on the new path, even when k is real, providing to not capture any pole during the deformation, knowing that the first singularities of integrands are for $|\text{Re}\alpha| \geq |\text{Re}(\theta_1)| \neq 0$. Therefore, we now show that, for this extension including k real, we can modify the path, following:

$$I = \int_{-i\infty}^{+i\infty} K(\alpha',\alpha)v(\alpha')d\alpha' \to I = \int_{C_\epsilon} K(\alpha',\alpha)v(\alpha')d\alpha', \tag{3.60}$$

by changing the initial path $\alpha' = ix'$ with $x' \in R$ for the path C_ϵ with $\alpha' = g_\epsilon(x')$ given by:

$$\begin{aligned} &C_\epsilon \equiv \{\alpha' = g_\epsilon(x') \text{ with } g_\epsilon(x') = ix' + \epsilon\tanh x',\ x' \in R,\ |\epsilon| < |\text{Re}(\theta_1)|\}, \\ &\text{as } 0 < |\epsilon| \leq \text{sign}(\epsilon)\arg(ik\Delta) < \pi/2 + |\epsilon|/N, \epsilon \in R, k\Delta \neq 0, N > 1. \end{aligned} \tag{3.61}$$

where the case $|\epsilon| = 0$ will be considered in the sense of the limit.

3.6.4.2. *Vanishing properties on C_ϵ and analysis of Re$(ik\cos(g_\epsilon(x)))$*

To show how $e^{-ik\Delta\cos\alpha'} \to 0$ at infinity on C_ϵ, we have to analyze the behavior of $\text{Re}(ik\cos(g_\epsilon(x)))$ as $|x| \to \infty$. We can write:

$$\begin{aligned} &ik\cos(g_\epsilon(x)) = |ik|(e^{-x+i(\arg(ik)+\epsilon\tanh x)} + e^{x+i(\arg(ik)-\epsilon\tanh x)})/2, \\ &\text{Re}(i\frac{k}{|k|}\cos(g_\epsilon(x))) = \cos(\arg(ik) - \epsilon\tanh|x|)e^{|x|}(\frac{1+e^{-2|x|}\frac{\cos(\arg(ik)+\epsilon\tanh|x|)}{\cos(\arg(ik)-\epsilon\tanh|x|)}}{2}). \end{aligned} \tag{3.62}$$

As $0 < |\epsilon| \leq \text{sign}(\epsilon)\arg(ik\Delta) \leq \pi/2$, we then note that we have:

$$\begin{aligned} &\cos(\arg(ik) - \epsilon\tanh|x|) > 0,\ \partial_{|x|}\left(\frac{\cos(\arg(ik) + \epsilon\tanh|x|)}{\cos(\arg(ik) - \epsilon\tanh|x|)}\right) \leq 0, \\ &\left|\frac{\cos(\arg(ik) + \epsilon\tanh|x|)}{\cos(\arg(ik) - \epsilon\tanh|x|)}\right| \leq 1, \end{aligned} \tag{3.63}$$

while, as $\pi/2 < \text{sign}(\epsilon)\arg(ik\Delta) < \pi/2 + |\epsilon|/N$, when $\tanh|x| \geq \tanh|x_{M,N}| = M/N$ with $1 < M < N$, we have:

$$\begin{aligned} &\cos(\arg(ik) - \epsilon\tanh|x|) > 0,\ \partial_{|x|}\left(\frac{\cos(\arg(ik) + \epsilon\tanh|x|)}{\cos(\arg(ik) - \epsilon\tanh|x|)}\right) > 0, \\ &\frac{\sin(\epsilon(1+M)/N)}{\sin(\epsilon(1-M)/N)} < \frac{\cos(\arg(ik) + \epsilon\frac{M}{N})}{\cos(\arg(ik) - \epsilon\frac{M}{N})} < \frac{\cos(\arg(ik) + \epsilon\tanh|x|)}{\cos(\arg(ik) - \epsilon\tanh|x|)} < -1, \\ &1 + e^{-2|x|}\frac{\cos(\arg(ik) + \epsilon\tanh|x|)}{\cos(\arg(ik) - \epsilon\tanh|x|)} > 1 - e^{-2|x|}\left|\frac{\sin(\epsilon(M+1)/N)}{\sin(\epsilon(M-1)/N)}\right|. \end{aligned} \tag{3.64}$$

Therefore, we can write in whole generality that, as $0 < |\epsilon| \leq \text{sign}(\epsilon)\arg(ik\Delta) < \pi/2 + |\epsilon|/N$, when $|x| \geq \max(|x_{M,N}|, \frac{1}{2}\ln|\frac{\sin(\epsilon(M+1)/N)}{\sin(\epsilon(M-1)/N)}|) = a_{\epsilon,N}$ with $1 < M < N$, we have:

$$\text{Re}(ik\cos(g_\epsilon(x))) > |k|\cos(\arg(ik) - \epsilon\tanh|x|)\frac{e^{|x|}}{2}(1 - e^{-2(|x|-a_{\epsilon,N})}) > 0$$

$$\text{with } \frac{e^{|x|}}{2}(1 - e^{-2(|x|-a_{\epsilon,N})}) = e^{a_{\epsilon,N}}\sinh(|x'|-a_{\epsilon,N}). \tag{3.65}$$

Finally, considering freely M and N such that $1 < M < N$, the term $e^{-ik\Delta\cos\alpha'}$ vanishes at infinity on the path $\mathcal{C}_\epsilon$ defined in (3.61), by verifying, as $|x'| \geq a_{\epsilon,N}$:

$$|e^{-ik\Delta\cos\alpha'}| \leq e^{-\gamma(k\Delta,x')e^{a_{\epsilon,N}}\sinh(|x'|-a_{\epsilon,N})},$$

$$\text{with } \gamma(k\Delta, x') = \text{Re}(ik\Delta e^{-i\epsilon\tanh|x'|}) > 0, \tag{3.66}$$

when ϵ and $\arg(ik\Delta)$ $(= \arg(ik))$ satisfy:

$$0 < |\epsilon| \leq \text{sign}(\epsilon)\arg(ik\Delta) < \pi/2 + |\epsilon|/N,$$

$$\text{with } N > 1,\ k\Delta \neq 0 \text{ and } 0 < |\epsilon| < |\text{Re}(\theta_1)|. \tag{3.67}$$

Correlatively, at N, M and ϵ fixed, for a $\mathcal{C}_\epsilon$ given, we define reciprocally:

$$\text{the set } \mathcal{T}_\epsilon \text{ of } k\Delta \neq 0 \text{ with } \arg(ik\Delta) \text{ verifying (3.67)}, \tag{3.68}$$

for which, the exponential term vanishes at infinity on $\mathcal{C}_\epsilon$ following (3.66). So defined, we notice that, for $k\Delta$ belonging to $\mathcal{T}_\epsilon$, we have:

$$\gamma(k\Delta, x') \text{ increases, as } |\arg(ik\Delta)| \text{ decreases or as } |x'| \text{ increases,}$$

$$\text{and as } |k\Delta| \text{ increases when } |x'| \geq a_{\epsilon,N}. \tag{3.69}$$

NOTE 3.4.– We note that, as $|\arg(ik)| \leq \pi/2$:

$$|e^{-ik\Delta\cos\alpha'}| \leq e^{-\gamma(k\Delta,x')\sinh|x'|} \text{ with } \gamma(k\Delta, x') = \text{Re}(ik\Delta e^{-i\epsilon\tanh|x'|}). \tag{3.70}$$

3.6.4.3. *Analyticity of integral expressions of f_{br} and f_{ar} with the new path $\mathcal{C}_\epsilon$*

The integrals (3.45) and (3.46) with new path of integration $\mathcal{C}_\epsilon$ give us expressions of:

$$\frac{f_{br}(\alpha)}{\Psi_{+1}(\alpha)} \text{ and } \frac{f_{ar}(\alpha)}{\Psi_{1-}(\alpha)}, \tag{3.71}$$

which are regular, respectively, as:

$$\text{Re}(g_\epsilon(\text{Im}(\alpha))) - \Phi_+ < \text{Re}(\alpha) < \text{Re}(g_\epsilon(\text{Im}(\alpha))) + 3\Phi_+,$$

$$\text{and } \text{Re}(g_\epsilon(\text{Im}(\alpha))) - 3\Phi_- < \text{Re}(\alpha) < \text{Re}(g_\epsilon(\text{Im}(\alpha))) + \Phi_-, \tag{3.72}$$

with $\text{Re}(g_\epsilon(\text{Im}(\alpha))) = \epsilon \tanh(\text{Im}(\alpha))$, $|\text{Re}(g_\epsilon(\text{Im}(\alpha)))| \leq |\epsilon|$, where we note that the functions $f_{br,ar}(\alpha)$ are $O(1)$ at infinity, with $f_{br,ar}(-i\infty) = -f_{br,ar}(+i\infty)$. These functions satisfy, by analytic continuation of integrals, the functional equations (3.35)–(3.38), and are meromorphic functions of the complex plane.

NOTE 3.5.– The continuation of complex integrals consists of adding residues terms when poles cross integration paths as parameters vary, and/or in deforming integration paths to avoid singularities. Functional equations can complete the continuation.

3.6.4.4. *Integral equations on* C_ϵ *as* $|\epsilon| \leq$*sign*$(\epsilon)\arg(ik\Delta) < \pi/2 + |\epsilon|/N$

Changing the path for C_ϵ, the integral term I in (3.60) now becomes:

$$I = \int_{-\infty}^{+\infty} (K(g_\epsilon(x'), g_\epsilon(x))v(\alpha(x'))\partial_{x'} g_\epsilon(x'))dx', \tag{3.73}$$

for $|\epsilon| \leq \text{sign}(\epsilon)\arg(ik\Delta) < \pi/2 + |\epsilon|/N$, $N > 1$, with $k\Delta \neq 0$ and $0 < |\epsilon| < |\text{Re}(\theta_1)|$. So, replacing the integration path by C_ϵ in the integral equations (3.53) and (3.54), we obtain:

$$\frac{S_a^+(g_\epsilon(x))}{(\sin(g_\epsilon(x)) + \sin\theta_1)} = \frac{-i}{8\Phi_+}\int_{-\infty}^{+\infty} \frac{S_b^-(g_\epsilon(x'))}{(\sin(g_\epsilon(x')) - \sin\theta_1)}$$

$$K_{b,a}(g_\epsilon(x'), g_\epsilon(x))\partial_{x'} g_\epsilon(x')dx' + \frac{1}{2}[\Psi_{+1}(g_\epsilon(x) + \frac{\pi}{2} - \frac{\Phi_b}{2})\chi_b^i(g_\epsilon(x) + \frac{\pi}{2} - \frac{\Phi_b}{2})$$

$$+ \frac{(\sin(g_\epsilon(x)) - \sin\theta_1)\Psi_{+1}(-\alpha + \frac{\pi}{2} - \frac{\Phi_b}{2})}{(\sin(g_\epsilon(x)) + \sin\theta_1)}\chi_b^i(-g_\epsilon(x) + \frac{\pi}{2} - \frac{\Phi_b}{2})], \tag{3.74}$$

and

$$\frac{S_b^-(g_\epsilon(x))}{(\sin(g_\epsilon(x)) - \sin\theta_1)} = \frac{i}{8\Phi_-}\int_{-\infty}^{+\infty} \frac{S_a^+(g_\epsilon(x'))}{(\sin(g_\epsilon(x')) + \sin\theta_1)}$$

$$K_{a,b}(g_\epsilon(x'), g_\epsilon(x))\partial_{x'} g_\epsilon(x')dx' + \frac{1}{2}[\Psi_{1-}(g_\epsilon(x) - \frac{\pi}{2} + \frac{\Phi_a}{2})\chi_a^i(g_\epsilon(x) - \frac{\pi}{2} + \frac{\Phi_a}{2})$$

$$+ \frac{(\sin(g_\epsilon(x)) + \sin\theta_1)\Psi_{1-}(-g_\epsilon(x) - \frac{\pi}{2} + \frac{\Phi_a}{2})}{(\sin(g_\epsilon(x)) - \sin\theta_1)}\chi_a^i(-g_\epsilon(x) - \frac{\pi}{2} + \frac{\Phi_a}{2})], \tag{3.75}$$

for $x \in R$, with $g_\epsilon(x) = ix + \epsilon \tanh x$. Following (3.57), the dependence on $k\Delta$ of kernels is only in the exponential term, and we can express them as:

$$K_{a,b}(g_\epsilon(x'), g_\epsilon(x))\partial_{x'} g_\epsilon(x') = k_{a,b}(g_\epsilon(x'), g_\epsilon(x))\partial_{x'} g_\epsilon(x')e^{-ik\Delta\cos(g_\epsilon(x'))}, \tag{3.76}$$

where the $k_{a,b}(g_\epsilon(x'), g_\epsilon(x))$ are independent of $k\Delta$, and the exponential term is an entire function of $k\Delta$. Considering (3.55) and (3.56), the $k_{a,b}$ are analytic functions of $k\Delta$, regular and bounded by constants independent of $k\Delta$, which verify:

$$|k_{a,b}(g_\epsilon(x'), g_\epsilon(x))\partial_{x'} g_\epsilon(x')| = O\Big(\frac{\Big(\frac{\cosh(a_1 x)\cosh(a_1 x')}{\cosh(a_1 x)+c\cosh(a_1 x')} + \frac{d\cosh(a_1 x)}{1+b_1\cosh(x)}\Big)}{\cosh(a_1 x) + c\cosh(a_1 x')}\Big), \tag{3.77}$$

as $|x| > A$ and $|x'| > A'$, for some positive real constants A, A', a_1, $b_1 > 0$, $c > 0$ and d independent of $k\Delta$, while we note that the source terms are $O(1/\cosh(ax))$ as $|x| \to \infty$.

3.6.4.5. *Fredholmness on $\mathcal{C}_\epsilon \times \mathcal{C}_\epsilon$, in $L^2(R) \times L^2(R)$*

Let us consider $\mathcal{T}_\epsilon$ defined with (3.68)–(3.69), the domain of $k\Delta$ where the integral equations can be considered on $C_\epsilon \times C_\epsilon$. We note that, for $k\Delta$ belonging to any closed subset T of $\mathcal{T}_\epsilon$, it exists two complex values $\xi_{1,2}$ of $\mathcal{T}_\epsilon$ with $0 < |\xi_1| < |\xi_2|$, $|\arg(i\xi_1)| = |\arg(i\xi_2)|$, such that:

$$|\arg(ik\Delta)| \leq |\arg(i\xi_1)|,\ |\xi_1| \leq |k\Delta| \leq |\xi_2|. \tag{3.78}$$

We can then write in T that, from (3.66) to (3.69), we have as $|x'| \geq a_{\epsilon,N}$:

$$|e^{-ik\Delta\cos(g_\epsilon(x'))}| \leq e^{-\gamma(k\Delta,x')e^{a_{\epsilon,N}}\sinh(|x'|-a_{\epsilon,N})} \leq e^{-\gamma(\xi_1,x')e^{a_{\epsilon,N}}\sinh(|x'|-a_{\epsilon,N})},$$

$$\gamma(\xi_1, x') = \text{Re}(i\xi_1 e^{-i\epsilon\tanh|x'|}) > 0, \gamma(\xi_1, x') > \gamma(\xi_1, a_{\epsilon,N}) > 0, \tag{3.79}$$

and, as $|x'| \leq a_{\epsilon,N}$,

$$|e^{-ik\Delta\cos(g_\epsilon(x'))}| \leq C_{\xi_{1,2}} \leq C_{\xi_{1,2}} e^{-\gamma(\xi_1,a_{\epsilon,N})e^{a_{\epsilon,N}}\sinh(|x'|-a_{\epsilon,N})}, \tag{3.80}$$

where $C_{\xi_{1,2}}$ only depends on ϵ, ξ_1, ξ_2 and $a_{\epsilon,N}$. Therefore, we obtain the following property:

$$|e^{-ik\Delta\cos(g_\epsilon(x'))}| \leq D_{\xi_{1,2}} e^{-\gamma_1\sinh(|x'|-a_{\epsilon,N})},$$
$$\text{with } D_{\xi_{1,2}} = \max(1, C_{\xi_{1,2}}), \gamma_1 = \gamma(\xi_1, a_{\epsilon,N})e^{a_{\epsilon,N}} > 0, \tag{3.81}$$

where $D_{\xi_{1,2}}$ and γ_1 depend only on the choice of the subset T.

After some elementary calculations, we then derive for $k\Delta$ belonging to T:

$$\int_{-\infty}^{+\infty}\int_{-\infty}^{+\infty} |K_{a,b}(g_\epsilon(x'), g_\epsilon(x))\partial_{x'} g(x')|^2 dx' dx \leq B_0 < \infty,$$

$$\int_{-\infty}^{+\infty} |K_{a,b}(g_\epsilon(x'), g_\epsilon(x))\partial_{x'} g(x')|^2 dx' \leq F_{1,a,b}(x) \leq F_0(x),$$

$$\int_{-\infty}^{+\infty} |K_{a,b}(g_\epsilon(x'), g_\epsilon(x))\partial_{x'} g(x')|^2 dx \leq F_{2,a,b}(x') \leq F_0(x'),$$

$$\int_{-\infty}^{+\infty} F_{1,a,b}(x)dx \leq B_0, \int_{-\infty}^{+\infty} F_{2,a,b}(x')dx' \leq B_0, \int_{-\infty}^{+\infty} F_0(x)dx = B_0, \tag{3.82}$$

where, considering (3.77) and using (3.81), we take $F_0 = \sum_{j,a,b} F_{j,a,b}$, with F_1 and F_2 the functions only depending on T, positive and definite on real axis, given by:

$$F_1(x) = D^2_{\xi_{1,2}} \int_{-\infty}^{+\infty} |k_{a,b}(g_\epsilon(x'), g_\epsilon(x))\partial_{x'} g(x')|^2 e^{-2\gamma_1 \sinh(|x'|-a_{\epsilon,N})} dx',$$

$$F_2(x') = D^2_{\xi_{1,2}} e^{-2\gamma_1 \sinh(|x'|-a_{\epsilon,N})} |\partial_{x'} g(x')|^2 \int_{-\infty}^{+\infty} |k_{a,b}(g_\epsilon(x'), g_\epsilon(x))|^2 dx. \tag{3.83}$$

We have then satisfied the conditions, from the theory of integral equations depending analytically on a parameter [27] to satisfying Tamarkin theorem [28], when extended to systems (note: $(Id - \mu A)^{-1} = Id + \mu(Id - \mu A)^{-1}A$ in general). This allows us to assume the meromorphicity for the resolvent, usually considered for Fredholm-type equations of second kind in $L^2(R) \times L^2(R)$ (here $\mathcal{C}_\epsilon \times \mathcal{C}_\epsilon$ for $\alpha = ix + \epsilon \tanh x$, $x \in R$), here for $k\Delta \in \mathcal{T}_\epsilon$ when $|\epsilon| \leq$sign$(\epsilon) \arg(ik\Delta) < \pi/2 + |\epsilon|/N$, with $0 < |\epsilon| < |\text{Re}(\theta_1)|$, $k\Delta \neq 0$ and $N > 1$, which includes in particular the case $k\Delta$ real.

3.7. Existence and uniqueness for the integral equations on $\mathcal{C}_\epsilon$

3.7.1. *Uniqueness for the integral equations as* $|\arg(ik\Delta)| \leq \pi/2$

Concerning uniqueness, we consider the solution of integral equations (3.74)–(3.75) in $L^2(R)$ along integration path $\mathcal{C}_\epsilon \equiv \{\alpha = ix + \epsilon \tanh x, x \in R\}$ when χ^i_a and χ^i_b are null (we have no source terms), with $|\epsilon| \leq$sign$(\epsilon) \arg(ik\Delta) \leq \pi/2$, as $0 < |\epsilon| < |\text{Re}(\theta_1)|$, $k\Delta \neq 0$.

3.7.1.1. f^a *and* f^b *from* $S^+_a(\alpha')$ *and* $S^-_b(\alpha')$ *when* $\chi^i_a = \chi^i_b = 0$

Let us take $S^+_a(\alpha')$ and $S^-_b(\alpha')$ derived from (3.74) to (3.75) when $\chi^i_a = \chi^i_b = 0$. Considering the same kernels as the ones used in (3.45)–(3.46), we now define two functions $f^{br}$and f^{ar} following:

$$\frac{f^{br}(\alpha)}{\Psi_{+1}(\alpha)} = \frac{i}{4\Phi_+} \int_{-i\infty-\epsilon}^{+i\infty+\epsilon} \frac{S^-_b(\alpha')e^{-ik\Delta\cos\alpha'} \tan(\frac{\pi}{4\Phi_+}(\alpha - \Phi_+ - \alpha'))}{(\sin\alpha' - \sin\theta_1)\Psi_{+1}(\alpha' - \Phi_+)} d\alpha'$$

$$= \frac{-i}{4\Phi_+} \int_{-i\infty-\epsilon}^{+i\infty+\epsilon} \frac{S^-_b(\alpha')/(\sin\alpha' - \sin\theta_1)}{\Psi_{+1}(\alpha' - \Phi_+)} \frac{e^{-ik\Delta\cos\alpha'} \sin(\frac{\pi\alpha'}{2\Phi_+})}{\cos(\frac{\pi}{2\Phi_+}(\alpha - \Phi_+)) + \cos(\frac{\pi\alpha'}{2\Phi_+})} d\alpha', \tag{3.84}$$

and

$$\frac{f^{ar}(\alpha)}{\Psi_{1-}(\alpha)} = \frac{-i}{4\Phi_-}\int_{-i\infty-\epsilon}^{+i\infty+\epsilon} \frac{S_a^+(\alpha')e^{-ik\Delta\cos\alpha'}\tan(\frac{\pi}{4\Phi_-}(\alpha+\Phi_- -\alpha'))}{(\sin\alpha'+\sin\theta_1)\Psi_{1-}(\alpha'+\Phi_-)}d\alpha'$$

$$= \frac{i}{4\Phi_-}\int_{-i\infty-\epsilon}^{+i\infty+\epsilon} \frac{S_a^+(\alpha')/(\sin\alpha'+\sin\theta_1)}{\Psi_{1-}(\alpha'+\Phi_-)} \frac{e^{-ik\Delta\cos\alpha'}\sin(\frac{\pi\alpha'}{2\Phi_-})}{\cos(\frac{\pi}{2\Phi_-}(\alpha+\Phi_-))+\cos(\frac{\pi\alpha'}{2\Phi_-})}d\alpha', \tag{3.85}$$

These expressions are definite in the respective bands:

$$\begin{aligned} &\mathrm{Re}(g_\epsilon(\mathrm{Im}(\alpha))) - \Phi_+ < \mathrm{Re}(\alpha) < \mathrm{Re}(g_\epsilon(\mathrm{Im}(\alpha))) + 3\Phi_+, \\ &\mathrm{Re}(g_\epsilon(\mathrm{Im}(\alpha))) - 3\Phi_- < \mathrm{Re}(\alpha) < \mathrm{Re}(g_\epsilon(\mathrm{Im}(\alpha))) + \Phi_-, \end{aligned} \tag{3.86}$$

where, from analytical properties of kernels, the functions f^{br} and f^{ar} are analytical functions, $O(1)$ at infinity and verifying $f^{ar(br)}(i\infty) + f^{ar(br)}(-i\infty) = 0$. Considering the integral equations (3.74)–(3.75) when χ_a^i and χ_b^i are null, satisfied by S_a^+ and S_b^-,, we have:

$$\begin{aligned} f^{ar}(\alpha - \frac{\pi}{2} + \frac{\Phi_a}{2}) + \frac{(\sin\alpha+\sin\theta_1)f^{ar}(-\alpha-\frac{\pi}{2}+\frac{\Phi_a}{2})}{(\sin\alpha-\sin\theta_1)} &= \frac{2S_b^-(\alpha)}{(\sin\alpha-\sin\theta_1)}, \\ f^{br}(\alpha + \frac{\pi}{2} - \frac{\Phi_b}{2}) + \frac{(\sin\alpha-\sin\theta_1)f^{br}(-\alpha+\frac{\pi}{2}-\frac{\Phi_b}{2})}{(\sin\alpha+\sin\theta_1)} &= \frac{2S_a^+(\alpha)}{(\sin\alpha+\sin\theta_1)}, \end{aligned} \tag{3.87}$$

which implies, from meromorphic character of kernels, that f^{ar} and f^{br} are also meromorphic functions of the complex plane (note: analytic continuation is done by adding residues terms associated with poles crossing the integration path as α varies). We can then consider functional equations attached to (3.45)–(3.46). We then deduce, after using elementary properties of kernels, that $f^b(\alpha) = f^{br}(\alpha - \frac{\Phi_b}{2})$ and $f^a(\alpha) = f^{ar}(\alpha + \frac{\Phi_a}{2})$, satisfy, in place of $f_b(\alpha) = f_{br}(\alpha - \frac{\Phi_b}{2})$ and $f_a(\alpha) = f_{ar}(\alpha + \frac{\Phi_a}{2})$, the functional equations (3.29)–(3.30) with (3.28), and thus the condition (3.34) which implies, in particular, that f^a and f^b refer to the same field. Considering the zone of regularity of integral expressions, their asymptotics and their analytical continuation, the conditions (a) and (b) are satisfied for f^a and f^b.

Therefore, defined from S_a^+ and S_b^-, given by (3.74)–(3.75) when $\chi_a^i = \chi_b^i = 0$, we then conclude that:

f^a and f^b are two spectral functions refering to the representation
of the same field u with two distinct origins of coordinates, which
satisfies the boundary conditions for the impedance three-part polygon,
and the elementary conditions (a'), (b') and (c') without illumination. (3.88)

3.7.1.2. *Vanishing property of f^a and f^b, then uniqueness*

We now consider the following properties:

(i) In the absence of source terms, from uniqueness theorem on the field (see appendix C), the field u and its normal derivative shall be null on the surface of impedance polygons.

(ii) Due to single-face expression of spectral functions (3.7)–(3.8), the functions f^a and f^b vanish when the field and its normal derivative are null on the surface of the polygon.

After use of (i) for (ii), we can assert that the spectral functions f^a and f^b verifying (3.88), and thus S_b^- and S_a^+ satisfying (3.87), have to be null, which leads us to conclude, by definition, that the solutions of our spectral integral equations (3.74)–(3.75) when $\chi_a^i = \chi_b^i = 0$ are too. This property implies by linearity that we have uniqueness of solutions for our system of spectral integral equations (3.74)–(3.75) in $L^2(R) \times L^2(R)$ as $|\arg(ik)| \leq \pi/2$, $\Delta \neq 0$.

NOTE 3.6.– The solution is also uniquely defined in the case $\Delta = 0$ (simple wedge) from Thuzhilin's theory [24], when conditions (a) and (b) are satisfied.

3.7.2. *Existence and uniqueness of solution*

Concerning existence, we remind that the kernels are analytic functions of $k\Delta$ on $\mathcal{T}_\epsilon$ verifying (3.82). Considering then the theory of integral equations depending on a parameter in $L^2(R)$ (from Smirnov [27] or Tamarkin [28]), the resolvent for our system of Fredholm integral equations of second kind in $L^2(R)\times L^2(R)$ can be written in the form of $\mathcal{R}(x, x', k\Delta) = \mathcal{D}_1(x, x', k\Delta, 1)/D_2(k\Delta,1)$, where $\mathcal{D}_1(x, x', k\Delta, \mu)$ and $D_2(k\Delta,\mu)$ are given by entire series of μ, which are regular with respect to $k\Delta$ in $\mathcal{T}_\epsilon$, $k\Delta \neq 0$. This defines a unique solution in $L^2(R) \times L^2(R)$, except if $D_2(k\Delta, 1) \equiv 0$ or if $D_2(k\Delta, 1) = 0$ for some discrete values of $k\Delta$. In both cases where the resolvent is singular, the solution is not unique and we have value of $k\Delta$ where the integral equations without source terms has a non-trivial (non null) solution.

Therefore, considering the uniqueness of our spectral integral equations on $C_\epsilon \times C_\epsilon$ obtained independently in the previous section, we deduce that a unique solution of (3.74)–(3.75) exists in $L^2(R) \times L^2(R)$, as $0 < |\epsilon| \leq \text{sign}(\epsilon)\arg(ik\Delta) \leq \pi/2$, with $|\epsilon| < |\text{Re}(\theta_1)|$. In other words, we have derived that:

$$\begin{aligned}&\text{the solution of (3.74)–(3.75)exists and is unique, when considered}\\&\text{in } L^2(R) \times L^2(R) \text{ on } C_\epsilon \times C_\epsilon, \text{ as } |\epsilon| \leq \text{sign}(\epsilon)\arg(ik\Delta) \leq \pi/2,\\&\text{with } 0 < |\epsilon| < |\text{Re}(\theta_1)|, k\Delta \neq 0, \text{ which includes the case } k \text{ real.}\end{aligned} \quad (3.89)$$

3.8. Some particular features of the system of integral equations for the three-part impedance polygon and their consequences

Our non-singular spectral integral equations can be solved numerically to derive the different elements of radiated fields as the scattering diagram [1] and [3] (see also appendix E), but some suitable manipulations also permit approximations and simplifications in some particular configurations. This then leads us to new types of integral equations with specific remarkable properties, as the decoupling for some particular geometries and the development of new kernels vanishing for small $k\Delta$, while the approximation of integrals for large $k\Delta$ allows the derivation of closed form asymptotics. These are all special features we are going to illustrate.

3.8.1. *Decoupling in the case of unsymmetric three-part impedance planes*

For three-part impedance planes, $\Phi_a = \Phi_b = 0$ and $\Phi_+ = \Phi_- = \frac{\pi}{2}$, and we can use parity of $[f_{ar}/\Psi_{1-}](\alpha - \pi/2)$ and $[f_{br}/\Psi_{+1}](\alpha + \pi/2)$. Considering then that:

$$\frac{\Psi_{1-}(\alpha - \frac{\pi}{2})}{2\Psi_{+1}(\alpha - \frac{\pi}{2})} - \frac{\Psi_{1-}(-\alpha - \frac{\pi}{2})}{2\Psi_{+1}(-\alpha - \frac{\pi}{2})} =$$

$$= \frac{\Psi_{1-}(\alpha - \frac{\pi}{2})}{\Psi_{+1}(\alpha - \frac{\pi}{2})} \frac{\sin\alpha(\sin\theta_- - \sin\theta_1)}{(\sin\alpha + \sin\theta_-)(\sin\alpha - \sin\theta_1)} = \frac{C(\alpha)\sin\alpha(\sin\theta_- - \sin\theta_1)}{\sin\alpha - \sin\theta_1},$$

$$\frac{\Psi_{+1}(\alpha + \frac{\pi}{2})}{2\Psi_{1-}(\alpha + \frac{\pi}{2})} - \frac{\Psi_{+1}(\frac{\pi}{2} - \alpha)}{2\Psi_{1-}(\frac{\pi}{2} - \alpha)} =$$

$$= \frac{\Psi_{+1}(\frac{\pi}{2} - \alpha)}{\Psi_{1-}(\frac{\pi}{2} - \alpha)} \frac{-\sin\alpha(\sin\theta_+ - \sin\theta_1)}{(\sin\alpha + \sin\theta_+)(\sin\alpha - \sin\theta_1)} = \frac{-C(\alpha)\sin\alpha(\sin\theta_+ - \sin\theta_1)}{\sin\alpha - \sin\theta_1}, \tag{3.90}$$

and letting $N(\alpha') = \frac{iC(\alpha')\sin\alpha'/2\pi}{\sin\alpha' - \sin\theta_1}$, we then derive:

$$\frac{(\sin\theta_1 - \sin\theta_+)^{1/2} f_{br}(\alpha + \frac{\pi}{2})}{\Psi_{+1}(\alpha + \frac{\pi}{2})}$$

$$= \eta_m \int_{-i\infty-\epsilon}^{+i\infty+\epsilon} N(\alpha') \frac{(\sin\theta_1 - \sin\theta_-)^{1/2} f_{ar}(\alpha' - \frac{\pi}{2})}{\Psi_{1-}(\alpha' - \frac{\pi}{2})}$$

$$\times \frac{e^{-ik\Delta\cos\alpha'}\sin\alpha'}{\cos\alpha + \cos\alpha'} d\alpha' + (\sin\theta_1 - \sin\theta_+)^{1/2} \chi_b^i(\alpha + \frac{\pi}{2}), \tag{3.91}$$

and

$$\frac{(\sin\theta_1 - \sin\theta_-)^{1/2} f_{ar}(\alpha - \frac{\pi}{2})}{\Psi_{1-}(\alpha - \frac{\pi}{2})}$$
$$= \eta_m \int_{-i\infty-\epsilon}^{+i\infty+\epsilon} N(\alpha') \frac{(\sin\theta_1 - \sin\theta_+)^{1/2} f_{br}(\alpha' + \frac{\pi}{2})}{\Psi_{+1}(\alpha' + \frac{\pi}{2})}$$
$$\times \frac{e^{-ik\Delta\cos\alpha'}\sin\alpha'}{\cos\alpha + \cos\alpha'} d\alpha' + (\sin\theta_1 - \sin\theta_-)^{1/2}\chi_a^i(\alpha - \frac{\pi}{2}), \tag{3.92}$$

where $\eta_m = (\sin\theta_1 - \sin\theta_+)^{1/2}(\sin\theta_1 - \sin\theta_-)^{1/2}$.

We obtain two uncoupled equations by simple addition (S) and substraction (D):

$$(S, D) = \eta_m \int_{-i\infty-\epsilon}^{+i\infty+\epsilon} N(\alpha')((S, -D)\frac{e^{-ik\Delta\cos\alpha'}\sin\alpha'}{\cos\alpha + \cos\alpha'} d\alpha' + (S^i, -D^i). \tag{3.93}$$

3.8.2. *Decoupling for symmetric three-parts polygons ($\Phi_b = \Phi_a$ and $\sin\theta_+ = \sin\theta_-$)*

In this case, the systems (3.74)–(3.75) can also be decoupled. For this, we first express the system of integral equations (3.74)–(3.75) in a new form. We observe that $\Psi_{-1}(\alpha, \Phi_-)$ (also denoted $\Psi_{-1}(\alpha)$) satisfies $\Psi_{-1}(\alpha, \Phi_-) = \Psi_{1-}(-\alpha, \Phi_-)$ (see appendix A). Considering (3.45)–(3.46), we can write:

$$\frac{f_{br}(-\alpha + \frac{\pi}{2} - \frac{\Phi_b}{2})}{\Psi_{+1}(-\alpha + \frac{\pi}{2} - \frac{\Phi_b}{2})} = \frac{-i}{4\Phi_+} \int_{-i\infty-\epsilon}^{+i\infty+\epsilon} \frac{f_{ar}(\alpha' - \frac{\pi}{2} + \frac{\Phi_a}{2})}{\Psi_{+1}(\alpha' - \Phi_+)}$$
$$\times \frac{e^{-ik\Delta\cos\alpha'}\sin(\frac{\pi\alpha'}{2\Phi_+})}{\cos(\frac{\pi}{2\Phi_+}(\alpha + \Phi_b)) + \cos(\frac{\pi\alpha'}{2\Phi_+})} d\alpha' + \chi_b^i(-\alpha + \frac{\pi}{2} - \frac{\Phi_b}{2}),$$

$$\frac{f_{ar}(\alpha - \frac{\pi}{2} + \frac{\Phi_a}{2})}{\Psi_{-1}(-\alpha + \frac{\pi}{2} - \frac{\Phi_a}{2})} = \frac{-i}{4\Phi_-} \int_{-i\infty-\epsilon}^{+i\infty+\epsilon} \frac{f_{br}(-\alpha' + \frac{\pi}{2} - \frac{\Phi_b}{2})}{\Psi_{-1}(\alpha' - \Phi_-)}$$
$$\times \frac{e^{-ik\Delta\cos\alpha'}\sin(\frac{\pi\alpha'}{2\Phi_-})}{\cos(\frac{\pi}{2\Phi_-}(\alpha + \Phi_a)) + \cos(\frac{\pi\alpha'}{2\Phi_-})} d\alpha' + \chi_a^i(\alpha - \frac{\pi}{2} + \frac{\Phi_a}{2}). \tag{3.94}$$

This form is particularly suitable for the case of a symmetric polygon. In this case, $\Phi_+ = \Phi_-$ (i.e. $\Phi_b = \Phi_a$) and $\sin\theta_+ = \sin\theta_-$, so that $\Psi_{+1} = \Psi_{-1}$ and the equations have the same kernels. Thus, by addition and substraction, we derive a system of decoupled equations:

$$\frac{(S,D)(\alpha)}{\Psi_{+1}(-\alpha+\frac{\pi}{2}-\frac{\Phi_b}{2})}=$$

$$=\frac{-i}{4\Phi_+}\int_{-i\infty-\epsilon}^{+i\infty+\epsilon}\frac{\frac{(S,-D)(\alpha')}{\Psi_{+1}(\alpha'-\Phi_+)}e^{-ik\Delta\cos\alpha'}\sin(\frac{\pi\alpha'}{2\Phi_+})}{\cos(\frac{\pi}{2\Phi_+}(\alpha+\Phi_b))+\cos(\frac{\pi\alpha'}{2\Phi_+})}d\alpha'+(S^i,D^i)(\alpha), \quad (3.95)$$

where

$$S(\alpha)=f_{br}(-\alpha+\frac{\pi}{2}-\frac{\Phi_b}{2})+f_{ar}(\alpha-\frac{\pi}{2}+\frac{\Phi_b}{2}),$$
$$D(\alpha)=f_{br}(-\alpha+\frac{\pi}{2}-\frac{\Phi_b}{2})-f_{ar}(\alpha-\frac{\pi}{2}+\frac{\Phi_b}{2}),$$
$$S^i(\alpha)=\chi^i_{br}(-\alpha+\frac{\pi}{2}-\frac{\Phi_b}{2})+\chi^i_{ar}(\alpha-\frac{\pi}{2}+\frac{\Phi_b}{2}),$$
$$D^i(\alpha)=\chi^i_{br}(-\alpha+\frac{\pi}{2}-\frac{\Phi_b}{2})-\chi^i_{ar}(\alpha-\frac{\pi}{2}+\frac{\Phi_b}{2}), \quad (3.96)$$

for $|\varphi_o|<\min(\frac{\pi}{2},\frac{\pi}{2}+\Phi_b)$.

3.8.3. *Partial inversion and new kernels for small* $k\Delta$

Considering the solution known for $k\Delta=0$ [18], we can apply the identity [1]:

$$e^{-ik\Delta\cos\alpha'}=(e^{-ik\Delta\cos\alpha'}-1)+1, \quad (3.97)$$

and invert the part corresponding to unit term, which results in equations with kernels vanishing as $k\Delta\to 0$, suitable for approximations. To simplify the reading, we will detail the method in appendix. In this reduction, we begin by defining the functions:

$$f_{ar0}(\alpha)=[f_{ar}(\alpha)-f_0^{ab}(\alpha-(\Phi_+-\frac{\pi}{2}))],$$
$$f_{br0}(\alpha)=[f_{br}(\alpha)-e^{ik\Delta\sin\varphi_o}f_0^{ab}(\alpha+(\Phi_--\frac{\pi}{2}))],$$
$$\text{with } f_0^{ab}(\alpha,\varphi_o)=\frac{\pi}{2\Phi_d}\frac{\Psi_{+-}(\alpha)\cos(\frac{\pi\varphi_{o,d}}{2\Phi_d})}{\Psi_{+-}(\varphi_{o,d})(\sin(\frac{\pi\alpha}{2\Phi_d})-\sin(\frac{\pi\varphi_{o,d}}{2\Phi_d}))}, \quad (3.98)$$

with $\Phi_+=\frac{\pi}{2}+\frac{\Phi_b}{2}$, $\Phi_-=\frac{\pi}{2}+\frac{\Phi_a}{2}$, where the function $f_0^{ab}(\alpha,\varphi_o)$ corresponds to the solution for $\Delta=0$ [18] (see appendix A), that is, for a simple wedge of exterior angle $2\Phi_d=\Phi_a+\Phi_b+\pi$, with $\varphi_{o,d}=\varphi_o+(\Phi_a-\Phi_b)/2$, $-\Phi_d<\varphi_{o,d}<\Phi_d$. After some developments (given in appendix B), it results in a system of coupled equations with unknowns, kernels and source terms vanishing as $k\Delta\to 0$, following:

$$f_{br0}(\alpha+\frac{\pi}{2}-\frac{\Phi_b}{2})=\frac{1}{8\pi^2}\int_{-i\infty}^{+i\infty}d\alpha'(\frac{\Psi_{1-}(\alpha'+\Phi_-)(B_{b0}(\alpha')-B_{b0}(-\alpha'))}{\sin\alpha'-\sin\theta_1}$$
$$\times M_+(\alpha,\alpha')-\frac{\Psi_{+1}(\alpha'-\Phi_+)(B_{a0}(\alpha')-B_{a0}(-\alpha'))}{\sin\alpha'+\sin\theta_1}M_-(\alpha,\alpha')),$$
$$f_{ar0}(\alpha-\frac{\pi}{2}+\frac{\Phi_a}{2})=\frac{1}{8\pi^2}\int_{-i\infty}^{+i\infty}d\alpha'(\frac{\Psi_{1-}(\alpha'+\Phi_-)(B_{b0}(\alpha')-B_{b0}(-\alpha'))}{\sin\alpha'-\sin\theta_1}$$
$$\times N_+(\alpha,\alpha')-\frac{\Psi_{+1}(\alpha'-\Phi_+)(B_{a0}(\alpha')-B_{a0}(-\alpha'))}{\sin\alpha'+\sin\theta_1}N_-(\alpha,\alpha')), \tag{3.99}$$

where we can write, $M_\pm(\alpha,\alpha')=L_\pm(\alpha+\frac{\pi}{2}+\frac{(\Phi_a-\Phi_b)}{2},\alpha')$, $N_\pm(\alpha,\alpha')=L_\pm(\alpha-\frac{\pi}{2}+\frac{(\Phi_a-\Phi_b)}{2},\alpha')$ with:

$$L_\pm(\alpha,\alpha')=\frac{\pi\sin\alpha'\Psi_{+-}(\alpha)}{2\Phi_d}\int_{-i\infty}^{i\infty}\frac{\cos(\frac{\pi(\varphi'+(\Phi_a-\Phi_b)/2)}{2\Phi_d})}{\Psi_{+-}(\varphi'+(\Phi_a-\Phi_b)/2)}$$
$$\times\frac{(\cos\varphi'+\sin\theta_1)}{\cos(\varphi'\pm\frac{\pi}{2})+\cos\alpha'}\frac{1}{(\sin(\frac{\pi\alpha}{2\Phi_d})-\sin(\frac{\pi(\varphi'+(\Phi_a-\Phi_b)/2)}{2\Phi_d}))}d\varphi', \tag{3.100}$$

while,

$$B_{a0}(\alpha')=\frac{f_{ar0}(\alpha'-\frac{\pi}{2}+\frac{\Phi_a}{2})}{\Psi_{+1}(\alpha'-\Phi_+)}(e^{-ik\Delta\cos\alpha'}-1)$$
$$+\frac{f_0^{ab}(\alpha'-\frac{\pi}{2}+(\Phi_a-\Phi_b)/2,\varphi_\circ)}{\Psi_{+1}(\alpha'-\Phi_+)}(e^{-ik\Delta(\cos\alpha'+\sin\varphi_\circ)}-1)e^{ik\Delta\sin\varphi_\circ},$$
$$B_{b0}(\alpha')=\frac{f_{br0}(\alpha'+\frac{\pi}{2}-\frac{\Phi_b}{2})}{\Psi_{1-}(\alpha'+\Phi_-)}(e^{-ik\Delta\cos\alpha'}-1)$$
$$+\frac{f_0^{ab}(\alpha'+\frac{\pi}{2}-(\Phi_b-\Phi_a)/2,\varphi_\circ)}{\Psi_{1-}(\alpha'+\Phi_-)}(e^{-ik\Delta(\cos\alpha'-\sin\varphi_\circ)}-1), \tag{3.101}$$

with $\Phi_+=\frac{\pi}{2}+\frac{\Phi_b}{2}$, $\Phi_-=\frac{\pi}{2}+\frac{\Phi_a}{2}$, and $\Phi_d=\frac{\pi}{2}+\frac{\Phi_a+\Phi_b}{2}$. In the particular case $\Phi_a=\Phi_b=0$, $\Phi_d=\frac{\pi}{2}$, the functions $L_\pm$ can be simplified so that we recover the expressions found in [2] for the three-part impedance plane.

3.8.4. Asymptotics for large $k\Delta$

We can write, from previous integral expressions in (3.45)–(3.46):

$$\mathcal{R}_b(\alpha)=\frac{-i}{4\Phi_+}m_1(\alpha)\int_{-i\infty}^{+i\infty}\mathcal{R}_a(\alpha')\frac{e^{-ik\Delta\cos\alpha'}\sin(\frac{\pi\alpha'}{2\Phi_+})}{\cos(\frac{\pi}{2\Phi_+}(\alpha-\Phi_b))+\cos(\frac{\pi\alpha'}{2\Phi_+})}d\alpha'$$
$$+m_1(\alpha)\chi_b^i(\alpha+\frac{\pi}{2}-\frac{\Phi_b}{2}), \tag{3.102}$$

as $-\Phi_+ <\text{Re}(\alpha + \frac{\pi}{2} - \frac{\Phi_b}{2}) < 3\Phi_+$, and,

$$\mathcal{R}_a(\alpha) = \frac{i}{4\Phi_-} m_2(\alpha) \int_{-i\infty}^{+i\infty} \mathcal{R}_b(\alpha') \frac{e^{-ik\Delta\cos\alpha'} \sin(\frac{\pi\alpha'}{2\Phi_-})}{\cos(\frac{\pi}{2\Phi_-}(\alpha + \Phi_a)) + \cos(\frac{\pi\alpha'}{2\Phi_-})} d\alpha'$$

$$+ m_2(\alpha)\chi_a^i(\alpha - \frac{\pi}{2} + \frac{\Phi_a}{2}), \tag{3.103}$$

as $-3\Phi_- <\text{Re}(\alpha - \frac{\pi}{2} + \frac{\Phi_a}{2}) < \Phi_-$, for $-\min(\frac{\pi}{2}, \frac{\pi}{2} + \Phi_a) < \varphi_\circ < \min(\frac{\pi}{2}, \frac{\pi}{2} + \Phi_b)$, where:

$$\mathcal{R}_b(\alpha) = \frac{f_{br}(\alpha + \frac{\pi}{2} - \frac{\Phi_b}{2})}{\Psi_{1-}(\alpha + \Phi_-)}, \mathcal{R}_a(\alpha) = \frac{f_{ar}(\alpha - \frac{\pi}{2} + \frac{\Phi_a}{2})}{\Psi_{+1}(\alpha - \Phi_+)},$$

$$m_1(\alpha) = \frac{\Psi_{+1}(\alpha + \frac{\pi}{2} - \frac{\Phi_b}{2})}{\Psi_{1-}(\alpha + \Phi_-)}, m_2(\alpha) = \frac{\Psi_{1-}(\alpha - \frac{\pi}{2} + \frac{\Phi_a}{2})}{\Psi_{+1}(\alpha - \Phi_+)}. \tag{3.104}$$

To consider asymptotic evaluations of $\mathcal{R}_{a,b}(\alpha')$ from the equations (3.102)–(3.104) as $k\Delta$ is large, we can choose to take account or to neglect the influence of complex poles of $\mathcal{R}_{a,b}(\alpha')$ (corresponding to guided waves) in vicinity of the integration path. To simplify the presentation, we develop here the second case and consider the first case in appendix D.

We begin by assuming that the principal contribution of $\mathcal{R}_{a,b}(\alpha')$ comes from the vicinity of the stationary phase point $\alpha' = 0$. Letting $\cos\alpha' = 1 - x^2/2$ with $x = 2\sin(\alpha'/2)$, and taking into account the parity of the integration path, we have:

$$\mathcal{R}_b(\alpha) \sim m_1(\alpha)\chi_b^i(\alpha + \frac{\pi}{2} - \frac{\Phi_b}{2}) + \Big(\frac{m_1(\alpha)(\partial_\alpha \mathcal{R}_a(\alpha)|_{\alpha=0})e^{-ik\Delta}}{2\pi i}$$

$$\times \int_{-i\infty}^{+i\infty} \frac{e^{\frac{ik\Delta x^2}{2}} x^2}{(\frac{\pi}{2\Phi_+})^{-2}(1 + \cos(\frac{\pi}{2\Phi_+}(\alpha - \Phi_b))) - x^2} dx \Big), \tag{3.105}$$

as $-\Phi_+ <\text{Re}(\alpha + \frac{\pi}{2} - \frac{\Phi_b}{2}) < 3\Phi_+$, and:

$$\mathcal{R}_a(\alpha) \sim m_2(\alpha)\chi_a^i(\alpha - \frac{\pi}{2} + \frac{\Phi_a}{2}) - \Big(\frac{m_2(\alpha)(\partial_\alpha \mathcal{R}_b(\alpha)|_{\alpha=0})e^{-ik\Delta}}{2\pi i}$$

$$\times \int_{-i\infty}^{+i\infty} \frac{e^{\frac{ik\Delta x^2}{2}} x^2}{(\frac{\pi}{2\Phi_-})^{-2}(1 + \cos(\frac{\pi}{2\Phi_-}(\alpha + \Phi_a))) - x^2} dx \Big), \tag{3.106}$$

as $-3\Phi_- <\text{Re}(\alpha - \frac{\pi}{2} + \frac{\Phi_a}{2}) < \Phi_-$. After derivation of previous equations, we obtain:

$$(\partial_\alpha \mathcal{R}_b(\alpha)|_{\alpha=0}) \sim \partial_\alpha(m_1(\alpha)\chi_b^i(\alpha + \frac{\pi}{2} - \frac{\Phi_b}{2}))|_{\alpha=0}),$$

$$(\partial_\alpha \mathcal{R}_a(\alpha)|_{\alpha=0}) \sim \partial_\alpha(m_2(\alpha)\chi_a^i(\alpha - \frac{\pi}{2} + \frac{\Phi_a}{2}))|_{\alpha=0}), \tag{3.107}$$

and we get

$$\mathcal{R}_b(\alpha) \sim \frac{m_1(\alpha)(\partial_\alpha(m_2(\alpha)\chi_a^i(\alpha - \frac{\pi}{2} + \frac{\Phi_a}{2}))|_{\alpha=0})e^{-ik\Delta}}{2\pi i}$$

$$\int_{-i\infty}^{+i\infty} \frac{e^{\frac{ik\Delta x^2}{2}} x^2}{(\frac{\pi}{2\Phi_+})^{-2}(1+\cos(\frac{\pi}{2\Phi_+}(\alpha - \Phi_b))) - x^2} dx + m_1(\alpha)\chi_b^i(\alpha + \frac{\pi}{2} - \frac{\Phi_b}{2}), \tag{3.108}$$

as $-\Phi_+ <\mathrm{Re}(\alpha + \frac{\pi}{2} - \frac{\Phi_b}{2}) < 3\Phi_+$, and,

$$\mathcal{R}_a(\alpha) \sim -\frac{m_2(\alpha)(\partial_\alpha(m_1(\alpha)\chi_b^i(\alpha + \frac{\pi}{2} - \frac{\Phi_b}{2}))|_{\alpha=0})e^{-ik\Delta}}{2\pi i}$$

$$\int_{-i\infty}^{+i\infty} \frac{e^{\frac{ik\Delta x^2}{2}} x^2}{(\frac{\pi}{2\Phi_-})^{-2}(1+\cos(\frac{\pi}{2\Phi_-}(\alpha + \Phi_a))) - x^2} dx + m_2(\alpha)\chi_a^i(\alpha - \frac{\pi}{2} + \frac{\Phi_a}{2}), \tag{3.109}$$

as $-3\Phi_- <\mathrm{Re}(\alpha - \frac{\pi}{2} + \frac{\Phi_a}{2}) < \Phi_-$, for $-\min(\frac{\pi}{2}, \frac{\pi}{2} + \Phi_a) < \varphi_o < \min(\frac{\pi}{2}, \frac{\pi}{2} + \Phi_b)$, $\Phi_- = \frac{\pi}{2} + \frac{\Phi_a}{2}$, $\Phi_+ = \frac{\pi}{2} + \frac{\Phi_b}{2}$.

Concerning the diffracted field $u_d^{a,b}(\rho_{a,b}, \varphi_{a,b})$, referring to origins at $\rho_{a,b} = 0$, we have,

$$u_d^{a,b}(\rho_{a,b}, \varphi_{a,b}) \sim \frac{-e^{-i\pi/4 - ik\rho_{a,b}}}{\sqrt{2\pi k \rho_{a,b}}}[f_{a,b}(\pi + \varphi_{a,b}) - f_{a,b}(-\pi + \varphi_{a,b})],$$

$$\text{with } \Psi_{1-}(\alpha + \Phi_-)\mathcal{R}_b(\alpha) = f_{br}(\alpha + \frac{\pi}{2} - \frac{\Phi_b}{2}) = f_b(\alpha + \frac{\pi}{2}),$$

$$\text{and } \Psi_{+1}(\alpha - \Phi_+)\mathcal{R}_a(\alpha) = f_{ar}(\alpha - \frac{\pi}{2} + \frac{\Phi_a}{2}) = f_a(\alpha - \frac{\pi}{2}). \tag{3.110}$$

The determination of $f_{a,b}(\pm\pi + \varphi_{a,b})$ can be reduced, after the use of functional difference equation on $f_{a,b}$, to evaluations of $\mathcal{R}_{a,b}(\alpha)$ for some real $\alpha = \varphi_r$ (see appendix E on scattering diagram), provided we are in the domains of regularity of integral expressions, respectively, $-\Phi_+ <\mathrm{Re}(\varphi_r + \frac{\pi}{2} - \frac{\Phi_b}{2}) < 3\Phi_+$ for $\mathcal{R}_b(\varphi_r)$, and $-3\Phi_- <\mathrm{Re}(\varphi_r - \frac{\pi}{2} + \frac{\Phi_a}{2}) < \Phi_-$ for $\mathcal{R}_a(\varphi_r)$.

For that, we can deform the path of integration to the steepest descent path, while neglecting the contribution of complex poles, and thus derive:

$$\mathcal{R}_b(\varphi_r) \sim \frac{m_1(\varphi_r)\partial_\alpha(m_2(\alpha)\chi_a^i(\alpha - \frac{\pi}{2} + \frac{\Phi_a}{2}))|_{\alpha=0}}{\sqrt{2ik\Delta}}$$

$$\times\, 2e^{-ik\Delta}(v\, e^{v^2}\mathrm{erfc}(v) - \frac{1}{\sqrt{\pi}}) + m_1(\varphi_r)\chi_b^i(\varphi_r + \frac{\pi}{2} - \frac{\Phi_b}{2}), \tag{3.111}$$

as $-\Phi_+ < (\varphi_r + \frac{\pi}{2} - \frac{\Phi_b}{2}) < 3\Phi_+$, $v^2 = ik\Delta(\frac{\pi}{2\Phi_+})^{-2}(1 + \cos(\frac{\pi}{2\Phi_+}(\varphi_r - \Phi_b)))$, and:

$$\mathcal{R}_a(\varphi_r) \sim -\frac{m_2(\varphi_r)\partial_\alpha(m_1(\alpha)\chi_b^i(\alpha + \frac{\pi}{2} - \frac{\Phi_b}{2}))|_{\alpha=0}}{\sqrt{2ik\Delta}}$$

$$\times\, 2e^{-ik\Delta}(w\, e^{w^2}\operatorname{erfc}(w) - \frac{1}{\sqrt{\pi}}) + m_2(\varphi_r)\chi_a^i(\varphi_r - \frac{\pi}{2} + \frac{\Phi_a}{2}), \qquad (3.112)$$

as $-3\Phi_- < (\varphi_r - \frac{\pi}{2} + \frac{\Phi_a}{2}) < \Phi_-$, $w^2 = ik\Delta(\frac{\pi}{2\Phi_-})^{-2}(1 + \cos(\frac{\pi}{2\Phi_-}(\varphi_r + \Phi_a)))$, for $-\min(\frac{\pi}{2}, \frac{\pi}{2} + \Phi_a) < \varphi_o < \min(\frac{\pi}{2}, \frac{\pi}{2} + \Phi_b)$, with $\Phi_+ = \frac{\pi}{2} + \frac{\Phi_b}{2}$ and $\Phi_- = \frac{\pi}{2} + \frac{\Phi_a}{2}$, where:

$$ae^{a^2}\operatorname{erfc}(a) - \frac{1}{\sqrt{\pi}} = \frac{1}{\pi}\int_{-\infty}^{\infty}\frac{-t^2e^{-t^2}}{a^2+t^2}dt = \frac{-1}{2\sqrt{\pi}a^2} + O(\frac{1}{a^4}), \qquad (3.113)$$

when $\mathrm{Re}(a) > 0$. In the right-handed sides, note that the first term corresponds to double diffraction, and the second term to single diffraction (note: see appendix D for contributions of complex poles).

3.9. Exact first order expressions for a small complex cavity in a step, when $\Phi_b = -\Phi_a$ and $\sin\theta_+ = \sin\theta_-$

We consider a step of length Δ, when we have $\Phi_b = -\Phi_a$ in the three-part polygon geometry, so that the right and the left-handed semi-infinite planes are parallel, then we assume a complex cavity in this step. This cavity is composed of n faces, of angles Φ_j and length d_j, between extreme edges 'a' and 'b' (see Figures 3.4 and 3.5).

We will search to express the diffracted part u_d of u for this complex cavity, given by:

$$u_d = \frac{-e^{-i\pi/4-ik\rho_a}}{\sqrt{2\pi k\rho_a}}[f_a(\pi + \varphi_a) - f_a(-\pi + \varphi_a)] + O(1/(k\rho)^{3/2}), \qquad (3.114)$$

when we have impedance boundary conditions and the cavity dimension is small. We will show that, when $\sin\theta_+ = \sin\theta_-$, the first order of u_d in power of k can be determined by only studying difference functional equations, in combination of single-face integral expression of spectral functions, with no need of integral equations. In what follows, we write ρ_a and φ_a without subscript as ρ and φ, and let $\varphi = \varphi' - \Phi_a$, $\varphi_o = \varphi'_o - \Phi_a$.

3.9.1. *Impedance boundary conditions and elementary equations for spectral functions*

The semi-infinite planes, respectively right and left-handed, with relative surface impedances $\sin\theta_-$ and $\sin\theta_+$, and, each face of angle Φ_j, with surface relative impedances $\sin\theta_j$, satisfy respective boundary conditions:

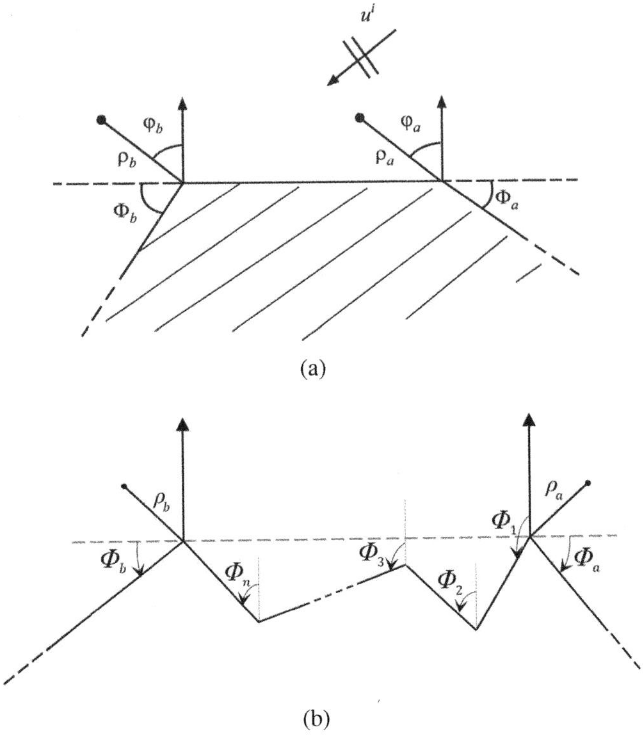

Figure 3.4. *Geometry of complex cavity in a step*

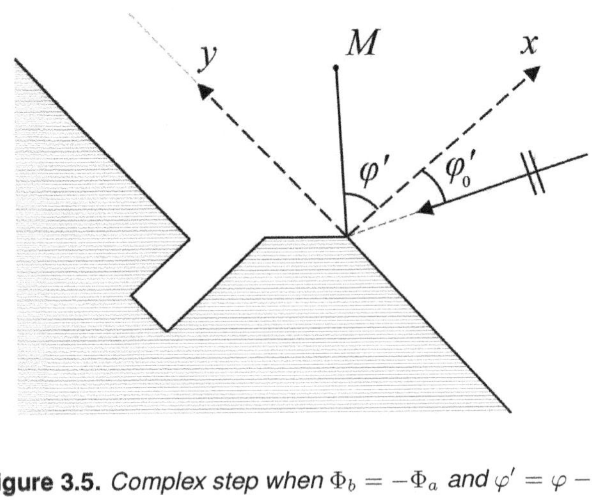

Figure 3.5. *Complex step when $\Phi_b = -\Phi_a$ and $\varphi' = \varphi - \Phi_b$*

$$(\frac{\partial}{\partial n} - ik\sin\theta_{\pm})u = 0, (\frac{\partial}{\partial n} - ik\sin\theta_j)u = 0. \tag{3.115}$$

Using single-face representation of spectral function (3.7) for spectral functions, and noting $f_e^- = f_a$, $f_e^+ = f_b$, we can then write:

$$\begin{aligned}
&(\sin\alpha + \sin\theta_+)f_b(\alpha + \pi/2 + \Phi_b) - (-\sin\alpha + \sin\theta_+)f_b(-\alpha + \pi/2 + \Phi_b) = 0,\\
&(\sin\alpha - \sin\theta_-)f_a(\alpha - \pi/2 - \Phi_a) - (-\sin\alpha - \sin\theta_-)f_a(-\alpha - \pi/2 - \Phi_a) = 0,
\end{aligned} \tag{3.116}$$

where,

$$\begin{aligned}
f_a(\alpha) = &\frac{ik}{2}\sum_{1\leq j\leq n} e^{-ik\sum_{1\leq i<j} d_i\cos(\alpha-\Phi_i)}\\
&\times\int_0^{d_j}(-\sin(\alpha-\Phi_j) + \sin\theta_j)u(\rho_j', \Phi_j)e^{-ik\rho_j'\cos(\alpha-\Phi_j)})d\rho_j'\\
&+ e^{-ik\sum_{1\leq i\leq n} d_i\cos(\alpha-\Phi_i)}f_b(\alpha) = f_1(\alpha) + e^{-ik\Delta\cos(\alpha-\pi/2)}f_b(\alpha),
\end{aligned} \tag{3.117}$$

with,

$$\sum_{1\leq j\leq n} d_j\cos\Phi_j = 0, \sum_{1\leq j\leq n} d_j\sin\Phi_j = \Delta, \tag{3.118}$$

Therefore, we have:

$$\begin{aligned}
&f_b(\alpha) = e^{ik\Delta\cos(\alpha-\pi/2)}(f_a(\alpha) - f_1(\alpha)),\\
&(\sin\alpha + \sin\theta_+)e^{ik\Delta\cos(\alpha+\Phi_b)}(f_a(\alpha + \pi/2 + \Phi_b) - f_1(\alpha + \pi/2 + \Phi_b))\\
&+ (\sin\alpha - \sin\theta_+)e^{ik\Delta\cos(\Phi_b-\alpha)}(f_a(\pi/2 - \alpha + \Phi_b) - f_1(\pi/2 - \alpha + \Phi_b)) = 0,\\
&(\sin\alpha - \sin\theta_-)f_a(\alpha - \pi/2 - \Phi_a) + (\sin\alpha + \sin\theta_-)f_a(-\alpha - \pi/2 - \Phi_a) = 0,
\end{aligned} \tag{3.119}$$

that we can formulate following:

$$\begin{aligned}
&f_a(\alpha + \pi/2 + \Phi_b)\\
&+ \frac{(\sin\alpha - \sin\theta_+)}{(\sin\alpha + \sin\theta_+)}e^{ik\Delta(\cos(\alpha-\Phi_b)-\cos(\alpha+\Phi_b))}f_a(-\alpha + \pi/2 + \Phi_b)\\
&= f_1(\alpha + \pi/2 + \Phi_b)\\
&+ \frac{(\sin\alpha - \sin\theta_+)}{(\sin\alpha + \sin\theta_+)}e^{ik\Delta(\cos(\alpha-\Phi_b)-\cos(\alpha+\Phi_b))}f_1(-\alpha + \pi/2 + \Phi_b),\\
&(\sin\alpha - \sin\theta_-)f_a(\alpha - \pi/2 - \Phi_a) + (\sin\alpha + \sin\theta_-)f_a(-\alpha - \pi/2 - \Phi_a) = 0.
\end{aligned} \tag{3.120}$$

3.9.2. *Reduction of far field function when* $\sin\theta_+ = \sin\theta_-$ *and* $\Phi_a = -\Phi_b$

Considering that $f_1(\alpha) = O(k\Delta)$, we then have:

$$f_a(\pi+\varphi) - \frac{(-\cos(\varphi-\Phi_b)+\sin\theta_+)}{(\cos(\varphi-\Phi_b)+\sin\theta_+)} e^{2ik\Delta\cos(\varphi-\Phi_b)\sin\Phi_b} f_a(-\varphi+2\Phi_b)$$
$$= f_1(\pi+\varphi) + \frac{(\cos(\varphi-\Phi_b)-\sin\theta_+)}{(\cos(\varphi-\Phi_b)+\sin\theta_+)} f_1(-\varphi+2\Phi_b) + O((k\Delta)^2),$$
$$f_a(-\pi+\varphi) - \frac{(\cos(\varphi+\Phi_a)-\sin\theta_-)}{(-\cos(\varphi+\Phi_a)-\sin\theta_-)} f_a(-\varphi-2\Phi_a) = 0, \tag{3.121}$$

and thus,

$$f_a(\pi+\varphi) - f_a(-\pi+\varphi) =$$
$$f_1(\pi+\varphi) + \frac{\cos(\varphi-\Phi_b)-\sin\theta_+}{\cos(\varphi-\Phi_b)+\sin\theta_+} f_1(-\varphi+2\Phi_b)$$
$$- \frac{\cos(\varphi-\Phi_b)-\sin\theta_+}{\cos(\varphi-\Phi_b)+\sin\theta_+} e^{2ik\Delta\cos(\varphi-\Phi_b)\sin\Phi_b} f_a(-\varphi+2\Phi_b)$$
$$+ \frac{\cos(\varphi+\Phi_a)-\sin\theta_-}{\cos(\varphi+\Phi_a)+\sin\theta_-} f_a(-\varphi-2\Phi_a) + O((k\Delta)^2). \tag{3.122}$$

For $\Phi_a = -\Phi_b$, $\sin\theta_+ = \sin\theta_-$, we can then write:

$$f_a(\pi+\varphi) - f_a(-\pi+\varphi)$$
$$= f_1(\pi+\varphi) + \frac{(\cos(\varphi-\Phi_b)-\sin\theta_+)}{(\cos(\varphi-\Phi_b)+\sin\theta_+)} f_1(-\varphi+2\Phi_b)$$
$$+ f_a(-\varphi+2\Phi_b)\frac{(\cos(\varphi-\Phi_b)-\sin\theta_+)}{(\cos(\varphi-\Phi_b)+\sin\theta_+)}(1 - e^{2ik\Delta\cos(\varphi-\Phi_b)\sin\Phi_b})$$
$$+ O((k\Delta)^2), \tag{3.123}$$

then, letting $\varphi' = \varphi - \Phi_b$,

$$f_a(\pi+\varphi'+\Phi_b) - f_a(-\pi+\varphi'+\Phi_b)$$
$$= f_1(\pi+\varphi'+\Phi_b) + \frac{(\cos\varphi'-\sin\theta_+)}{(\cos\varphi'+\sin\theta_+)} f_1(-\varphi'+\Phi_b) + D_u, \tag{3.124}$$

with,

$$D_u = -f_a(-\varphi' + \Phi_b)\frac{(\cos\varphi' - \sin\theta_+)}{(\cos\varphi' + \sin\theta_+)}2ik\Delta\cos\varphi'\sin\Phi_b + o(k\Delta),$$

$$f_a(-\varphi' + \Phi_b) = f_0(-\varphi' + \Phi_b + (\Phi_a - \Phi_b)/2, \varphi_0) + O(k\Delta)$$
$$= f_0(-\varphi', \varphi_0) + O(k\Delta), \tag{3.125}$$

where f_0 is the spectral function for $k\Delta = 0$ given by (3.41) with, in our case, $\Phi_d = \pi/2$ and $\sin\theta_+ = \sin\theta_-$, which gives us:

$$f_0(\alpha, \varphi_\circ) = \frac{(\cos\alpha + \sin\theta_+)\cos\varphi'_\circ}{(\cos\varphi'_\circ + \sin\theta_+)(\sin\alpha - \sin\varphi'_\circ)}. \tag{3.126}$$

This leads us to the following expression of the far field function:

$$f_a(\pi + \varphi' + \Phi_b) - f_a(-\pi + \varphi' + \Phi_b) = \frac{(\cos\varphi' - \sin\theta_+)}{(\cos\varphi' + \sin\theta_+)}f_1(-\varphi' + \Phi_b)$$
$$+ f_1(\pi + \varphi' + \Phi_b) + \frac{(\cos\varphi' - \sin\theta_+)\cos\varphi'_\circ\cos\varphi'}{(\cos\varphi'_\circ + \sin\theta_+)(\sin\varphi' + \sin\varphi'_\circ)}2ik\Delta\sin\Phi_b + o(k\Delta). \tag{3.127}$$

3.9.3. *Exact first-order expressions for f_1 and for the far field function*

We can write:

$$f_1(\alpha) = \frac{ik}{2}\sum_{1\leq j\leq n} e^{-ik\sum_{1\leq i<j} d_i\cos(\alpha-\Phi_i)}\int_0^{d_j}(-\sin(\alpha - \Phi_j)$$
$$+ \sin\theta_j)u(\rho'_j, \Phi_j)e^{-ik\rho'_j\cos(\alpha-\Phi_j)})d\rho'_j \sim \frac{ikM_u}{2}\sum_j d_j(-\sin(\alpha - \Phi_j) + \sin\theta_j), \tag{3.128}$$

for small $k\Delta$, where M_u is a mean value of u for small $k\Delta$. Considering the solution for $k\Delta = 0$, we have $M_u \sim M_u^0 = \frac{2\cos\varphi'_\circ}{(\cos\varphi'_\circ + \sin\theta_+)}$ as $\sin\theta_+ = \sin\theta_-$, and using $\sum_{1\leq j\leq n} d_j\cos\Phi_j = 0$ and $\sum_{1\leq j\leq n} d_j\sin\Phi_j = \Delta$, we have:

$$f_1(\alpha) \sim \frac{ik\cos\varphi'_\circ}{(\cos\varphi'_\circ + \sin\theta_+)}(\Delta\cos\alpha + \sum_j d_j\sin\theta_j). \tag{3.129}$$

We then obtain for (3.127):

$$\begin{aligned}
&f_a(\pi+\varphi'+\Phi_b) - f_a(-\pi+\varphi'+\Phi_b) \\
&\sim \frac{2ik\cos\varphi'_\circ\cos\varphi'}{(\cos\varphi'_\circ+\sin\theta_+)(\cos\varphi'+\sin\theta_+)}(-\Delta\cos\Phi_b\sin\theta_+ + \sin\Phi_b\Delta\sin\varphi' \\
&+\sum_j d_j\sin\theta_j) + \frac{2(\cos\varphi'-\sin\theta_+)\cos\varphi'_\circ\cos\varphi'}{(\cos\varphi'_\circ+\sin\theta_+)(\sin\varphi'+\sin\varphi'_\circ)} ik\Delta\sin\Phi_b \\
&\sim \frac{2ik\cos\varphi'_\circ\cos\varphi'}{(\cos\varphi'_\circ+\sin\theta_+)(\cos\varphi'+\sin\theta_+)}(-\Delta\cos\Phi_b\sin\theta_+ + \sum_j d_j\sin\theta_j) \\
&+\frac{2ik\Delta\sin\Phi_b\cos\varphi'_\circ\cos\varphi'}{(\cos\varphi'_\circ+\sin\theta_+)(\cos\varphi'+\sin\theta_+)(\sin\varphi'+\sin\varphi'_\circ)} \\
&\times(\sin\varphi'(\sin\varphi'+\sin\varphi'_\circ)+(\cos^2\varphi'-\sin^2\theta_+)),
\end{aligned} \tag{3.130}$$

and thus,

$$\begin{aligned}
&f_a(\pi+\varphi'+\Phi_b) - f_a(-\pi+\varphi'+\Phi_b) \sim \frac{2ik\cos\varphi'_\circ\cos\varphi'}{(\cos\varphi'_\circ+\sin\theta_+)(\cos\varphi'+\sin\theta_+)} \\
&\times\left(-\Delta\cos\Phi_b\sin\theta_+ + \sum_j d_j\sin\theta_j + \frac{\Delta\sin\Phi_b(\cos^2\theta_+ + \sin\varphi'\sin\varphi'_\circ)}{(\sin\varphi'+\sin\varphi'_\circ)}\right),
\end{aligned} \tag{3.131}$$

or more generally, letting $\sum_j d_j\sin\theta_j = \int_{O_a}^{O_b}\sin\theta(l)dl$:

$$\begin{aligned}
&f_a(\pi+\varphi'+\Phi_b) - f_a(-\pi+\varphi'+\Phi_b) \sim \frac{2ik\cos\varphi'_\circ\cos\varphi'}{(\cos\varphi'_\circ+\sin\theta_+)(\cos\varphi'+\sin\theta_+)} \\
&\times\left(-\Delta\cos\Phi_b\sin\theta_+ + \int_{O_a}^{O_b}\sin\theta(l)dl + \frac{\Delta\sin\Phi_b(\cos^2\theta_+ + \sin\varphi'\sin\varphi'_\circ)}{(\sin\varphi'+\sin\varphi'_\circ)}\right).
\end{aligned} \tag{3.132}$$

3.9.4. *Exact first-order expression for the diffracted field*

Using (3.132) in (3.114), we then obtain for the diffracted field at large distance:

$$\begin{aligned}
u_d &\sim \frac{-2ik\Delta\cos\varphi'_\circ\cos\varphi' e^{-i\pi/4-ik\rho_a}}{\sqrt{2\pi k\rho_a}(\cos\varphi'_\circ+\sin\theta_+)(\cos\varphi'+\sin\theta_+)}(-\cos\Phi_b\sin\theta_+ + \sin\theta'_1 \\
&+\frac{\sin\Phi_b}{(\sin\varphi'+\sin\varphi'_\circ)}(\cos^2\theta_+ + \sin\varphi'\sin\varphi'_\circ)),
\end{aligned} \tag{3.133}$$

where,

$$l_1 = \sum_j d_j = \int_{O_a}^{O_b} dl, \ \sin\theta_1' = \int_{O_a}^{O_b} \sin\theta(l)dl/\Delta,$$

$$\varphi_a = \varphi' + \Phi_b, \varphi_\circ = \varphi_\circ' + \Phi_b, \Phi_b = -\Phi_a, \sin\theta_+ = \sin\theta_-. \qquad (3.134)$$

It is worth noticing that this expression is the exact first order of the asymptotics for small $k\Delta$, and that it satisfies the reciprocity theorem, that is, it remains the same when we change φ' for $\varphi_\circ'$ (in contrary to the one that we would obtain by physical optics approximation).

3.10. References

[1] Bernard, J.M.L. (2006). A spectral approach for scattering by impedance polygons. *Quart. J Mech. and Appl. Math.*, 59(4), 517–550.

[2] Bernard, J.M.L. (2005). Scattering by a three-part impedance plane: A new spectral approach. *Quart. J. Mech. and Appl. Math.*, 58(3), 383–418.

[3] Lyalinov, M. and Zhu, N.Y. (2021). Scattering of a surface wave in a polygonal domain with impedance boundary. *Algebra i analiz*, 33(2), 98–135.

[4] Fokas, A.S. (2001). Two dimensional linear PDE's in a convex polygon. *Proceedings of the Royal Society A*, 457, 371–393.

[5] Daniele, V. and Lombardi, G. (2020). The Wiener-Hopf theory for the scattering by an impenetrable polygonal structure. *Proceedings of the IEEE International Symposium on Antennas and Propagation*, Toronto.

[6] Shanin, A.V. (2001). Three theorems concerning diffraction by a strip or a slit. *Q. Jl. Mech. Appl. Math.*, 54(1), 107–137.

[7] Craster, R.V., Shanin, A.V., Doubravsky, A.V. (2003). Embedding formulae in diffraction theory. *Proc. R. Soc. Lond. A*, 459, 2475–2496.

[8] Idemen, M. and Alkumru, A. (2003). On a class of functional equations of the Wiener-Hopf type and their applications in n-part scattering problems. *IMA J. Appl. Math.*, 68, 563–586.

[9] Vasil'ev, E.N., Solodukhov, V.V., Makkaveeva, V.F. (1996). Numerical solving of a new class of canonical problems and their application. *Radio Science*, 31(6), 1853–1861.

[10] Noble, B. (1958). *Methods Based on the Wiener-Hopf Techniques*. Pergamon, Oxford.

[11] Craster, R.V. and Shanin, A.V. (2005). Embedding formulae for diffraction by rational wedge and angular geometries. *Proc. Royal Soc. A*, 461(2059), 2227–2242.

[12] Kamotskii, V.V. (2005). An application of the method of spectral function to the problem of scattering by two wedges. *Zap. Nauchn. Sem. S.-Petersburg, Otdel. Mat. Inst. Steklov. (POMI)*, 324, 61–76.

[13] Borovikov, V.A. (1962). Two-dimensional problem of diffraction on a polygon. *Sov. Phys. Dokl.*, 7(6), 499–501.

[14] Budaev, B.V. and Bogy, D.B. (2006). Diffraction by a convex polygon with side-wise constant impedance. *Wave Motion*, 43, 631–645.

[15] Jones, D.S. (1964). *The Theory of Electromagnetism*. Pergamon Press, London.

[16] Felsen, L.B. and Marcuvitz, N. (1973). *Radiation and Scattering of Waves*. Prentice Hall, Englewood Cliffs, NJ.

[17] Maliuzhinets, G.D. (1958). Inversion formula for the Sommerfeld integral. *Sov. Phys. Dokl.*, 3, 52–56.

[18] Maliuzhinets, G.D. (1958). Excitation, reflection and emission of surface waves from a wedge with given face impedances. *Sov. Phys. Dokl.*, 3, 752–755.

[19] Tuzhilin, A.A. (1973). Diffraction of plane sound wave in an angular domain with absolutely hard and slippery face bounded by thin elastic plates. *Diff. Urav.*, 9(10), 1875–1888.

[20] Bernard, J.M.L. (1998). Diffraction at skew incidence by an anisotropic impedance wedge in electromagnetism theory: A new class of canonical cases. *J. Phys. A: Math. Gen.*, 31, 595–613.

[21] Bernard, J.M.L. (2013). On the scattering by a cavity in an impedance plane in three dimensions: Boundary integral equations and novel Green's function. *Jl. Mech. Appl. Math.*, 66(4), 609–639.

[22] Tuzhilin, A.A. (1973). The theory of Maliuzhinets inhomogeneous functional equations. *Diff. Urav.*, 9(11), 2058–2064.

[23] Sakharova, M.P. and Filippov, A.F. (1967). The solution of non-stationary problem of the diffraction at an impedance wedge using tabulated functions. *Zh. vychisl. Mat. Fiz.*, 7(3), 568–579.

[24] Tuzhilin, A.A. (1970). The theory of Maliuzhinets functional equations I. Homogenous functional equations, general properties of solutions, special cases. *Diff. Urav.*, 6, 692–704.

[25] Gradshteyn, I.S. and Ryzhik, I.M. (1980). *Table of Integrals, Series and Products*. Academic Press, London.

[26] Abramowitz, M. and Stegun, I. *Handbook of Mathematical Functions*. Dover Publications Inc., New York.

[27] Smirnov, V.I. (1964). *A Course of Higher Mathematics*. Pergamon Press, Oxford.

[28] Tamarkin, J.D. (1926). On Fredholm's integral equations, whose kernels are analytic in a parameter. *Annals of Math.*, 28(1), 127–152.

[29] Bernard, J.M.L. and Lyalinov, M.A. (2004). Electromagnetic scattering by a smooth convex impedance cone. *IMA Journal of Applied Mathematics*, 69, 285–333.

[30] Levine, L.M. (1964). A uniqueness theorem for the reduced wave equation. *Comm. on Pure and Appl. Math.*, 17, 147–176.

[31] Levine, L.M. (1961). *Some Uniqueness Theorems for the Reduced Wave Equation*. Mathematics Division, Air Force Office of Scientific Research, New York.

[32] Jones, D.S. (1953). The eigenvalues of $\Delta u + k^2 u = 0$ when the boundary conditions are given on semi-infinite domains. *Math. Proc. Cambridge Phil. Soc.*, 49, 668–684.

[33] Colton, D. (1970). Cauchy's problem for almost linear elliptic equations in two independent variables. *J. Approximation Theory*, 3, 66–71.

3.A. Appendix A: About $\Psi_{+-}(\alpha, \Phi)$ and the solution for an impedance wedge

3.A.1. $\Psi_{+-}(\alpha)$ *in passive case*

The function $\Psi_{+-}(\alpha)$, also denoted $\Psi_{+-}(\alpha, \Phi)$, has been defined by Maliuzhinets [18], for the spectral function $f_{+-}(\alpha) = \frac{\Psi_{+-}(\alpha)}{\Psi_{+-}(\varphi')}\sigma(\alpha)$ attached to the representation of the field, in presence of a passive impedance wedge illuminated by a unit plane wave of direction φ'. The spectral function f_{+-} is the solution of the equations:

$$(\sin\alpha + \sin\theta_+)f_{+-}(\alpha + \Phi) - (-\sin\alpha + \sin\theta_+)f_{+-}(-\alpha + \Phi) = 0,$$
$$(\sin\alpha - \sin\theta_-)f_{+-}(\alpha - \Phi) - (-\sin\alpha - \sin\theta_-)f_{+-}(-\alpha - \Phi) = 0, \qquad (3.135)$$

regular in the strip $|\mathrm{Re}\alpha| \leq \Phi$, except for the pole with unit residue at $\alpha = \varphi'$, and $O(1)$ at infinity in this band, where $\sin\theta_+$ (respectively, $\sin\theta_-$) corresponds to the relative impedance attached to the face $\varphi = +\Phi$ (respectively, $\varphi = -\Phi$). The function $\Psi_{+-}(\alpha)$ is the solution of (3.135) without poles or zeros as $|\mathrm{Re}\alpha| \leq \Phi$, when $\mathrm{Re}(\sin\theta_{+,-}) > 0$ with $0 < \mathrm{Re}(\theta_\pm) \leq \pi/2$ (passive case), $O(\cos(\pi\alpha/2\Phi))$ in this band, and $\sigma(\alpha) = \frac{\pi}{2\Phi}\cos(\frac{\pi\varphi'}{2\Phi})/(\sin(\frac{\pi\alpha}{2\Phi}) - \sin(\frac{\pi\varphi'}{2\Phi}))$ has the unit pole for incident field at $\alpha = \varphi'$ in the band $|\mathrm{Re}\alpha| \leq \Phi$ [18].

The function $\Psi_{+-}(\alpha)$ has numerous properties. We have:

$$\begin{aligned}\Psi_{+-}(\alpha) &= A\Psi_\Phi(\alpha + \Phi + (\frac{\pi}{2} - \theta_+))\Psi_\Phi(\alpha - \Phi - (\frac{\pi}{2} - \theta_-))\\ &\times \Psi_\Phi(\alpha + \Phi - (\frac{\pi}{2} - \theta_+))\Psi_\Phi(\alpha - \Phi + (\frac{\pi}{2} - \theta_-)),\\ &= A\Psi_\Phi^4(\frac{\pi}{2})(\cos(\frac{\pi}{4\Phi}(\alpha + \Phi - \theta_+))\cos(\frac{\pi}{4\Phi}(\alpha - \Phi + \theta_-))\\ &\times \frac{\Psi_\Phi(\alpha + \Phi - \frac{\pi}{2} + \theta_+)\Psi_\Phi(\alpha - \Phi + \frac{\pi}{2} - \theta_-)}{\Psi_\Phi(\alpha + \Phi - \frac{\pi}{2} - \theta_+)\Psi_\Phi(\alpha - \Phi + \frac{\pi}{2} + \theta_-)},\end{aligned} \qquad (3.136)$$

where Ψ_Φ is the Maliuzhinets function [18], $0 < \mathrm{Re}(\theta_\pm) \leq \frac{\pi}{2}$, A is an arbitrary constant. This even function satisfies $\Psi_\Phi(\alpha) = A'_0 e^{\mp i\pi\alpha/8\Phi}(1 + O(\alpha^\nu|e^{\pm i\mu'\alpha}|))$ when $\mathrm{Im}\alpha \to \pm\infty$, where A'_0 and ν are constants, $\mu' = \min(\frac{\pi}{2\Phi}, 1)$ (see [23] for more details), and we can write $\Psi_{+-}(\alpha) = A_0\cos(\pi\alpha/2\Phi)(1 + O(\alpha^\nu e^{-\mu'|\mathrm{Im}\alpha|}))$ for $|\mathrm{Im}\alpha|$ large, A_0 a constant. Since $\Psi_\Phi(\alpha) = \Psi_\Phi(-\alpha)$ and $\Psi_\Phi(\alpha + \frac{\pi}{2})\Psi_\Phi(\alpha - \frac{\pi}{2}) = \Psi_\Phi^2(\frac{\pi}{2})\cos(\pi\alpha/4\Phi)$, we have $\Psi_{+-}(-\alpha) = \Psi_{-+}(\alpha)$ and:

$$4\Psi_{+-}(\alpha+\frac{\pi}{2})\Psi_{+-}(\alpha-\frac{\pi}{2}) = A^2\Psi_\Phi^8(\frac{\pi}{2})(\cos(\pi(\frac{\pi}{2}-\theta_+)/2\Phi)-\sin(\pi\alpha/2\Phi))$$
$$\times(\cos(\pi(\frac{\pi}{2}-\theta_-)/2\Phi)+\sin(\pi\alpha/2\Phi)), \tag{3.137}$$

with $A\Psi_\Phi^4(\frac{\pi}{2}) = 2A_0$. Besides, from [27], we can write, for $|\mathrm{Re}\alpha| \leq \Phi + (N+1)\pi$:

$$\Psi_{+-}(\alpha) = B_N \prod_{\pm}\Big(\prod_{m=0}^{N}[\Gamma(\frac{1}{2}+\frac{1}{4\Phi}(\alpha\pm\Phi+\theta_\pm+m\pi))$$
$$\Gamma(\frac{1}{2}-\frac{1}{4\Phi}(\alpha\pm\Phi-(\theta_\pm+m\pi)))\Gamma(\frac{1}{2}+\frac{1}{4\Phi}(\alpha\pm\Phi+\pi-\theta_\pm+m\pi))$$
$$\Gamma(\frac{1}{2}-\frac{1}{4\Phi}(\alpha\pm\Phi-(\pi-\theta_\pm+m\pi)))]^{(-1)^{m+1}}$$
$$\times\exp\Big(\int_0^\infty(-e^{-\nu\pi})^{N+1}\frac{e^{-\nu\theta_\pm}+e^{-\nu(\pi-\theta_\pm)}}{(1+e^{-\nu\pi})}\frac{(1-\cosh(\nu(\alpha\pm\Phi)))}{\nu\sinh(2\nu\Phi)}d\nu\Big)\Big), \tag{3.138}$$

where N is an arbitrary positive integer and B_N is a constant. For the applications, we choose N fixed and define Ψ_{+-} with $B_N = 1$. This expression is suitable for numerical calculus (with $N = 1$ or 2) or to derive the analytical properties of $\Psi_{+-}(\alpha)$ (with $N \to \infty$) from those of Γ.

The zeros which are the closest to the imaginary axis are $\alpha = \Phi + \theta_+$ and $\alpha = -\Phi - \theta_-$, and the closest poles are $\alpha = \Phi + \theta_+ + \pi$ and $\alpha = -\Phi - \theta_- - \pi$.

NOTE 3.7.– Using (3.135) in (3.137), we deduce that:

$$\Psi_{+-}(\alpha\mp(\Phi-\frac{\pi}{2}))\Psi_{+-}(-\alpha\mp(\Phi-\frac{\pi}{2}))$$
$$= A_0^2(\cos(\pi\alpha/2\Phi)+\cos(\pi(\frac{\pi}{2}-\theta_\pm)/2\Phi))(\cos\alpha+\sin\theta_\mp)$$
$$\times\frac{(\cos(\pi\alpha/2\Phi)-\cos(\pi(\frac{\pi}{2}-\theta_\mp)/2\Phi))}{(\cos\alpha-\sin\theta_\mp)}, \tag{3.139}$$

and,

$$\Psi_{+-}(-\alpha\mp(\Phi-\pi)) =$$
$$= A_0^2(\cos(\pi(\alpha\mp\pi/2)/2\Phi)+\cos(\pi(\frac{\pi}{2}-\theta_\pm)/2\Phi))(\pm\sin\alpha+\sin\theta_\mp)$$
$$\times\frac{(\cos(\pi(\alpha\mp\pi/2)/2\Phi)-\cos(\pi(\frac{\pi}{2}-\theta_\mp)/2\Phi))}{\pm\Psi_{+-}(\alpha\mp\Phi)(\sin\alpha\mp\sin\theta_\mp)}, \tag{3.140}$$

then,

$$\frac{\Psi_{+-}(-\alpha \mp (\Phi - \pi))}{\Psi_{+-}(\alpha \mp (\Phi - \pi))} = \frac{(\cos(\pi(\alpha \mp \pi/2)/2\Phi) + \cos(\pi(\frac{\pi}{2} - \theta_{\pm})/2\Phi))}{(\cos(\pi(\alpha \pm \pi/2)/2\Phi) + \cos(\pi(\frac{\pi}{2} - \theta_{\pm})/2\Phi))}$$
$$\times \frac{(\cos(\pi(\alpha \mp \pi/2)/2\Phi) - \cos(\pi(\frac{\pi}{2} - \theta_{\mp})/2\Phi))}{(\cos(\pi(\alpha \pm \pi/2)/2\Phi) - \cos(\pi(\frac{\pi}{2} - \theta_{\mp})/2\Phi))} \frac{(\sin\alpha \pm \sin\theta_{\mp})}{(-\sin\alpha \pm \sin\theta_{\mp})}. \quad (3.141)$$

NOTE 3.8.– From [18], the zeros of Ψ_Φ which are the closest to $\alpha = 0$ and the corresponding poles are the points $\alpha = \pm(\frac{\pi}{2} + 2\Phi)$ and $\alpha = \pm(\frac{3\pi}{2} + 2\Phi)$. In other respects, Ψ_Φ satisfies:

$$\frac{\Psi_\Phi(\alpha + 2\Phi)}{\Psi_\Phi(\alpha - 2\Phi)} = \cot(\frac{\alpha + \frac{\pi}{2}}{2}), \Psi_\Phi(\alpha + \Phi)\Psi_\Phi(\alpha - \Phi) = \Psi_\Phi^2(\Phi)\Psi_{\Phi/2}(\alpha). \quad (3.142)$$

3.A.2. *Some miscellaneous properties in general case (passive or active)*

The solution is more complex when we want to consider arbitrary face constant impedances in whole generality, passive ($0 < \mathrm{Re}(\theta^\pm) \leq \pi/2$) or active ($0 < -\mathrm{Re}(\theta^\pm) \leq \pi/2$). For that, we begin by letting:

$$\Psi_{+-}(\alpha, \sin\theta^+, \sin\theta^-) = \Psi^+(\alpha, \sin\theta^+)\Psi^-(\alpha, \sin\theta^-), \quad (3.143)$$

where the $\Psi^\pm$ verifies:

$$(\sin\alpha \pm \sin\theta_\pm)\Psi^\pm(\alpha \pm \Phi) - (-\sin\alpha \pm \sin\theta_\pm)\Psi^\pm(-\alpha \pm \Phi) = 0,$$
$$\Psi^\pm(\alpha \mp \Phi) - \Psi^\pm(-\alpha \mp \Phi) = 0, \quad (3.144)$$

with Ψ_{+-} remaining without any zero or pole in the band $|\mathrm{Re}\alpha| \leq \Phi$, and thus:

$$\Psi^\pm(\alpha, \sin\theta^\pm) = 1/\Psi^\pm(\alpha, -\sin\theta^\pm) \text{ as } \mathrm{Re}(\theta^\pm) \neq 0. \quad (3.145)$$

The function $\Psi^\pm$ with an active $\sin\theta^\pm$ then directly derives from $\Psi^\pm$ with passive one, and, noting $s^\pm = \mathrm{sign}(\sin\theta^\pm)$, we remark that, when $s^\pm = 1$ (respectively, $s^\pm = -1$), $\Psi^\pm(\alpha,\sin\theta^\pm)$ has zeros (respectively, poles):

$$\alpha = \mp\Phi + \epsilon(s^\pm\theta^\pm + (4p+2)\Phi + 2m\pi),$$
$$\text{and } \alpha = \mp\Phi + \epsilon(\pi - s^\pm\theta^\pm + (4p+2)\Phi + 2m\pi), \quad (3.146)$$

and poles (respectively, zeros) at:

$$\alpha = \mp\Phi + \epsilon(\pi + s^\pm\theta^\pm + (4p+2)\Phi + 2m\pi),$$
$$\text{and } \alpha = \mp\Phi + \epsilon(2\pi - s^\pm\theta^\pm + (4p+2)\Phi + 2m\pi), \quad (3.147)$$

for $m \geq 0, p \geq 0, \epsilon = 1$ and -1, while we have:

$$\Psi^\pm(\alpha) = A_\pm(\cos(\frac{\pi\alpha}{4\Phi}))^{s^\pm}(1 + O(\alpha^\nu e^{-\mu'|\mathrm{Im}\alpha|})), \quad (3.148)$$

for large $|\mathrm{Im}\alpha|$, with constants $A_\pm$ and $\mu' > 0$.

3.A.3. *The solution for the diffraction by a wedge with passive or active impedances*

We can express the spectral solution $f_0(\alpha)$ for the diffraction by a wedge with passive or active impedance faces with the following general definition:

$$
\begin{aligned}
&f_0(\alpha) = \chi_0(\alpha) - \frac{1}{2}(\chi_0(i\infty) + \chi_0(-i\infty)),\\
&\chi_0(\alpha) = \Psi_{+-}(\alpha, \sin\theta^+, \sin\theta^-)P_n(\sin(\mu\alpha))\sigma(\alpha)/\Psi_{+-}(\varphi', \sin\theta^+, \sin\theta^-),\\
&\sigma(\alpha) = \mu\cos(\mu\varphi')/(\sin(\mu\alpha) - \sin(\mu\varphi')),
\end{aligned}
\tag{3.149}
$$

where P_n is a polynomial of order n, $\mu = \pi/2\Phi$, and φ' is the incidence angle. Since we have $f_0(\alpha) \sim 1/(\alpha - \varphi')$ in vicinity of incidence pole $\alpha = \varphi'$, and, from unique definition of u at $\rho = 0$, $f_0(\alpha) = O(1)$ as $|\text{Im}\alpha| \to \infty$, we can write:

$$
P_n(\sin(\mu\varphi')) = 1,\ n = \frac{1}{2}\sum_{\pm}(1 - s^\pm),
\tag{3.150}
$$

with $s^\pm$ =sign(Re(sin$\theta^\pm$)), and thus $n = 0$ when both impedances are passive. In addition, the n zeros of $P_n(\sin(\mu\alpha))$ have to suppress, for each active face with $s^\pm = -1$, the pole $\alpha_s^\pm$ of $\Psi^\pm(\alpha,\sin\theta^\pm)$ with $\pm$sign(arg(ik))Im$\alpha_s^\pm > 0$, among $\alpha_1^\pm = \pm(s^\pm\theta^\pm + \Phi)$ and $\alpha_2^\pm = \pm(\pi - s^\pm\theta^\pm + \Phi)$, that give poles of $f_0(\alpha + \varphi)$ within the region $|\text{Re}\alpha| < \pi$ with $\text{Re}(ik\cos\alpha) > 0$, as $|\text{Re}\varphi| < \Phi$, which leads to cancel non-physical contribution $e^{ik\rho\cos(\alpha_s^\pm - \varphi)}$ that can diverge at infinity as $|\text{Re}\varphi| < \Phi$. We then have:

$$
P_n(\sin(\mu\alpha)) = \prod_{\pm}\left(\frac{(\sin(\mu\alpha) - \sin(\mu\alpha_s^\pm))}{(\sin(\mu\varphi') - \sin(\mu\alpha_s^\pm))}\right)^{\frac{1-s^\pm}{2}},
\tag{3.151}
$$

which completes the definition of f_0 for arbitrary impedance (passive or active).

3.B. Appendix B: principle of semi-inversion for our system of integral equations

We can modify equations and derive integral equations with kernels vanishing as $k\Delta \to 0$ for the three-part semi-infinite impedance polygon [1], which allows approximations when $k\Delta$ is small. For this, we begin with changing the unknowns in the equations, considering the functions, vanishing when $\Delta = 0$:

$$
\begin{aligned}
&f_{ar0}(\alpha) = [f_{ar}(\alpha) - f_0^{ab}(\alpha - (\Phi_+ - \frac{\pi}{2}))],\\
&f_{br0}(\alpha) = [f_{br}(\alpha) - e^{ik\Delta\sin\varphi_o} f_0^{ab}(\alpha + (\Phi_- - \frac{\pi}{2}))],
\end{aligned}
\tag{3.152}
$$

with $\Phi_+ = \frac{\pi}{2} + \frac{\Phi_b}{2}$, $\Phi_- = \frac{\pi}{2} + \frac{\Phi_a}{2}$. The function $f_0^{ab}(\alpha, \varphi_\circ)$ corresponds to the solution for $\Delta = 0$, that is, for a simple wedge of exterior angle $2\Phi_d = \Phi_a + \Phi_b + \pi$, given by:

$$f_0^{ab}(\alpha, \varphi_\circ) = \frac{\pi}{2\Phi_d} \frac{\Psi_{+-}(\alpha)\cos(\frac{\pi\varphi_{\circ,d}}{2\Phi_d})}{\Psi_{+-}(\varphi_{\circ,d})(\sin(\frac{\pi\alpha}{2\Phi_d}) - \sin(\frac{\pi\varphi_{\circ,d}}{2\Phi_d}))}, \tag{3.153}$$

with $\varphi_{\circ,d} = \varphi_\circ + (\Phi_a - \Phi_b)/2$, $-\Phi_d < \varphi_{\circ,d} < \Phi_d$; this, we note, satisfies a system of coupled equations as $\Delta = 0$, deduced from the integral expressions (3.45)–(3.46) for $\Delta = 0$. Multiplying the first equation by$e^{ik\Delta\sin\varphi_\circ}$ and the second one by 1, we can substract them from initial equations derived from (3.45)–(3.46), and then deduce a new system. Therefore, we obtain that the functions f_{ar0} and f_{br0}, which vanish as $\Delta = 0$, satisfy:

$$\begin{aligned}\frac{f_{br0}(\alpha + \frac{\pi}{2} - \frac{\Phi_b}{2})}{\Psi_{+1}(\alpha + \frac{\pi}{2} - \frac{\Phi_b}{2})} &= \frac{-i}{4\Phi_+}\Bigg(\int_{-i\infty}^{+i\infty} \frac{f_{ar0}(\alpha' - \frac{\pi}{2} + \frac{\Phi_a}{2})}{\Psi_{+1}(\alpha' - \Phi_+)} \\ &\times \frac{\sin(\frac{\pi\alpha'}{2\Phi_+})}{\cos(\frac{\pi}{2\Phi_+}(\alpha - \Phi_b)) + \cos(\frac{\pi\alpha'}{2\Phi_+})} d\alpha' \\ &+ \int_{-i\infty}^{+i\infty} \frac{B_{a0}(\alpha')\sin(\frac{\pi\alpha'}{2\Phi_+})}{\cos(\frac{\pi}{2\Phi_+}(\alpha - \Phi_b)) + \cos(\frac{\pi\alpha'}{2\Phi_+})} d\alpha'\Bigg),\end{aligned} \tag{3.154}$$

where,

$$\begin{aligned}B_{a0}(\alpha') &= \frac{f_{ar0}(\alpha' - \frac{\pi}{2} + \frac{\Phi_a}{2})}{\Psi_{+1}(\alpha' - \Phi_+)}(e^{-ik\Delta\cos\alpha'} - 1) \\ &+ \frac{f_0^{ab}(\alpha' - \frac{\pi}{2} + (\Phi_a - \Phi_b)/2, \varphi_\circ)}{\Psi_{+1}(\alpha' - \Phi_+)}(e^{-ik\Delta(\cos\alpha' + \sin\varphi_\circ)} - 1)e^{ik\Delta\sin\varphi_\circ},\end{aligned} \tag{3.155}$$

as $-\Phi_+ <\text{Re}(\alpha + \frac{\pi}{2} - \frac{\Phi_b}{2}) < 3\Phi_+$, and

$$\begin{aligned}\frac{f_{ar0}(\alpha - \frac{\pi}{2} + \frac{\Phi_a}{2})}{\Psi_{1-}(\alpha - \frac{\pi}{2} + \frac{\Phi_a}{2})} &= \frac{i}{4\Phi_-}\Bigg(\int_{-i\infty}^{+i\infty} \frac{f_{br0}(\alpha' + \frac{\pi}{2} - \frac{\Phi_b}{2})}{\Psi_{1-}(\alpha' + \Phi_-)} \\ &\times \frac{\sin(\frac{\pi\alpha'}{2\Phi_-})}{\cos(\frac{\pi}{2\Phi_-}(\alpha + \Phi_a)) + \cos(\frac{\pi\alpha'}{2\Phi_-})} d\alpha' \\ &+ \int_{-i\infty}^{+i\infty} \frac{B_{b0}(\alpha')\sin(\frac{\pi\alpha'}{2\Phi_-})}{\cos(\frac{\pi}{2\Phi_-}(\alpha + \Phi_a)) + \cos(\frac{\pi\alpha'}{2\Phi_-})} d\alpha'\Bigg),\end{aligned} \tag{3.156}$$

where,

$$B_{b0}(\alpha') = \frac{f_{br0}(\alpha' + \frac{\pi}{2} - \frac{\Phi_b}{2})}{\Psi_{1-}(\alpha' + \Phi_-)}(e^{-ik\Delta\cos\alpha'} - 1)$$

$$+ \frac{f_0^{ab}(\alpha' + \frac{\pi}{2} - (\Phi_b - \Phi_a)/2, \varphi_\circ)}{\Psi_{1-}(\alpha' + \Phi_-)}(e^{-ik\Delta(\cos\alpha' - \sin\varphi_\circ)} - 1), \qquad (3.157)$$

as $-3\Phi_- <\text{Re}(\alpha - \frac{\pi}{2} + \frac{\Phi_a}{2}) < \Phi_-$, for $-\min(\frac{\pi}{2}, \frac{\pi}{2} + \Phi_a) < \varphi_\circ < \min(\frac{\pi}{2}, \frac{\pi}{2} + \Phi_b)$, with $\Phi_+ = \frac{\pi}{2} + \frac{\Phi_b}{2}$ and $\Phi_- = \frac{\pi}{2} + \frac{\Phi_a}{2}$.

We then notice some similarity with the equations satisfied when $\Delta = 0$, related to spectral function f_0^{ab} for an impedance wedge. Thus, we let:

$$f'_{br0}(\alpha + \frac{\pi}{2} - \frac{\Phi_b}{2}) = \int_{-i\infty}^{+i\infty} G(\varphi') f_0^{ab}(\alpha + \frac{\pi}{2} + (\Phi_a - \Phi_b)/2, \varphi') d\varphi',$$

$$f'_{ar0}(\alpha - \frac{\pi}{2} + \frac{\Phi_a}{2}) = \int_{-i\infty}^{+i\infty} G(\varphi') f_0^{ab}(\alpha - \frac{\pi}{2} + (\Phi_a - \Phi_b)/2, \varphi') d\varphi', \quad (3.158)$$

as $|\text{Re}(\alpha)| < \frac{\pi}{2}$. The functions $f_0^{ab}(\alpha \pm \frac{\pi}{2} + (\Phi_a - \Phi_b)/2, \varphi')$ are regular and $O(1/\cos(\pi\varphi'/2\Phi_d))$ on the imaginary axis, and a pole at $\varphi' = \alpha \pm \frac{\pi}{2}$ ensures that we have, even if $\Phi_a = \Phi_b = 0$, generally $f'_{br0} \neq f'_{ar0}$.

We search to define $G(\varphi')$, regular in the band $|\text{Re}(\varphi')| \leq \frac{\pi}{2}$, so that $f'_{br0} \equiv f_{br0}$ and $f'_{ar0} \equiv f_{ar0}$ verifies (3.154)–(3.157) for arbitrary Δ. For that, we consider the integral expressions (3.45)–(3.46) for f_{ar} and f_{br} as $\Delta = 0$, then multiply by $G(\varphi')$, integrate following (3.158), and we obtain:

$$\frac{f'_{br0}(\alpha + \frac{\pi}{2} - \frac{\Phi_b}{2})}{\Psi_{+1}(\alpha + \frac{\pi}{2} - \frac{\Phi_b}{2})} = \int_{-i\infty}^{+i\infty} \frac{\frac{-i}{4\Phi_+} f'_{ar0}(\alpha' - \frac{\pi}{2} + \frac{\Phi_a}{2}) \sin(\frac{\pi\alpha'}{2\Phi_+})}{\Psi_{+1}(\alpha' - \Phi_+)(\cos(\frac{\pi}{2\Phi_+}(\alpha - \Phi_b)) + \cos(\frac{\pi\alpha'}{2\Phi_+}))} d\alpha'$$

$$+ \frac{\pi}{2\Phi_+} \int_{-i\infty}^{+i\infty} \frac{G(\varphi')}{\Psi_{+1}(\varphi' - \frac{\Phi_b}{2})} \frac{\sin\frac{\pi(\varphi' + \frac{\pi}{2})}{2\Phi_+}}{(\cos\frac{\pi(\alpha - \Phi_b)}{2\Phi_+} + \cos\frac{\pi(\varphi' + \frac{\pi}{2})}{2\Phi_+})} d\varphi', \qquad (3.159)$$

and,

$$\frac{f'_{ar0}(\alpha - \frac{\pi}{2} + \frac{\Phi_a}{2})}{\Psi_{1-}(\alpha - \frac{\pi}{2} + \frac{\Phi_a}{2})} = \int_{-i\infty}^{+i\infty} \frac{\frac{i}{4\Phi_-} f'_{br0}(\alpha' + \frac{\pi}{2} - \frac{\Phi_b}{2}) \sin(\frac{\pi\alpha'}{2\Phi_-})}{\Psi_{1-}(\alpha' + \Phi_-)(\cos(\frac{\pi}{2\Phi_-}(\alpha + \Phi_a)) + \cos(\frac{\pi\alpha'}{2\Phi_-}))} d\alpha'$$

$$+ \frac{\pi}{2\Phi_-} \int_{-i\infty}^{i\infty} \frac{G(\varphi')}{\Psi_{1-}(\varphi' + \frac{\Phi_a}{2})} \frac{\sin\frac{\pi(\varphi' - \frac{\pi}{2})}{2\Phi_-}}{(\cos\frac{\pi(\alpha + \Phi_a)}{2\Phi_-} + \cos\frac{\pi(\varphi' - \frac{\pi}{2})}{2\Phi_-})} d\varphi'. \qquad (3.160)$$

In the case where $G(\varphi')$ is regular in the band $|\mathrm{Re}(\varphi')| \leq \frac{\pi}{2}$, we can shift the integral paths in the integrals containing $G(\varphi')$. Comparing (3.154)–(3.157) with (3.159)–(3.160), we notice that (f'_{br0}, f'_{ar0}) is a solution of the system of equations (3.154)–(3.157) if G satisfies the conditions:

$$\frac{G(\alpha'+\frac{\pi}{2})}{\Psi_{1-}(\alpha'+\Phi_-)} - \frac{G(-\alpha'+\frac{\pi}{2})}{\Psi_{1-}(-\alpha'+\Phi_-)} = \frac{i}{2\pi}(B_{b0}(\alpha') - B_{b0}(-\alpha')),$$
$$\frac{G(\alpha'-\frac{\pi}{2})}{\Psi_{+1}(\alpha'-\Phi_+)} - \frac{G(-\alpha'-\frac{\pi}{2})}{\Psi_{+1}(-\alpha'-\Phi_+)} = \frac{-i}{2\pi}(B_{a0}(\alpha') - B_{a0}(-\alpha')). \qquad (3.161)$$

Taking account of the properties of Ψ_{+1} and Ψ_{1-} (see appendix A), and letting $G(\alpha') = (\cos\alpha' + \sin\theta_1)g(\alpha')$, (3.161) can be written as:

$$g(\alpha'+\frac{\pi}{2}) - g(-\alpha'+\frac{\pi}{2}) = \frac{i\Psi_{1-}(\alpha'+\Phi_-)(B_{b0}(\alpha') - B_{b0}(-\alpha'))}{2\pi(-\sin\alpha' + \sin\theta_1)},$$
$$g(\alpha'-\frac{\pi}{2}) - g(-\alpha'-\frac{\pi}{2}) = \frac{i\Psi_{+1}(\alpha'-\Phi_+)(B_{a0}(\alpha') - B_{a0}(-\alpha'))}{2\pi(-\sin\alpha' - \sin\theta_1)}. \qquad (3.162)$$

Since $G(\alpha')$ is regular in the band $|\mathrm{Re}\alpha'| \leq \frac{\pi}{2}$ and $\mathrm{Re}(\sin\theta_1) > 0$, $g(\alpha')$ is regular in this band. We can then use (3.42)–(3.43), and write, as $|\mathrm{Re}\alpha| < \frac{\pi}{2}$:

$$g(\alpha) = \frac{i}{4\pi}\int_{-i\infty}^{+i\infty}\Big(\frac{i\Psi_{1-}(\alpha'+\Phi_-)(B_{b0}(\alpha') - B_{b0}(-\alpha'))}{2\pi(\sin\alpha' - \sin\theta_1)}\tan(\frac{(\alpha+\frac{\pi}{2}-\alpha')}{2})$$
$$- \frac{i\Psi_{+1}(\alpha'-\Phi_+)(B_{a0}(\alpha') - B_{a0}(-\alpha'))}{2\pi(\sin\alpha' + \sin\theta_1)}\tan(\frac{(\alpha-\frac{\pi}{2}-\alpha')}{2})\Big)d\alpha', \qquad (3.163)$$

Using (3.158) and (3.163), we obtain the equations with kernels vanishing as $k\Delta \to 0$:

$$f_{br0}(\alpha+\frac{\pi}{2}-\frac{\Phi_b}{2}) = \int_{-i\infty}^{+i\infty}\Big(\frac{\frac{\Psi_{1-}(\alpha'+\Phi_-)}{8\pi^2}(B_{b0}(\alpha') - B_{b0}(-\alpha'))}{\sin\alpha' - \sin\theta_1}M_+(\alpha,\alpha')$$
$$- \frac{\frac{\Psi_{+1}(\alpha'-\Phi_+)}{8\pi^2}(B_{a0}(\alpha') - B_{a0}(-\alpha'))}{\sin\alpha' + \sin\theta_1}M_-(\alpha,\alpha')\Big)d\alpha', \qquad (3.164)$$

and,

$$f_{ar0}(\alpha-\frac{\pi}{2}+\frac{\Phi_a}{2}) = \int_{-i\infty}^{+i\infty}\Big(\frac{\frac{\Psi_{1-}(\alpha'+\Phi_-)}{8\pi^2}(B_{b0}(\alpha') - B_{b0}(-\alpha'))}{\sin\alpha' - \sin\theta_1}N_+(\alpha,\alpha')$$
$$- \frac{\frac{\Psi_{+1}(\alpha'-\Phi_+)}{8\pi^2}(B_{a0}(\alpha') - B_{a0}(-\alpha'))}{\sin\alpha' + \sin\theta_1}N_-(\alpha,\alpha')\Big)d\alpha', \qquad (3.165)$$

where we can write $M_{\pm}(\alpha,\alpha') = L_{\pm}(\alpha + \frac{\pi}{2} + \frac{(\Phi_a - \Phi_b)}{2}, \alpha')$, $N_{\pm}(\alpha,\alpha') = L_{\pm}(\alpha - \frac{\pi}{2} + \frac{(\Phi_a-\Phi_b)}{2}, \alpha')$ with:

$$L_{\pm}(\alpha,\alpha') = \frac{\pi \sin\alpha' \Psi_{+-}(\alpha)}{2\Phi_d} \int_{-i\infty}^{+i\infty} \frac{\cos(\frac{\pi(\varphi' + (\Phi_a - \Phi_b)/2)}{2\Phi_d})}{\Psi_{+-}(\varphi' + (\Phi_a - \Phi_b)/2)}$$

$$\times \frac{(\cos\varphi' + \sin\theta_1)}{\cos(\varphi' \pm \frac{\pi}{2}) + \cos\alpha'} \frac{1}{(\sin(\frac{\pi\alpha}{2\Phi_d}) - \sin(\frac{\pi(\varphi' + (\Phi_a - \Phi_b)/2)}{2\Phi_d}))} d\varphi'. \quad (3.166)$$

In the particular case $\Phi_a = \Phi_b = 0$, $\Phi_d = \frac{\pi}{2}$, the functions $L_{\pm}$ can be simplified so that we recover the expressions found in [2] for the three-part impedance plane (see note 3.9).

NOTE 3.9.– It exists analytical expressions of $L_{\pm}$ when $\Phi_a = \Phi_b = 0$. For this, we consider Ψ_{+-} for $\Phi = \pi/2$, where we have:

$$\frac{(-\sin\alpha + \sin\theta_-)}{\Psi_{+-}(\alpha + \frac{\pi}{2})} - \frac{(\sin\alpha + \sin\theta_-)}{\Psi_{+-}(-\alpha + \frac{\pi}{2})}$$

$$= (\sin\theta_- - \sin\theta_+)(\frac{1}{\Psi_{+-}(\alpha + \frac{\pi}{2})} - \frac{1}{\Psi_{+-}(-\alpha + \frac{\pi}{2})}),$$

$$\frac{(\sin\alpha + \sin\theta_-)}{\Psi_{+-}(\alpha - \frac{\pi}{2})} - \frac{(-\sin\alpha + \sin\theta_-)}{\Psi_{+-}(-\alpha - \frac{\pi}{2})} = 0. \quad (3.167)$$

Using that $\Psi_{+-}(-\alpha)\Psi_{+-}(\alpha) = A_0^2(\cos\alpha + \sin\theta_+)(\cos\alpha + \sin\theta_-)$, and $\Psi_{+-}(\alpha)/\cos\alpha \to A_0$ as $\alpha \to i\infty$, we can let $\chi(\alpha) = \frac{A_0(\cos\alpha + \sin\theta_-)}{\Psi_{+-}(\alpha)}$ in (3.42)–(3.43), and derive:

$$\frac{(\cos\alpha + \sin\theta_-)}{\Psi_{+-}(\alpha)} - \frac{1}{A_0} = \frac{(\sin\theta_- - \sin\theta_+)}{4\pi i} \int_{-i\infty}^{+i\infty} H(\alpha') \tan\Big(\frac{\alpha + \frac{\pi}{2} - \alpha'}{2}\Big) d\alpha',$$

$$\frac{\Psi_{+-}(-\alpha)/A_0^2}{(\cos\alpha + \sin\theta_+)} - \frac{1}{A_0} = \frac{(\sin\theta_- - \sin\theta_+)}{4\pi i} \int_{-i\infty}^{+i\infty} H(\alpha') \tan\Big(\frac{\alpha + \frac{\pi}{2} - \alpha'}{2}\Big) d\alpha', \quad (3.168)$$

for $\frac{-3\pi}{2} < \mathrm{Re}(\alpha) < \frac{\pi}{2}$, where $H(\alpha') = 2\sin\alpha'/(\Psi_{+-}(\alpha' + \frac{\pi}{2})(\sin\alpha' + \sin\theta_+))$ and:

$$H(\alpha) = \frac{1}{\Psi_{+-}(\alpha + \frac{\pi}{2})} - \frac{1}{\Psi_{+-}(-\alpha + \frac{\pi}{2})}$$

$$= \frac{\sin\alpha/\sin\theta_+}{\Psi_{+-}(\alpha + \frac{\pi}{2})} + \frac{\sin\alpha/\sin\theta_+}{\Psi_{+-}(-\alpha + \frac{\pi}{2})}. \quad (3.169)$$

From oddness of $H(\alpha)$, $\tan(\frac{1}{2}(\alpha+\frac{\pi}{2}-\alpha'))$ can be replaced by $\frac{-\sin\alpha'}{(\cos(\alpha+\frac{\pi}{2})+\cos\alpha')}$, and we note that, after deformation of integration path, we have:

$$\int_{-i\infty}^{+i\infty}\frac{1}{\Psi_{+-}(\varphi')}\frac{\cos\varphi'}{\cos\alpha+\sin\varphi'}d\varphi' = -2\pi i\frac{\frac{\sin\alpha+\sin\theta_-}{\Psi(\alpha-\frac{\pi}{2})}-\frac{1}{A_0}}{(\sin\theta_+-\sin\theta_-)},$$

$$\int_{-i\infty}^{+i\infty}\frac{\cos\varphi'}{\Psi_{+-}(\varphi')}\frac{\cos\varphi'}{\cos\alpha+\sin\varphi'}d\varphi' = 2\pi i\sin\theta_+\frac{\frac{\sin\alpha+\sin\theta_-}{\Psi(\alpha-\frac{\pi}{2})}-\frac{1}{A_0}}{(\sin\theta_+-\sin\theta_-)}+\frac{i\pi}{A_0}, \quad (3.170)$$

as $|\mathrm{Re}\alpha| < \frac{\pi}{2}$. Similar expressions with $-\sin\varphi'$ in place of $\sin\varphi'$ can be obtained from $\Psi_{+-}(-\alpha) \rightarrow \Psi_{-+}(\alpha)$ when $\theta_\pm \rightarrow \theta_\mp$ (or by continuation with capture of poles). After noticing that we have:

$$\frac{1}{(z+\sin\varphi')(z'+\sin\varphi')} = (\frac{1}{z+\sin\varphi'}-\frac{1}{z'+\sin\varphi'})\frac{1}{z'-z}, \quad (3.171)$$

the equalities in (3.170) permits us, by elementary combinations, to obtain analytical expressions of $L_\pm$ terms when $\Phi_a = \Phi_b = 0$.

3.C. Appendix C: uniqueness of fields when impedance boundary conditions on piecewise regular geometry

3.C.1. *First Green theorem for uniqueness: nullity of field outside S' when $|\arg(ik)| < \pi/2$, and of field and its normal derivative on regular parts of S' when $|\arg(ik)| = \pi/2$*

We analyze here the uniqueness of solutions for the boundary value problem, that we previously used in [21], in the domain Ω outside a piecewise regular boundary S' in 2D and 3D (finite or infinite), when the field is a real-analytic function of coordinates satisfying the Helmholtz equation in Ω, except possibly at discontinuity of S'. In [30], Levine develops an uniqueness theorem for the scalar acoustic field p, that is, a proof that $p_{inc} \equiv 0$ implies $p \equiv 0$, in the case of a scatterer with impedance boundary conditions on S', when no primary source is present. He considers piecewise $C^{(2+\lambda)}$ boundary S' (with no zero exterior angle), $\lambda > 0$, without auxiliary "edge conditions" at edges or corner points, except that p is continuous. He first studies bounded scatterers, but he also gives, in section 3.7, the elements to generalize his results to scatterers with infinite boundaries and general conical points (detailed in [31]), in particular by the use of Jones' uniqueness theorem [32].

We begin by noticing first that the conditions given by Levine to apply the Green's first theorem are as follows: the scatterer surface is piecewise analytic (with no zero exterior angle), the field is continuous at discontinuities of S' and satisfies impedance boundary conditions:

$$(\partial p/\partial n - ikg(s)p)|_{s\in S'} = 0, \quad (3.172)$$

with Re$(g) \geq 0$, and Re$(g) \neq 0$ on a regular part S_0 of S', the conditions at infinity, no sources outside S', and harmonic dependence on time $e^{i\omega t}$. From Green's first theorem, we can write:

$$\int_{\Omega} (p^*(r)\Delta p(r) + \text{grad}p^*(r)\text{grad}p(r))d\Omega = -\int_{S'} p^*(r)(\widehat{n}\,\text{grad}(p(r)))dS +$$

$$+ \lim_{a\to\infty} \int_{|r|=a,\, r\in\Omega} p^*(r)(\frac{\partial p(r)}{\partial |r|})dS, \tag{3.173}$$

where Ω is the infinite domain outside S' and $\widehat{n}$ is the inward normal to Ω. Besides, from the impedance boundary conditions, we have:

$$\text{Re}(\int_{\Omega} -ik|p(r)|^2 + \frac{|\text{grad}p(r)|^2}{-ik}d\Omega) = \int_{S'} \text{Re}(g)|p(r)|^2 dS + I_\infty, \tag{3.174}$$

where,

$$I_\infty = \text{Re}(\lim_{a\to\infty} I_a) \text{ with } I_a = \int_{|r|=a,\, r\in\Omega} p^*(r)(\frac{\partial p(r)}{-ik\partial |r|})dS \tag{3.175}$$

As $|\arg(ik)| < \pi/2$ (space with losses), we can assume:

$$I_\infty = \text{Re}(\lim_{a\to\infty} O(e^{-\delta a})) = 0 \text{ as } |\arg(ik)| < \pi/2. \tag{3.176}$$

As $|\arg(ik)| = \pi/2$ (without losses), the field at large distance (when no primary source) is principally the sum of a diffracted term radiating longitudinally from origin and a term p_s which includes possible guided wave along S', following:

$$p(r) = F(\frac{r}{|r|})\frac{e^{-ik|r|}}{|r|^{(m-1)/2}} + p_s(\frac{r}{|r|}, |r|) \text{ with } F(\frac{r}{|r|}) = O(1) \text{ in } \Omega,$$

$$|r^{(m-1)/2}p_s(r)| \text{ and } |r^{(m-1)/2}\partial_{|r|}p_s(r)| \to 0 \text{ as } |r| \to \infty \text{ in } \Omega\backslash S',$$

$$|p_s(r)| \text{ and } |\partial_{|r|}p_s(r)| = O(1) \text{ on } S' \text{ as } |r| > a_0, \tag{3.177}$$

with $m = 2$ in 2D and $m = 3$ in 3D, and thus:

$$I_\infty = \lim_{a\to\infty} \int_{|r|=a,\, r\in\Omega} |p(r)|^2 dS > 0. \tag{3.178}$$

For Re$(g) \geq 0$ and $|\arg(ik)| \leq \pi/2$, the left-hand term of (3.174) is negative since Re$(ik) \geq 0$, while the right-handed term is positive, and thus both terms vanish. Consequently, we have, when $|\arg(ik)| < \pi/2$:

$$p(r) = 0 \text{ in } \Omega, \text{ for Re}(g) \geq 0, \tag{3.179}$$

while, when $|\arg(ik)| = \pi/2$, the far field radiation function F like I_∞ is null, the term $\text{Re}(g)|p(r)|^2$ vanishes on S', and thus:

$$\partial_n p(r) = ikg\, p(r) = 0 \text{ on any regular part } S_0 \text{ of } S' \text{ with } \text{Re}(g) > 0. \qquad (3.180)$$

In the latter case, we can use the second Green theorem to conclude that p vanishes in Ω (see section 3.C.2) to complete the proof of uniqueness. We could also use when $S_0 \equiv S'$, as suggested by Levine, the Jones' uniqueness theorem [32] for surfaces conical at infinity (conical outside a sphere of large radius), when the Neumann boundary condition ($\partial_n p(r)|_{S'} = 0$) is satisfied, which implies $p \equiv 0$ in the entire domain Ω. This uniqueness proof is also possible for the electromagnetic field, from the use of expressions with surface tangential components (note: the use of Cauchy–Kovalevskaia theorem, the Colton results for elliptic operator [33], or the demonstration given in [3] lead to the same result).

3.C.2. *Uniqueness for* $|\arg(ik)| \leq \pi/2$ *comes from nullity of field and its normal derivative on a regular part of boundaries*

Let a field p' satisfying $\Delta p' + k^2 p' = 0$ in a subdomain $\Omega_1 \subset \Omega_2$ without sources, bounded by a surface $\partial\Omega_1$, and a Green function $\mathcal{G}$ verifying:

$$(\Delta + k^2)\mathcal{G}(k|r - r'|) = \delta(r - r'), \qquad (3.181)$$

like $\mathcal{G}(k|r|) = -\frac{e^{-ik|r|}}{4\pi|r|}$ as $|r| = \sqrt{x^2+y^2+z^2}$ and $\delta(r) \equiv \delta(x)\delta(y)\delta(z)$ in 3D for $|\arg(ik)| \leq \pi/2$, or like $\mathcal{G}(k|r|) = \frac{i}{4}H^{(2)}(k|r|)$ as $|r| = \sqrt{x^2+y^2}$ and $\delta(r) \equiv \delta(x)\delta(y)$ in 2D. We can then use the second Green theorem (or identity) to express the field p' as a combination of simple and double layer expressions on $\partial\Omega_1$ following:

$$\int_{\Omega_1} (p'\Delta\mathcal{G} - \mathcal{G}\Delta p')d\Omega = -\int_{\partial\Omega_1} (p'\frac{\partial\mathcal{G}}{\partial n} - \mathcal{G}\frac{\partial p'}{\partial n})dS = \left| \begin{matrix} p'(r'), r' \in \Omega_1 \\ 0, r' \notin \Omega_1 \end{matrix} \right., \qquad (3.182)$$

with the normal vector here directed inside $\partial\Omega_1$, where it is worth noticing the nullity outside $\partial\Omega_1$. If we have $\partial_n p(r) = p(r) = 0$ on an analytic part $\mathcal{H}$ of $\partial\Omega_1$, we have a hole in the surface field expression, and we can write:

$$-\int_{\partial\Omega_1 \backslash \mathcal{H}} (p'\frac{\partial\mathcal{G}}{\partial n} - \mathcal{G}\frac{\partial p'}{\partial n})dS = \left| \begin{matrix} p'(r'), r' \in \Omega_1 \\ 0, r' \notin \Omega_1 \end{matrix} \right., \qquad (3.183)$$

which implies by analyticity that nullity outside $\partial\Omega_1$ continues inside $\partial\Omega_1$, that is, $p' = 0$ in Ω_1 and thus everywhere in Ω_2.

3.C.3. *Some extension in 3D electromagnetism*

In electromagnetism, we have similar results. Let us consider fields E and H verifying:

$$\text{rot}(H) = i\omega\epsilon E, \ \text{rot}(E) = -i\omega\mu H,$$
$$(\Delta + k^2)H = (\Delta + k^2)E = 0, \text{div}(E) = \text{div}(H) = 0, \tag{3.184}$$

in a domain Ω_2 without sources, with homogeneous isotropic media of permittivity ϵ and permeability μ, $k = \omega\sqrt{\epsilon\mu}$, and a subdomain $\Omega_1 \subset \Omega_2$. Considering the Helmholtz equations alone, we could write:

$$\int_{\Omega_1} (|_H^E \Delta\mathcal{G} - \mathcal{G}\Delta|_H^E)d\Omega = -\int_{\partial\Omega_1} (|_H^E \frac{\partial\mathcal{G}}{\partial n} - \mathcal{G}\frac{\partial|_H^E}{\partial n})dS_r = \left| \begin{matrix} |_H^E(r'), r' \in \Omega_1 \\ 0, r' \notin \Omega_1 \end{matrix} \right., \tag{3.185}$$

with the normal vector here directed inside $\partial\Omega_1$, but we can also obtain expressions only related to tangential surface components. For that, we use the electromagnetic Green functions E_G and H_G radiated by dirac sources $j_0\delta$ and $m_0\delta$ in Ω_2 that satisfy:

$$\text{rot}(H_G) = i\omega\epsilon E_G + j_0\delta(r - r'), \ \text{rot}(E_G) = -i\omega\mu H_G - m_0\delta(r - r'), \tag{3.186}$$

then the identities $\text{div}(e_l \wedge h_q) = h_q.\text{rot}e_l - e_l.\text{rot}h_q$ and $\int_\Omega \text{div}(a)dV = \int_{\partial\Omega} n_s.a \, dS$, and we obtain:

$$\int_{\partial\Omega_1} (E_G.(n_s \wedge H) + H_G.(n_s \wedge E))dS =$$
$$\int_{\Omega_1} (m_0\delta.H - j_0\delta.E) \, d\Omega = \left| \begin{matrix} m_0.H(r') - j_0.E(r'), r' \in \Omega_1 \\ 0, r' \notin \Omega_1 \end{matrix} \right., \tag{3.187}$$

where n_s is the unit outward normal to Ω_1 ($\widehat{n} = -n_s$ is the unit inward normal to Ω_1). In a similar manner to a scalar case, we then obtain that if the tangential components of E and H vanishes on a regular part of $\partial\Omega_1$, we have a hole in expression (3.187) on $\partial\Omega_1$, and by analyticity, the fields continue to be zero from outside to inside $\partial\Omega_1$, and thus are null in all parts of Ω_2.

NOTE 3.10.– When E_G and H_G are the free space Green functions in electromagnetism, we note that we have:

$$(\Delta + k^2)E_G = (\text{grad}(\text{div}(\frac{ij_0(.)}{\epsilon\omega})) + i\omega\mu j_0(.))\delta + \text{rot}(m_0(.))\delta,$$
$$(\Delta + k^2)H_G = (\text{grad}(\text{div}(\frac{im_0(.)}{\mu\omega})) + i\omega\epsilon m_0(.))\delta - \text{rot}(j_0(.))\delta, \tag{3.188}$$

This implies, after using $\mathcal{G}$ verifying $(\Delta + k^2)\mathcal{G}(k|r - r'|) = \delta(r - r')$, that:

$$E_G = [(\text{grad}_r(\text{div}_r(\frac{ij_0(.)}{\epsilon\omega})) + i\omega\mu j_0(.)) + \text{rot}_r(m_0(.))]\mathcal{G}(k|r - r'|),$$

$$j(r).E_G = [\frac{ij_0}{\epsilon\omega}.(\text{grad}_{r'}(\text{div}_{r'}(j(r)(.))) + k^2 j(r))$$

$$+ m_0.\text{rot}_{r'}(j(r)(.))]\mathcal{G}(k|r - r'|),$$

$$H_G = [(\text{grad}_r(\text{div}_r(\frac{im_0(.)}{\mu\omega})) + i\omega\epsilon m_0(.)) - \text{rot}_r(j_0(.))]\mathcal{G}(k|r - r'|),$$

$$m(r).H_G = [\frac{im_0}{\mu\omega}.(\text{grad}_{r'}(\text{div}_{r'}(m(r)(.)) + k^2 m(r)).$$

$$- j_0[\text{rot}_{r'}(m(r)(.))]\mathcal{G}(k|r - r'|) \qquad (3.189)$$

3.D. Appendix D: asymptotics for $\mathcal{R}_a(\alpha)$ and $\mathcal{R}_b(\alpha)$ taking account of complex poles

When $k\Delta$ is large, the highly oscillatory nature of the exponential terms in kernels leads to simple approximations. We consider (3.102)–(3.104) and use the method of steepest descent path, taking, this time, account of the influence of complex poles of $\mathcal{R}_{a,b}(\alpha')$. During the deformation of the initial path of integration along imaginary axis to SDP, where $\text{Im}(ik(\sin(\alpha'/2))^2) = 0$ and $\text{Re}(ik(\sin(\alpha'/2))^2) \leq 0$, poles of $\mathcal{R}_a(\alpha')$ and $\mathcal{R}_b(\alpha')$ can be captured, and we have:

$$\mathcal{R}_b(\alpha) = \frac{-i}{4\Phi_+} m_1(\alpha)\Big(\int_{SDP} \mathcal{R}_a(\alpha') \frac{e^{-ik\Delta\cos\alpha'}\sin(\frac{\pi\alpha'}{2\Phi_+})}{\cos(\frac{\pi}{2\Phi_+}(\alpha - \Phi_b)) + \cos(\frac{\pi\alpha'}{2\Phi_+})} d\alpha'$$

$$+ 2\pi i \sum_s \frac{e^{-ik\Delta\cos\alpha'_s}\sin(\frac{\pi\alpha'_s}{2\Phi_+})}{\cos(\frac{\pi}{2\Phi_+}(\alpha - \Phi_b)) + \cos(\frac{\pi\alpha'_s}{2\Phi_+})} \epsilon_{a,s}\text{Residue}(\mathcal{R}_a(\alpha'))|_{\alpha' = \alpha_s, \text{Re}\alpha_s < 0}\Big)$$

$$+ m_1(\alpha)\chi_b^i(\alpha + \pi/2 - \Phi_b/2), \qquad (3.190)$$

as $\alpha + \pi/2 - \Phi_b/2$ is in the domain limited by $SDP - \Phi_+$ and $SDP + 3\Phi_+$, and:

$$\mathcal{R}_a(\alpha) = \frac{i}{4\Phi_-} m_2(\alpha)\Big(\int_{SDP} \mathcal{R}_b(\alpha') \frac{e^{-ik\Delta\cos\alpha'}\sin(\frac{\pi\alpha'}{2\Phi_-})}{\cos(\frac{\pi}{2\Phi_-}(\alpha + \Phi_a)) + \cos(\frac{\pi\alpha'}{2\Phi_-})} d\alpha'$$

$$+ 2\pi i \sum_s \frac{e^{-ik\Delta\cos\alpha'_s}\sin(\frac{\pi\alpha'_s}{2\Phi_-})}{\cos(\frac{\pi}{2\Phi_-}(\alpha + \Phi_a)) + \cos(\frac{\pi\alpha'_s}{2\Phi_-})} \epsilon_{b,s}\text{Residue}(\mathcal{R}_b(\alpha'))|_{\alpha' = \alpha_s, \text{Re}\alpha_s > 0}\Big)$$

$$+ m_2(\alpha)\chi_a^i(\alpha - \pi/2 + \Phi_a/2), \qquad (3.191)$$

as $\alpha - \pi/2 + \Phi_a/2$ is in the domain limited by $SDP - 3\Phi_-$ and $SDP + \Phi_-$, where:

$$\mathcal{R}_b(\alpha) = (\frac{f_{br}(\alpha + \pi/2 - \Phi_b/2)}{\Psi_{1-}(\alpha + \Phi_-)}),\ \mathcal{R}_a(\alpha) = (\frac{f_{ar}(\alpha - \pi/2 + \Phi_a/2)}{\Psi_{+1}(\alpha - \Phi_+)}),$$

$$m_1(\alpha) = \frac{\Psi_{+1}(\alpha + \pi/2 - \Phi_b/2)}{\Psi_{1-}(\alpha + \Phi_-)},\ m_2(\alpha) = \frac{\Psi_{1-}(\alpha - \pi/2 + \Phi_a/2)}{\Psi_{+1}(\alpha - \Phi_+)},$$

$$\chi_b^i(\alpha) = \frac{\pi}{2\Phi_+}(\frac{e^{ik\Delta \sin\varphi_o} \cos\frac{\pi\varphi_{o,b}}{2\Phi_+}}{\Psi_{+1}(\varphi_{o,b})(\sin\frac{\pi\alpha}{2\Phi_+} - \sin\frac{\pi\varphi_{o,b}}{2\Phi_+})}),$$

$$\chi_a^i(\alpha) = \frac{\pi}{2\Phi_-}(\frac{\cos\frac{\pi\varphi_{o,a}}{2\Phi_-}}{\Psi_{1-}(\varphi_{o,a})(\sin\frac{\pi\alpha}{2\Phi_-} - \sin\frac{\pi\varphi_{o,a}}{2\Phi_-})}), \tag{3.192}$$

with $\varphi_{o,b} = \varphi_o - \Phi_b/2$, $\varphi_{o,a} = \varphi_o + \Phi_a/2$, $\Phi_- = \frac{\pi}{2} + \frac{\Phi_a}{2}$, $\Phi_+ = \frac{\pi}{2} + \frac{\Phi_b}{2}$, and $\epsilon_{a(b),s} = 1$ or -1 if the pole s is captured, $\epsilon_{a(b),s} = 0$ else.

For large $k\Delta$, the residue terms are principally given by:

$$\text{Residue}(\mathcal{R}_a(\alpha'))|_{\alpha'=\alpha_s} \sim \text{Residue}(m_2(\alpha')\chi_a^i(\alpha' - \pi/2 + \Phi_a/2))|_{\alpha'=-\theta_1},$$
$$\text{Residue}(\mathcal{R}_b(\alpha'))|_{\alpha'=\alpha_s} \sim \text{Residue}(m_1(\alpha')\chi_b^i(\alpha' + \pi/2 - \Phi_b/2))|_{\alpha'=\theta_1}, \tag{3.193}$$

where $\alpha = -\theta_1$ and $\alpha = \theta_1$ are the zeros of $\Psi_{+1}(\alpha - \Phi_+)$ and $\Psi_{1-}(\alpha + \Phi_-)$, the closest to the imaginary axis, that can be captured. Letting $t = -i\sqrt{2ik\Delta}\sin(\alpha'/2)$, we then derive:

$$\mathcal{R}_b(\alpha) \sim \frac{m_1(\alpha)(\partial_\alpha \mathcal{R}_a(\alpha)|_{\alpha=0})}{\sqrt{2ik\Delta}} 2e^{-ik\Delta}(v\, e^{v^2}\text{erfc}(v) - \frac{1}{\sqrt{\pi}})$$
$$+ \frac{2\pi}{4\Phi_+}\frac{e^{-ik\Delta\cos\theta_1}\sin(\frac{-\pi\theta_1}{2\Phi_+})\chi_a^i(-\theta_1 - \frac{\pi}{2} + \frac{\Phi_a}{2})}{\cos(\frac{\pi}{2\Phi_+}(\alpha - \Phi_b)) + \cos(\frac{\pi\theta_1}{2\Phi_+})}\epsilon_a \text{Residue}(m_2(\alpha'))|_{\alpha'=-\theta_1}$$
$$s$$
$$+ m_1(\alpha)\chi_b^i(\alpha + \pi/2 - \Phi_b/2), \tag{3.194}$$

with $v^2 = ik\Delta(\frac{\pi}{2\Phi_+})^{-2}(1 + \cos(\frac{\pi}{2\Phi_+}(\alpha - \Phi_b)))$, $\alpha + \pi/2 - \Phi_b/2$ in the domain limited by $SDP - \Phi_+$ and $SDP + 3\Phi_+$, and:

$$\mathcal{R}_a(\alpha) \sim -\frac{m_2(\alpha)(\partial_\alpha \mathcal{R}_b(\alpha)|_{\alpha=0})}{\sqrt{2ik\Delta}} 2e^{-ik\Delta}(w\, e^{w^2}\text{erfc}(w) - \frac{1}{\sqrt{\pi}})$$
$$- \frac{2\pi}{4\Phi_-}\frac{e^{-ik\Delta\cos\alpha_s'}\sin(\frac{\pi\theta_1}{2\Phi_-})\chi_b^i(\theta_1 + \frac{\pi}{2} - \frac{\Phi_b}{2})}{\cos(\frac{\pi}{2\Phi_-}(\alpha + \Phi_a)) + \cos(\frac{\pi\theta_1}{2\Phi_-})}\epsilon_b \text{Residue}(m_1(\alpha'))|_{\alpha'=\theta_1}$$
$$+ m_2(\alpha)\chi_a^i(\alpha - \pi/2 + \Phi_a/2), \tag{3.195}$$

with $w^2 = ik\Delta(\frac{\pi}{2\Phi_-})^{-2}(1+\cos(\frac{\pi}{2\Phi_-}(\alpha+\Phi_a)))$, $\alpha - \pi/2 + \Phi_a/2$ in the domain limited by $SDP{-}3\Phi_-$ and $SDP{+}\Phi_-$, for $-\min(\pi/2, \pi/2{+}\Phi_a) < \varphi_\circ < \min(\pi/2, \pi/2 + \Phi_b)$.

To obtain $\partial_\alpha \mathcal{R}_b(\alpha)|_{\alpha=0}$ and $\partial_\alpha \mathcal{R}_a(\alpha)|_{\alpha=0}$, we derivate the previous expressions and solve the linear system of equations. We obtain, for large $k\Delta$:

$$\begin{aligned}\partial_\alpha \mathcal{R}_b(\alpha)|_{\alpha=0} &\sim \partial_\alpha(m_1(\alpha)\chi_b^i(\alpha+\pi/2-\Phi_b/2))|_{\alpha=0},\\ \partial_\alpha \mathcal{R}_a(\alpha)|_{\alpha=0} &\sim \partial_\alpha(m_2(\alpha)\chi_a^i(\alpha-\pi/2+\Phi_a/2))|_{\alpha=0}.\end{aligned} \tag{3.196}$$

In consequence, we can write:

$$\begin{aligned}\mathcal{R}_b(\alpha) \sim &\frac{m_1(\alpha)\partial_\alpha(m_2(\alpha)\chi_a^i(\alpha-\frac{\pi}{2}+\frac{\Phi_a}{2}))|_{\alpha=0}}{\sqrt{ik\Delta/2}} e^{-ik\Delta}(v\, e^{v^2}\mathrm{erfc}(v) - \frac{1}{\sqrt{\pi}})\\ &+ \frac{2\pi}{4\Phi_+}\frac{e^{-ik\Delta\cos\theta_1}\sin(\frac{-\pi\theta_1}{2\Phi_+})\chi_a^i(-\theta_1-\frac{\pi}{2}+\frac{\Phi_a}{2})}{\cos(\frac{\pi}{2\Phi_+}(\alpha-\Phi_b))+\cos(\frac{\pi\theta_1}{2\Phi_+})}\epsilon_a \mathrm{Residue}(m_2(\alpha'))|_{\alpha'=-\theta_1}\\ &+ m_1(\alpha)\chi_b^i(\alpha+\pi/2-\Phi_b/2),\end{aligned} \tag{3.197}$$

with $v^2 = ik\Delta(\frac{\pi}{2\Phi_+})^{-2}(1+\cos(\frac{\pi}{2\Phi_+}(\alpha-\Phi_b)))$, $\alpha+\pi/2-\Phi_b/2$ in the domain limited by $SDP - \Phi_+$ and $SDP + 3\Phi_+$, and,

$$\begin{aligned}\mathcal{R}_a(\alpha) \sim &-\frac{m_2(\alpha)\partial_\alpha(m_1(\alpha)\chi_b^i(\alpha+\frac{\pi}{2}-\frac{\Phi_b}{2}))|_{\alpha=0}}{\sqrt{ik\Delta/2}} e^{-ik\Delta}(w\, e^{w^2}\mathrm{erfc}(w) - \frac{1}{\sqrt{\pi}})\\ &- \frac{2\pi}{4\Phi_-}\frac{e^{-ik\Delta\cos\alpha_s'}\sin(\frac{\pi\theta_1}{2\Phi_-})\chi_b^i(\theta_1+\frac{\pi}{2}-\frac{\Phi_b}{2})}{\cos(\frac{\pi}{2\Phi_-}(\alpha+\Phi_a))+\cos(\frac{\pi\theta_1}{2\Phi_-})}\epsilon_b \mathrm{Residue}(m_1(\alpha'))|_{\alpha'=\theta_1}\\ &+ m_2(\alpha)\chi_a^i(\alpha-\pi/2+\Phi_a/2),\end{aligned} \tag{3.198}$$

with $w^2 = ik\Delta(\frac{\pi}{2\Phi_-})^{-2}(1+\cos(\frac{\pi}{2\Phi_-}(\alpha+\Phi_a)))$, $\alpha - \pi/2 + \Phi_a/2$ in the domain limited by $SDP - 3\Phi_-$ and $SDP + \Phi_-$, while $-\min(\pi/2, \pi/2 + \Phi_a) < \varphi_\circ < \min(\pi/2, \pi/2 + \Phi_b)$.

In right-handed sides of previous expressions, the first term corresponds to the double diffraction mechanism, the second term to the influence of guided waves and the third term to the single diffraction mechanism.

NOTE 3.11.–

The integral term is calculated as follows:

$$\frac{i}{\pi}\int_{-\infty}^{\infty}\frac{t^2e^{-t^2}}{z^2-t^2}dt = \frac{i}{\pi}\int_{-\infty}^{\infty}\frac{z^2e^{-t^2}}{z^2-t^2}dt - \frac{i}{\sqrt{\pi}} = ze^{-z^2}\mathrm{erfc}(-iz) - \frac{i}{\sqrt{\pi}}, \tag{3.199}$$

as Im$z > 0$, with erfc$(a) = 1-$erf$(a) \sim e^{-a^2}(1/a - 1/(2a^3))/\sqrt{\pi}$ for large a, erf(a) and erfc(a) being, respectively, the error and the complementary error functions [25], [26].

3.E. Appendix E: the scattering diagram from the solutions of the integral equations

In a previous section, we have reduced the problem to a system of non-singular integral equations. When the solutions of the integral equations are known, the different elements of the decomposition of the field can be evaluated from the integral expressions (3.45)–(3.46), where the integral terms can be considered as smooth coupling terms between both edges. In this case, the difference functional equations (3.28)–(3.30) have to be used to reduce the calculus in the band of validity of integral expressions.

Here, we study the reduction of the field term with radial dependence $\frac{\exp(-ik\rho_{a,b})}{\sqrt{2\pi k\rho_{a,b}}}$ for large $\rho_{a,b}$. Its angular dependence $\mathcal{F}_{a,b}$ in the direction $\varphi_a = \varphi_b = \varphi$ is commonly called the scattering diagram (or directivity) [15][16][29], given from (3.5) by:

$$\mathcal{F}_{a,b}(\varphi) = -e^{-i\pi/4}(f_{a,b}(\pi + \varphi) - f_{a,b}(-\pi + \varphi)), \tag{3.200}$$

where, from (3.25), $\mathcal{F}_a(\varphi) = e^{ik\Delta\sin\varphi}\mathcal{F}_b(\varphi)$.

We illustrate the development by the one of $\mathcal{F}_b(\varphi_b)$ in the case where $\Phi_b > -\frac{\pi}{2}$, $\Phi_a > 0$, and $-\min(\frac{\pi}{2}, \frac{\pi}{2} + \Phi_a) < \varphi_o < \min(\frac{\pi}{2}, \frac{\pi}{2} + \Phi_b)$, when $-\frac{\pi}{2} < \varphi_b < \frac{\pi}{2} + \Phi_b$ (similar to $\mathcal{F}_a(\varphi_a)$ when $-\frac{\pi}{2} < -\varphi_a < \frac{\pi}{2} + \Phi_a$, $\Phi_a > -\frac{\pi}{2}$, $\Phi_b > 0$), as in [1].

We first consider the reduction of $f_b(\pi + \varphi_b)$ in (3.200). If $\Phi_b > 0$, then $-\frac{\pi}{2} < \pi + \varphi_b < \frac{3\pi}{2} + 2\Phi_b$, and we are in the domain of validity of (3.45)–(3.46). If $\Phi_b < 0$, then we have a concave part and there are two cases. If $-\frac{\pi}{2} < \pi + \varphi_b < \frac{3\pi}{2} + 2\Phi_b$, (3.45)–(3.46) directly applies. On the contrary, if $\frac{3\pi}{2} + 2\Phi_b < \pi + \varphi_b < \frac{3\pi}{2} + \Phi_b$, we use (3.30) and write:

$$f_b(\pi + \varphi_b) = \frac{(-\sin(\frac{\pi}{2} - \Phi_b + \varphi_b) + \sin\theta_+)}{(\sin(\frac{\pi}{2} - \Phi_b + \varphi_b) + \sin\theta_+)} f_b(-\varphi_b + 2\Phi_b), \tag{3.201}$$

for which we use (3.28)–(3.29), and thus:

$$f_b(-\varphi_b + 2\Phi_b) = \frac{(\sin(\frac{\pi}{2} - \varphi_b + 2\Phi_b) + \sin\theta_1)}{(-\sin(\frac{\pi}{2} - \varphi_b + 2\Phi_b) + \sin\theta_1)}(f_b(-\pi + \varphi_b - 2\Phi_b)$$
$$- e^{ik\Delta\sin(2\Phi_b - \varphi_b)} f_a(-\pi + \varphi_b - 2\Phi_b)) + e^{ik\Delta\sin(2\Phi_b - \varphi_b)} f_a(-\varphi_b + 2\Phi_b), \tag{3.202}$$

where the terms in the right-handed side of (3.202) can be expressed by the use of (3.45) and (3.46).

Then, we reduce $f_b(-\pi+\varphi_b)$ in (3.200). If $-\frac{\pi}{2}<-\pi+\varphi_b<\frac{3\pi}{2}+2\Phi_b$, that is, $\varphi_b>\frac{\pi}{2}$, which implies $\Phi_b>0$, we are in the domain of validity of (3.45)–(3.46). On the contrary, when $-\frac{\pi}{2}<\varphi_b<\frac{\pi}{2}$, which is obliged when $\Phi_b<0$, we first have to use (3.28)–(3.29):

$$f_b(-\pi+\varphi_b)=\frac{(\sin(\frac{\pi}{2}-\varphi_b)-\sin\theta_1)}{(-\sin(\frac{\pi}{2}-\varphi_b)-\sin\theta_1)}$$
$$\times\,(f_b(-\varphi_b)-e^{-ik\Delta\sin\varphi_b}f_a(-\varphi_b))+e^{-ik\Delta\sin\varphi_b}f_a(-\pi+\varphi_b). \qquad (3.203)$$

In this case, since $-\frac{3\pi}{2}-2\Phi_a<-\varphi_b<\frac{\pi}{2}$ and $-\frac{3\pi}{2}-2\Phi_a<-\pi+\varphi_b<\frac{\pi}{2}$ when $\Phi_a>0$, and $-\frac{\pi}{2}<-\varphi_b<\frac{3\pi}{2}+2\Phi_b$ when $\Phi_b>-\frac{\pi}{2}$, $f_a(-\pi+\varphi_b)$ and $f_a(-\varphi_b)$ can be expressed with (3.46), and $f_b(-\varphi_b)$ with (3.45), which ends the reduction.

3.F. Appendix F: approximated second-order ameliorations for small cavity

3.F.1. *Amelioration of equivalent cavity impedance* $\sin\theta_1'$

Using [21], we can add a second order to $\sin\theta_1'$, following:

$$\sin\theta_1'=\sum_j d_j\sin\theta_j/\Delta+ik(S_+-S_-)/\Delta, \qquad (3.204)$$

where S_+ is the total surface of cavities below the straight line between the exterior discontinuities 'a' and 'b', while S_- is the total surface of cavities above the straight line. In the case where some elements would be filled with materials of impedance $Z_j=\sqrt{\epsilon_2\mu_2}/\epsilon_2$ (mode H when $u\equiv H_z$) or admittance $Y_j=\sqrt{\epsilon_2\mu_2}/\mu_2$ (mode E when $u\equiv E_z$), and with wave number k_m, we can write:

$$\sin\theta_1'|_{\text{mode } H}=\sum_j d_j\sin\theta_j/\Delta+\sum_{\text{m}} i\frac{Z_m}{Z_0}k_m(S_{m+}-S_{m-})/\Delta,$$

$$\sin\theta_1'|_{\text{mode } E}=\sum_j d_j\sin\theta_j/\Delta+\sum_{\text{m}} i\frac{Y_m}{Y_0}k_m(S_{m+}-S_{m-})/\Delta, \qquad (3.205)$$

where the $\sin\theta_j$ remain the relative surface in free space.

3.F.2. *A change of* M_u

We can refine in (3.128), from [1], [2], [21], the expression of the mean value M_u:

$$M_u \sim \frac{(1 - \frac{i}{\pi} B_0 \sin\varphi'_\circ \sin\Phi_b)}{1 + \frac{i}{\pi} B_0(\sin\theta'_1 - \sin\theta_+ \cos\Phi_b)} \frac{2\cos\varphi'_\circ}{(\cos\varphi'_\circ + \sin\theta_+)} = D_0 M_u^0$$

$$B_0 = \int_{-i\infty}^{+i\infty} d\alpha' \frac{(e^{-ik\Delta\cos\alpha'} - 1)}{2\cos\alpha'} \sim -k\Delta(\ln(k\Delta) + \gamma_0 - 1 + i\pi/2)$$

$$D_0 = \frac{(1 - \frac{i}{\pi} B_0 \sin\varphi'_\circ \sin\Phi_b)}{1 + \frac{i}{\pi} B_0(\sin\theta'_1 - \sin\theta_+ \cos\Phi_b)}, \gamma_0 \approx .577, \quad (3.206)$$

and then write, in place of (3.132):

$$f_a(\pi + \varphi' + \Phi_b) - f_a(-\pi + \varphi' + \Phi_b)$$
$$\sim \frac{2ik\Delta\cos\varphi'_\circ \cos\varphi'}{(\cos\varphi'_\circ + \sin\theta_+)(\cos\varphi' + \sin\theta_+)}(-\cos\Phi_b \sin\theta_+ + \sin\Phi_b \sin\varphi'$$
$$+ \sin\theta'_1) \times D_0 + \frac{2ik\Delta\sin\Phi_b(\cos\varphi' - \sin\theta_+)\cos\varphi' \cos\varphi'_\circ}{(\cos\varphi'_\circ + \sin\theta_+)(\sin\varphi' + \sin\varphi'_\circ)}. \quad (3.207)$$

A more refined expression could be found by partial inversion (see section 3.8.3 and appendix B devoted to semi-inversion).

4

Advanced Properties of Spectral Functions in Frequency and Time Domains for Diffraction by a Wedge-shaped Region

4.1. Introduction

The scattering by objects with singular geometries is a particularly delicate problem [1]–[20], [23]–[31], and the Sommerfeld–Maliuzhinets representation of fields, when it is associated with Maliuzhinets inversion [3]–[4] and single-face representation [7]–[8] methods, is a powerful tool to investigate the acoustic and electromagnetic waves diffraction by a complex wedge-shaped region, in particular based on remarkable properties of spectral functions in this representation.

In first step of this chapter, we review some general expressions and properties in scalar case for these problems: general properties in complex plane, spectral representation of Green function, single-face representation of spectral function and consequences, and determination of spectral function from far field (diffraction) coefficient are analyzed for scatterers with imperfectly reflecting surfaces that can be finite or infinite.

In a second step, we analyze more specifically the case where the scatterer is a wedge with straight semi-infinite faces. We begin by considering the general solution for a wedge with faces impedances of arbitrary signs (passive or active case), then the solution for the diffraction of a skew incident plane wave by a passive anisotropic impedance wedge of any angle in vectorial case [9] (2D 1/2 electromagnetism problems), and we detail efficient expansions of special functions used for them and their properties.

Later, an explicitly causal representation of fields in time domain is developed, taking both account of passive and active poles of reflection coefficients, which generalizes previous known works [12]–[14] on impedance wedges, and applies on a large domain of validity, including the general case of a dispersive wedge with multimode boundary conditions.

4.2. Basic properties of spectral function in Sommerfeld–Maliuzhinets representation

4.2.1. *Sommerfeld–Maliuzhinets representation for scattering by a wedge-shaped sector*

Let us consider the case of diffraction in free space of a scalar incident plane wave:

$$u^i(\rho, \varphi) = U^i(\omega) e^{ik\rho\cos(\varphi - \varphi_o)}, \tag{4.1}$$

by a scatterer (finite or infinite) enclosed in a wedge-shaped region, defined in cylindrical coordinates (ρ, φ) as the domain outside the free space angular sector $-\Phi_r \leq \varphi \leq \Phi_l$ with origin O, while z axis is defined as along the edge of the sector. The angle of illumination φ_o is considered with $-\Phi_r \leq \varphi_o \leq \Phi_l$, and the plane wave field u^i verifies the Helmholtz equation in whole space. An implicit harmonic dependence on time $e^{i\omega t}$ is understood and henceforth suppressed with $|\arg(i\omega)| \leq \frac{\pi}{2}$, and $k = \frac{\omega}{c}$ denotes the wavenumber of the exterior medium with $|\arg(ik)| < \frac{\pi}{2}$. Generally, $|\arg(ik)| < \frac{\pi}{2}$ is most often associated with some losses in free space, while $|\arg(ik)| = \frac{\pi}{2}$ can be assumed as a limit case.

The term $U^i(\omega)$ is considered as an analytical function, regular for $\mathrm{Im}\omega \leq 0$ and $O(\omega^{-b})$ with $b > 1$ as $|\mathrm{Re}\omega| \to \infty$. The scattered field u_{sc} can be then assumed with the same behavior, as analytical for $\mathrm{Im}\omega \leq 0$ and $O(\omega^{-v})$ with $v > 1$ as $|\mathrm{Re}\omega| \to \infty$. The characteristics of the scatterer are supposed to be independent of z coordinate and the total field in the free space region, $u = u_{sc} + u^i$, satisfies the Helmholtz equation [1]–[2]:

$$(\Delta + k^2)u(\rho, \varphi) = 0, \tag{4.2}$$

in the free space, as $|\arg(ik)| < \frac{\pi}{2}$ (note: $|\arg(ik)| = \frac{\pi}{2}$ is considered in the sense of the limit). It is analytic with respect to ρ, φ and φ_o, except possibly at the origin, and it exists a constant s_o such that $\int_0^\infty |u(\rho, \varphi)e^{-s_o\rho}|d\rho < \infty$. The problem can be considered in acoustic, but also in electromagnetism when we consider u as the z-component of the electric or the magnetic field parallel to the edge of the wedge.

We can then search the total field u for $-\Phi_r \leq \varphi \leq \Phi_l$ as a Sommerfeld–Maliuzhinets integral [3],[8]:

$$u(\rho, \varphi) = \frac{1}{2\pi i}\int_\gamma f(\alpha + \varphi, \varphi_o) e^{ik\rho\cos\alpha} d\alpha, \tag{4.3}$$

which satisfies the Helmholtz equation. In this representation, f is an analytic function and the path γ consists of two branches: one, named γ_+, going from $(i\infty+\arg(ik) + (a_1 + \frac{\pi}{2}))$ to $(i\infty+\arg(ik) - (a_2 + \frac{\pi}{2}))$ with $0 < a_{1,2} < \pi$, as Im$\alpha \geq d > 0$, above all the singularities of the integrand, and the other, named γ_-, obtained by inversion of γ_+ with respect to $\alpha = 0$.

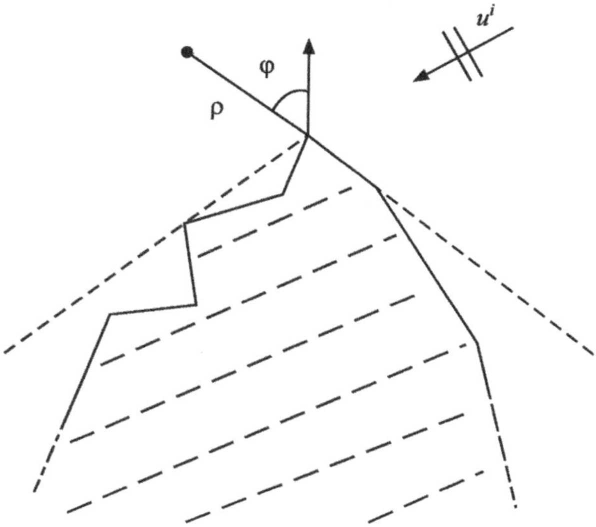

Figure 4.1. *Geometry for a wedge-shaped region*

We can develop the expression, considering a deformation of γ to the steepest descent path SDP for large $k\rho$, composed of two paths $SDP_\pm$ centered on the stationary phase points $\alpha = \pm\pi$. Assuming that, between γ and SDP, $f(\alpha + \varphi, \varphi_o)e^{ik\rho\cos\alpha}$ is meromorphic and vanishes at infinity, we can write for $-\Phi_r \leq \varphi \leq \Phi_l$:

$$u(\rho,\varphi) = u_i + \sum u_r^\pm + \sum u_{tr}^\pm + \sum u_s^\pm$$
$$+ \frac{e^{-ik\rho}}{2\pi i} \sum_\pm \int_{SDP_\pm} f(\alpha + \varphi, \varphi_o)e^{ik\rho(\cos\alpha+1)}d\alpha, \tag{4.4}$$

where:

– the terms u_i, $u_r^\pm$ and $u_{tr}^\pm$, for incident, reflected and transmitted geometrical optics (GO) plane waves (if any), correspond to real poles of $f(\alpha+\varphi)$ captured during deformation of γ to SDP, that is, the respective real poles α, in the band $|\text{Re}\alpha| < \pi$, verifying in particular:

$$\text{for } u_i : \alpha + \varphi = \varphi_\circ, \text{ i.e. } \alpha = \varphi_\circ - \varphi,$$
$$\text{for } u_r^\pm : \alpha + \varphi = -\varphi_\circ + 2\Phi_\pm, \text{ i.e. } \alpha = (\Phi_\pm - \varphi) + (\Phi_\pm - \varphi_\circ), \tag{4.5}$$

for $-\Phi_r \leq \varphi \leq \Phi_l$. The first term gives the incident field in the illuminated zone, while the second term expresses the reflection by the semi-infinite plane parts at $\varphi = \Phi_\pm$ (if any) of the scatterer enclosed by face $\pm$;

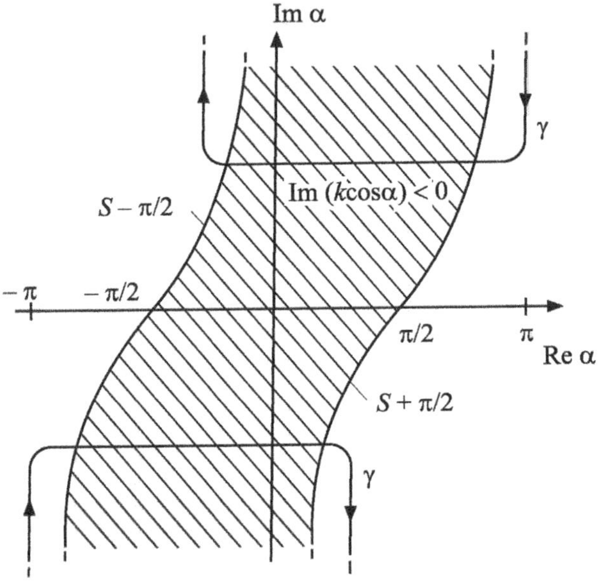

Figure 4.2. *Complex path of integration*

– the terms $u_s^\pm$ correspond to complex poles of $f(\alpha + \varphi)$, if any, captured during deformation of γ to SDP, for terms of complex plane guided waves excited by the edge (note: these terms are different from creeping waves guided by faces if they are curved);

– the last term, named u_d, is principally radiated cylindrically when the observation point M at (ρ, φ) is far, that is, $k\rho = k|OM| \to \infty$. Approximating $f(\alpha + \varphi)$ on $SDP_\pm$ in vicinity of $\alpha = \pm\pi$, we obtain:

$$u_d(M) \sim \frac{-e^{-i\pi/4 - ik\rho}}{\sqrt{2\pi k\rho}} F(\varphi, \varphi_\circ), \tag{4.6}$$

where

$$F(\varphi, \varphi_\circ) = f(\pi + \varphi, \varphi_\circ) - f(-\pi + \varphi, \varphi_\circ), \tag{4.7}$$

is the diffraction (or far field) coefficient attached to the wedge-shaped region relatively to the origin.

NOTE 4.1.– Let us notice that we can add a constant or any 2π-periodic function to f without changing F (and thus u_d). The field u associated with any finite order polynomial of $\cos(\alpha)$ and $\sin(\alpha)$ is null, and thus, some conditions at infinity on f and on its singularities are generally added to enforce its unique determination, as in the inversion theorem of Maliuzhinets [3].

NOTE 4.2.– Considering the Green's surface integral expression of the scattered field [1]–[2] and the behavior of incident field at infinity as $|\arg(ik)| < \frac{\pi}{2}$, we note that, for $\Phi_r = \Phi_l = \Phi > \pi/2$, and the conditions $\frac{\pi}{2} < \Phi \mp \varphi_o < \frac{3\pi}{2}$, $\frac{\pi}{2} < \Phi \mp \varphi < \frac{3\pi}{2}$ satisfied simultaneously for convergence of integrals, we have:

$$
\begin{aligned}
F(\varphi, \varphi_o) &= \frac{-1}{2} \sum_{\pm} \int_{L^{\pm}_{0,\infty}} (u \nabla_M(e^{ik\widehat{\tau}.OM}) - \nabla_M(u) e^{ik\widehat{\tau}.OM}).\widehat{n}^{\pm} ds_M, \\
0 &= \frac{-1}{2} \sum_{\pm} \int_{L^{\pm}_{0,\infty}} (u^i \nabla_M(e^{ik\widehat{\tau}.OM}) - \nabla_M(u^i) e^{ik\widehat{\tau}.OM}).\widehat{n}^{\pm} ds_M,
\end{aligned}
\tag{4.8}
$$

where $L^+_{0,\infty} \cup L^-_{0,\infty}$ encloses both faces of the physical surface of the scatterer, within the wedge-shaped region $|\varphi| < \Phi$, $\widehat{\tau}$ refers to the direction of observation of angle φ and $\widehat{n}^{\pm}$ is the outward normal to the face $L^{\pm}_{0,\infty}$ on each smooth facet (and semi-infinite plane) composing each face.

4.2.2. Basic properties of total field u and spectral function f

Some elementary properties can be assumed to hold:

– (a') the only incoming plane wave, from the free space sector $-\Phi_r \leq \varphi \leq \Phi_l$, is the incident field.

– (b') as $\rho \to 0$, the limit of the field u is finite and does not depend on φ with $u(0, \varphi) \equiv u(0) < u_0$, while the derivatives $\partial_\rho u$ and $\partial_\varphi u/\rho$ are $O(\rho^{c_0 - 1}(\ln \rho)^{d_0})$, where u_0, d_0 and $c_0 > 0$ can be chosen as three constants independent of the pulsation ω (note: $\partial_\rho u$ and $\partial_\varphi u/\rho$ are then locally summable with respect to ρ in the vicinity of the origin). This property applies for an origin taken at any point out of or upon the scatterer.

– (c') the field, except possibly its GO part when $\mathrm{Im}(k) \neq 0$, does not grow at infinity. In addition, some bounds on the far field are assumed, and for large $k\rho$, as ω or ρ vary, with $|\arg(ik)| \leq \frac{\pi}{2}$, the field is in general $O(1 + |e^{ik\rho\cos(\varphi - \varphi_o)}|)$ when the scattering object satisfies passive boundary conditions, and more particularly, is $O(e^{ik\rho\cos(\varphi - \varphi_o)})$ for a scatterer with passive impedance surface [3], [7]–[8].

These properties lead us to assume some elementary conditions on f [3]–[4], [8]–[9]:

a) $(f(\alpha) - u^i(O)/(\alpha - \varphi_o))$ is regular at points with Re(α) belonging to the free space angular sector with origin O, $-\Phi_r \leq$Re$(\alpha) \leq \Phi_l$, which ensures (a').

b) There exists some constants $g^{\pm}$, some analytic function h and some Maliuzhinets contour γ such that $|f(\alpha+\varphi) \mp f(-\alpha+\varphi) - g^{\pm}| <| h(\alpha) |$ on and inside the loop formed by the upper branch γ_+ of γ, when $-\Phi_r \leq \varphi \leq \Phi_l$, the function h being summable on γ_+, regular on and within it. Since γ is odd, we can add a constant to f without changing u, which implies that we can choose, from now, to define f with $g^- = 0$. We then note that we can write, for $-\Phi_r \leq \varphi \leq \Phi_l$:

$$f(i|\ln\rho|+\varphi) - f(-i|\ln\rho|+\varphi) = -iu(0) + O(\frac{\rho\partial u(\rho,\varphi)}{\partial\rho}),$$

$$f(i|\ln\rho|+\varphi) + f(-i|\ln\rho|+\varphi) = O(\frac{\partial u(\rho,\varphi)}{\partial\varphi}),$$

$$\text{with } \frac{\rho\partial u}{\partial\rho} = O(e^{c_1\ln\rho}(\ln\rho)^{d_1}), \frac{\partial u(\rho,\varphi)}{\partial\varphi} = O(e^{c_2\ln\rho}(\ln\rho)^{d_2}), \tag{4.9}$$

as $\rho \to 0$, with $u(0) = ig^+$, $c_{1,2} \geq c_0 > 0$, $d_0 \geq d_{1,2} > 0$, and that we have, for α belonging to the semi-infinite domain bounded by the loop γ_+:

$$\lim_{\text{Im}\alpha\to\infty} f(\alpha+\varphi) = -\lim_{\text{Im}\alpha\to\infty} f(-\alpha+\varphi) = -\frac{i}{2}u(0), \tag{4.10}$$

that is, from definition of γ_+, for arg$(ik) - \frac{\pi}{2} \leq$Re$\alpha \leq$arg$(ik) + \frac{\pi}{2}$. This ensures (b').

c) $f(\alpha+\varphi)$ has no singularity, except possibly those associated with incident, reflected, transmitted plane waves not vanishing at infinity, or bounded guided waves, in the zone with Re$(ik\text{cos}\alpha) \geq 0$ as $|\text{Re}(\alpha)| \leq \pi$, $-\Phi_r \leq \varphi \leq \Phi_l$, $|\arg(ik)| \leq \frac{\pi}{2}$. This ensures (c').

NOTE 4.3.– Letting $\Phi_r = \Phi_l = \Phi$, we note from (4.9) to(4.10) that, we have, as arg$(ik) - \frac{\pi}{2} - \Phi \leq \pm$Re$\alpha' \leq$arg$(ik) + \frac{\pi}{2} + \Phi$:

$$f(\alpha') = \mp\frac{i}{2}u(0) + O(e^{\pm ic_\pm\alpha}\alpha^{d_\pm}) \text{ with } c_\pm > 0,\ d_\pm > 0, \text{ as } \pm\,\text{Im}\alpha' \to \infty. \tag{4.11}$$

The presence of $\alpha^{d_\pm}$ terms (respectively, $(\ln\rho)^{d_\pm}$ terms for u) in the asymptotics explains why singularities of derivatives continue to remain for a discontinuity of material on a plane (see our sections on complete asymptotics of spectral function for an impedance wedge with distinct faces impedance parameters).

NOTE 4.4.– Using (c') when $|\arg(ik)| < \frac{\pi}{2}$, $f(\alpha+\varphi)$ has no singularity in the region with Re$(ik\text{cos}\alpha) > 0$ as $|\text{Re}(\alpha)| < \pi$ when Re$(ik\cos(\varphi-\varphi_o)) < 0$, that is, $\frac{\pi}{2} < |\varphi - \varphi_o| < \frac{3\pi}{2}$.

4.2.3. *Higher order expressions with Sommerfeld–Maliuzhinets integrals*

4.2.3.1. *Basic equalities for derivatives of integral representation of u*

Considering previous properties, in particular convergence at infinity of $f(\alpha+\varphi)$ for large imaginary argument, we can obtain different reduced expressions for a differential operator applied to u, in particular, from the use of integration by parts. So, for positive or null integers n, m, k, p, l and $\delta = 0$ or 1, we have the basic equality:

$$
\begin{aligned}
&(ik\rho)^n(\frac{\partial}{ik\partial\rho})^m(ik\rho)^k(\frac{\partial}{ik\rho\partial\varphi})^p(\frac{\partial}{ik\partial\rho})^l(\frac{\partial}{ik\rho\partial\varphi})^\delta(u(\rho,\varphi)) = \\
&= \frac{1}{2\pi i}\int_\gamma e^{ik\rho\cos\alpha}(\frac{\partial}{\partial\alpha}(\frac{(.)}{\sin\alpha}))^n\Big((\cos\alpha)^m(\frac{\partial}{\partial\alpha}(\frac{(.)}{\sin\alpha}))^{k-p} \\
&[\frac{\partial^p f(\alpha+\varphi,\varphi_\circ)}{\partial\varphi^p}(\cos\alpha)^l(\sin\alpha)^\delta]\Big)d\alpha \\
&= \frac{1}{2\pi i}\int_{\gamma_+}\Big[(\frac{\partial}{\partial\alpha}(\frac{(.)}{\sin\alpha}))^n\Big((\cos\alpha)^m(\frac{\partial}{\partial\alpha}(\frac{(.)}{\sin\alpha}))^{k-p} \\
&[(\cos\alpha)^l(\sin\alpha)^\delta(\frac{\partial}{\partial\alpha})^p f(\alpha+\varphi,\varphi_\circ)]\Big) \\
&-(\frac{\partial}{\partial\alpha}(\frac{(.)}{\sin\alpha}))^n\Big((\cos\alpha)^m(\frac{\partial}{\partial\alpha}(\frac{(.)}{\sin\alpha}))^{k-p} \\
&[(\cos\alpha)^l(-\sin\alpha)^\delta(\frac{-\partial}{\partial\alpha})^p f(-\alpha+\varphi,\varphi_\circ)]\Big)\Big]e^{ik\rho\cos\alpha}d\alpha,
\end{aligned} \tag{4.12}
$$

where we note $(\frac{\partial}{\partial\alpha}(\frac{(.)}{\sin\alpha}))^n \equiv \frac{\partial}{\partial\alpha}(\frac{(.)}{\sin\alpha})\frac{\partial}{\partial\alpha}(\frac{(.)}{\sin\alpha})\ldots\frac{\partial}{\partial\alpha}(\frac{(.)}{\sin\alpha})$ n times.

By using (4.12), we remark in particular, that we have:

$$
\begin{aligned}
&(\rho/ik)(\Delta+k^2)u = (ik\rho(\frac{\partial}{ik\partial\rho})^2 + \frac{\partial}{ik\partial\rho} + \frac{\partial}{ik\rho\partial\varphi}\frac{\partial}{\partial\varphi} - ik\rho)u \\
&= \frac{1}{2\pi i}\int_\gamma[\frac{\partial}{\partial\alpha}(\frac{(.)}{\sin\alpha})(f(\alpha+\varphi,\varphi_\circ)(1-(\sin\alpha)^2)) + f(\alpha+\varphi,\varphi_\circ)\cos\alpha \\
&+\sin\alpha\frac{\partial}{\partial\alpha}f(\alpha+\varphi,\varphi_\circ) - \frac{\partial}{\partial\alpha}(\frac{(.)}{\sin\alpha})f(\alpha+\varphi,\varphi_\circ)]e^{ik\rho\cos\alpha}d\alpha = 0.
\end{aligned} \tag{4.13}
$$

which is in accordance with the Helmholtz equation verified by u, and implies:

$$
(\frac{\partial}{\rho\partial\rho} + (\frac{\partial}{\rho\partial\varphi})^2)u = -(k^2 + (\frac{\partial}{\partial\rho})^2)u. \tag{4.14}
$$

In complements, considering Cartesian coordinates $(x^\pm, y^\pm)$ attached to faces $\varphi = \pm\Phi$, with $y^\pm$ axes directed as the outward normal to faces $\pm$, we shall also note that:

$$\frac{\partial u}{\partial y^{\pm}} = \frac{\mp\partial u}{\rho\partial\varphi}, (\frac{\partial}{\partial y^{\pm}})^2 u = -(k^2 + (\frac{\partial}{\partial\rho})^2)u \text{ as } \varphi = \pm\Phi,$$

$$(\frac{\partial u}{ik\partial y^{\pm}})^n|_{\varphi=\pm\Phi} = \frac{1}{2\pi i}\int_\gamma (\mp\sin\alpha)^n f(\alpha\pm\Phi,\varphi_\circ)e^{ik\rho\cos\alpha}d\alpha,$$

$$(\frac{\partial u}{ik\partial x^{\pm}})^n|_{\varphi=\pm\Phi} = \frac{1}{2\pi i}\int_\gamma (\mp\cos\alpha)^n f(\alpha\pm\Phi,\varphi_\circ)e^{ik\rho\cos\alpha}d\alpha. \quad (4.15)$$

Therefore, we can exploit these reductions for derivatives of fields, in conjunction with the Maliuzhinets inversion theorem, to determine functional equations on f from differential boundary equations.

NOTE 4.5.– Considering $y^{\pm} = \mp\rho\sin(\varphi\mp\Phi)$ and $x^{\pm} = \mp\rho\cos(\varphi\mp\Phi)$, we have:

$$\frac{\partial}{\partial y^{\pm}} = \mp\cos(\varphi\mp\Phi)\frac{\partial}{\rho\partial\varphi} \mp\sin(\varphi\mp\Phi)\frac{\partial}{\partial\rho},$$

$$\frac{\partial}{\partial x^{\pm}} = \pm\sin(\varphi\mp\Phi)\frac{\partial}{\rho\partial\varphi} \mp\cos(\varphi\mp\Phi)\frac{\partial}{\partial\rho}. \quad (4.16)$$

4.2.3.2. *Basic example of determination of functional equations on f*

For example of use of (4.12), we can consider a wedge with general high-order boundary conditions on plane faces $\varphi = \pm\Phi$ (taking $\Phi_r = \Phi_l = \Phi$) that are combinations of higher order derivatives of fields [16]–[18], following:

$$\sum_{n,m,k,p,\delta} a^{\pm}_{n,m,k,p,\delta}((ik\rho)^n(\frac{\partial}{ik\partial\rho})^m(ik\rho)^k(\frac{\partial}{ik\rho\partial\varphi})^p(\frac{\partial}{ik\rho\partial\varphi})^\delta u(\rho,\varphi))|_{\varphi=\pm\Phi} = 0, \quad (4.17)$$

where n, m, k, and p are positive or null integers, and $\delta = 0$ or 1. The coefficients $a^{\pm}_{n,m,k,p,\delta}$ are generally obtain from asymptotic considerations on characteristics of materials and geometry near the edge. Using an approach similar (slightly modified) to the one used by Maliuzhinets in [3] for its inversion theorem, we can exploit (4.12) and then derive functional equations on $f(\alpha)$ in the complex plane.

For that, considering each sum in (4.17) as $O(\rho^{-Q^{\pm}-d})$ in vicinity of the origin, with $0 \leq d < 1$ and $Q^{\pm}$ integer, we multiply (4.17) by $\rho^{Q^{\pm}+1}$, and using (4.12), apply a Laplace transform on ρ. Considering the behavior of integrand in the loop γ_+, we can close the loop γ_+ at infinity, and derive, from the residue theorem, the functional equations on f:

$$(\frac{\partial}{\partial\alpha}(\frac{(.)}{\sin\alpha}))^{Q^{\pm}+1}\sum_{n,m,k,p,\delta}^{N,M,K,P,1} a^{\pm}_l(D^{\pm}_{l,\alpha}(f(\alpha\pm\Phi)) - D^{\pm}_{l,-\alpha}(f(-\alpha\pm\Phi)) = 0, \quad (4.18)$$

where l denotes a set $\{n, m, k, p, \delta\}$ of positive or null integers (with $\delta \leq 1$), while

$$D^{\pm}_{l,\alpha}(f(\alpha \pm \Phi)) =$$

$$(\frac{\partial}{\partial\alpha}(\frac{(.)}{\sin\alpha}))^n[(\cos\alpha)^m(\frac{\partial}{\partial\alpha}(\frac{(.)}{\sin\alpha}))^{k-p}((\sin\alpha)^\delta(\frac{\partial}{\partial\alpha})^p f(\alpha \pm \Phi, \varphi_o))]. \quad (4.19)$$

This implies, after integration:

$$\sum_{n,m,k,p,\delta}^{N,M,K,P,1} a^{\pm}_{n,m,k,p,\delta}(D^{\pm}_{\alpha}(f(\alpha \pm \Phi)) - D^{\pm}_{-\alpha}(f(-\alpha \pm \Phi))$$

$$= \sum_{q=1}^{Q^{\pm}+1} c^{\pm}_q \sin\alpha(\cos\alpha)^{q-1}, \quad (4.20)$$

Considering u finite at the origin, we can use (4.11) in (4.20), and comparing each side for Im$\alpha \to \infty$, we can take $Q^{\pm} + 1 = \max_{a^{\pm}_{n,m,k,p,\delta} \neq 0}(m + p - k - n + \delta)$.

This type of (difference) functional equations, encountered for wedges with plane faces in [4], [16]–[18], can also be derived asymptotically from conditions on curved faces. By analytical continuation, these equations permit a determination everywhere from the one in the band $|\text{Re}\alpha| \leq \Phi$. Furthermore, the constants $c^{\pm}_q$ in (4.20) can be determined from additional conditions on u at the edge (related to conditions at infinity on f), but also from reciprocity principle (if it applies) [17]-[19], [25]-[26] (see also appendix D).

NOTE 4.6.– If we use that $f(\alpha \pm \Phi) + f(-\alpha \pm \Phi) \to 0$ for $|\text{Im}\alpha| \to \infty$ on and inside the loops $\gamma_{\pm}$ of γ, we notice that $c^{\pm}_{Q^{\pm}+1} = 0$ if $N = K = P = 0$.

4.3. Spectral function attached to radiation of a single face and properties

4.3.1. *Basic integral expression of radiation of one face*

We now apply Green's theorems [1], [2], when applied to Helmholtz equation, to derive important properties on representation of spectral functions attached to the diffraction of a 2D polygonal scatterer. To simplify the notation without losing generality, we take from now $\Phi_r = \Phi_l = \Phi$. From the properties (b'), (c') on u, the scattered field in free space $u_{sc} = u - u^i$ for $|\varphi| < \Phi$ can be written as the sum $u_+ + u_-$ of the radiations of equivalent surface currents carried by the faces $\varphi = +\Phi$

and $-\Phi$, following:

$$u_{\pm}(\rho,\varphi) = \frac{-i}{4}\int_{0+}^{\infty}(u(\rho',\varphi')\frac{\partial H_0^{(2)}(kR)}{\partial n} - \frac{\partial u(\rho',\varphi')}{\partial n}H_0^{(2)}(kR))|_{\varphi'=\pm\Phi}d\rho'$$

$$= \frac{-i}{4}\int_{0+}^{\infty}(u(\rho',\varphi')\nabla H_0^{(2)}(kR) - \nabla(u(\rho',\varphi'))H_0^{(2)}(kR))|_{\varphi'=\pm\Phi}.\widehat{n}\,d\rho', \quad (4.21)$$

with $R = \sqrt{\rho^2+\rho'^2-2\rho\rho'\cos(\varphi-\varphi')}$, $\partial(.)/\partial n = \widehat{n}\nabla(.) = \mp\partial(.)/\rho'\partial\varphi'$, $\widehat{n}$ the outward normal to the face $\varphi' = \pm\Phi$, $|\varphi| < \Phi$, $|\arg(ik)| < \frac{\pi}{2}$, and where $H_0^{(2)}$ is the Hankel function of second kind [21].

We will show that it is possible, after expressing the spectral function corresponding to $u_{\pm}$ with f, to express f for the total field from the radiation of a single face in far field. This expression will then have the remarkable property, from Green's theorem, to have a path of integration that can be deformed without changing the result. For this, we begin by using a Sommerfeld–Maliuzhinets representation of $H_0^{(2)}(kR)$.

NOTE 4.7.– The possibility to deform integration paths without change of result is well known in Cauchy theorem for complex plane and in Green's theorem for 2D/3D spaces.

4.3.2. *Spectral function associated with* $H_0^{(2)}(kR)$

As shown in [7], [8], [13], the spectral function associated with Sommerfeld–Maliuzhinets representation of $H_0^{(2)}(kR)$ is given by:

$$f_{H_0^{(2)}}(\alpha) = \frac{-1}{2\pi}\int_S e^{ik\rho'\cos(\alpha'-\varphi')}(\tan(\frac{1}{2}(\alpha-\alpha')) - g_{\circ}(\alpha'))d\alpha', \quad (4.22)$$

where S is the path from $-i\infty-\arg(ik)$ to $i\infty+\arg(ik)$ with $\mathrm{Im}(k\sin\alpha) = 0$, as $R = \sqrt{\rho^2+\rho'^2-2\rho\rho'\cos(\varphi-\varphi')}$, $|\varphi'| > \frac{\pi}{2}$, $|\varphi| < |\varphi'|$. The terms $g_{\circ}$, normally unnecessary because γ is odd, can be chosen as $g_{\circ}(\alpha) = -\tan(\frac{\alpha}{2})$ in order to ameliorate the convergence of the integral. The expression (4.22) is a determination for $|\varphi'| > \frac{\pi}{2}$, regular as $\alpha \in]S-\pi, S+\pi[$ (i.e. the domain in complex plane limited by $S-\pi$ and $S+\pi$), and we verify (after capture of poles of $\tan(\frac{1}{2}(\alpha-\alpha'))$ as α varies):

$$f_{H_0^{(2)}}(\pi+\alpha') - f_{H_0^{(2)}}(-\pi+\alpha') = -2ie^{ik\rho'\cos(\alpha'-\varphi')}, \quad (4.23)$$

which is a functional equation that can be considered, by argument of analyticity, in complement of (4.22) for a determination of $f_{H_0^{(2)}}$ everywhere as α varies.

The path S can be also deformed continuously, as long as the integrand remains bounded in (4.22), without changing $f_{H_0^{(2)}}$, and this expression can be then continued for $|\varphi'| \leq \frac{\pi}{2}$ from shift of S and summation of residues due to poles captured by S during this shift.

4.3.3. *Expression of spectral function associated with* $u_\pm$ *and properties*

Therefore, we can write $f_{H_0^{(2)}}$ corresponding to $H_0^{(2)}(kR)$ for arbitrary φ'. By letting $\varphi' = \epsilon\Phi$, with $\epsilon \equiv +\text{or}-$ and $\Phi > 0$, a real positive constant, we can write the spectral function $f_{H_0^{(2)}} \equiv f^\epsilon_{H_0^{(2)}}$, associated with $H_0^{(2)}(kR_\epsilon)$ with:

$$R_\epsilon = \sqrt{\rho^2 + \rho'^2 - 2\rho\rho' \cos(\varphi - \epsilon\Phi)}, \tag{4.24}$$

following

$$f^\epsilon_{H_0^{(2)}}(\alpha) = \frac{-1}{2\pi} \int_{\mathcal{S}_\epsilon} e^{ik\rho' \cos(\alpha' - \epsilon\Phi)} (\tan(\frac{1}{2}(\alpha - \alpha')) - g_\circ(\alpha'))d\alpha', \tag{4.25}$$

where $\mathcal{S}_\epsilon = \mathcal{S} + a_\epsilon$, with a_ϵ a constant satisfying $\Phi - \frac{3\pi}{2} < \epsilon a_\epsilon < \Phi - \frac{\pi}{2}$. This gives us a regular value of $f^\epsilon_{H_0^{(2)}}(\alpha)$ between $\mathcal{S} + \epsilon(\Phi - \frac{5\pi}{2})$ and $\mathcal{S} + \epsilon(\Phi + \frac{\pi}{2})$, bounded at infinity in this domain if $\rho' \neq 0$, verifying everywhere from (4.23):

$$f^\epsilon_{H_0^{(2)}}(\pi + \alpha') - f^\epsilon_{H_0^{(2)}}(-\pi + \alpha') = -2ie^{ik\rho' \cos(\alpha' - \epsilon\Phi)}, \tag{4.26}$$

which allows an explicit Sommerfeld–Maliuzhinets integral representation for $H_0^{(2)}(kR_\epsilon)$ as $|\varphi| < \Phi$.

Using the expression (4.25) for $H_0^{(2)}(kR_\epsilon)$ and $\partial_n H_0^{(2)}(kR_\epsilon)$ in (4.21), we then obtain a Sommerfeld–Maliuzhinets representation of $u_\pm$ with f [7], [8]:

$$u_\pm(\rho, \varphi) = \frac{1}{2\pi i} \int_\gamma f_\pm(\alpha + \varphi) e^{ik\rho \cos\alpha} d\alpha, \tag{4.27}$$

where, the function f_ϵ, with $\epsilon \equiv +\text{or}-$, satisfies:

$$\begin{aligned} f_\epsilon(\alpha) &= \frac{1}{4\pi i} \int_{\mathcal{S}_\epsilon} \epsilon f(\epsilon\pi + \alpha')(\tan(\frac{1}{2}(\alpha - \alpha')) - g_\circ(\alpha'))d\alpha', \\ f_\epsilon(\alpha + \varphi) - f_\epsilon(-\alpha + \varphi) &= \frac{1}{2\pi i} \int_{\mathcal{S}_\epsilon} \frac{\epsilon f(\epsilon\pi + \alpha') \sin\alpha}{\cos\alpha + \cos(\varphi - \alpha')} d\alpha', \end{aligned} \tag{4.28}$$

for α between $\mathcal{S}_\epsilon - \pi$ and $\mathcal{S}_\epsilon + \pi$, provided $\frac{\pi}{2} < \Phi - \epsilon a_\epsilon < \frac{3\pi}{2}$, $\frac{\pi}{2} < \Phi - \epsilon\varphi_\circ < \frac{3\pi}{2}$.

Since f is an analytic function, f_ϵ can be analytically continued in the whole complex plane. We note, in particular, that, taking account of the poles of $\tan(\frac{1}{2}(\alpha - \alpha'))$ which can be captured by $\mathcal{S}_\epsilon$ as α varies, $f_\pm$ satisfies:

$$f_\pm(\pi + \alpha) - f_\pm(-\pi + \alpha) = \pm f(\pm\pi + \alpha). \tag{4.29}$$

Concerning the dependence on $\varphi_\circ$ (or Φ), the expression (4.28) has been determined for $\frac{\pi}{2} < \Phi - \epsilon\varphi_\circ < \frac{3\pi}{2}$, but we can consider $f_\epsilon(\alpha)$ outside of this domain by analytical continuation on $\varphi_\circ$ (or Φ), which corresponds to take account of the contribution of any singularity that would go through $\mathcal{S}_\epsilon$ as $\varphi_\circ$ (or Φ) goes into these regions.

NOTE 4.8.– As $\sum_\pm f(\epsilon\pi + \epsilon a_\epsilon \pm (\alpha' - \epsilon a_\epsilon))$ exponentially vanishes for large α' on $\mathcal{S}_\epsilon$ (see property (b) considered on f and (4.11)), we note that $f_\epsilon(\alpha+\varphi) - f_\epsilon(-\alpha+\varphi)$ is bounded for large α on γ, even if $f_\epsilon(\alpha + \varphi)$ is not when $f(\pm i\infty) \neq 0$. This property is in accordance with the expression $u_\pm$ given by (4.21), which is finite as $\rho \to 0$, while its normal derivative is not when u is non null at the origin.

4.4. Far-field radiation of one face and single-face expression of spectral function f

4.4.1. *The spectral function f derived from fields on a semi-straight line and properties*

The steepest descent path method [1]–[2] can be applied to (4.27) to find the far-field radiation of the face $\varphi = \epsilon\Phi$, also denoted as $\varphi = \pm\Phi$. From the regularity of f_ϵ, we can deform γ to stationary phase points $\alpha = +\pi$ and $-\pi$, when $\frac{\pi}{2} < \Phi - \epsilon\varphi_\circ < \frac{3\pi}{2}$ and $\frac{\pi}{2} < \Phi - \epsilon\varphi < \frac{3\pi}{2}$, and thus, out of the reflected and shadowed regions $2\Phi - \epsilon(\varphi + \varphi_\circ) < \pi$ and $|\varphi - \varphi_\circ| > \pi$. In this case, equation (4.27) then gives us:

$$u_\pm(\rho,\varphi) = \frac{-e^{-i\pi/4}}{\sqrt{2\pi k\rho}} e^{-ik\rho}(f_\pm(\pi+\varphi) - f_\pm(-\pi+\varphi) + O(1/(k\rho))), \tag{4.30}$$

which, from (4.29), can be written as:

$$u_\pm(\rho,\varphi) = \frac{-e^{-i\pi/4}}{\sqrt{2\pi k\rho}} e^{-ik\rho}(\pm f(\pm\pi+\varphi) + O(1/(k\rho))). \tag{4.31}$$

At this step, let us notice that this result is a generalization of Michaeli's result [20] obtained for Dirichlet boundary conditions. In other respects, we can consider the far field derived from (4.21), using formulas [21]:

$$\begin{aligned} &H_0^{(2)}(kR) = \sqrt{2/(\pi kR)}\; e^{-ikR+i\pi/4}(1 + O(1/(kR))), \\ &\partial_n H_0^{(2)}(kR) = -i\sqrt{2/(\pi kR)}\; e^{-ikR+i\pi/4}\partial_n(kR)(1 + O(1/(kR))), \end{aligned} \tag{4.32}$$

for large R, with $\partial_n(kR) = \mp\partial(kR)/\rho'\partial\varphi'|_{\varphi'=\pm\Phi} = (\pm k\rho\sin(\varphi-\varphi')/R)|_{\varphi'=\pm\Phi}$, $R = \sqrt{\rho^2 + \rho'^2 - 2\rho\rho'\cos(\varphi-\varphi')}$. Taking into account that:

$$R = \rho - \rho'\cos(\varphi \mp \Phi) + \frac{(\rho'\sin(\varphi\mp\Phi))^2}{R + \rho - \rho'\cos(\varphi\mp\Phi)}, \tag{4.33}$$

when $R + \rho - \rho' \cos(\varphi \mp \Phi) \neq 0$, and considering the properties (b')-(c') on u, we then obtain the expression for the far field from (4.21) which, compared with (4.31), gives us:

$$f(\pm\pi + \varphi) = \frac{1}{2}\int_{0+}^{\infty}(iku(\rho', \pm\Phi)\sin(\varphi \mp \Phi)$$
$$\pm \frac{\partial u}{\partial n}(\rho', \pm\Phi))e^{ik\rho'\cos(\varphi\mp\Phi)}d\rho', \tag{4.34}$$

as $\frac{\pi}{2} < \Phi \mp \varphi_\circ < \frac{3\pi}{2}$ and $\frac{\pi}{2} < \Phi \mp \varphi < \frac{3\pi}{2}$ with $|\arg(ik)| < \frac{\pi}{2}$, that is, as $e^{ik\rho'\cos(\varphi\mp\Phi)}$ and $e^{ik\rho'\cos(\varphi_\circ\mp\Phi)}$ vanish at infinity. We can also consider the field along the line $\varphi = \pm\Phi_e^\pm$ with $0 < \Phi_e^\pm \leq \Phi$, letting $\pm\pi + \varphi = \alpha \pm \Phi_e^\pm$. In this case, we can write, when $\frac{\pi}{2} < \Phi_e^\pm \mp \varphi_\circ < \frac{3\pi}{2}$ and $\frac{\pi}{2} < \mp\alpha + \pi < \frac{3\pi}{2}$:

$$f(\alpha \pm \Phi_e^\pm) = \frac{1}{2}\int_{0+}^{\infty}(-iku(\rho', \pm\Phi_e^\pm)\sin\alpha \pm \frac{\partial u}{\partial n}(\rho', \pm\Phi_e^\pm))e^{-ik\rho'\cos\alpha}d\rho'$$
$$= \frac{1}{2}\int_{0+}^{\infty}(-iku(\rho', \pm\Phi_e^\pm)\sin\alpha - \frac{\partial u}{\rho\partial\varphi}(\rho', \pm\Phi_e^\pm))e^{-ik\rho'\cos\alpha}d\rho', \tag{4.35}$$

which is also valid, by analytic continuation, as $\mathrm{Re}(ik(\cos\alpha - \cos(\Phi_e^\pm \mp \varphi_\circ))) > 0$, $|\mathrm{Re}\alpha| < \pi$, and by continuation for complex $\varphi_\circ$. This expression implies that:

$$f(ix \pm \Phi_e^\pm) - f(-ix \pm \Phi_e^\pm) = -iu(0, \pm\Phi_e^\pm) + O(\frac{\partial u(\rho, \pm\Phi_e^\pm)}{\partial \ln \rho}|_{|\ln\rho|=x}),$$
$$f(ix \pm \Phi_e^\pm) + f(-ix \pm \Phi_e^\pm) = O(\frac{\partial u(\rho, \varphi)}{\partial\varphi}|_{|\ln\rho|=x, \varphi=\pm\Phi_e^\pm}), \tag{4.36}$$

as $\rho \to 0$ and $x = |\ln\rho| \to \infty$, which is in accordance with the property (b) of the spectral function f attached to the field u.

Thus, the definition of the field and its normal derivatives on one single face (a semi-line) leads us to a (explicit) definition of a unique analytic function f attached to the total radiated field (uniqueness and existence), with $f(ix \pm \Phi_e^\pm) + f(-ix \pm \Phi_e^\pm)$ vanishing at infinity as $\frac{\pi}{2} < \Phi_e^\pm \mp \varphi_\circ < \frac{3\pi}{2}$, $0 < \Phi_e^\pm \leq \Phi$, in accordance with condition (b).

NOTE 4.9.– From (4.35), valid as $\mathrm{Re}(ik(\cos\alpha - \cos(\Phi_e^\pm \mp \varphi_\circ))) > 0$, in the band $|\mathrm{Re}\alpha| < \pi$, for $0 < \Phi_e^\pm \leq \Phi$, we can write in this band, as $\mathrm{Re}(ik\cos\alpha) > \mathrm{Re}(ik)$, $|\varphi| \leq \Phi$,

$$f(\alpha + \varphi) = \frac{1}{2}\int_{0+}^{\infty}(-iku(\rho, \varphi)\sin\alpha - \frac{\partial u}{\rho\partial\varphi}(\rho, \varphi))e^{-ik\rho\cos\alpha}d\rho. \tag{4.37}$$

NOTE 4.10.– It is possible to recover (4.35) by another way, as already noticed in [7]. For this, we can use the inversion theorem of Maliuzhinets [3] (taking care about time convention) on $u(\rho, \pm\Phi)$ and $\partial u/\partial n(\rho, \pm\Phi)$ to obtain:

$$f(\alpha+\varphi) - f(-\alpha+\varphi) = -ik\sin\alpha \int_{0+}^{\infty} u(\rho,\varphi)e^{-ik\rho\cos\alpha}d\rho,$$

$$f(\alpha+\varphi) + f(-\alpha+\varphi) = -\int_{0+}^{\infty} \frac{\partial u}{\rho\partial\varphi}(\rho,\varphi)e^{-ik\rho\cos\alpha}d\rho, \tag{4.38}$$

for Re$(ik\rho\cos\alpha)$ >Re$(ik\rho)$ as $|\text{Re}\alpha| < \pi$. Elementary combinations and analyticity then permit us to recover (4.35). We prefer to not use this way here because it does not allow to simply relate (4.35) to Green's theorem, which is a key point to derive the remarkable property of deformation of the integration path given in next section.

4.4.2. *The spectral function f derived from fields on a piecewise smooth semi-line*

4.4.2.1. *Single-face expressions of f along $L^{\pm}_{0,\infty}$ with the use of tangent angles*

Using the properties of Green's theorem [1], it is worth noticing [7]–[8] that the contour of integration along $\varphi = \pm\Phi$ in (4.21) can be deformed into any path $L^{\pm}_{0,\infty}$ without changing the field $u_{\pm}(M)$, provided that $L^{\pm}_{0,\infty}$ is a piecewise smooth semi-line, and that, during the deformation, the integral remains definite, M is not captured by the path during its deformation, and no source passes through the path. Afterwards we can write (4.34) in a more general form, with integration along the new general path $L^{\pm}_{0,\infty}$, a piecewise smooth semi-line from the origin to infinity, following:

$$f(\pm\pi+\varphi) = \frac{1}{2}\int_{L^{\pm}_{0,\infty}} (iku(\rho',\varphi')\sin(\varphi-\varphi'_t)$$

$$\pm \frac{\partial u}{\partial n}(\rho',\varphi'))e^{ik\rho'\cos(\varphi-\varphi')}dl'(\rho',\varphi'), \tag{4.39}$$

where $dl'(\rho',\varphi')$ is the element of length, $\frac{\partial}{\partial n}$ is the outward normal derivative, and φ'_t is the tangent angle on each smooth segment composing $L^{\pm}_{0,\infty}$.

If we divide the piecewise smooth semi-line $L^{\pm}_{0,\infty}$ (deriving from a deformation of the faces $\varphi = \pm\Phi$ enclosing the scatterer, described above) into $L^{\pm}_{0,\Delta^{\pm}}$ (i.e. $0 < l' < \Delta^{\pm}$) and $L^{\pm}_{\Delta,\infty}$ (i.e. $l' > \Delta^{\pm}$), we have:

$$f(\pm\pi+\varphi) = \frac{1}{2}\int_{L^{\pm}_{0,\Delta^{\pm}}} (iku(\rho',\varphi')\sin(\varphi-\varphi'_t)$$

$$\pm \frac{\partial u}{\partial n}(\rho',\varphi'))e^{ik\rho'\cos(\varphi-\varphi')}dl'(\rho',\varphi') + f_{L^{\pm}_{\Delta^{\pm},\infty}}(\pm\pi+\varphi), \tag{4.40}$$

where $f_{L^{\pm}_{\Delta^{\pm},\infty}}(\alpha) = e^{-ik\rho_{\Delta^{\pm}}\cos(\alpha-\varphi_{\Delta^{\pm}})} f_e^{\pm}(\alpha)$, $f_e^{\pm}(\alpha)$ is the spectral function related to the Sommerfeld–Maliuzhinets representation of the field in coordinates with origin at points $(\rho_{\Delta^{\pm}}, \varphi_{\Delta^{\pm}})$ with $l' = \Delta^{\pm}$. We can then write, by analytic continuation:

$$f(\alpha) = \frac{1}{2}\int_{L^{\pm}_{0,\Delta^{\pm}}} (-iku(\rho', \varphi')\sin(\alpha - \varphi'_t)$$

$$\pm \frac{\partial u}{\partial n}(\rho', \varphi'))e^{-ik\rho'\cos(\alpha-\varphi')}dl'(\rho', \varphi') + f_{L^{\pm}_{\Delta^{\pm},\infty}}(\alpha). \tag{4.41}$$

4.4.2.2. *Single-face expressions of f along $L^{\pm}_{0,\infty}$ with the use of gradient*

Using the expression of normal derivative with gradient, we derive another form of the expression of the spectral function associated with total field, in presence of a scatterer enclosed by a surface with piecewise smooth faces $L^{\pm}_{0,\infty}$, ended by semi-infinite planes with tangent angles $\pm\Phi^{\pm}$ (note: in the case of a closed scatterer, we can take $\Phi^{\pm} = \pi$). Therefore, the spectral function f satisfies the single-face expressions, as $\frac{\pi}{2} < \Phi^{\pm} \mp \varphi_{\circ} < \frac{3\pi}{2}$ and $\frac{\pi}{2} < \Phi^{\pm} \mp \varphi < \frac{3\pi}{2}$ [8], following:

$$f(\pm\pi + \varphi) = \frac{1}{2}\int_{L^{\pm}_{0,\infty}} \mp(u(M)\nabla_M(e^{ik\widehat{\tau}.\overline{OM}})$$

$$- \nabla_M(u(M))e^{ik\widehat{\tau}.\overline{OM}}).\widehat{n}^{\pm}ds(M), \tag{4.42}$$

where $\widehat{\tau}(\varphi) = \cos\varphi\,\widehat{x} + \sin\varphi\,\widehat{y}$ refers to the direction of observation of angle φ, and $\widehat{n}^{\pm}ds = \mp(\widehat{z}\wedge d(\overline{OM}))$ on each smooth segment composing $L^{\pm}_{0,\infty}$. This expression is in agreement with (4.39) and the expression (4.8) of $F(\varphi, \varphi_{\circ}) = f(\pi+\varphi) - f(-\pi+\varphi)$. As for (4.41), we then derive:

$$f(\alpha) = \frac{1}{2}\int_{L^{\pm}_{0,\Delta^{\pm}}} \mp(u(M)\nabla_M(e^{-ik\widehat{\tau}_e.\overline{OM}})$$

$$- \nabla_M(u(M))e^{-ik\widehat{\tau}_e.\overline{OM}}).\widehat{n}^{\pm}ds(M) + f_{L^{\pm}_{\Delta^{\pm},\infty}}(\alpha), \tag{4.43}$$

where $\widehat{\tau}_e$ refers to the complex angle α, with $\widehat{\tau}_e = \cos\alpha\,\widehat{x}+\sin\alpha\,\widehat{y}$ verifying $\widehat{\tau}_e.\widehat{\tau}_e = 1$ (note: we use $\widehat{\tau}_e = -\widehat{\tau}(\mp\pi + \alpha)$, with scalar product of complex vectors defined by analytic continuation (no Hermitian product and no conjugate)).

NOTE 4.11.– When $L^{\pm}_{0,\infty}$ has a succession of singularities at points M_n, we can consider $L^{\pm}_{0,\infty}$ as the addition of curved line segments without M_n, that is, the integral as the addition of improper (or limit) integrals on open paths, in the sense that the ends of paths are not strictly considered at M_n but as $M \to M_n$.

4.5. Expression of f from the far field function F and its consequences

4.5.1. *Integral expression of f relatively to F*

Considering (4.27), we can write for the spectral function f attached to the total field:

$$f(\alpha) = f_+(\alpha) + f_-(\alpha) + f_i(\alpha) - A_F, \tag{4.44}$$

where $f_i(\alpha) = \frac{\text{Res}f|_{\alpha=\varphi_\circ}}{2}\cot(\frac{1}{2}(\alpha - \varphi_\circ))$, the $f_\pm$ verify (4.28) and (4.29), while:

$$A_F = \frac{1}{2}\sum_{\pm} \lim_{\text{Im}\alpha\to\pm\infty}(f_+(\alpha) + f_-(\alpha)), \tag{4.45}$$

so that $f(i\infty) = -f(i\infty)$ as specified in the condition (b). The constant A_F does not change the field since γ is odd, and is only here to define uniquely f with $f(i\infty) = -f(-i\infty)$.

When both conditions $\frac{\pi}{2} < \Phi \mp \varphi_\circ < \frac{3\pi}{2}$ are satisfied (permitted with $\pi > \Phi > \pi/2$), we then obtain, from (4.28) and (4.29), for $\alpha \in]\mathcal{S} - \pi, \mathcal{S} + \pi[$:

$$f(\alpha, \varphi_\circ) = \frac{1}{4\pi i}\int_{\mathcal{S}} F(\alpha', \varphi_\circ)(\tan(\frac{1}{2}(\alpha - \alpha')) - g_\circ(\alpha'))d\alpha'$$

$$+ \frac{\text{Res}f|_{\alpha=\varphi_\circ}}{2}\cot(\frac{1}{2}(\alpha - \varphi_\circ)) - A_F, \tag{4.46}$$

with,

$$F(\alpha', \varphi_\circ) = f(\pi + \alpha', \varphi_\circ) - f(-\pi + \alpha', \varphi_\circ),$$

$$A_F = \frac{-1}{4\pi i}\int_{\mathcal{S}} F(\alpha', \varphi_\circ)g_\circ(\alpha')d\alpha', \tag{4.47}$$

where the far field function (also known as the diffraction coefficient at the edge in (4.7)) vanishes at infinity on $\mathcal{S}$, and $f(i\infty) = -f(i\infty)$ is verified.

As a first consequence, we conclude the existence of f from F as $\Phi > \pi/2$, defined from (4.46) and its analytical continuation with (4.47), which implies the definition of representation (4.3) for the field u. Moreover, the expression (4.46) can be analytically continued on $\varphi_\circ$ for finite or infinite scatterers. In the particular case of a finite scatterer, F is an entire function and (4.46) directly applies for any $\varphi_\circ$.

The expression (4.46) can be also modified by analytical continuation for semi-infinite scatterer, when the pole of incident field of f is not 2π-periodic. For that, we begin to rewrite (4.46) in the form:

$$f(\alpha,\varphi_\circ) = \frac{-1}{4\pi i}\Big(\int_{\mathcal{S}+\pi} - \int_{\mathcal{S}-\pi}\Big) f(\alpha',\varphi_\circ)(\cot(\frac{1}{2}(\alpha-\alpha')) + \cot(\frac{\alpha'}{2}))d\alpha'$$

$$+ \frac{\mathrm{Res}f|_{\alpha=\varphi_\circ}}{2}\cot(\frac{1}{2}(\alpha-\varphi_\circ)) - A_F, \tag{4.48}$$

We can assume that the nearest singularities of $f(\alpha,\varphi_\circ)$ on real axis are the incident pole at $\alpha=\varphi_\circ$ and the GO reflection poles at $\alpha = 2\Phi_\pm - \varphi_\circ = \pm 2\Phi^\pm - \varphi_\circ$, where the constant $\Phi^\pm \geq \Phi$ refers to the angle of semi-infinite plane composing each physical face, while the singularities apart of real axis, which are independent of $\varphi_\circ$, are with $|\mathrm{Im}\alpha| > \delta > 0$. We can then use Cauchy theorem, and deform $\mathcal{S}$ to a new path $\mathcal{S}_0^\pm$, with capture of poles on real axis [13]. When F has the pole of incident field (i.e. if the scatterer is not finite), this leads us to write:

$$f(\alpha,\varphi_\circ) = -\Big(\int_{\mathcal{S}_0^\pm+\pi} - \int_{\mathcal{S}_0^\pm-\pi}\Big) f(\alpha',\varphi_\circ)(\cot(\frac{\alpha-\alpha'}{2}) + \cot(\frac{\alpha'}{2}))\frac{d\alpha'}{4\pi i} - A_F$$

$$= \frac{1}{4\pi i}\int_{\mathcal{S}_0^\pm} F(\alpha',\varphi_\circ)(\tan(\frac{\alpha-\alpha'}{2}) - g_\circ(\alpha'))d\alpha' - A_F, \tag{4.49}$$

as $\alpha \in]\mathcal{S}_0^\pm - \pi, \mathcal{S}_0^\pm + \pi[$, for $\Phi \geq \pm\varphi_\circ > \pi/2$. For $|\arg(ik)| = \pi/2 - 0^+$, let us note that we can consider that $\mathcal{S}_0^\pm \equiv \mathcal{S} \pm i\delta\mathrm{sign}(\arg(ik))$, and that (4.49) remains valid by continuation for complex $\varphi_\circ$ in the domain $\Omega_{\delta,\Phi}^\pm$ defined by $\Phi \geq \pm\mathrm{Re}(\varphi_\circ) > \pi/2$ with $|\mathrm{Im}(\varphi_\circ)| < \delta$. Therefore, we can use:

the expression (4.46) of f from F for any scatterers (finite or not),

and the expression (4.49) from F for semi-infinite scatterers. (4.50)

NOTE 4.12.– Both conditions $\frac{\pi}{2} < \Phi \mp \varphi_\circ < \frac{3\pi}{2}$, considered to apply (4.46), are satisfied only if $2\Phi > \pi$. With these conditions, all poles of $f(\alpha) - \frac{\mathrm{Res}f|_{\alpha=\varphi_\circ}}{2}\cot(\frac{1}{2}(\alpha - \varphi_\circ))$ are outside the band $]\mathcal{S}-\pi, \mathcal{S}+\pi[$, in particular the GO reflection poles at $\alpha_\pm = -\varphi_\circ \pm 2\Phi^\pm$ with $\Phi^\pm \geq \Phi$, since the conditions $\frac{\pi}{2} < \Phi \mp \varphi_\circ < \frac{3\pi}{2}$ implies $2\Phi \mp \varphi_\circ > \frac{\pi}{2} + \Phi > \pi$ and thus $2\Phi^\pm \mp \varphi_\circ > \pi$.

4.5.2. *Equations on f from functional equations on F and consequences*

The form (4.49) for a semi-infinite scatterer has important consequence to deduce new analytical properties of $f(\alpha,\beta)$ on incident variable β. For that, let us take the example of $F(\alpha',\beta)$ satisfying the analytical functional equations, relatively to variable β:

$$\mathcal{G}_\beta^\pm(F(\alpha',\beta)) = 0, \text{ as } \alpha' \in \mathcal{S}_0^\pm, \beta \in \Omega_{\delta,\Phi}^\pm, \tag{4.51}$$

where $\mathcal{G}_\beta^\pm$ is an analytical operator, linear relatively to F. We directly deduce from (4.49):

$$\mathcal{G}_\beta^\pm(f(\alpha,\beta)) = 0, \tag{4.52}$$

for $\beta \in \Omega_{\delta,\Phi}^\pm$ as $\alpha \in]\mathcal{S}_0^\pm - \pi, \mathcal{S}_0^\pm + \pi[$, then everywhere by analytical continuation.

It is worth noticing that when the reciprocity principle applies, we have $F(\alpha',\varphi_\circ) = F(\varphi_\circ,\alpha')$, and functional equations for F on variable $\varphi_\circ$ derive from those on α'. Thus, (4.52) is a remarkable way, while $f(\alpha',\varphi_\circ) \neq f(\varphi_\circ,\alpha')$, to exploit reciprocity and obtain equations on second variable of f.

The property (4.52) was first noted and illustrated in [13] in the particular case where $\mathcal{G}_\beta^\pm(F(\alpha',\beta)) = 0$ are difference functional equations derived from boundary conditions on the faces of a wedge, when reciprocity applies and $F(\alpha',\varphi_\circ) = F(\varphi_\circ,\alpha')$. In this type of problems, we can generally write, from boundary conditions on the field for each face of a wedge $\varphi = \pm\Phi$, functional difference equations on f [4], [16]–[18], following:

$$\mathcal{A}_\alpha^\pm f(\alpha \pm \Phi,\beta) - \mathcal{A}_{-\alpha}^\pm f(-\alpha \pm \Phi,\beta) = \sin\alpha \sum_{q=1}^{Q^\pm} a_q^\pm (\cos\alpha)^{q-1}, \tag{4.53}$$

with $\mathcal{A}_\alpha = \mathcal{A}_{\alpha+2\pi}$, and thus deduce

$$\mathcal{A}_{\pi+\alpha}^\pm F(\alpha \pm \Phi,\beta) + \mathcal{A}_{\pi-\alpha}^\pm F(-\alpha \pm \Phi,\beta) = 0, \tag{4.54}$$

which gives us, from $F(\alpha,\alpha') = F(\alpha',\alpha)$ when reciprocity applies:

$$\mathcal{A}_{\pi+\beta}^\pm F(\alpha,\beta \pm \Phi) + \mathcal{A}_{\pi-\beta}^\pm F(\alpha,-\beta \pm \Phi) = 0. \tag{4.55}$$

This equation permits us, from the application of (4.52), to obtain functional equations relatively to the second variable on f, following:

$$\mathcal{A}_{\pi+\beta}^\pm f(\alpha,\beta \pm \Phi) + \mathcal{A}_{\pi-\beta}^\pm f(\alpha,-\beta \pm \Phi) = 0, \tag{4.56}$$

valid, by analytical continuation, for any α and β.

4.5.3. Expression of f from F with a shift of origin

Considering the expression of u_d in (4.6), we can express, relatively to new coordinates (ρ_1,φ_1) when the origin is shifted at the point $O_1 \equiv (\rho_s,\varphi_s)$, the far field function F_1, as follows:

$$\begin{aligned} F_1(\varphi_1,\varphi_\circ) &= -\lim_{\rho_1\to\infty}(\sqrt{2\pi k\rho_1}e^{i\pi/4+ik\rho_1}u_d(M)) \\ &= \lim_{\rho_1\to\infty}(\sqrt{(\rho_1/\rho)}e^{ik(\rho_1-\rho)}F(\varphi,\varphi_\circ)) \\ &= e^{-ik\rho_s\cos(\varphi_1-\varphi_s)}F(\varphi_1,\varphi_\circ), \end{aligned} \tag{4.57}$$

since $\varphi \sim \varphi_1$ and $\rho = |OM| \sim \rho_1 + \frac{OO_1.O_1M}{|O_1M|}$ as $\rho_1 \to \infty$ [1]. We can then use (4.46) and (4.49) for the spectral function f_1, attached to Sommerfeld–Maliuzhinets representation for coordinates (ρ_1, φ_1) and origin O_1, with its far field function F_1, and obtain:

$$f_1(\alpha, \varphi_o) = \frac{1}{4\pi i}\int_S F_1(\alpha', \varphi_o)(\tan(\frac{1}{2}(\alpha - \alpha')) - g_o(\alpha'))d\alpha'$$

$$+ e^{ik\rho_s \cos(\varphi_o - \varphi_s)}\frac{\mathrm{Res} f|_{\alpha=\varphi_o}}{2}\cot(\frac{1}{2}(\alpha - \varphi_o)) - A_F, \tag{4.58}$$

for finite or semi-infinite scatterers, as $\alpha \in]\mathcal{S}_0 - \pi, \mathcal{S}_0 + \pi[$, and

$$f_1(\alpha, \varphi_o) = \frac{1}{4\pi i}\int_{\mathcal{S}_0^\pm} F_1(\alpha', \varphi_o)(\tan(\frac{1}{2}(\alpha - \alpha')) - g_o(\alpha'))d\alpha' - A_{F_1}, \tag{4.59}$$

for semi-infinite scatterers, as $\alpha \in]\mathcal{S}_0^\pm - \pi, \mathcal{S}_0^\pm + \pi[$, where we take, $\Phi \geq \pm\varphi_o > \pi/2$, $|\varphi_s| < \pi/2$, and $F_1(\alpha', \varphi_o) = e^{-ik\rho_s \cos(\alpha' - \varphi_s)}F(\alpha', \varphi_o)$. This expression of f_1 can be analytically continued for $|\varphi_s| > \pi/2$ by shift of $\mathcal{S}_0^\pm$, as long as the integrand continues to vanish at infinity, taking account of singularities captured during it (note: we can also write $-\cos(\alpha' - \varphi_s) = \cos(\alpha' - \varphi_s')$ with $\pi/2 < |\varphi_s'| \leq \pi$).

NOTE 4.13.– We note that

$$e^{-ik\rho_s \cos(\alpha' - \varphi_s)}F(\alpha', \varphi_o) =$$

$$= e^{ik\rho_s \cos(\pi + \alpha' - \varphi_s)}f(\pi + \alpha', \varphi_o) - e^{ik\rho_s \cos(-\pi + \alpha' - \varphi_s)}f(-\pi + \alpha', \varphi_o), \tag{4.60}$$

but the reader will notice that we have generally

$$f_1(\alpha', \varphi_o) \neq e^{ik\rho_s \cos(\alpha' - \varphi_s)}f(\alpha', \varphi_o). \tag{4.61}$$

Indeed, in the Sommerfeld–Maliuzhinets integral representation, the spectral function has to be $O(\alpha^{b'}(\cos\alpha)^{a'})$ at infinity on and inside the loops $\gamma_\pm$ of γ, where a' and b' are two constants, and we then generally need (4.59) to express f_1 from f.

NOTE 4.14.– The expression (4.1) of the incident field u^i for a shift of origin can be used with:

$$e^{iku_0.(OM)} = e^{iku_0.(OO_1 + O_1M)} = (e^{ik\rho_s \cos(\varphi_o - \varphi_s)})e^{ik\rho_1 \cos(\varphi_1 - \varphi_o)}. \tag{4.62}$$

4.6. The spectral function $f_0(\alpha)$ for the diffraction by a wedge, with passive or active impedance faces, and an illumination normal to the edge

4.6.1. *Definition of $f_0(\alpha)$ in general case, with passive or active faces*

The boundary conditions, for the 2D diffraction by a constant impedance wedge with straight faces $\varphi = \pm\Phi$, are in general:

$$(\frac{\partial}{\partial n} - ik\sin\theta^{\pm})u(\rho,\varphi)|_{\varphi=\pm\Phi} = 0, \text{i.e.} (\frac{\mp\partial}{ik\rho\partial\varphi} - \sin\theta^{\pm})u(\rho,\varphi)|_{\varphi=\pm\Phi} = 0, \tag{4.63}$$

where $\sin\theta^{\pm}$ are the relative impedances of the faces $\varphi = \pm\Phi$, with $|\mathrm{Re}\theta^{\pm}| \leq \pi/2$. These conditions are a particular case of (4.17), and the spectral function f_0 attached to the total field u then satisfies:

$$\begin{aligned}&(\sin\alpha + \sin\theta^{+})f_0(\alpha + \Phi) - (-\sin\alpha + \sin\theta^{+})f_0(-\alpha + \Phi) = 0,\\&(\sin\alpha - \sin\theta^{-})f_0(\alpha - \Phi) - (-\sin\alpha - \sin\theta^{-})f_0(-\alpha - \Phi) = 0,\end{aligned} \tag{4.64}$$

with $f_0(i\infty) = -f_0(-i\infty)$. In whole generality, we consider $s^{\pm}$ =sign(Re(sin$\theta^{\pm}$)), with Re(sin$\theta^{\pm}$) $\neq 0$, that can be positive (passive impedance face) or negative (active impedance face). In letting $\mu = \pi/2\Phi$ and φ_o the angle of illumination, we have:

$$\begin{aligned}&f_0(\alpha) = \chi_0(\alpha) - (\chi_0(i\infty) + \chi_0(-i\infty))/2,\\&\chi_0(\alpha) = \Psi(\alpha, \sin\theta^{+}, \sin\theta^{-})P_n(\sin(\mu\alpha))\sigma(\alpha)/\Psi(\varphi_o, \sin\theta^{+}, \sin\theta^{-}),\\&\sigma(\alpha) = \mu\cos(\mu\varphi_o)/(\sin(\mu\alpha) - \sin(\mu\varphi_o)),\\&\Psi(\alpha, \sin\theta^{+}, \sin\theta^{-}) = \Psi^{+}(\alpha, \sin\theta^{+})\Psi^{-}(\alpha, \sin\theta^{-}),\end{aligned} \tag{4.65}$$

where P_n is a polynomial of order n, and $f_0(i\infty) = -f_0(-i\infty)$. The function $P_n(\sin(\mu\alpha))\sigma(\alpha)$, that has the pole of incidence $\alpha = \varphi_o$ as Re$\alpha < \Phi$, verifies:

$$P_n(\sin(\mu(\alpha \pm \Phi)))\sigma(\alpha \pm \Phi) = P_n(\sin(\mu(-\alpha \pm \Phi)))\sigma(-\alpha \pm \Phi), \tag{4.66}$$

while $\Psi(\alpha)$, without any zero or pole as Re$\alpha < \Phi$, is the solution of:

$$\begin{aligned}&(\sin\alpha + \sin\theta^{+})\Psi(\alpha + \Phi) - (-\sin\alpha + \sin\theta^{+})\Psi(-\alpha + \Phi) = 0,\\&(\sin\alpha - \sin\theta^{-})\Psi(\alpha - \Phi) - (-\sin\alpha - \sin\theta^{-})\Psi(-\alpha - \Phi) = 0,\end{aligned} \tag{4.67}$$

so that f_0 verifies (4.64). The polynomial P_n will be defined so that f_0 satisfies the conditions (a)–(b)–(c) on the spectral function for a scatterer in a wedge-shaped sector illuminated by a unit plane wave.

4.6.2. *General exact derivation of* $\Psi(\alpha) = \Psi^+(\alpha)\Psi^-(\alpha)$

4.6.2.1. *Efficient expansion of* $\Psi(\alpha) = \Psi^+(\alpha)\Psi^-(\alpha)$ *with* Γ

The function $\Psi(\alpha)$ can be derived from the developments given in (4.193)–(4.194) of appendix A. From (4.67), we have:

$$\partial_\alpha \ln \frac{\Psi(\alpha \pm \Phi)}{\Psi(-\alpha \pm \Phi)} = \pm\partial_\alpha \ln \frac{(\sin\alpha - \sin\theta^\pm)}{(\sin\alpha + \sin\theta^\pm)},$$

$$\frac{\Psi'}{\Psi}(\alpha \pm \Phi) + \frac{\Psi'}{\Psi}(-\alpha \pm \Phi) = S^\pm(\alpha),$$

$$S^\pm(\alpha) = \pm s^\pm \partial_\alpha \ln \frac{(\sin\alpha - s^\pm \sin\theta^\pm)}{(\sin\alpha + s^\pm \sin\theta^\pm)}, \tag{4.68}$$

where $s^\pm$ =sign(Re($\sin\theta^\pm$)) with $0 < |\mathrm{Re}\theta^\pm| \leq \pi/2$. We can then consider $s(\alpha) = \partial \ln(\Psi(\alpha))$ as the solution of:

$$s(\alpha \pm \Phi) + s(-\alpha \pm \Phi) = S^\pm(\alpha),$$

$$S^\pm(\alpha) = \frac{\pm 2\sin\theta^\pm \cos\alpha}{(\sin\alpha + s^\pm \sin\theta^\pm)(\sin\alpha - s^\pm \sin\theta^\pm)}$$

$$= \frac{\pm s^\pm \cos\alpha}{(\sin\alpha - s^\pm \sin\theta^\pm)} - \frac{\pm s^\pm \cos\alpha}{(\sin\alpha + s^\pm \sin\theta^\pm)}. \tag{4.69}$$

which can be solved following methods described in appendix A. To use (4.193)–(4.194), we begin by expressing, using residue theorem:

$$r_\pm(\omega) = \int_{-i\infty}^{+i\infty} S^\pm(\alpha')e^{i\omega\alpha'}d\alpha' = \pm(-2\pi i)s^\pm \frac{e^{i\omega s^\pm\theta^\pm} + e^{i\omega(-s^\pm\theta^\pm+\pi)}}{1+e^{i\omega\pi}},$$

$$R_\pm(i\nu) = \frac{-i}{2\sqrt{2\pi}} r_\pm(i\nu) = \mp\frac{s^\pm\sqrt{2\pi}}{2}\frac{e^{-\nu s^\pm\theta^\pm} + e^{-\nu(-s^\pm\theta^\pm+\pi)}}{1+e^{-\nu\pi}}, \tag{4.70}$$

where $0 < s^\pm\mathrm{Re}\theta^\pm \leq \pi/2$, and we derive

$$\Psi(\alpha) = \Psi^+(\alpha, \sin\theta^+)\Psi^-(\alpha, \sin\theta^-),$$

$$\frac{\Psi^{\pm\prime}}{\Psi^\pm}(\alpha) = is^\pm \int_0^{i\infty} \frac{e^{i\omega s^\pm\theta^\pm} + e^{i\omega(-s^\pm\theta^\pm+\pi)}}{(1+e^{i\omega\pi})\sin(2\omega\Phi)} \sin(\omega(\alpha \pm \Phi))d\omega$$

$$= -s^\pm \int_0^\infty \frac{e^{-\nu s^\pm\theta^\pm} + e^{-\nu(-s^\pm\theta^\pm+\pi)}}{(1+e^{-\nu\pi})} \frac{\sinh(\nu(\alpha \pm \Phi))}{\sinh(2\nu\Phi)} d\nu$$

$$= -s^\pm \int_0^\infty \frac{\cosh(\nu(\pi/2 - s^\pm\theta^\pm))}{\cosh(\nu\pi/2)} \frac{\sinh(\nu(\alpha \pm \Phi))}{\sinh(2\nu\Phi)} d\nu, \tag{4.71}$$

as $|\text{Re}\alpha| \leq \Phi$, with $\Psi^{\pm}(\alpha) \equiv \Psi^{\pm}(\alpha, \sin\theta^{\pm})$. After integration, we then obtain:

$$\Psi^{\pm}(\alpha) = B^{\pm} \exp\Big(s^{\pm} \int_0^{\infty} \frac{e^{-\nu s^{\pm}\theta^{\pm}} + e^{-\nu(\pi - s^{\pm}\theta^{\pm})}}{(1+e^{-\nu\pi})} \frac{(1-\cosh(\nu(\alpha \pm \Phi)))}{\nu \sinh(2\nu\Phi)} d\nu\Big), \tag{4.72}$$

as $|\text{Re}\alpha| \leq \Phi$, where $B^{\pm} = \Psi^{\pm}(\mp\Phi)$ is a free constant, that we could take equal to unity without changing the expression of f_0. Considering that we have [21]:

$$\begin{aligned}
&(1+e^{i\omega\pi})^{-1} = \sum_{m=0}^{N} (-e^{i\omega\pi})^m + \frac{(-e^{i\omega\pi})^{N+1}}{1+e^{i\omega\pi}}, \\
&\int_0^{\infty} e^{-\mu x} \frac{\sinh(\beta x)}{\sinh(bx)} dx = \frac{1}{2b}(\partial \ln(\Gamma)(\frac{1}{2} + \frac{\mu+\beta}{2b}) - \partial \ln(\Gamma)(\frac{1}{2} + \frac{\mu-\beta}{2b})) \\
&= \partial_{\beta}(\ln(\Gamma(\frac{1}{2} + \frac{(\mu+\beta)}{2b})\Gamma(\frac{1}{2} + \frac{(\mu-\beta)}{2b}))), \\
&\int_0^{\infty} e^{-\mu x} \frac{1-\cosh(\beta x)}{x \sinh(bx)} dx = -\ln(\frac{(\Gamma(\frac{1}{2} + \frac{(\mu+\beta)}{2b})\Gamma(\frac{1}{2} + \frac{(\mu-\beta)}{2b})}{\Gamma(\frac{1}{2} + \frac{\mu}{2b})\Gamma(\frac{1}{2} + \frac{\mu}{2b})}),
\end{aligned} \tag{4.73}$$

as $|\text{Re}(\mu \pm \beta)| < |\text{Re}(b)|$, leads us to expand (4.72), as $|\text{Re}\alpha| \leq \Phi + (N+1)\pi$, following:

$$\begin{aligned}
\Psi^{\pm}(\alpha) = {} & B_N^{\pm} \prod_{m=0}^{N} [\Gamma(\frac{1}{2} + \frac{1}{4\Phi}(\alpha \pm \Phi + s^{\pm}\theta^{\pm} + m\pi)) \\
& \Gamma(\frac{1}{2} - \frac{1}{4\Phi}(\alpha \pm \Phi - (s^{\pm}\theta^{\pm} + m\pi)))\Gamma(\frac{1}{2} + \frac{1}{4\Phi}(\alpha \pm \Phi + \pi - s^{\pm}\theta^{\pm} + m\pi)) \\
& \Gamma(\frac{1}{2} - \frac{1}{4\Phi}(\alpha \pm \Phi - (\pi - s^{\pm}\theta^{\pm} + m\pi)))]^{s^{\pm}(-1)^{m+1}} \\
& \times \exp\Big(s^{\pm} \int_0^{\infty} (-e^{-\nu\pi})^{N+1} \frac{e^{-\nu s^{\pm}\theta^{\pm}} + e^{-\nu(\pi - s^{\pm}\theta^{\pm})}}{(1+e^{-\nu\pi})} \frac{(1-\cosh(\nu(\alpha \pm \Phi)))}{\nu \sinh(2\nu\Phi)} d\nu\Big),
\end{aligned} \tag{4.74}$$

where N is an arbitrary positive integer, $s^{\pm}$ =sign(Re($\sin\theta^{\pm}$)), $0 < |\text{Re}\theta^{\pm}| \leq \pi/2$. The $B_N^{\pm}$ are constants depending on N, while $\Psi^{\pm}(\alpha)/\Psi^{\pm}(\varphi_o)$ does not depend on it.

Since $\mathrm{e}^{-\nu(N+1)\pi}$ rapidly decreases for large ν or N, this expression has the advantage of giving an efficient calculation for $N \sim 1$, and all analytic properties from those of Γ function when we consider $N \to \infty$. It is then particularly simple to note that $\Psi(\alpha)$ is a meromorphic function of the complex plane.

4.6.2.2. *Efficient expansion of Maliuzhinets function Ψ_Φ with Γ*

We note, from (4.72) and the use of $2\cos(a)\cos(b) = \cos(a+b) + \cos(a-b)$, that we have:

$$\frac{\Psi(\alpha)}{\Psi(\varphi')} = \prod_{\pm} \exp\Big(s^{\pm}\int_0^{\infty}\sum_{\epsilon'=-1,1}\Big(\frac{\cosh(\nu(\varphi' \pm \Phi + \epsilon'(s^{\pm}\theta_1^{\pm} - \frac{\pi}{2}))) - 1}{2\nu\sinh(2\nu\Phi)\cosh(\frac{\nu\pi}{2})}$$

$$+ \frac{1 - \cosh(\nu(\alpha \pm \Phi + \epsilon'(s^{\pm}\theta_1^{\pm} - \frac{\pi}{2})))}{2\nu\sinh(2\nu\Phi)\cosh(\frac{\nu\pi}{2})}\Big)d\nu\Big), \qquad (4.75)$$

for $|\mathrm{Re}\alpha| < \Phi$. We can then write Ψ, for $|\mathrm{Re}\alpha + \epsilon'(s^{\pm}\theta_1^{\pm} - \frac{\pi}{2})| < \Phi + (N+1)\pi$ as $\epsilon' = -1$ and $+1$:

$$\Psi(\alpha) = C_N \prod_{\epsilon'=-1,1}\Big(\prod_{\pm}\prod_{m\geq 0}^{N}[\Gamma(\frac{1}{2} + \frac{(\alpha \pm \Phi + \epsilon'(s^{\pm}\theta_1^{\pm} - \frac{\pi}{2})) + \frac{\pi}{2} + m\pi}{4\Phi})$$

$$\times \Gamma(\frac{1}{2} + \frac{-(\alpha \pm \Phi + \epsilon'(s^{\pm}\theta_1^{\pm} - \frac{\pi}{2})) + \frac{\pi}{2} + m\pi}{4\Phi})]^{s^{\pm}(-1)^{m+1}}$$

$$\times \exp\Big(s^{\pm}\int_0^{\infty}(-e^{-\nu\pi})^{N+1}\frac{1 - \cosh(\nu(\alpha \pm \Phi + \epsilon'(s^{\pm}\theta_1^{\pm} - \frac{\pi}{2})))}{2\nu\sinh(2\nu\Phi)\cosh(\frac{\nu\pi}{2})}d\nu\Big)\Big)$$

$$= C_N \prod_{\epsilon'=-1,1}\prod_{\pm}(\Psi_{\Phi,N}(\alpha \pm \Phi + \epsilon'(s^{\pm}\theta_1^{\pm} - \frac{\pi}{2})))^{s^{\pm}}, \qquad (4.76)$$

where C_N is a constant, and $\Psi_{\Phi,N}(\alpha)$ is given, for $|\mathrm{Re}\alpha| < 2\Phi + (N+1)\pi + \frac{\pi}{2}$, by:

$$\Psi_{\Phi,N}(\alpha) = \Big(\prod_{m\geq 0}^{N}[\Gamma(\frac{1}{2} + \frac{\alpha + \frac{\pi}{2} + m\pi}{4\Phi})\Gamma(\frac{1}{2} + \frac{-\alpha + \frac{\pi}{2} + m\pi}{4\Phi})]^{(-1)^{m+1}}\Big)$$

$$\times \exp\Big(\int_0^{\infty}(-e^{-\nu\pi})^{N+1}e^{-\nu\frac{\pi}{2}}\frac{1 - \cosh(\nu\alpha)}{\nu\sinh(2\nu\Phi)(1 + e^{-\nu\pi})}d\nu\Big), \qquad (4.77)$$

As for the expansion of $\Psi^{\pm}(\alpha)$ with Γ function, $\Psi_{\Phi,N}(\alpha)/\Psi_{\Phi,N}(0)$ is independent of N. Comparing this expression with the one given for the Maliuzhinets function Ψ_Φ given in [30], we remark that:

$$\Psi_{\Phi,N=-1}(\alpha) = \Psi_\Phi(\alpha),\ \Psi_{\Phi,N=-1}(0) = 1, \qquad (4.78)$$

and finally,

$$\Psi_{\Phi,N}(\alpha)/\Psi_{\Phi,N}(0) = \Psi_\Phi(\alpha), \qquad (4.79)$$

which represents a new efficient expansion of $\Psi_\Phi(\alpha)$ with Γ function.

NOTE 4.15.– We also note that $(\Psi_\Phi(\alpha))^2 = \frac{\Psi^{\pm}(\alpha\mp\Phi, \sin\theta^{\pm}=1)}{\Psi^{\pm}(\mp\Phi, \sin\theta^{\pm}=1)}$.

NOTE 4.16.– For $|\text{Im}\alpha| \to \infty$, the principal contribution, in the integral expressions (4.72), comes from the vicinity of $\nu = 0$. We then note that:

$$\int_0^{\infty} \nu^{2m} \frac{\sinh(\beta\nu)}{\sinh(b\nu)} d\nu = \partial_{\beta^{1+2m}} (\ln(\frac{\pi}{\cos(\frac{\pi}{2b}\beta)})),$$

$$\partial_\alpha(\ln(\Psi^{\pm}(\alpha)) = s^{\pm}\partial_\alpha(\ln(\cos(\frac{\pi(\alpha \pm \Phi)}{4\Phi}))) + O(1/\alpha^2), \tag{4.80}$$

and thus,

$$\Psi^{\pm}(\alpha) \sim C_{\pm}(\cos(\frac{\pi}{4\Phi}(\alpha \pm \Phi)))^{s^{\pm}} \text{ with } C_{\pm} \text{ a constant.} \tag{4.81}$$

A complete asymptotics for $\Psi^{\pm}$ will be given in a further section.

4.6.2.3. *Additional expressions for* $\Psi^{\pm}$ *and* Ψ_{Φ}

Other expressions of $\Psi^{\pm}$ and Maliuzhinets function Ψ_{Φ} can be derived from, this time, the use of (4.195) in appendix A. Considering $s(\alpha) = \partial_\alpha \ln \Psi^{\pm}(\alpha, \sin\theta^{\pm})$, $S^{\pm}(\alpha) = \pm s^{\pm}\partial_\alpha \ln \frac{(\sin\alpha - s^{\pm}\sin\theta^{\pm})}{(\sin\alpha + s^{\pm}\sin\theta^{\pm})}$ and $S^{\mp}(\alpha) = 0$ with $s^{\pm}$ =sign(Re($\sin\theta^{\pm}$)), we have, from (4.195), as $|\text{Re}\alpha| < \Phi$, with $\text{Re}(\sin\theta^{\pm}) \neq 0$:

$$S^{\pm}(\alpha) = \pm s^{\pm}\partial_\alpha(\ln\left(\tan\left(\frac{(\alpha - s^{\pm}\theta^{\pm})}{2}\right)\right) - \ln\left(\tan\left(\frac{(\alpha + s^{\pm}\theta^{\pm})}{2}\right)\right))$$

$$= \mp s^{\pm}\left(\frac{1}{\cos(\alpha + (s^{\pm}\theta^{\pm} - \frac{\pi}{2}))} + \frac{1}{\cos(\alpha - (s^{\pm}\theta^{\pm} - \frac{\pi}{2}))}\right),$$

$$\partial_\alpha \ln \Psi^{\pm}(\alpha, \sin\theta^{\pm}) = \frac{-i}{8\Phi}\int_{-i\infty}^{+i\infty} \pm S^{+}(\alpha')\tan(\frac{\pi}{4\Phi}(\alpha \pm \Phi - \alpha'))d\alpha'$$

$$= \frac{is^{\pm}}{8\Phi}\int_{-i\infty}^{+i\infty} \sum_{\epsilon=-1,1} \frac{1}{\cos(\alpha' + \epsilon(s^{\pm}\theta^{\pm} - \frac{\pi}{2}))}\tan(\frac{\pi}{4\Phi}(\alpha \pm \Phi - \alpha'))d\alpha', \tag{4.82}$$

where, by definition, $0 < \text{Re}(s^{\pm}\theta^{\pm}) \leq \pi/2$. After integration, we can choose to add a constant, in the form of a suitable integral, and write as $|\text{Re}\alpha| < 2\Phi$:

$$\ln \Psi^{\pm}(\alpha \mp \Phi, \sin\theta^{\pm}) = \frac{-is^{\pm}}{2\pi}\int_{-i\infty}^{+i\infty} \sum_{\epsilon=-1,1} \frac{\ln\left(\frac{\cos(\frac{\pi}{4\Phi}(\alpha-\alpha'))}{\cos(\frac{\pi}{4\Phi}\alpha')}\right)}{\cos(\alpha' + \epsilon(s^{\pm}\theta^{\pm} - \frac{\pi}{2}))}d\alpha' + \ln B^{\pm}$$

$$= \frac{-is^{\pm}}{2\pi}\ln(\cos(\frac{\pi}{4\Phi}\alpha))\int_{-i\infty}^{+i\infty} \sum_{\epsilon=-1,1} \frac{1}{\cos(\alpha' + \epsilon(s^{\pm}\theta^{\pm} - \frac{\pi}{2}))}d\alpha' + \ln B^{\pm}$$

$$+ \frac{-is^{\pm}}{2\pi}\int_{-i\infty}^{+i\infty} \sum_{\epsilon=-1,1} \frac{1}{\cos(\alpha' + \epsilon(s^{\pm}\theta^{\pm} - \frac{\pi}{2}))}\ln\left(\frac{\cos(\frac{\pi}{4\Phi}(\alpha - \alpha'))}{\cos(\frac{\pi}{4\Phi}\alpha)\cos(\frac{\pi}{4\Phi}\alpha')}\right)d\alpha', \tag{4.83}$$

where we notice that $B^{\pm} = \Psi^{\pm}(\mp\Phi, \sin\theta^{\pm})$.

In a similar manner, but, this time, when $|\mathrm{Re}(\alpha+\epsilon(s^{\pm}\theta^{\pm}-\frac{\pi}{2}))|<2\Phi$ for $\epsilon=-1$ and $\epsilon=1$, we can write, after integration and addition of another constant integral:

$$\ln\Psi^{\pm}(\alpha\mp\Phi,\sin\theta^{\pm})=\frac{-is^{\pm}}{2\pi}\int_{-i\infty}^{+i\infty}\sum_{\epsilon=-1,1}\frac{\ln\left(\frac{\cos(\frac{\pi}{4\Phi}(\alpha-\alpha'))}{\cos(\frac{\pi}{4\Phi}\alpha'+\epsilon(s^{\pm}\theta^{\pm}-\frac{\pi}{2}))}\right)}{\cos(\alpha'+\epsilon(s^{\pm}\theta^{\pm}-\frac{\pi}{2}))}d\alpha'$$

$$+\ln C^{\pm}=\frac{-is^{\pm}}{2\pi}\int_{-i\infty}^{+i\infty}\sum_{\epsilon=-1,1}\frac{\ln\left(\frac{\cos(\frac{\pi}{4\Phi}(\alpha+\epsilon(s^{\pm}\theta^{\pm}-\frac{\pi}{2})-\alpha'))}{\cos(\frac{\pi}{4\Phi}\alpha')}\right)}{\cos\alpha'}d\alpha'+\ln C^{\pm}$$

$$=s^{\pm}\sum_{\epsilon=-1,1}\ln(\cos(\frac{\pi}{4\Phi}\alpha+\epsilon(s^{\pm}\theta^{\pm}-\frac{\pi}{2})))(\frac{-i}{2\pi}\int_{-i\infty}^{+i\infty}\frac{d\alpha'}{\cos\alpha'})+\ln C^{\pm}$$

$$+\frac{-is^{\pm}}{2\pi}\int_{-i\infty}^{+i\infty}\sum_{\epsilon=-1,1}\frac{\ln\left(\frac{\cos(\frac{\pi}{4\Phi}(\alpha+\epsilon(s^{\pm}\theta^{\pm}-\frac{\pi}{2})-\alpha'))}{\cos(\frac{\pi}{4\Phi}(\alpha+\epsilon(s^{\pm}\theta^{\pm}-\frac{\pi}{2})))\cos(\frac{\pi}{4\Phi}\alpha')}\right)}{\cos\alpha'}d\alpha',\tag{4.84}$$

where we have used a shift of the integration path, permitted from $0\leq\mathrm{Re}(\pi/2-s^{\pm}\theta^{\pm})<\pi/2$.

Therefore, considering that:

$$\int_{-i\infty}^{+i\infty}\frac{-i}{\cos(\alpha'+\epsilon(s^{\pm}\theta^{\pm}-\frac{\pi}{2}))}d\alpha'=\int_{-i\infty}^{+i\infty}\frac{-i}{\cos\alpha'}d\alpha'=\pi.\tag{4.85}$$

we obtain, from (4.83), as $|\mathrm{Re}\alpha|<2\Phi$:

$$\Psi^{\pm}(\alpha\mp\Phi,\sin\theta^{\pm})=B^{\pm}\exp\Big(s^{\pm}\ln(\cos(\frac{\pi}{4\Phi}\alpha))$$

$$+\frac{-is^{\pm}}{2\pi}\sum_{\epsilon=-1,1}\int_{-i\infty}^{+i\infty}\frac{\ln(1+\tan(\frac{\pi}{4\Phi}\alpha')\tan(\frac{\pi}{4\Phi}\alpha))d\alpha'}{\cos(\alpha'+\epsilon(s^{\pm}\theta^{\pm}-\frac{\pi}{2}))}\Big)$$

$$=B^{\pm}(\cos(\frac{\pi}{4\Phi}\alpha))^{s^{\pm}}\prod_{\epsilon=-1,1}\exp\Big(s^{\pm}\int_{-\infty}^{+\infty}\frac{\ln(1+i\tanh(\frac{\pi}{4\Phi}\nu)\tan(\frac{\pi}{4\Phi}\alpha))}{2\pi\cos(i\nu+\epsilon(s^{\pm}\theta^{\pm}-\frac{\pi}{2}))}d\nu\Big),\tag{4.86}$$

while, from (4.84), when $|\mathrm{Re}(\alpha+\epsilon(s^{\pm}\theta^{\pm}-\frac{\pi}{2}))|<2\Phi$ are both satisfied as $\epsilon=-1$ and $\epsilon=1$, we get:

$$\Psi^{\pm}(\alpha\mp\Phi,\sin\theta^{+})=C^{\pm}\prod_{\epsilon=-1,1}\Big((\cos(\frac{\pi}{4\Phi}(\alpha+\epsilon(s^{\pm}\theta^{\pm}-\frac{\pi}{2}))))^{\frac{s^{\pm}}{2}}$$

$$\exp\Big(s^{\pm}\int_{-\infty}^{+\infty}\frac{\ln(1+i\tanh(\frac{\pi}{4\Phi}\nu)\tan(\frac{\pi}{4\Phi}(\alpha+\epsilon(s^{\pm}\theta^{\pm}-\frac{\pi}{2}))))}{2\pi\cosh\nu}d\nu\Big)\Big).\tag{4.87}$$

The terms $B^{\pm}$ and $C^{\pm}$ are free constants that we can take arbitrary without changing the spectral function, while we can let $\tanh(\frac{\pi}{4\Phi}\nu)$ for $-\tanh(\frac{\pi}{4\Phi}\nu)$ in (4.86) and (4.87) without changing the expressions, so that we can note that $\Psi^{\pm}(\alpha \mp \Phi, \sin\theta^{\pm}) = \Psi^{\pm}(-\alpha \mp \Phi, \sin\theta^{\pm})$ is well verified.

Considering the definition of $\Psi^{\pm}$ with function Ψ_{Φ} (as used in (4.76)–(4.78)) following:

$$\Psi^{\pm}(\alpha \mp \Phi) = C^{\pm} \prod_{\epsilon'=-1,1} \Psi_{\Phi}(\alpha + \epsilon'(s^{\pm}\theta_1^{\pm} - \frac{\pi}{2}))^{s^{\pm}} \text{ with } \Psi_{\Phi}(0) = 1, \tag{4.88}$$

we compare it with (4.87), and obtain, in a original manner, another expression of the Maliuzhinets function Ψ_{Φ} for $|\mathrm{Re}(\alpha)| < 2\Phi$:

$$\Psi_{\Phi}(\alpha) = (\cos(\frac{\pi}{4\Phi}\alpha))^{\frac{1}{2}} \exp\Big(\int_{-\infty}^{+\infty} \frac{\ln(1 - i\tanh(\frac{\pi}{4\Phi}\nu)\tan(\frac{\pi}{4\Phi}\alpha))}{2\pi\cosh\nu} d\nu\Big). \tag{4.89}$$

This expression, previously quoted in [12], is useful to give the asymptotics of Ψ_{Φ}, in particular its first order, as $|\mathrm{Im}\alpha| \to \infty$.

NOTE 4.17.– Let us note that we can write:

$$\Psi^{\pm}(\alpha \mp \Phi, \sin\theta^{\pm}) \sim B^{\pm} \text{ as } |\mathrm{Im}(\theta^{\pm})| \to \infty. \tag{4.90}$$

4.6.3. *Complete asymptotic expressions of* $\frac{\Psi^{\pm\prime}}{\Psi^{\pm}}$ *and* $\Psi^{\pm}$

4.6.3.1. *Asymptotic expressions*

For a complete asymptotic expansions of $\frac{\Psi^{\pm\prime}}{\Psi^{\pm}}$ and $\Psi^{\pm}$ for large $\mathrm{Im}\alpha$, we can transform (4.71) and (4.72) expressions, using residue theorem. Considering $\frac{\Psi^{\pm\prime}}{\Psi^{\pm}}$, for $\mathrm{Im}\alpha \neq 0$ with $\epsilon = \mathrm{sign}(\mathrm{Im}\alpha)$ and $|\mathrm{Re}\alpha| < \Phi$, we develop (4.71), following:

$$\begin{aligned}
\frac{\Psi^{\pm\prime}}{\Psi^{\pm}}(\alpha) &= is^{\pm} \int_0^{i\infty} \frac{e^{i\omega s^{\pm}\theta^{\pm}} + e^{i\omega(-s^{\pm}\theta^{\pm}+\pi)}}{(1+e^{i\omega\pi})\sin(2\omega\Phi)} \sin(\omega(\alpha \pm \Phi))d\omega \\
&= \frac{\epsilon s^{\pm}}{4} \int_{-i\infty}^{i\infty} \frac{\cos(\omega(\pi/2 - s^{\pm}\theta^{\pm}))}{\cos(\omega\pi/2)\sin(2\omega\Phi)} (e^{i\omega\epsilon(\alpha\pm\Phi)} - e^{-i\omega\epsilon(\alpha\pm\Phi)})d\omega \\
&= -\pi i(\frac{\epsilon s^{\pm}}{4\Phi}) + \frac{\epsilon s^{\pm}}{4} \int_{0^+-i\infty}^{0^++i\infty} \frac{\cos(\omega(\pi/2 - s^{\pm}\theta^{\pm}))}{\cos(\omega\pi/2)\sin(2\omega\Phi)} e^{i\omega\epsilon(\alpha\pm\Phi)} d\omega \\
&\quad - \frac{\epsilon s^{\pm}}{4} \int_{-0^+-i\infty}^{-0^++i\infty} \frac{\cos(\omega(\pi/2 - s^{\pm}\theta^{\pm}))}{\cos(\omega\pi/2)\sin(2\omega\Phi)} e^{-i\omega\epsilon(\alpha\pm\Phi)} d\omega \\
&= -\pi i(\frac{\epsilon s^{\pm}}{4\Phi}) + 2 \times \frac{\epsilon s^{\pm}}{4} \int_{0^+-i\infty}^{0^++i\infty} \frac{\cos(\omega(\pi/2 - s^{\pm}\theta^{\pm}))}{\cos(\omega\pi/2)\sin(2\omega\Phi)} e^{i\omega\epsilon(\alpha\pm\Phi)} d\omega.
\end{aligned} \tag{4.91}$$

Shifting the path to $\mathrm{Re}\omega = \delta > 0$, while considering residues (first- and second-order ones) due to the capture of poles between $\mathrm{Re}\omega = 0^+$ and $\mathrm{Re}\omega = \delta$, we obtain:

$$
\begin{aligned}
&\frac{\Psi^{\pm\prime}}{\Psi^{\pm}}(\alpha) = -i\epsilon s^{\pm}(\frac{\pi}{4\Phi}) + \frac{\epsilon s^{\pm}}{2}(-2\pi i) \\
&\times \Bigg[\Big(\sum_{m\geq 1, \frac{m\pi}{2\Phi}\neq 2p+1}^{\frac{m\pi}{2\Phi}<\delta} (-1)^m \frac{1}{2\Phi}\frac{\cos(\frac{m\pi(\pi/2-s^{\pm}\theta^{\pm})}{2\Phi})}{\cos(\frac{m\pi(\pi/2)}{2\Phi})} e^{i\frac{m\pi}{2\Phi}\epsilon(\alpha\pm\Phi)}\Big) \\
&+\Big(\sum_{p\geq 0, 2p+1\neq\frac{m\pi}{2\Phi}}^{2p+1<\delta} (-1)^{p+1}\frac{2}{\pi}\frac{\cos((2p+1)(\pi/2-s^{\pm}\theta^{\pm}))}{\sin(2(2p+1)\Phi)} e^{i(2p+1)\epsilon(\alpha\pm\Phi)}\Big) \\
&+\Big(\sum_{p\geq 0, 2p+1=\frac{m\pi}{2\Phi}}^{2p+1<\delta} \frac{(-1)^{m+p+1}\frac{2}{\pi}}{2\Phi}(-(\pi/2-s^{\pm}\theta^{\pm})\sin(\frac{m\pi(\pi/2-s^{\pm}\theta^{\pm})}{2\Phi}) \\
&+ i\epsilon(\alpha\pm\Phi)\cos(\frac{m\pi(\pi/2-s^{\pm}\theta^{\pm})}{2\Phi}))e^{i\frac{m\pi}{2\Phi}\epsilon(\alpha\pm\Phi)}\Big)\Bigg] \\
&+\frac{\epsilon s^{\pm}}{2}e^{i\delta\epsilon(\alpha\pm\Phi)}\int_{\delta-i\infty}^{\delta+i\infty}\frac{\cos(\omega(\pi/2-s^{\pm}\theta^{\pm}))}{\cos(\omega\pi/2)\sin(2\omega\Phi)}e^{i(\omega-\delta)\epsilon(\alpha\pm\Phi)}d\omega,
\end{aligned}
\tag{4.92}
$$

taking care that $\frac{\delta 2\Phi}{\pi}$ and $\frac{\delta-1}{2}$ are not integers. After integration of (4.92), we have:

$$
\begin{aligned}
&\Psi^{\pm}(\alpha) = B^{\pm}e^{-i\epsilon s^{\pm}\frac{\pi}{4\Phi}(\alpha\pm\Phi)} \\
&\times\exp\left[\frac{s^{\pm}}{2}\int_{0^+-i\infty}^{0^++i\infty}\frac{\cos(\omega(\pi/2-s^{\pm}\theta^{\pm}))}{\cos(\omega\pi/2)\sin(2\omega\Phi)}\frac{e^{i\epsilon\omega(\alpha\pm\Phi)}-1}{i\omega}d\omega\right] \\
&= B^{\pm}e^{D_{\pm}}e^{-i\epsilon s^{\pm}\frac{\pi}{4\Phi}(\alpha\pm\Phi)} \\
&\times\exp\Bigg[-\pi s^{\pm}\Bigg(\Big(\sum_{m\geq 1,\frac{m\pi}{2\Phi}\neq 2p+1}^{\frac{m\pi}{2\Phi}<\delta}\frac{(-1)^m\frac{1}{2\Phi}\cos(\frac{m\pi(\pi/2-s^{\pm}\theta^{\pm})}{2\Phi})}{\frac{m\pi}{2\Phi}\cos(\frac{m\pi(\pi/2)}{2\Phi})}e^{i\epsilon\frac{m\pi}{2\Phi}(\alpha\pm\Phi)}\Big) \\
&+\Big(\sum_{p\geq 0,2p+1\neq\frac{m\pi}{2\Phi}}^{2p+1<\delta}\frac{(-1)^{p+1}\frac{2}{\pi}\cos((2p+1)(\pi/2-s^{\pm}\theta^{\pm}))}{(2p+1)\sin(2(2p+1)\Phi)}e^{i\epsilon(2p+1)(\alpha\pm\Phi)}\Big) \\
&+\Big(\sum_{p\geq 0,2p+1=\frac{m\pi}{2\Phi}}^{2p+1<\delta}\frac{(-1)^{m+p+1}\frac{2}{\pi}}{2\Phi}\frac{2\Phi}{m\pi}\Big(-(\frac{\pi}{2}-s^{\pm}\theta^{\pm})\sin(\frac{m\pi(\frac{\pi}{2}-s^{\pm}\theta^{\pm})}{2\Phi}) \\
&+\cos(\frac{m\pi(\frac{\pi}{2}-s^{\pm}\theta^{\pm})}{2\Phi})(i\epsilon(\alpha\pm\Phi)-\frac{2\Phi}{m\pi})\Big)e^{i\frac{m\pi}{2\Phi}\epsilon(\alpha\pm\Phi)}\Big)\Bigg)
\end{aligned}
$$

$$+\frac{s^{\pm}e^{i\epsilon\delta(\alpha\pm\Phi)}}{2\delta}\int_{\delta-i\infty}^{\delta+i\infty}\frac{\delta\cos(\omega(\pi/2-s^{\pm}\theta^{\pm}))}{i\omega\cos(\omega\pi/2)\sin(2\omega\Phi)}e^{i\epsilon(\omega-\delta)(\alpha\pm\Phi)}d\omega\Bigg], \tag{4.93}$$

where we note that $B^{\pm}=\Psi^{\pm}(\mp\Phi,\sin\theta^{\pm})$, while:

$$D_{\pm}=\frac{-s^{\pm}}{2i}\int_{0^{+}-i\infty}^{0^{+}+i\infty}\frac{\cos(\omega(\pi/2-s^{\pm}\theta^{\pm}))}{\cos(\omega\pi/2)\sin(2\omega\Phi)}\frac{1}{\omega}d\omega$$

$$=\frac{-s^{\pm}}{2i}\int_{-i\infty}^{+i\infty}\partial_{\omega}((\frac{\cos(\omega\pi/2)\frac{\sin(2\omega\Phi)}{\omega}}{\cos(\omega(\pi/2-s^{\pm}\theta^{\pm}))})^{-1})\frac{1}{\omega}d\omega. \tag{4.94}$$

The last integral terms in (4.92) and (4.93) being $O(1)$, and considering that:

$$|\exp(z)-\sum_{m=0}^{M}\frac{z^{m}}{m!}|=|\int_{0}^{z}\frac{t^{M}}{M!}e^{t}dt|\leq\frac{|z|^{M+1}\max(1,|e^{z}|)}{(M+1)!},$$

$$\exp(z_1+z_2+...)=\exp(z_1)\times\exp(z_2)\times ..., \tag{4.95}$$

we have obtained, thanks to (4.92) and (4.93) with δ arbitrarily large, the complete asymptotics of $\frac{\Psi^{\pm\prime}}{\Psi^{\pm}}(\alpha)$ and $\Psi^{\pm}(\alpha)$ for large $|\mathrm{Im}\alpha|$, as $\epsilon=\mathrm{sign}(\mathrm{Im}\alpha)$, $|\mathrm{Re}\alpha|<\Phi$.

NOTE 4.18.– Second-order residue in (4.92) and (4.93) concerns the case $2p+1=\frac{m\pi}{2\Phi}$. It implies that we have $\frac{\pi}{2\Phi}=\frac{2p_0+1}{m_0}$, $2p+1=(2q+1)(2p_0+1)$, and $m=(2q+1)m_0$, with $q=0,1,2,3,\ ...,$ when $2p_0+1$ and m_0 are mutually prime.

4.6.3.2. *Some complements on the determination of* $D_{\pm}$

For the calculus of $D_{\pm}$, we can also write:

$$D_{\pm}=\frac{is^{\pm}}{2}\int_{-i\infty}^{+i\infty}\frac{(\frac{\cos(\omega(\pi/2-s^{\pm}\theta^{\pm}))}{\cos(\omega\pi/2)}-1)}{\omega\sin(2\omega\Phi)}d\omega+\frac{is^{\pm}}{2}\int_{0^{+}-i\infty}^{0^{+}+i\infty}\frac{1}{\sin(2\omega\Phi)}\frac{1}{\omega}d\omega. \tag{4.96}$$

We can then use a method of integration by parts, and shifts of integration path, to reduce the first and second integrals. We note first that we have:

$$\Big(\frac{\cos(\omega a)}{\cos(\omega\pi/2)}-1\Big)\frac{1}{\omega}=\frac{1}{\pi i}\sum_{\epsilon'=+1,-1}\Big(\int_{-i\infty}^{+i\infty}\frac{(\cos(\omega(\alpha+\epsilon' a))-1)}{2\omega\cos(\alpha)}d\alpha\Big)$$

$$=\frac{1}{\pi i}\sum_{\epsilon'=+1,-1}\Big(\int_{-i\infty}^{+i\infty}\frac{(\cos(\omega\alpha)-1)}{2\omega\cos(\alpha+\epsilon' a)}d\alpha\Big)$$

$$=\frac{1}{\pi i}\sum_{\epsilon'=+1,-1}\int_{0}^{i\infty}\sin(\omega\alpha)\ln\Big(\frac{1}{i}\tan(\frac{\alpha+\epsilon' a+\frac{\pi}{2}}{2})\Big)d\alpha, \tag{4.97}$$

for $|\mathrm{Re}\omega| < 1$, $|\mathrm{Re}a| < \pi/2$, while we remark that:

$$\begin{aligned}
&\frac{i}{2}\int_{-i\infty+0^+}^{+i\infty+0^+} \frac{1}{\omega\sin(2\omega\Phi)}d\omega = \frac{i}{2}\int_{-i\infty+0^+}^{+i\infty+0^+} \frac{1}{\omega\sin(\omega\pi/2)}d\omega \\
&= \frac{i}{2}\int_{-i\infty}^{+i\infty} \frac{1}{(1+\omega)\cos(\omega\pi/2)}d\omega = i\int_{0}^{+i\infty} \frac{1}{(1-\omega^2)\cos(\omega\pi/2)}d\omega \\
&= -\int_{0}^{\infty} \frac{1}{(1+\nu^2)\cosh(\nu\pi/2)}d\nu = -\ln(2). \qquad (4.98)
\end{aligned}$$

Considering these results, we can then write:

$$\begin{aligned}
&D_\pm + s^\pm \ln(2) = \frac{i}{2}\frac{2s^\pm}{\pi i}\sum_{\epsilon'=+1,-1}\int_0^{i\infty} \ln\Big(\frac{1}{i}\tan\Big(\frac{\alpha+\epsilon'(\pi/2-s^\pm\theta^\pm)+\frac{\pi}{2}}{2}\Big)\Big) \\
&\times \int_0^{i\infty} \frac{\sin(\omega\alpha)}{\sin(2\omega\Phi)}d\omega d\alpha \\
&= \frac{s^\pm}{\pi}\sum_{\epsilon'=+1,-1}\int_0^{i\infty} \ln\Big(\frac{1}{i}\tan\Big(\frac{\alpha+\epsilon'(\pi/2-s^\pm\theta^\pm)+\frac{\pi}{2}}{2}\Big)\Big)\frac{\pi i}{4\Phi}\tan\Big(\frac{\pi}{4\Phi}\alpha\Big)d\alpha \\
&= \frac{-1}{i}\frac{s^\pm}{2\pi}\sum_{\epsilon'=+1,-1}\int_{-i\infty}^{+i\infty} \frac{\ln(\cos(\frac{\pi}{4\Phi}\alpha))}{\cos(\alpha+\epsilon'(\pi/2-s^\pm\theta^\pm))}d\alpha, \qquad (4.99)
\end{aligned}$$

after using integration by parts. With this reduction of $D_\pm$, it is worth noticing that the first order of this asymptotics is in perfect agreement with the one given by (4.86).

4.6.4. *Links between $\Psi^\pm$ for passive and active impedances, and positions of poles*

4.6.4.1. *The link between $\Psi^\pm$ for passive and active impedances*

For arbitrary sign of $\mathrm{Re}(\sin\theta^\pm)$, $\mathrm{Re}(\sin\theta^\pm) > 0$ (passive) or $\mathrm{Re}(\sin\theta^\pm) < 0$ (active), we note that the functions $\Psi^\pm$ in the expression of $\Psi(\alpha) = \Psi^+(\alpha)\Psi^-(\alpha)$ verify:

$$\begin{aligned}
&(\sin\alpha \pm \sin\theta^\pm)\Psi^\pm(\alpha\pm\Phi) - (-\sin\alpha \pm \sin\theta^\pm)\Psi^\pm(-\alpha\pm\Phi) = 0, \\
&\Psi^\pm(\alpha\mp\Phi) - \Psi^\pm(-\alpha\mp\Phi) = 0, \qquad (4.100)
\end{aligned}$$

that is, when $\mathrm{Re}(\sin\theta^\pm) \neq 0$,

$$\begin{aligned}
&\frac{\Psi^{\pm\prime}}{\Psi^\pm}(\alpha\pm\Phi) + \frac{\Psi^{\pm\prime}}{\Psi^\pm}(-\alpha\pm\Phi) = \pm\partial_\alpha \ln\frac{(\sin\alpha - \sin\theta^\pm)}{(\sin\alpha + \sin\theta^\pm)}, \\
&\frac{\Psi^{\pm\prime}}{\Psi^\pm}(\alpha\mp\Phi) + \frac{\Psi^{\pm\prime}}{\Psi^\pm}(-\alpha\mp\Phi) = 0. \qquad (4.101)
\end{aligned}$$

Since we have:

$$\partial_\alpha \ln \frac{(\sin\alpha - \sin\theta^\pm)}{(\sin\alpha + \sin\theta^\pm)} = -\partial_\alpha \ln \frac{(\sin\alpha + \sin\theta^\pm)}{(\sin\alpha - \sin\theta^\pm)}, \tag{4.102}$$

and the function $\Psi^\pm$ has to remain without any zero or pole in the band $|\text{Re}\alpha| \leq \Phi$, we can, letting $\Psi^\pm(\alpha,\sin\theta^\pm) \equiv \Psi^\pm(\alpha)$ for impedance $\sin\theta^\pm$, derive $\Psi^\pm$ with an active impedance face directly from $\Psi^\pm$ with passive impedance face, following:

$$\Psi^\pm(\alpha, \sin\theta^\pm) = B_0/\Psi^\pm(\alpha, -\sin\theta^\pm) \text{ as } \text{Re}(\sin\theta^\pm) \neq 0, \tag{4.103}$$

with B_0 a constant, as we can yet notice from general expression (4.74).

4.6.4.2. *Positions of poles and zeros of* $\Psi^\pm(\alpha)$

Letting $s^\pm$ =sign(Re($\sin\theta^\pm$)) and using (4.74), we directly derive zeros and poles of $\Psi^\pm$ from poles of the function $\Gamma(z)$. The poles and zeros of $\Psi^\pm(\alpha) \equiv \Psi^\pm(\alpha, \sin\theta^\pm)$, respectively, in passive and active cases, are linked, and we note that, when $s^\pm = 1$ (respectively, $s^\pm = -1$), $\Psi^\pm(\alpha, \sin\theta^\pm)$ has zeros (respectively, poles) at:

$$\begin{aligned} &\alpha = \mp\Phi + \epsilon(s^\pm\theta^\pm + (4p+2)\Phi + 2m\pi), \\ &\text{and } \alpha = \mp\Phi + \epsilon(\pi - s^\pm\theta^\pm + (4p+2)\Phi + 2m\pi), \end{aligned} \tag{4.104}$$

and poles (respectively, zeros) at

$$\begin{aligned} &\alpha = \mp\Phi + \epsilon(\pi + s^\pm\theta^\pm + (4p+2)\Phi + 2m\pi), \\ &\text{and } \alpha = \mp\Phi + \epsilon(2\pi - s^\pm\theta^\pm + (4p+2)\Phi + 2m\pi), \end{aligned} \tag{4.105}$$

for $m \geq 0, p \geq 0, \epsilon = 1$, and -1.

4.6.5. *Determination of* $P_n(\sin(\mu\alpha))$ *for passive or active faces*

The polynomial P_n permits us to complete the satisfaction of conditions (a)–(b)–(c) on f_0. Considering a unit incident plane wave, we have $f_0(\alpha) \sim 1/(\alpha - \varphi_o)$ in vicinity of incidence pole $\alpha = \varphi_o$. This implies $P_n(\sin(\mu\varphi_o))\sigma(\alpha) \sim 1/(\alpha - \varphi_o)$, and thus:

$$P_n(\sin(\mu\varphi_o)) = 1. \tag{4.106}$$

From unique definition of u at $\rho = 0$, we have to satisfy $f_0(\alpha) = O(1)$ as $|\text{Im}\alpha| \to \infty$, and considering (4.93) (see also (4.86) or (4.81)), we can write:

$$f_0(\alpha) \sim C_0 \frac{\prod_{\pm} \cos(\frac{\mu}{2}(\alpha \pm \Phi)))^{s^{\pm}}}{(\sin(\mu\alpha))^{1-n}} = O(1),$$

$$\text{which implies } (n-1)\mu + \sum_{\pm} \frac{s^{\pm}}{2}\mu = 0,$$

$$\text{i.e. } n = \sum_{\pm}(1-s^{\pm})/2 \; = \text{the number of active faces.} \tag{4.107}$$

For $|\text{Re}\varphi| < \Phi$ and passive impedance wedge with $\text{Re}(\sin\theta^{\pm}) > 0$, the poles of $\Psi(\alpha+\varphi)$ in the expression of $f_0(\alpha+\varphi)$ are outside the strip $|\text{Re}\alpha| < \pi$, but it is no more the case for active impedance faces. In this latter case, the n zeros of $P_n(\sin(\mu\alpha))$ have to suppress, for each active face with $s^{\pm} = -1$, the nonphysical pole $\alpha_s^{\pm}$ of $\Psi^{\pm}(\alpha,\sin\theta^{\pm})$ as $|\text{Re}\alpha| < \pi + \Phi$ that can give poles of $f_0(\alpha+\varphi)$ within the region $|\text{Re}\alpha| < \pi$ with $\text{Re}(ik\cos\alpha) > 0$ as $|\text{Re}\varphi| < \Phi$, which leads to cancel nonphysical contribution $e^{ik\rho\cos(\alpha_s^{\pm}-\varphi)}$ that can diverge at infinity as $|\text{Re}\varphi| < \Phi$ (note: for real k, notice that the nonphysical pole $\alpha_s^{\pm}$ is the one with $\pm\text{sign}(\arg(ik))\text{Im}\alpha_s^{\pm} > 0$, among $\alpha_1^{\pm} = \pm(s^{\pm}\theta^{\pm}+\Phi)$ and $\alpha_2^{\pm} = \pm(\pi - s^{\pm}\theta^{\pm}+\Phi)$). We then have:

$$P_n(\sin(\mu\alpha)) = \prod_{\pm}\left(\frac{\sin(\mu\alpha)-\sin(\mu\alpha_s^{\pm})}{\sin(\mu\varphi_o)-\sin(\mu\alpha_s^{\pm})}\right)^{\frac{1-s^{\pm}}{2}},$$

$$s^{\pm} = \text{sign}(\text{Re}(\sin\theta^{\pm})), \text{Re}(\sin\theta^{\pm}) \neq 0, \tag{4.108}$$

which completes the definition of f_0 for arbitrary impedances (passive or active).

4.7. Analysis for a wedge at skew incidence (2D1/2) and associated special functions

We study in this section the special functions used for the electromagnetic diffraction of a skew incident plane wave by a passive anisotropic wedge of any angle, their expansions and their properties. The problem is then no more scalar, as in strict 2D problem, but more generally vectorial.

We present here our solution [9], [23] for balanced hybrid conditions (when the product of diagonal terms of impedance matrix is equal to unity), and study its relation with Bobrovnikov–Fisanov function used in the solution of Lyalinov and Zhu in [24] and with Maliuzhinets function. On this occasion, we insist on the common links with the Γ special function [21], as in (4.74) for $\Psi^{\pm}$, which directly reveals important analytic properties and permits efficient calculus.

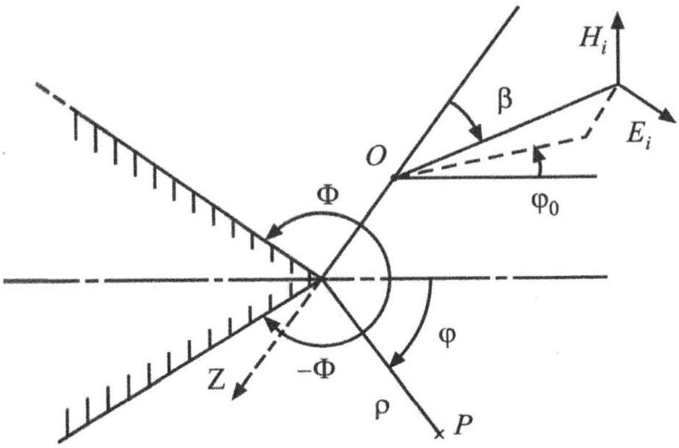

Figure 4.3. *Geometry at skew incidence*

4.7.1. *Our solution given in [9]*

The plane wave incident on the wedge is vectorial. It is characterized by the z-components of the electric and magnetic fields along the edge, respectively, E_z^i and H_z^i, given by:

$$\begin{bmatrix} E_z^i(\rho,\varphi,z) \\ Z_o H_z^i(\rho,\varphi,z) \end{bmatrix} = \begin{bmatrix} D_1 \\ D_2 \end{bmatrix} e^{ik[\rho\sin\beta\cos(\varphi-\varphi')-z\cos\beta]}, \tag{4.109}$$

In this expression, k ($\text{Im}k \leq 0$) and Z_o denote, respectively, the exterior medium wavenumber and impedance, and β is the angle of the incident direction with the edge of the wedge.

The z-components of the field, in cylindrical coordinates (ρ,φ, z), are searched in the form of Sommerfeld–Maliuzhinets integrals [9], following:

$$\begin{bmatrix} E_z(\rho,\varphi,z) \\ Z_o H_z(\rho,\varphi,z) \end{bmatrix} = \frac{e^{-ikz\cos\beta}}{2\pi i}\int_\gamma \begin{bmatrix} f_1(\alpha+\varphi) \\ f_2(\alpha+\varphi) \end{bmatrix} e^{ik\rho\sin\beta\cos\alpha} d\alpha, \tag{4.110}$$

which satisfy the Helmholtz equation $(\Delta+k^2)u = 0$ on each component. The contour γ is odd and composed of two symmetric loops as for 2D problems. Usual conditions at the edge and at infinity are assumed for z-components of fields, as previously for u in 2D, and we consider an anisotropic constant impedance boundary condition $(E - n^{\pm}(n^{\pm}E)$=$(Z^{\pm})(n^{\pm} \wedge H)$ on each face $\varphi = \pm\Phi$, where $n^{\pm}$ is the unit vector along the outward-pointing normal to the face, and $(Z^{\pm})/Z_o$ a relative constant impedance tensor. This implies coupled functional equations:

$$A_\alpha^{\pm} \begin{bmatrix} f_1(\alpha\pm\Phi) \\ f_2(\alpha\pm\Phi) \end{bmatrix} - A_{-\alpha}^{\pm} \begin{bmatrix} f_1(-\alpha\pm\Phi) \\ f_2(-\alpha\pm\Phi) \end{bmatrix} = \sin\alpha \begin{bmatrix} c_1 \\ c_2 \end{bmatrix}, \tag{4.111}$$

with

$$A_{\alpha}^{\pm} = \begin{bmatrix} \cos\beta\cos\alpha & \sin\alpha \pm \eta_h^{\pm}\sin\beta \\ \sin\alpha \pm \sin\beta/\eta_e^{\pm} & -\cos\beta\cos\alpha \end{bmatrix}, \tag{4.112}$$

when $(Z^{\pm})/Z_{\circ}$ is characterized by a class of diagonal relative impedance matrices with elements $\eta_h^{\pm}, \eta_e^{\pm}$.

Considering the oddness of γ, we note that we can choose to add some constants to spectral functions $f_{1,2}$ without changing the field, and the constants $c_{1,2}$ have no influence on it (see details in [9]). In these circumstances, we note that the previous sections on general properties of spectral functions, in Sommerfeld–Maliuzhinets representation of 2D fields, can apply providing that we let:

$$f(\alpha) \equiv e^{-ikz\cos\beta}\Big(f_{1,2}(\alpha) - \frac{f_{1,2}(i\infty) + f_{1,2}(-i\infty)}{2}\Big), \tag{4.113}$$

as $u \equiv E_z(\rho,\varphi,z)$ or $Z_{\circ}H_z(\rho,\varphi,z)$, with $\varphi_{\circ} \equiv \varphi'$, and $e^{ik\rho\sin\beta\cos\alpha}$ in place of $e^{ik\rho\cos\alpha}$ in integral representation of u.

By employing an original factorization technique for coupled equations [9], we found a closed-form solution when the products of these elements are equal to unity, that is, $\eta_h^{\pm}\eta_e^{\pm} = 1$. This case of anisotropy is important. It corresponds to a GO reflection coefficient, which is independent of the polarization of the incident field.

We let $\eta_h^{\pm} = 1/\eta_e^{\pm} = \sin\theta_1^{\pm}$ with $0 < \mathrm{Re}\theta_1^{\pm} \leq \pi/2$, $\delta = \ln(\tan(\beta/2))$, and obtain [9], [23]:

$$\begin{bmatrix} f_1(\alpha) \\ f_2(\alpha) \end{bmatrix} = \frac{\Psi_{an}(\alpha)}{\Psi_{an}(\varphi')} \begin{bmatrix} \cos\Delta(\alpha) & \sin\Delta(\alpha) \\ -\sin\Delta(\alpha) & \cos\Delta(\alpha) \end{bmatrix} \begin{bmatrix} D_1 \\ D_2 \end{bmatrix} \sigma(\alpha), \tag{4.114}$$

where $\Delta(\alpha) = v_{an}(\alpha) - v_{an}(\varphi')$, $\sigma(\alpha) = \mu\cos(\mu\varphi')(\sin(\mu\alpha) - \sin(\mu\varphi'))^{-1}$, $\mu = \pi/2\Phi$, and the functions Ψ_{an} and v_{an} verify:

$$\frac{\Psi_{an}(\alpha \pm \Phi)}{\Psi_{an}(-\alpha \pm \Phi)} = \Big(\frac{\det A_{-\alpha}^{\pm}}{\det A_{\alpha}^{\pm}}\Big)^{\frac{1}{2}},$$

$$v_{an}(\alpha \pm \Phi) - v_{an}(-\alpha \pm \Phi) = \arctan(\frac{n(\alpha)}{l^{\pm}(\alpha)}), \tag{4.115}$$

with

$$\begin{aligned} &\det(A_{\alpha}^{\pm}) = -\sin^2\beta\,(\sin\alpha \pm \sin(\theta_1^{\pm} - i\delta)) \\ &\times (\sin\alpha \pm \sin(\theta_1^{\pm} + i\delta)), \\ &\det(A_{\alpha}^{\pm})\det(A_{-\alpha}^{\pm}) = (n(\alpha))^2 + (l^{\pm}(\alpha))^2, \\ &l^{\pm}(\alpha) = \cos^2\beta\,\sin^2\alpha - \cos^2\alpha - ((\eta_h^{\pm})^2 - 1)\sin^2\beta, \\ &n(\alpha) = \cos\beta\,\sin(2\alpha), \end{aligned} \tag{4.116}$$

and $\sin(\beta)\sin(\theta_1^\pm - \epsilon i\delta) = (\eta_h^\pm + \epsilon\cos\beta\sqrt{(\eta_h^\pm)^2 - 1})$ (note: $(\arctan\frac{n}{l})'$ is detailed in appendix E). The explicit closed-form expressions of the two special functions $\Psi_{an}(\alpha)$ and $v_{an}(\alpha)$ is derived [9],[23], for $|\mathrm{Re}\alpha| < \Phi + (N+1)\pi$, following:

$$
\begin{aligned}
\Psi_{an}(\alpha) = &\prod_{\epsilon=-1,1}\prod_{\pm} \\
&\Big(\prod_{l\geq 0}^{N}[\,\Gamma(\frac{1}{2} + \frac{1}{4\Phi}((\alpha\pm\Phi) + \theta_1^\pm - \epsilon i\delta + l\pi))^{-\frac{1}{2}} \\
&\Gamma(\frac{1}{2} + \frac{1}{4\Phi}(-(\alpha\pm\Phi) + \theta_1^\pm - \epsilon i\delta + l\pi))^{-\frac{1}{2}} \\
&\Gamma(\frac{1}{2} + \frac{1}{4\Phi}((\alpha\pm\Phi) + \pi - \theta_1^\pm + \epsilon i\delta + l\pi))^{-\frac{1}{2}} \\
&\Gamma(\frac{1}{2} + \frac{1}{4\Phi}(-(\alpha\pm\Phi) + \pi - \theta_1^\pm + \epsilon i\delta + l\pi))^{-\frac{1}{2}}\,]^{(-1)^l} \\
&\times \exp\Big(\int_0^\infty (-1)^{N+1} \times \frac{e^{-\nu(\theta_1^\pm - \epsilon i\delta)} + e^{-\nu(\pi - \theta_1^\pm + \epsilon i\delta)}}{2\times(1 + e^{-\nu\pi})} \\
&\times e^{-\nu(N+1)\pi}\frac{1 - \cosh(\nu(\alpha\pm\Phi))}{\nu\sinh(2\nu\Phi)}d\nu\Big)\Big),
\end{aligned}
\tag{4.117}
$$

and

$$
\begin{aligned}
v_{an}(\alpha) = i&\sum_{\pm}\sum_{\epsilon=-1,+1} \\
&\Big(\sum_{l\geq 0}^{N}\ln([\Gamma(\frac{1}{2} + \frac{1}{4\Phi}(-(\alpha\pm\Phi) + \theta_1^\pm - \epsilon i\delta + l\pi))^{\mp\frac{1}{2}} \\
&\Gamma(\frac{1}{2} + \frac{1}{4\Phi}((\alpha\pm\Phi) + \theta_1^\pm - \epsilon i\delta + l\pi))^{\mp\frac{1}{2}} \\
&\Gamma(\frac{1}{2} + \frac{1}{4\Phi}((\alpha\pm\Phi) + \pi - \theta_1^\pm + \epsilon i\delta + l\pi))^{\pm\frac{1}{2}} \\
&\Gamma(\frac{1}{2} + \frac{1}{4\Phi}(-(\alpha\pm\Phi) + \pi - \theta_1^\pm + \epsilon i\delta + l\pi))^{\pm\frac{1}{2}}\,]^{\epsilon}) \\
&+\epsilon\int_0^\infty \frac{e^{-\nu(\theta_1^\pm - \epsilon i\delta)} - e^{-\nu(\pi - \theta_1^\pm + \epsilon i\delta)}}{2\times(1 - e^{-\nu\pi})} \\
&\times e^{-\nu(N+1)\pi}\frac{\mp(\cosh(\nu(\alpha\pm\Phi)) - 1)}{\nu\sinh(2\nu\Phi)}d\nu\Big),
\end{aligned}
\tag{4.118}
$$

for arbitrary $N \geq -1$, where we have used $\frac{1}{1\pm x} = \sum_{l\geq 0}^{N}(\mp x)^l + \frac{(\mp x)^{N+1}}{1\pm x}$ as in [25]–[26] (note: $\prod_{l\geq 0}^{-1} = 1$ and $\sum_{l\geq 0}^{-1}() = 0$). Since $\mathrm{e}^{-\nu(N+1)\pi}$ rapidly decreases for large ν or N, these expressions have the advantage to give an efficient calculation for $N \sim 1$, and all analytic properties from those of Γ function when we consider $N \to \infty$.

Besides, it is worth noticing that it is then particularly simple to realize that $\Psi_{an}(\alpha)e^{\pm i v_{an}(\alpha)}$ is a meromorphic function of the complex plane, for which we can identify all complex zeros and poles. For the particular case of normal incidence (i.e. for $\delta = 0$ when we have no more coupling between TM and TE waves), or for isotropic unit impedance ($\theta_1^{\pm} = \pi/2$), we recover the expressions given in [4], [12] and [25]–[26].

NOTE 4.19.– Among all analytical properties of Γ, we can quote the remarkable identities:

$$\Gamma(z+1) = z\Gamma(z), -z\Gamma(-z)\Gamma(z) = \pi/\sin(\pi z), \Gamma(1) = 1,$$

$$\Gamma(nz) = (2\pi)^{\frac{1}{2}(1-n)} n^{nz-\frac{1}{2}} \prod_{k=0}^{n-1} \Gamma(z + \frac{k}{n}). \tag{4.119}$$

4.7.2. *The relation with the solution given by Lyalinov and Zhu*

We now exhibit the relations of the special functions Ψ_{an} and v_{an} of our solution with the Bobrovnikov–Fisanov function $\chi_{\Phi}(\alpha)$, used by Lyalinov and Zhu in [24]. For that, we let $N = -1$ in (4.117) and (4.118), and note that:

$$\begin{aligned}
\frac{\Psi_{an}(\alpha)}{\Psi_{an}(\varphi')} &= \prod_{\epsilon=-1,1} \prod_{\pm} \\
&\exp\Bigl(\int_0^{\infty} \frac{\cosh(\nu(\pi/2 - \theta_1^{\pm} + \epsilon i\delta))}{2 \times \cosh(\nu(\pi/2))} \\
&\times \frac{\cosh(\nu(\varphi' \pm \Phi)) - \cosh(\nu(\alpha \pm \Phi))}{\nu \sinh(2\nu\Phi)} d\nu\Bigr) \\
&= \prod_{\epsilon=-1,1} \prod_{\pm} \\
&\exp\Bigl(\int_0^{\infty} \frac{\sinh(\nu(\pi - \theta_1^{\pm} + \epsilon i\delta)) - \sinh(\nu(-\theta_1^{\pm} + \epsilon i\delta))}{2\sinh(\nu\pi)} \\
&\times \frac{\cosh(\nu(\varphi' \pm \Phi)) - \cosh(\nu(\alpha \pm \Phi))}{\nu \sinh(2\nu\Phi)} d\nu\Bigr),
\end{aligned} \tag{4.120}$$

and

$$v_{an}(\alpha) - v_{an}(\varphi') = i \sum_{\pm} \sum_{\epsilon=-1,+1}$$

$$\epsilon \int_0^\infty \frac{\sinh(\nu(\pi/2 - \theta_1^\mp + \epsilon i\delta))}{2\sinh(\nu\pi/2)}$$

$$\times \frac{\mp(\cosh(\nu(\alpha \pm \Phi)) - \cosh(\nu(\varphi' \pm \Phi)))}{\nu \sinh(2\nu\Phi)} d\nu$$

$$= i \sum_{\pm} \sum_{\epsilon=-1,+1}$$

$$\epsilon \int_0^\infty \frac{\sinh(\nu(\pi - \theta_1^\mp + \epsilon i\delta)) + \sinh(\nu(-\theta_1^\mp + \epsilon i\delta))}{2\sinh(\nu\pi)}$$

$$\times \frac{\mp(\cosh(\nu(\alpha \pm \Phi)) - \cosh(\nu(\varphi' \pm \Phi)))}{\nu \sinh(2\nu\Phi)} d\nu, \tag{4.121}$$

and then compare with the integral expression of $\chi_\Phi(\alpha)$ given in [27]:

$$\chi_\Phi(\alpha) = \exp\Big(\frac{1}{2}\int_0^\infty \frac{1}{\nu\sinh(\nu\pi)}\Big(\frac{\alpha}{2\Phi} - \frac{\sinh(\nu\alpha)}{\sinh(2\nu\Phi)}\Big)d\nu\Big). \tag{4.122}$$

Since, we have:

$$2(\sinh(\nu(\pi + x))\cosh(\nu y) - \sinh(\nu(x))\cosh(\nu y)) =$$

$$= (\sinh(\nu(y + \pi + x)) - \sinh(\nu(y - \pi - x)) -$$

$$- (\sinh(\nu(y + x)) - \sinh(\nu(y - x)))$$

$$= (\sinh(\nu(\pi + y + x)) - \sinh(\nu(y + x)) +$$

$$+ (\sinh(\nu(y - x)) - \sinh(\nu(-\pi + y - x))), \tag{4.123}$$

we can express the function $\frac{\Psi_{an}(\alpha)}{\Psi_{an}(\varphi')}$ as a combination of elementary complex functions of the type, $(\frac{\chi_\Phi(\pi+z+\alpha)\chi_\Phi(z'+\alpha)}{\chi_\Phi(z+\alpha)\chi_\Phi(-\pi+z'+\alpha)})^{\frac{1}{2}}(\frac{\chi_\Phi(\pi+z+\varphi')\chi_\Phi(z'+\varphi')}{\chi_\Phi(z+\varphi')\chi_\Phi(-\pi+z'+\varphi')})^{-\frac{1}{2}}$, and in a similar manner, since

$$2(\sinh(\nu(\pi + x))\cosh(\nu y) + \sinh(\nu x)\cosh(\nu y)) =$$

$$= (\sinh(\nu(y + \pi + x)) - \sinh(\nu(y - \pi - x)) +$$

$$+ (\sinh(\nu(y + x)) - \sinh(\nu(y - x)))$$

$$= (\sinh(\nu(\pi + y + x)) + (\sinh(\nu(y + x))) -$$

$$- (\sinh(\nu(y - x)) + \sinh(\nu(-\pi + y - x))), \tag{4.124}$$

we can express $\cos(\upsilon_{an}(\alpha)-\upsilon_{an}(\varphi'))$ and $\sin(\upsilon_{an}(\alpha)-\upsilon_{an}(\varphi'))$ as combinations of functions $(\frac{\chi_\Phi(\pi+z+\alpha)\chi_\Phi(z+\alpha)}{\chi_\Phi(z'+\alpha)\chi_\Phi(-\pi+z'+\alpha)})^{\pm\frac{1}{2}}$ $(\frac{\chi_\Phi(\pi+z+\varphi')\chi_\Phi(z+\varphi')}{\chi_\Phi(z'+\varphi')\chi_\Phi(-\pi+z'+\varphi')})^{\mp\frac{1}{2}}$, which implies that $\frac{\Psi_{an}(\alpha)\cos(\upsilon_{an}(\alpha)-\upsilon_{an}(\varphi'))}{\Psi_{an}(\varphi')}$ and $\frac{\Psi_{an}(\alpha)\sin(\upsilon_{an}(\alpha)-\upsilon_{an}(\varphi'))}{\Psi_{an}(\varphi')}$ are meromorphic functions in α which express as combinations of $\frac{\chi_\Phi(\pi+z+\alpha)}{\chi_\Phi(-\pi+z'+\alpha)}(\frac{\chi_\Phi(\pi+z+\varphi')}{\chi_\Phi(-\pi+z'+\varphi')})^{-1}$ and $\frac{\chi_\Phi(z'+\alpha)}{\chi_\Phi(z+\alpha)}(\frac{\chi_\Phi(z'+\varphi')}{\chi_\Phi(z+\varphi')})^{-1}$ terms. This result permits us to recover the expressions obtained by Zhu and Lyalinov [24].

4.7.3. Ψ_{an} *in terms of* Ψ_Φ *functions and expressions of* Ψ_Φ

4.7.3.1. $\Psi_\Phi(\alpha)$ *from the expansion of* Ψ_{an} *with* Γ *function*

From (4.117), the expression of $\frac{\Psi_{an}(\alpha)}{\Psi_{an}(\varphi')}$ is given by:

$$\frac{\Psi_{an}(\alpha)}{\Psi_{an}(\varphi')}=\prod_{\epsilon=-1,1}\Big(\prod_{\pm}$$

$$\exp\Big(\int_0^\infty\sum_{\epsilon'=-1,1}\frac{\cosh(\nu(\varphi'\pm\Phi+\epsilon'(\theta_1^\mp-\epsilon i\delta-\frac{\pi}{2})))-1}{4\nu\ \sinh(2\nu\Phi)\cosh(\frac{\nu\pi}{2})}$$

$$+\frac{1-\cosh(\nu(\alpha\pm\Phi+\epsilon'(\theta_1^\mp-\epsilon i\delta-\frac{\pi}{2})))}{4\nu\sinh(2\nu\Phi)\cosh(\frac{\nu\pi}{2})}d\nu\Big)\Big),\tag{4.125}$$

for $|\text{Re}\alpha|<\Phi$. Considering that [21]:

$$\int_0^\infty e^{-\mu x}\frac{\sinh(\beta x)}{\sinh(bx)}dx=\partial_\beta(\ln(\Gamma(\frac{1}{2}+\frac{(\mu+\beta)}{2b})\Gamma(\frac{1}{2}+\frac{(\mu-\beta)}{2b}))),$$

$$\int_0^\infty e^{-\mu x}\frac{1-\cosh(\beta x)}{x\sinh(bx)}dx=-\ln(\frac{(\Gamma(\frac{1}{2}+\frac{(\mu+\beta)}{2b})\Gamma(\frac{1}{2}+\frac{(\mu-\beta)}{2b})}{\Gamma(\frac{1}{2}+\frac{\mu}{2b})\Gamma(\frac{1}{2}+\frac{\mu}{2b})}),$$

$$\frac{(-e^{-\nu\pi})^{N+1}}{(1+e^{-\nu\pi})}=(-e^{-\nu\pi})^{N+1}+\frac{(-e^{-\nu\pi})^{N+2}}{(1+e^{-\nu\pi})},\tag{4.126}$$

we can write, for $|\text{Re}\alpha+\epsilon'(\theta_1^\mp-\epsilon i\delta-\frac{\pi}{2})|<\Phi+(N+1)\pi$ as $\epsilon'=-1$ and 1:

$$\Psi_{an}(\alpha)=C_N\prod_{\epsilon=-1,1}\Big(\prod_{\epsilon'=-1,1}\Big(\prod_{\pm}$$

$$\prod_{l\geq 0}^{N}[\Gamma(\frac{1}{2}+\frac{(\alpha\pm\Phi+\epsilon'(\theta_1^\mp-\epsilon i\delta-\frac{\pi}{2}))+\frac{\pi}{2}+l\pi}{4\Phi})^{-\frac{1}{2}}$$

$$\Gamma(\frac{1}{2}+\frac{-(\alpha\pm\Phi+\epsilon'(\theta_1^\mp-\epsilon i\delta-\frac{\pi}{2}))+\frac{\pi}{2}+l\pi}{4\Phi})^{-\frac{1}{2}}]^{(-1)^l}$$

$$\times \exp\Big(\int_0^\infty (-e^{-\nu\pi})^{N+1} \frac{1-\cosh(\nu(\alpha \pm \Phi + \epsilon'(\theta_1^\pm - \epsilon i\delta - \frac{\pi}{2})))}{4\nu \sinh(2\nu\Phi)\cosh(\frac{\nu\pi}{2})} d\nu\Big)\Big)\Big)$$

$$= C_N \prod_{\epsilon=-1,1} \Big(\prod_{\epsilon'=-1,1} \Big(\prod_{\pm} \Big(\Psi_{\Phi,N}\Big(\alpha \pm \Phi + \epsilon'\Big(\theta_1^\pm - \epsilon i\delta - \frac{\pi}{2}\Big)\Big)\Big)^{\frac{1}{2}}\Big)\Big), \quad (4.127)$$

where C_N is a constant, and $\Psi_{\Phi,N}(\alpha)$ is given, for $|\mathrm{Re}\alpha| < 2\Phi + (N+1)\pi + \frac{\pi}{2}$, by:

$$\Psi_{\Phi,N}(\alpha) = \prod_{l\geq 0}^{N}$$

$$[\Gamma(\frac{1}{2} + \frac{\alpha + \frac{\pi}{2} + l\pi}{4\Phi})\Gamma(\frac{1}{2} + \frac{-\alpha + \frac{\pi}{2} + l\pi}{4\Phi})]^{(-1)^{l+1}}$$

$$\times \exp\Big(\int_0^\infty (-e^{-\nu\pi})^{N+1} e^{-\nu\frac{\pi}{2}} \frac{1-\cosh(\nu\alpha)}{\nu\sinh(2\nu\Phi)(1+e^{-\nu\pi})} d\nu\Big), \quad (4.128)$$

for $N \geq -1$, and $\prod_{l\geq 0}^{-1}() = 1$, which is in perfect agreement with (4.77). Comparing this expression with the one given for Ψ_Φ in [30], we conclude, as in (4.78), that we have:

$$\Psi_\Phi(\alpha) = \Psi_{\Phi,N}(\alpha)/\Psi_{\Phi,N}(0) \text{ with } \Psi_{\Phi,N=-1}(0) = 1, \quad (4.129)$$

where Ψ_Φ is the Maliuzhinets function, first defined in [4] for the diffraction by an impedance wedge, and verifying $\frac{\Psi_\Phi(\alpha+2\Phi)}{\Psi_\Phi(\alpha-2\Phi)} = \cot(\frac{1}{2}(\alpha + \frac{\pi}{2}))$.

The expression (4.129) is an efficient expansion to calculate the function $\Psi_\Phi(\alpha)$, when N is finite. It also permits us to recover all analytical properties if we let tend N to infinity, in particular from well-known remarkable identities satisfied by Γ in the complex plane.

Besides, it is worth noticing that the constants, C_N and $\Psi_{\Phi,N}(0)$, have no influence on the solution, and thus, we can take $C_N = 1$ and Ψ_Φ or $\Psi_{\Phi,N}$ indifferently, without changing the field.

4.7.3.2. *Complete asymptotic expression of* Ψ_Φ

For an asymptotic expression of Ψ_Φ for large $\mathrm{Im}\alpha$, we can transform the expression given by (4.129) as $N = -1$ using residue theorem. Letting $\epsilon = \mathrm{sign}(\mathrm{Im}\alpha)$, we can then write, as $\mathrm{Im}\alpha \neq 0$ and $|\mathrm{Re}\alpha| < 2\Phi + \frac{\pi}{2}$:

$$\Psi_\Phi(\alpha) = \exp\left[\frac{i}{4}\int_{-i\infty}^{i\infty} \frac{1-\cos(\omega\alpha)}{\omega\cos(\omega\pi/2)\sin(2\omega\Phi)} d\omega\right]$$

$$= \exp\left[\frac{1}{8i}\int_{-i\infty}^{i\infty} \frac{(e^{i\epsilon\omega\alpha} - i\epsilon\omega\alpha - 1) + (e^{-i\epsilon\omega\alpha} + i\epsilon\omega\alpha - 1)}{\omega\cos(\omega\pi/2)\sin(2\omega\Phi)} d\omega\right]$$

$$= \exp\left[\frac{(-\pi i)}{8i}(\frac{i\epsilon\alpha}{2\Phi}) + \frac{1}{8}\int_{0^+-i\infty}^{0^++i\infty}\frac{1}{i\omega}\frac{e^{i\epsilon\omega\alpha}-1}{\cos(\omega\pi/2)\sin(2\omega\Phi)}d\omega\right.$$

$$\left.+\frac{(\pi i)}{8i}(\frac{-i\epsilon\alpha}{2\Phi}) + \frac{1}{8}\int_{-0^+-i\infty}^{-0^++i\infty}\frac{1}{i\omega}\frac{e^{-i\epsilon\omega\alpha}-1}{\cos(\omega\pi/2)\sin(2\omega\Phi)}d\omega\right]$$

$$= e^{-i\epsilon\frac{\pi}{8\Phi}\alpha}\times\exp\left[\frac{1}{8}\int_{0^+-i\infty}^{0^++i\infty}\frac{1}{i\omega}\frac{e^{i\epsilon\omega\alpha}-1}{\cos(\omega\pi/2)\sin(2\omega\Phi)}d\omega\right.$$

$$\left.+\frac{1}{8}\int_{-0^+-i\infty}^{-0^++i\infty}\frac{1}{i\omega}\frac{e^{-i\epsilon\omega\alpha}-1}{\cos(\omega\pi/2)\sin(2\omega\Phi)}d\omega\right]$$

$$= e^{-i\epsilon\frac{\pi}{8\Phi}\alpha}\times\exp\left[\frac{1}{8}\int_{0^+-i\infty}^{0^++i\infty}\frac{1}{i\omega}\frac{e^{i\epsilon\omega\alpha}-1}{\cos(\omega\pi/2)\sin(2\omega\Phi)}d\omega\right.$$

$$\left.+\frac{1}{8}\int_{0^+-i\infty}^{0^++i\infty}\frac{1}{i\omega}\frac{e^{i\epsilon\omega\alpha}-1}{\cos(\omega\pi/2)\sin(2\omega\Phi)}d\omega\right]$$

$$= e^{-i\epsilon\frac{\pi}{8\Phi}\alpha}\exp\left[\frac{1}{4}\int_{0^+-i\infty}^{0^++i\infty}\frac{1}{\cos(\omega\pi/2)\sin(2\omega\Phi)}\frac{-1}{i\omega}d\omega\right]$$

$$\times\exp\left[\frac{1}{4}\int_{0^+-i\infty}^{0^++i\infty}\frac{1}{\cos(\omega\pi/2)\sin(2\omega\Phi)}\frac{e^{i\epsilon\omega\alpha}}{i\omega}d\omega\right]. \tag{4.130}$$

In the last exponential term, we can then shift the path to Re$\alpha = \delta > 0$, taking care that $\frac{\delta 2\Phi}{\pi}$ and $\frac{\delta-1}{2}$ are not integers, while adding residues terms (first and second order), due to the capture of poles, between 0^+ and the constant δ, during the shifting. We then derive, as $|\mathrm{Re}\alpha| < 2\Phi + \frac{\pi}{2}$ and Im$\alpha \neq 0$:

$$\Psi_\Phi(\alpha) = e^B e^{-i\epsilon\frac{\pi}{8\Phi}\alpha}$$

$$\times\exp\left[-\frac{\pi}{2}\left(\left(\sum_{m\geq 1,\frac{m\pi}{2\Phi}\neq 2p+1}^{\frac{m\pi}{2\Phi}<\delta}\frac{(-1)^m\frac{1}{2\Phi}}{\frac{m\pi}{2\Phi}\cos(\frac{m\pi(\pi/2)}{2\Phi})}e^{i\epsilon\frac{m\pi}{2\Phi}\alpha}\right)\right.\right.$$

$$+\left(\sum_{p\geq 0,2p+1\neq\frac{m\pi}{2\Phi}}^{2p+1<\delta}\frac{(-1)^{p+1}\frac{2}{\pi}}{(2p+1)\sin(2(2p+1)\Phi)}e^{i\epsilon(2p+1)\alpha}\right)$$

$$\left.+\left(\sum_{p\geq 0,2p+1=\frac{m\pi}{2\Phi}}^{2p+1<\delta}\frac{(-1)^{m+p+1}}{2\Phi}\frac{2}{\pi}\frac{2\Phi}{m\pi}(i\epsilon\alpha-\frac{2\Phi}{m\pi})e^{i\frac{m\pi}{2\Phi}\epsilon\alpha}\right)\right)$$

$$\left.+\frac{e^{i\epsilon\delta\alpha}}{4\delta}\int_{\delta-i\infty}^{\delta+i\infty}\frac{\delta}{i\omega\cos(\omega\pi/2)\sin(2\omega\Phi)}e^{i\epsilon(\omega-\delta)\alpha}d\omega\right], \tag{4.131}$$

where

$$B=\frac{i}{4}\int_{0^+-i\infty}^{0^++i\infty}\frac{1}{\cos(\omega\pi/2)\sin(2\omega\Phi)}\frac{1}{\omega}d\omega=\frac{i}{4}\int_{-i\infty}^{+i\infty}\frac{\partial_\omega\frac{\omega}{\cos(\omega\pi/2)\sin(2\omega\Phi)}}{\omega}d\omega. \tag{4.132}$$

The last integral term in (4.131) being $O(1)$, we have obtained a complete asymptotic expansion for the function $\Psi_\Phi(\alpha)$ as $|\mathrm{Im}\alpha|\to\infty$, which completes, from a distinct method, the expression given in [12].

4.7.3.3. *Some complements on the determination of* B

For the calculus of B, we can also write:

$$B=\frac{i}{4}\int_{-i\infty}^{+i\infty}\frac{(\frac{1}{\cos(\omega\pi/2)}-1)}{\sin(2\omega\Phi)\omega}d\omega+\frac{i}{4}\int_{0^+-i\infty}^{0^++i\infty}\frac{1}{\sin(2\omega\Phi)}\frac{1}{\omega}d\omega. \tag{4.133}$$

We then use (4.97)–(4.98) and the method of integration by parts to simplify the first and second integrals. From (4.97)–(4.98), we note first that:

$$\begin{aligned}&(\frac{1}{\cos(\frac{\pi}{2}\omega)}-1)\frac{1}{\omega}=\frac{2}{\pi i}\int_0^{i\infty}\frac{(\cos(\omega\alpha)-1)}{\omega\cos(\alpha)}d\alpha\\&=\frac{2}{\pi i}\int_0^{i\infty}\sin(\omega\alpha)\ln(\frac{1}{i}\tan(\frac{\alpha+\frac{\pi}{2}}{2}))d\alpha,\ |\mathrm{Re}\omega|<1,\\&\frac{i}{2}\int_{-i\infty+0^+}^{+i\infty+0^+}\frac{1}{\omega\sin(2\omega\Phi)}d\omega=-\ln(2),\end{aligned} \tag{4.134}$$

and we can then write:

$$\begin{aligned}B+\frac{\ln(2)}{2}&=\frac{i}{2}\frac{2}{\pi i}\int_0^{i\infty}\ln(\frac{1}{i}\tan(\frac{\alpha+\frac{\pi}{2}}{2}))\int_0^{i\infty}\frac{\sin\omega\alpha}{\sin(2\omega\Phi)}d\omega d\alpha\\&=\frac{1}{\pi}\int_0^{i\infty}\ln(\frac{1}{i}\tan(\frac{\alpha+\frac{\pi}{2}}{2}))\frac{\pi i}{4\Phi}\tan(\frac{\pi}{4\Phi}\alpha)d\alpha\\&=\frac{-1}{i}\frac{1}{2\pi}\int_{-i\infty}^{+i\infty}\frac{\ln(\cos(\frac{\pi}{4\Phi}\alpha))}{\cos(\alpha)}d\alpha=\frac{-1}{2\pi}\int_{-\infty}^{+\infty}\frac{\ln(\cosh(\frac{\pi}{4\Phi}\nu))}{\cosh(\nu)}d\nu.\end{aligned} \tag{4.135}$$

With this reduction of B, it is worth noticing that the first order of this asymptotic is in perfect agreement with the one given by (4.89).

4.7.4. *Expressions for the Bobrovnikov–Fisanov function* χ_Φ

4.7.4.1. *An efficient expansion of the function* χ_Φ *with* Γ *function*

The Bobrovnikov–Fisanov function χ_Φ, used in [24] and [27], can be developed in an original manner from the special function Γ. We write the second-order derivative of (4.122), for $|\mathrm{Re}(\alpha)| < \pi + 2\Phi$, following:

$$\frac{\partial^2}{\partial\alpha^2}\ln(\chi_\Phi(\alpha)) = -\frac{1}{2}\int_0^\infty \frac{\nu}{\sinh(\nu\pi)}\frac{\sinh(\nu\alpha)}{\sinh(2\nu\Phi)}d\nu$$

$$= -\frac{1}{2}\int_0^\infty (2e^{-\nu\pi}\sum_{l\geq 0}^{N} e^{-2l\pi\nu} + \frac{e^{-\nu(N+1)2\pi}}{\sinh(\nu\pi)})\frac{\nu\sinh(\nu\alpha)}{\sinh(2\nu\Phi)}d\nu. \tag{4.136}$$

Considering then that we have [21]:

$$\int_0^\infty e^{-\mu\nu}\frac{\nu\sinh(\nu\alpha)}{\sinh(2\Phi\nu)}d\nu = \frac{-1}{4\Phi}\partial_\mu((\partial\ln\Gamma)(\frac{1}{2}+\frac{(\mu+\alpha)}{4\Phi})-$$

$$-(\partial\ln\Gamma)(\frac{1}{2}+\frac{(\mu-\alpha)}{4\Phi}))$$

$$= -\frac{\partial^2}{\partial\alpha^2}(\ln(\Gamma(\frac{1}{2}+\frac{(\mu+\alpha)}{4\Phi})) - \ln(\Gamma(\frac{1}{2}+\frac{(\mu-\alpha)}{4\Phi}))), \tag{4.137}$$

we can write, for $|\mathrm{Re}(\alpha)| < (N+1)2\pi + \pi + 2\Phi$,

$$\ln(\chi_\Phi(\alpha)) = (A_N + B_N\alpha)+$$

$$+\sum_{l\geq 0}^{N}\ln\left(\frac{\Gamma(\frac{1}{2}+\frac{1}{4\Phi}((2l+1)\pi+\alpha))}{\Gamma(\frac{1}{2}+\frac{1}{4\Phi}((2l+1)\pi-\alpha))}\right)$$

$$+\int_0^\infty \frac{e^{-\nu(N+1)2\pi}}{2\sinh(\nu\pi)}\frac{(\alpha-\frac{\sinh(\nu\alpha)}{\nu})}{\sinh(2\nu\Phi)}d\nu. \tag{4.138}$$

Thus, since $\chi_\Phi(0) = 1$, we obtain:

$$\chi_\Phi(\alpha) = e^{B_N\alpha}\chi_{\Phi,N}(\alpha), \tag{4.139}$$

with

$$\chi_{\Phi,N}(\alpha) = \prod_{l\geq 0}^{N}\left(\frac{\Gamma(\frac{1}{2}+\frac{1}{4\Phi}((2l+1)\pi+\alpha))}{\Gamma(\frac{1}{2}+\frac{1}{4\Phi}((2l+1)\pi-\alpha))}\right)$$

$$\times\exp\left(\int_0^\infty e^{-\nu(N+1)2\pi}\frac{(\nu\alpha-\sinh(\nu\alpha))}{2\nu\sinh(\nu\pi)\sinh(2\nu\Phi)}d\nu\right), \tag{4.140}$$

for $|\mathrm{Re}(\alpha)| < (N+1)2\pi+\pi+2\Phi$, $N \geq -1$, $\prod_{l\geq 0}^{-1} = 1$. Considering the derivative of the integral term in the expression of $\ln(\chi_{\Phi,N}(\alpha))$, null at $\alpha = 0$, and the derivative of $\ln(\chi_{\Phi}(\alpha))$ from (4.122), the constant B_N is given by:

$$B_N = \frac{1}{2}\int_0^{\infty} \frac{\left(\frac{1}{2\Phi} - \frac{\nu}{\sinh(2\nu\Phi)}\right)}{\nu\sinh(\nu\pi)} d\nu - 2\sum_{l\geq 0}^{N}(\partial \ln\Gamma)(\frac{1}{2} + \frac{1}{4\Phi}((2l+1)\pi)), \quad (4.141)$$

but also, from $\chi_{\Phi}(\pi) = 2^{\frac{1}{2}-\frac{\pi}{4\Phi}}$ [27], following:

$$B_N = \frac{1}{\pi}\ln\Big(\frac{2^{\frac{1}{2}-\frac{\pi}{4\Phi}}}{\chi_{\Phi,N}(\pi)}\Big). \quad (4.142)$$

The expressions (4.139) and (4.140) are efficient tools to calculate the functions χ_{Φ} and $\chi_{\Phi,N}$ when N is finite, permitting us to recover all the analytical properties from remarkable identities of Γ when we let tend N to infinity.

It is important to notice that, considering the wedge problem with anisotropic impedance condition, the constant B_N has no influence on the solution, and thus, that we can choose $\chi_{\Phi,N}$ in place of χ_{Φ}, at N arbitrarily fixed, without changing the final expression of the field.

NOTE 4.20.– $\chi_{\Phi,N}$ and χ_{Φ} are meromorphic functions with the same poles and zeros, and we have $\chi_{\Phi,N}(\alpha + 2\Phi) = (e^{-B_N 4\Phi}\cos(\frac{\alpha}{2}))\,\chi_{\Phi,N}(\alpha - 2\Phi)$.

4.7.4.2. *About asymptotic expansions of functions $\chi_{\Phi,N}$ and χ_{Φ}*

Let $N = -1$ in (4.140). Using a manner similar to the one use for Ψ_{Φ}, we can develop the complete asymptotics for $\ln\chi_{\Phi,N}$ and $\chi_{\Phi,N}$. For that, we let ϵ =sign(Im(α)), use residue theorem and write:

$$\ln\chi_{\Phi,N}(\alpha) = \frac{i\epsilon}{2}\int_{-i\infty}^{+i\infty} \frac{(\omega\epsilon\alpha - \sin(\omega\epsilon\alpha))}{2\omega\sin(\omega\pi)\sin(2\omega\Phi)} d\omega$$

$$= \frac{-\pi i\epsilon\alpha^2}{16\pi\Phi} - \frac{\epsilon}{4}\int_{-i\infty+0^+}^{+i\infty+0^+} \frac{e^{i\epsilon\omega\alpha} - i\epsilon\omega\alpha - 1}{2\omega\sin(\omega\pi)\sin(2\omega\Phi)} d\omega$$

$$+ \frac{\epsilon}{4}\int_{-i\infty-0^+}^{+i\infty-0^+} \frac{e^{-i\epsilon\omega\alpha} + i\epsilon\omega\alpha - 1}{2\omega\sin(\omega\pi)\sin(2\omega\Phi)} d\omega$$

$$= \frac{-i\epsilon\alpha^2}{16\Phi} + \epsilon B_0' + B_1'\alpha - \frac{\epsilon}{4}\int_{-i\infty+0^+}^{+i\infty+0^+} \frac{e^{i\epsilon\omega\alpha}}{\omega\sin(\omega\pi)\sin(2\omega\Phi)} d\omega, \quad (4.143)$$

where

$$B'_0 = \int_{-i\infty+0^+}^{+i\infty+0^+} \frac{d\omega/4}{\omega\sin(\omega\pi)\sin(2\omega\Phi)} = \int_{-i\infty}^{+i\infty} \partial_{\omega^3}(\frac{-\omega^2/8}{\sin(\omega\pi)\sin(2\omega\Phi)})\ln(\omega)d\omega$$

$$= -\frac{i\pi}{2}\int_0^{+i\infty} \partial_{\omega^3}(\frac{\omega^2/4}{\sin(\omega\pi)\sin(2\omega\Phi)})d\omega = \frac{i\pi}{2}\partial_{\omega^2}(\frac{\omega^2/4}{\sin(\omega\pi)\sin(2\omega\Phi)})|_{\omega=0}$$

$$= \frac{i\pi}{2}\frac{1}{4}\frac{1}{\pi 2\Phi}(\frac{(\pi)^2}{3} + \frac{(2\Phi)^2}{3}) = \frac{i}{12}(\frac{(\pi)^2}{4\Phi} + \Phi),$$

$$B'_1 = \int_{-i\infty+0^+}^{+i\infty+0^+} \frac{i/4}{\sin(\omega\pi)\sin(2\omega\Phi)}d\omega = \int_0^{+\infty} \partial_\nu(\frac{\nu^2/2}{\sinh(\nu\pi)\sinh(2\nu\Phi)})\frac{1}{\nu}d\nu. \tag{4.144}$$

Shifting the path of integration to Re$\omega = \delta > 0$ in the last integral term of (4.143), and using residue theorem (first and second order) for poles captured during it, we obtain the complete asymptotic expansion of $\ln \chi_{\Phi,N}(\alpha)$, then of $\chi_{\Phi,N}(\alpha)$, and thus, of $\chi_\Phi(\alpha)$ from (4.139).

NOTE 4.21.– We note that, using (4.98) and (4.141), we can also simplify B'_1 following:

$$B'_1 = \frac{i}{4}\int_{-i\infty}^{+i\infty} \frac{(\frac{\omega}{\sin(2\omega\Phi)} - \frac{1}{2\Phi})}{\omega\sin(\omega\pi)}d\omega + \frac{i}{8\Phi}\int_{-i\infty+0^+}^{+i\infty+0^+} \frac{1}{\omega\sin(\omega\pi)}d\omega$$

$$= \frac{1}{2}\int_0^{+\infty} \frac{(\frac{\nu}{\sinh(2\nu\Phi)} - \frac{1}{2\Phi})}{\nu\sinh(\nu\pi)}d\nu - \frac{\ln(2)}{4\Phi} = -B_{-1} - \frac{\ln(2)}{4\Phi}, \tag{4.145}$$

from which $B'_1 + B_{-1}$, that refers to χ_Φ from (4.139), can be derived in the closed form.

4.8. Explicitly causal expression in time domain for a dispersive wedge-shaped region

4.8.1. *Basic elements on causality and integral expressions*

Let us consider the total field $u(\rho, \varphi; i\omega)$ in presence of a 2D scatterer, enclosed in the region outside the free space sector $|\varphi| < \Phi$, when it is illuminated by a plane wave:

$$u^i(\rho, \varphi; i\omega) = U^i(i\omega)e^{ik\rho\cos(\varphi-\varphi_\circ)} \text{ with } ik = \frac{i\omega}{c}. \tag{4.146}$$

We assume that, in the domain $|\arg(i\omega)| \leq \pi/2$, $U^i(i\omega)$ and ik are holomorphic functions of $i\omega$, and $|\arg(ik)| \leq \pi/2$. The function $U^i(i\omega)$ is $O(\omega^{-b})$ as $|\text{Re}\omega| \to \infty$,

$b > 1$, while $u(\rho, \varphi; i\omega)$ satisfies the conditions (a'), (b') and (c'). The time domain response $\overline{u}^i(\rho, \varphi; t)$ (or $\overline{u}(t)$) is then given by:

$$\overline{u}^i(\rho, \varphi; t) = \lim_{A \to \infty} \frac{1}{2\pi i} \int_{-iA+\sigma}^{+iA+\sigma} u^i(\rho, \varphi; s)\, e^{st} ds,\ \sigma \geq 0. \tag{4.147}$$

Closing the contour of integration at infinity, we note that:

$$\overline{u}^i(\rho, \varphi; t) = 0 \ \text{ for } t < \mathcal{T}(\rho, \varphi) = \frac{-\rho\cos(\varphi - \varphi_\circ)}{c_\infty}, \tag{4.148}$$

where $c_\infty = \lim_{|\omega| \to \infty} c > 0$, and we can write:

$$u^i(\rho, \varphi; s) = \int_{\mathcal{T}(\rho,\varphi)}^{\infty} \overline{u}^i(\rho, \varphi; t) e^{-st} dt. \tag{4.149}$$

The time domain field $\overline{u}(\rho, \varphi; t)$ is real. It is here assumed to be bounded, and, in the presence of the wedge-shaped region illuminated by $\overline{u}^i(\rho, \varphi; t)$, it is causal with respect to the front of the incident wave, so that we have:

$$\overline{u}(\rho, \varphi; t) = 0 \text{ as } t < \mathcal{T}(\rho, \varphi). \tag{4.150}$$

Noticing that $\mathcal{T}(\rho, \varphi) > \mathcal{T}_0(\rho) = -\rho/c_\infty$, we can add that we have:

$$\overline{u}(\rho, \varphi; t) = \overline{u}^i(\rho, \varphi; t) = 0 \text{ as } t < \mathcal{T}_0(\rho) = -\rho/c_\infty. \tag{4.151}$$

Consequently, we have:

$$u(\rho, \varphi; s) = \int_{-\rho/c_\infty}^{\infty} \overline{u}(t) e^{-st} dt, \tag{4.152}$$

which, we note, implies:

$$\begin{aligned} &u(\rho, \varphi; s^*) = (u(\rho, \varphi; s))^*, \\ &u(\rho, \varphi; s) = O(e^{s\rho/c_\infty}) \text{ for } \mathrm{Re}(s) > 0, \end{aligned} \tag{4.153}$$

and thus no singularity of $u(\rho, \varphi; s)$ as $\mathrm{Re}(s) > 0$. Therefore, we can write:

$$\begin{aligned} \overline{u}(t) &= \frac{1}{2\pi i} \int_{-i\infty+\sigma}^{+i\infty+\sigma} u(\rho, \varphi; s)\, e^{st} ds,\ \sigma \geq 0, \\ \overline{u}(t) &= \frac{1}{2\pi} \Big(\int_0^\infty u(\rho, \varphi; i\omega)\, e^{i\omega t}\, d\omega + \int_{-\infty}^0 u(\rho, \varphi; i\omega)\, e^{i\omega t}\, d\omega \Big) \\ &= \frac{1}{\pi} \mathrm{Re}\Big(\int_0^\infty u(\rho, \varphi; i\omega)\, e^{i\omega t}\, d\omega \Big). \end{aligned} \tag{4.154}$$

Rather to express causality from analyticity of $u(\rho,\varphi;s)$ for complex $s=i\omega$, we now aim to reduce $\overline{u}(t)$ to an explicitly causal expression depending on position, that is, an integral on a "finite continuous" path $\mathcal{H}$ whose ends depends on (ρ,φ) and t.

For that, we will transform the expressions in (4.154) from the integral representation of u with spectral function $f(\alpha+\varphi,i\omega)$. After analyzing the properties of $f(\alpha+\varphi,i\omega)$, in particular those concerning the positions of their singularities in α for real ω, we will define and use some transforms of it that presents some intricate causality properties. Dispersive multilayered wedges with higher order boundary conditions can be then considered. To simplify, we assume in what follows that, in the free space:

$$c=\frac{\omega}{k} \text{ is a real positive constant independent of } \omega. \tag{4.155}$$

NOTE 4.22.– The problem when $c=\frac{\omega}{k}$ depends on ω, that is when the free space is a homogeneous dispersive medium, is studied in [13]–[14] for the diffraction by a passive impedance wedge, and we note in general that, from (4.151):

$$\int_{-i\infty+\sigma}^{+i\infty+\sigma} u(\rho',\varphi;s)\,e^{-s\rho/c+s\tau}ds=0 \text{ as } -\frac{\rho}{c_\infty}+\tau<-\frac{\rho'}{c_\infty}. \tag{4.156}$$

4.8.2. *Properties of spectral function f and trajectories of poles*

We consider that the spectral function $f(\alpha+\varphi;i\omega)$, in Sommerfeld–Maliuzhinets representation (4.3) of field u, has poles of GO on real axis whose positions are independent of ω, in particular those related to incident and reflected angles, while its complex singularities, whose positions depend analytically on $i\omega$, generally derive from singularities of reflections coefficients $R^{\pm}(\beta;i\omega)$ for both faces.

For now, we assume that each face is composed of passive materials (and metamaterials), characterized by complex relative permittivity ϵ_r and permeability μ_r with $|\arg(i\epsilon_r)|<\pi/2$ and $|\arg(i\mu_r)|<\pi/2$, which cover non-penetrable surfaces verifying passive impedance boundary conditions. As ϵ_r-1 and μ_r-1 tend to zero for $\omega\to\infty$ from causality principle [22], the contribution of materials vanishes at high frequencies while the one of impedance surface can continue to be present.

The function $f(\alpha';i\omega)$ is meromorphic in the domain $|\text{Re}(\alpha')|\leq\pi+\Phi$, with real angle poles (independent of ω), and possible complex angle poles (depending analytically on $i\omega$), while, at fixed ω, $f(\alpha';i\omega)$ is bounded at infinity in this domain. Concerning the trajectories $\alpha'=\alpha_s(i\omega)$ for complex poles, we consider the following properties:

Each trajectory described by complex poles of $f(\alpha'; i\omega)$,

$$\alpha' = \alpha_s(i\omega), \text{ numbered } s \in N, \text{ when } \omega \text{ varies from } \pm 0^+ \text{ to } \pm A,$$

can cross the domain with $|\text{Re}(\alpha')| < \pi + \Phi - \epsilon_A$, with $\epsilon_A > 0$,

but these trajectories have no end points in any closed subdomain of it

(with $\epsilon_A \to 0^+$ as $A \to \infty$); in other words, we suppose that

the end points are outside the band $|\text{Re}(\alpha')| < \pi + \Phi - \epsilon_A$, or at infinity. (4.157)

Moreover, we take precaution that we have no cusp ($\partial_\omega(\alpha_s(i\omega)) \neq 0$).

Besides, using the single-face expression (4.35) of f, we note that:

$$f(\alpha^* + \varphi; (i\omega)^*) = (f(\alpha + \varphi; i\omega))^*, \tag{4.158}$$

and thus, we can add that α_s verifies:

$$\alpha_s((i\omega)^*) = (\alpha_s(i\omega))^*. \tag{4.159}$$

The general assumption given in (4.157) has no use for a wedge with constant passive impedance boundaries, since, in this case, we have no pole depending on ω in the band $|\text{Re}(\alpha)| \leq \pi + \Phi$ [13]–[14], but must be considered when the faces are composed of a multilayer (see appendix B), and thus, $f(\alpha; i\omega)$ has possible active poles depending on ω in the band $|\text{Re}(\alpha)| \leq \pi + \Phi$.

4.8.3. *Elementary transform $\mathcal{F}_A^{\pm}$ of f and its analytical continuation $\mathcal{F}_{a,A}^{\pm}$*

4.8.3.1. *Definitions of domains $\Omega^{\pm}$ and of transforms $\mathcal{F}_A^{\pm}$, for real $\pm\omega > 0$*

The spectral function $f(\alpha + \varphi; i\omega)$ verifies the property (b), and thus we have:

for real ω, $f(\alpha'; i\omega)$ is bounded and regular in $\{\alpha' = \alpha + \varphi, \alpha \in \Omega, |\varphi| \leq \Phi\}$

where Ω is the region of the strip $|\text{Re}(\alpha)| \leq \pi$ with $\text{Re}(i\omega \cos\alpha) > 0$,

that is, the region where the function $e^{i\omega\rho\cos\alpha/c}$ diverges as $\rho \to \infty$. (4.160)

For real $\pm\omega > 0$, the domain Ω is denoted $\Omega^{\pm}$. It is composed of the subdomains $\Omega^{\pm}_{\epsilon=1}$ and $\Omega^{\pm}_{\epsilon=-1}$. Each $\Omega^{\pm}_{\epsilon}$ is bounded by the loops $\mathcal{L}^{\pm}_{\epsilon}$ defined following:

$$\begin{aligned}&\mathcal{L}^{+}_{\epsilon} \equiv]\epsilon(i\infty + \pi^+), \epsilon(\pi + i0^+)] \cup [\epsilon(\pi + i0^+), \epsilon i0^+] \cup [\epsilon i0^+, \epsilon i\infty[,\\&\mathcal{L}^{-}_{\epsilon} \equiv [\epsilon(i\infty), \epsilon i0^+[\cup[\epsilon i0^+, \epsilon(i0^+ - \pi)] \cup [\epsilon(i0^+ - \pi), \epsilon(-\pi + i\infty)[,\end{aligned} \tag{4.161}$$

where we notice that $\mathcal{L}^{+}_{\epsilon} \equiv (\mathcal{L}^{-}_{-\epsilon})^*$, and thus, $\Omega^{\pm}_{\epsilon}$ is the conjugate of $\Omega^{\mp}_{-\epsilon}$. In this domain of regularity, $\Omega^{\pm} \equiv \Omega^{\pm}_{\epsilon=1} \cup \Omega^{\pm}_{\epsilon=-1}$ limited by $\mathcal{L}^{\pm} \equiv \mathcal{L}^{\pm}_{\epsilon=1} \cup \mathcal{L}^{\pm}_{\epsilon=-1}$, f is from

(4.35), $O(\omega^{-v})$ as $|\text{Re}\omega| \to \infty$ for any finite $\text{Im}\omega \leq 0$, with $v > 1$. We then define, for any τ real:

$$\mathcal{F}_A^+(\alpha+\varphi,\tau) = \frac{1}{2\pi}\int_0^{+A} f(\alpha+\varphi;i\omega)e^{i\omega\tau}d\omega \text{ as } \alpha \in \Omega^+,$$

$$\mathcal{F}_A^-(\alpha+\varphi,\tau) = \frac{1}{2\pi}\int_{-A}^{0} f(\alpha+\varphi;i\omega)e^{i\omega\tau}d\omega \text{ as } \alpha \in \Omega^-, \tag{4.162}$$

with A a positive real constant. These expressions also permit us to consider, $\tau \equiv \tau_e(\alpha) = \rho\cos\alpha/c + t$, with α belonging to the frontiers $\mathcal{L}^\pm \equiv \mathcal{L}^\pm_{\epsilon=1} \cup \mathcal{L}^\pm_{\epsilon=-1}$, where $\cos\alpha$ is real, and to write:

$$\mathcal{F}_A^+(\alpha+\varphi,\tau_e(\alpha)) = \frac{1}{2\pi}\int_0^{+A} f(\alpha+\varphi;i\omega)e^{i\omega\tau_e(\alpha)}d\omega \text{ as } \alpha \in \mathcal{L}^+,$$

$$\mathcal{F}_A^-(\alpha+\varphi,\tau_e(\alpha)) = \frac{1}{2\pi}\int_{-A}^{0} f(\alpha+\varphi;i\omega)e^{i\omega\tau_e(\alpha)}d\omega \text{ as } \alpha \in \mathcal{L}^-. \tag{4.163}$$

We note that, so defined, the functions $\mathcal{F}_A^\pm$ verify:

$$(\mathcal{F}_A^\pm(\alpha+\varphi,\tau))^* = \mathcal{F}_A^\mp(\alpha^*+\varphi,\tau),$$

$$(\mathcal{F}_A^\pm(\alpha+\varphi,\tau_e(\alpha)))^* = \mathcal{F}_A^\mp(\alpha^*+\varphi,\tau_e(\alpha^*)). \tag{4.164}$$

4.8.3.2. *Definition of analytical continuation $\mathcal{F}^\pm_{a,A}$ of $\mathcal{F}^\pm_A$ in the domain with poles of f*

We can then consider the analytical continuation $\mathcal{F}^\pm_{a,A}(\alpha+\varphi,\tau_e(\alpha))$ (respectively, $\mathcal{F}^\pm_{a,A}(\alpha+\varphi,\tau)$) of $\mathcal{F}^\pm_A(\alpha+\varphi,\tau_e(\alpha))$ (respectively, $\mathcal{F}^\pm_A(\alpha+\varphi,\tau)$) in the domain with poles of f, complementary to $\Omega^\pm$ in the band $|\text{Re}(\alpha)| \leq \pi$, which corresponds to the part of the domain $|\text{Re}(\alpha)| \leq \pi$ where $e^{i\omega\tau_e(\alpha)} = O(1)$ as $\pm\omega > 0$. The frontiers, from which $\mathcal{F}^\pm_{a,A}(\alpha+\varphi,\tau_e(\alpha))$ (respectively, $\mathcal{F}^\pm_{a,A}(\alpha+\varphi,\tau)$) can be continued, are the lines $\{\alpha = ix \mp 0^+\text{sign}(x), x \in R\}$ and the real axis, where $\mathcal{F}^\pm_A(\alpha+\varphi,\tau_e(\alpha))$ (respectively, $\mathcal{F}^\pm_{a,A}(\alpha+\varphi,\tau)$) is known.

Because of (4.157), this continuation can be done without any singularity and cut in it, and on its boundary, as $|\varphi| < \Phi - \epsilon_A$, even when $A \to \infty$. The function $\mathcal{F}^\pm_{a,A}(\alpha+\varphi,\tau_e(\alpha))$ (respectively, $\mathcal{F}^\pm_{a,A}(\alpha+\varphi,\tau)$) will be then regular in the domain $\Omega^{\pm *} \equiv \Omega^\mp$ of α, where $e^{i\omega\tau_e(\alpha)} = O(1)$, bounded by $\mathcal{L}^{\pm *} \equiv \mathcal{L}^\mp$, in complement of the loop $\mathcal{L}^\pm \equiv \mathcal{L}^\pm_{\epsilon=1} \cup \mathcal{L}^\pm_{\epsilon=-1}$ on which $\mathcal{F}^\pm_A(\alpha+\varphi,\tau_e(\alpha))$ (respectively, $\mathcal{F}^\pm_{a,A}(\alpha+\varphi,\tau)$) was initially definite.

NOTE 4.23.– As mentioned earlier, for the particular case of the diffraction by a passive impedance wedge, $f(\alpha+\varphi;i\omega)$ is regular as $|\text{Re}(\alpha+\varphi)| \leq \pi + \Phi$ except at real GO poles, and thus in $\Omega^{\pm *}$, and we can directly consider the integral expression of $\mathcal{F}^\pm_A$ for $\mathcal{F}^\pm_{a,A}$ in it, but it is no more the case for a multilayered wedge.

4.8.3.3. *An example of analytical continuation for an integral with a continuous line of poles*

Let us consider the basic example of the exponential integral E_1 defined with:

$$E_1(z) = e^{-z} \int_{-\infty}^{0} \frac{e^s}{z-s} ds, \text{ as } |\arg(z)| < \pi,$$

$$E_1(z) = -\gamma - \ln(z) - \sum_{n \geq 1} \frac{(-z)^n}{nn!}. \tag{4.165}$$

The trajectory of poles in the integral is a straight semi-line from 0 to $-\infty$, which gives us the cut of $E_1(z)$, the same than for $\ln z$. An analytical continuation of $E_1(z)$ through this cut changes its determination.

By definition, we continue an analytical function by successive use of Taylor expansions, whose domains of convergence are limited by the radius of convergence (determined by circle with first singularity encountered), which amounts to say that we increase continuously the domain of holomorphy, and thus deform in our case the cut, to avoid z to cross it. The new cut is a curved semi-line $L_{-\infty,0}$, with same ends points 0 and $-\infty$, from which the continuation can be named $E_{1,L_{-\infty,0}}(z)$.

Now, from the use of Cauchy theorem, this continuation can directly be done from the integral representation (4.165) after "continuous" deformation of the integration path, and thus, by a change of the initial path to $L_{-\infty,0}$, which gives us:

$$E_{1,L_{-\infty,0}}(z) = e^{-z} \int_{L_{-\infty,0}} \frac{e^s}{z-s} ds, \text{ as } z \notin L_{-\infty,0},$$

$$E_{1,L_{-\infty,0}}(z) = -\gamma - \ln_L(z) - \sum_{n \geq 1} \frac{(-z)^n}{nn!} \text{ with } \ln_L \neq \ln, \tag{4.166}$$

This kind of developments will be used to take account of complex poles of f depending continuously on ω, for the determination of the continuation $\mathcal{F}^{\pm}_{a,A}$ of $\mathcal{F}^{\pm}_{A}$, provided (4.157) is satisfied.

4.8.3.4. *Expression of $\mathcal{F}^{\pm}_{a,A}$ in vicinity of real poles of f independent of ω*

On the path $\mathcal{L}^{\pm} = \mathcal{L}^{\pm}_{\epsilon=1} \cup \mathcal{L}^{\pm}_{\epsilon=-1}$ of the strip $|\text{Re}(\alpha)| \leq \pi$, we have real GO singularities of $f(\alpha+\varphi; i\omega)$ at $\alpha+\varphi = \alpha_r$ that have the particularity to be independent of ω (for incident and reflected GO waves), as $|\varphi| < \Phi$. To take account of them for the continuation on $\mathcal{L}^{\pm}$, we can write:

$$\mathcal{F}^{\pm}_{a,A}(\alpha+\varphi, \tau_e(\alpha)) = \frac{\pm 1}{2\pi} \int_0^{\pm A} \Big(f(\alpha+\varphi; i\omega) e^{i\omega\tau_e(\alpha)}$$

$$- \sum_r \frac{f_r(i\omega) e^{i\omega\tau_e(\alpha)}}{\alpha+\varphi-\alpha_r} \Big) d\omega + \frac{1}{2\pi} \sum_{\alpha_r} g_r(\alpha+\varphi, \tau_e(\alpha)), \tag{4.167}$$

where

$$f_r(i\omega) = \frac{1}{[\partial_\alpha(1/f)](\alpha_r, i\omega)},\ g_r^\pm(\alpha+\varphi, \tau_e(\alpha)) = \frac{\pm\int_0^{\pm A} f_r(i\omega)e^{i\omega\tau_e(\alpha)}d\omega}{\alpha+\varphi-\alpha_r}, \tag{4.168}$$

with $|\text{Re}(\alpha)| < \pi + \Phi$. This expression can also be considered for $\mathcal{F}_{a,A}^\pm(\alpha+\varphi, \tau)$.

4.8.3.5. *Analytical continuation $\mathcal{F}_{a,A}^\pm$ of $\mathcal{F}_A^\pm$ for complex poles of f depending on ω*

In the case of a multilayered wedge, we have to consider the presence of poles $\alpha_s(\omega)$ of $f(\alpha; i\omega)$ crossing the domain $|\text{Re}(\alpha)| < \pi + \Phi$ and verifying (4.157). By definition, $\mathcal{F}_{a,A}^\pm(\alpha+\varphi, \tau_e(\alpha))$ is the regular continuation of $\mathcal{F}_A^\pm(\alpha+\varphi, \tau_e(\alpha))$ in $\Omega^{\pm*}$, bounded by $\mathcal{L}^{\pm*}$, where $e^{i\omega\tau_e(\alpha)} = O(1)$ as $\pm\omega > 0$, with no branch point and cut in $\Omega^{\pm*}$ and on $\mathcal{L}^{\pm*}$ as $|\varphi| < \Phi - \epsilon_A$ (note: $\epsilon_A \to 0^+$ as $A \to \infty$). For that, we write:

$$\mathcal{F}_{a,A}^\pm(\alpha+\varphi, \tau_e(\alpha)) = \frac{\pm 1}{2\pi}\int_0^{\pm A}\Big(f(\alpha+\varphi; i\omega)e^{i\omega\tau_e(\alpha)}$$

$$-\sum_s \frac{f_s(i\omega)e^{i\omega\tau_e(\alpha)}}{\alpha+\varphi-\alpha_s(i\omega)}\Big)d\omega + \frac{1}{2\pi}\sum_s g_s^\pm(\alpha+\varphi, \tau_e(\alpha)), \tag{4.169}$$

where we have:

$$f_s(-i\omega^*) = (f_s(i\omega))^*,\ \alpha_s(-i\omega^*) = (\alpha_s(i\omega))^*,$$

$$f_s(i\omega) = \frac{1}{[\partial_\alpha(1/f)](\alpha_s(i\omega); i\omega)}, \tag{4.170}$$

and

$$g_s^\pm(\alpha+\varphi, \tau_e(\alpha)) = \pm\int_{\pm 0^+}^{\pm A} \frac{f_s(i\omega)e^{i\omega\tau_e(\alpha)}}{\alpha+\varphi-\alpha_s(i\omega)}d\omega,$$

$$= \pm\int_{\mathcal{T}_A^\pm} \frac{f_s(i\omega)e^{i\omega\tau_e(\alpha)}}{\alpha+\varphi-\alpha_s(i\omega)}\frac{1}{i\partial_{i\omega}(\alpha_s(i\omega))}d\alpha_s(i\omega),$$

$$= \pm\int_{\alpha_{s,0\pm}}^{\alpha_{s,\pm A}} \frac{f_s(i\omega)e^{\tau_e(\alpha)(\alpha_s^{(-1)}(\alpha_s))}}{\alpha+\varphi-\alpha_s}(-i\partial_v(\alpha_s^{(-1)}(v))|_{v=\alpha_s})d\alpha_s, \tag{4.171}$$

with $\mathcal{T}_A^\pm = \{\alpha_s(i\omega),\ \omega \in]0^\pm, \pm A[\}$, $i\omega = \alpha_s^{(-1)}\alpha_s(i\omega)$, $\alpha_{s,\pm A} = \alpha_s(\pm iA)$ and $\alpha_{s,0^\pm} = \alpha_s(i0^\pm)$. From (4.157), $\partial_\omega(\alpha_s(i\omega)) \neq 0$ (no cusp), and the endpoints value $\alpha_{s,0^\pm}$ and $\alpha_{s,\pm A}$ can be considered outside the band $|\text{Re}(\alpha)| < \pi + \Phi - \epsilon_A$.

In this hypothesis, g_s can be analytically continued as in (4.166), from α on the imaginary axis to α in $\Omega^{\pm *}$, by continuously deforming $\mathcal{T}_A^{\pm}$ as $\alpha + \varphi$ approaches the trajectory of α_s in order to avoid to cross it.

Thereafter, taking account of the continuation with (4.167) and (4.169) in $\Omega^{\pm *}$ where $e^{i\omega\tau_e(\alpha)} = O(1)$ as $\pm\omega > 0$, we have in general:

$$\mathcal{F}^{\pm}_{a,A}(\alpha + \varphi, \tau_e(\alpha)) \text{ is regular in } \Omega^{\pm *} \text{ and on } \mathcal{L}^{\pm} \cup \mathcal{L}^{\pm *},$$

$$\text{even for } A \to \infty, \text{ as } \tau_e(\alpha) = \rho \cos\alpha/c + t \text{ with arbitrary real } t,\ |\varphi| < \Phi. \quad (4.172)$$

These results can also be considered for $\mathcal{F}^{\pm}_{a,A}(\alpha + \varphi, \tau)$, as τ is real.

NOTE 4.24.– Considering (4.164) and the definition of $\mathcal{F}^{\pm}_{a,A}$, we note that:

$$(\mathcal{F}^{\pm}_{a,A}(\alpha + \varphi, \tau_e(\alpha)))^* = \mathcal{F}^{\mp}_{a,A}(\alpha^* + \varphi, \tau_e(\alpha^*)). \quad (4.173)$$

4.8.4. *Fourier transform of f and vanishing property due to causality*

After noting the regularity of $f(\alpha' + \varphi; i\omega)$ on imaginary axis for $|\text{Im}\alpha'| > 0$, we define on it:

$$\mathcal{F}_A(\alpha' + \varphi, \tau) = \frac{1}{2\pi} \int_{-A-i0^+}^{+A-i0^+} f(\alpha' + \varphi, i\omega) e^{i\omega\tau} d\omega, \quad (4.174)$$

for any τ real. Considering the single-face expression (4.35), which is valid for $\text{Re}(ik(\cos\alpha - \cos(\Phi_e^{\pm} \mp \varphi_o))) > 0$, $0 < \Phi_e^{\pm} < \Phi$, and thus, the expression (4.37) for $f(\alpha + \varphi)$ as $\text{Re}(ik\cos\alpha) > \text{Re}(ik)$, $|\varphi| \leq \Phi$, we can, for $\text{Re}\alpha' = 0$ as $\alpha' \neq 0$, interchange the order of the integration (permitted even for $A \to \infty$) and write:

$$\mathcal{F}_A(\alpha' + \varphi, \tau) = \frac{1}{4\pi} \int_{0^+}^{\infty} \int_{-A}^{+A} (-iku(\rho, \varphi) \sin\alpha'$$

$$- \frac{\partial u}{\rho \partial \varphi}(\rho, \varphi)) e^{i\omega t'} d\omega d\rho, \quad (4.175)$$

as $t' = -\rho' \cos\alpha'/c + \tau$, where u satisfies (4.152) and the conditions (a'), (b') and (c').

Since $\overline{u}(t)$ is causal with respect to the front of the incident plane wave, the contour of integration in ω can be closed at infinity for $\text{Im}(\omega) < 0$ when $t' < -\rho'/c$, and we can write:

$$\mathcal{F}_{A\to\infty}(\alpha + \varphi, \tau) \to 0 \text{ for } \text{Re}\alpha = 0,\ |\varphi| \leq \Phi,\ \alpha \neq 0 \text{ at any fixed } \tau < 0. \quad (4.176)$$

We then consider the analytical continuation $\mathcal{F}_{a,A}(\alpha+\varphi,\tau)$ of $\mathcal{F}_{A}(\alpha+\varphi,\tau)$, for any τ real, with no cut and no singularities as $|\mathrm{Re}(\alpha+\varphi)|<\pi+\Phi-\epsilon_A$, except the poles of GO on real axis. For this, we exploit that, in our construction of $\mathcal{F}^{\pm}_{a,A}$ with (4.167) and (4.169), $\tau_e(\alpha)$ remains an independent function that we are free to take as a constant, and note the regularity of $\mathcal{F}^{\pm}_{a,A}(\alpha+\varphi,\tau)$ and $\mathcal{F}^{\mp}_{A}(\alpha+\varphi,\tau)$ in $\Omega^{\pm *}\equiv\Omega^{\mp}$. Therefore, we can write:

$$\mathcal{F}_{a,A}(\alpha+\varphi,\tau)=\mathcal{F}^{+}_{a,A}(\alpha+\varphi,\tau)+\mathcal{F}^{-}_{A}(\alpha+\varphi,\tau)\text{ in }\Omega^{+*}\equiv\Omega^{-},$$
$$\mathcal{F}_{a,A}(\alpha+\varphi,\tau)=\mathcal{F}^{+}_{A}(\alpha+\varphi,\tau)+\mathcal{F}^{-}_{a,A}(\alpha+\varphi,\tau)\text{ in }\Omega^{-*}\equiv\Omega^{+},\qquad(4.177)$$

with $\mathcal{F}^{\pm}_{a,A}(\alpha+\varphi,\tau)=\mathcal{F}^{\pm}_{A}(\alpha+\varphi,\tau)$ in $\Omega^{\mp *}\equiv\Omega^{\pm}$ and on their frontiers $\mathcal{L}^{\mp *}\equiv\mathcal{L}^{\pm}$.

The regularity of $\mathcal{F}_{a,A}$ in α allows us to extend (4.176), and we can now assume a vanishing property for $\mathcal{F}_{a,A\to\infty}$, as follows:

$$\mathcal{F}_{a,A\to\infty}(\alpha+\varphi,\tau)\to 0\text{ at any fixed }\tau<0\text{ as }|\mathrm{Re}(\alpha+\varphi)|<\pi+\Phi.\qquad(4.178)$$

The factor $\tau_e(\alpha)$ being real on $\mathcal{L}^{\pm}\cup\mathcal{L}^{\pm *}$, we can infer that, for $\alpha\in\mathcal{L}^{\pm}\cup\mathcal{L}^{\pm *}$:

$$\mathcal{F}_{a,A}(\alpha+\varphi,\tau_e(\alpha))\equiv\mathcal{F}^{+}_{a,A}(\alpha+\varphi,\tau_e(\alpha))+\mathcal{F}^{-}_{a,A}(\alpha+\varphi,\tau_e(\alpha)),\qquad(4.179)$$

verifies the causality property:

$$\mathcal{F}_{a,A\to\infty}(\alpha+\varphi,\tau_e(\alpha))\to 0\text{ for }\tau_e(\alpha)<0,\ \text{ as }\alpha\in\mathcal{L}^{\pm}\cup\ \mathcal{L}^{\pm *},\ |\varphi|<\Phi.\quad(4.180)$$

NOTE 4.25.– Considering the problem when $c=\frac{\omega}{k}$ depends on ω, we note in general from (4.156) that:

$$\int_{-\infty-i0+}^{+\infty-i0^{+}}u(\rho',\varphi;i\omega)\,e^{-s\rho'\cos\alpha'/c+s\tau}d\omega=0\text{ as }\tau<0\text{ when }\mathrm{Re}\alpha'=0,\qquad(4.181)$$

and thus, that (4.176) and (4.178) apply in whole generality, when the free space is a homogeneous dispersive medium.

4.8.5. *Explicitly causal expression of fields in time domain with* $\mathcal{F}_{a,A\to\infty}$

From the expression (4.3) of u, and the domain of regularity of $\mathcal{F}^{\pm}_{a,A}(\alpha+\varphi,\tau_e(\alpha))$, the field in time domain is given by:

$$\begin{aligned}\overline{u}(t)&=\lim_{A\to\infty}\frac{1}{2\pi}\Big(\int_{0}^{A}u(\rho,\varphi;i\omega)\,e^{i\omega t}\,d\omega+\int_{-A}^{0}u(\rho,\varphi;i\omega)\,e^{i\omega t}\,d\omega\Big)\\&=\lim_{A\to\infty}\frac{1}{2\pi i}\Big(\int_{\mathcal{L}+}\mathcal{F}^{+}_{a,A}(\alpha+\varphi,\tau_e(\alpha))d\alpha+\int_{\mathcal{L}^{-}}\mathcal{F}^{-}_{a,A}(\alpha+\varphi,\tau_e(\alpha))d\alpha\Big),\end{aligned}\qquad(4.182)$$

with $\tau_e(\alpha) = \rho\cos\alpha/c + t$, where the double loops $\mathcal{L}^{\pm}$, as previously defined, correspond to the path of integration γ for (4.3) as $\pm k > 0$, $a_{1,2} = 0$ and $d = 0^+$.

Considering the analyticity of $\mathcal{F}^{+}_{a,A}(\alpha + \varphi, \tau_e(\alpha))$ and $\mathcal{F}^{-}_{a,A}(\alpha + \varphi, \tau_e(\alpha))$, we can deform in (4.182) both paths $\mathcal{L}^+$ and $\mathcal{L}^-$ (oriented) to the common path $\mathcal{D} \equiv \mathcal{D}_+ \cup \mathcal{D}_-$ independent of ω, where the loops $\mathcal{D}_{\pm}$, similar to loops $\gamma_{\pm}$ when we take $a_1 = \pi - \arg(ik) - \frac{\pi}{2}$, $a_2 = \pi + \arg(ik) - \frac{\pi}{2}$ and $d = 0^+$ in (4.3), verify:

$$\mathcal{D}_{\pm} \equiv]\pm(i\infty + \pi), \pm(\pi + i0^+)] \cup [\pm(\pi + i0^+), \pm(-\pi + i0^+)]$$
$$\cup [\pm(-\pi + i0^+), \pm(-\pi + i\infty)[. \tag{4.183}$$

Therefore, we can write, from (4.179):

$$\overline{u}(t) = \frac{1}{2\pi i}\int_{\mathcal{D}} \mathcal{F}_{a,A\to\infty}(\alpha + \varphi, \tau_e(\alpha)))d\alpha, \tag{4.184}$$

with $\tau_e(\alpha)$ real on $\mathcal{D}$.

We then use the causality property (4.180) for $\mathcal{F}_{a,A\to\infty}$ and obtain an expression for $\overline{u}(t)$, explicitly causal, on account of a finite path depending on time and position, following:

$$\overline{u}(t) = \frac{1}{2\pi i}\int_{\mathcal{H}(\rho,t)} \mathcal{F}_{a,A\to\infty}(\alpha + \varphi, \tau_e(\alpha))d\alpha, \tag{4.185}$$

where $\mathcal{H} = \mathcal{H}_+ \cup \mathcal{H}_-$ is the finite subset of $\mathcal{D}$ with $\tau_e(\alpha) = \rho\cos\alpha/c + t > 0$, and $\mathcal{F}_{a,A}$ is given by (4.179) on $\mathcal{H}$. For this expression, we can let:

$$iv(\rho, t) = \left| \begin{matrix} i\ln(ct/\rho + \sqrt{(ct/\rho)^2 - 1}), ct \geq \rho \\ \\ -\arccos(ct/\rho), -\rho < ct < \rho \end{matrix} \right. , \tag{4.186}$$

and note that the finite path $\mathcal{H}_+$ (oriented as $\mathcal{D}_+$) can be described:

$$\text{as } |ct| < \rho : [\pi + iv(\rho, t) + i0^+, -\pi - iv(\rho, t) + i0^+],$$
$$\text{as } ct \geq \rho :]\pi + iv(\rho, t), \pi + i0^+] \cup [\pi + i0^+, -\pi + i0^+]$$
$$\cup [-\pi + i0^+, iv(\rho, t) - \pi[, \tag{4.187}$$

above all of the real singularities of the integrand as $|\text{Re}(\alpha)| < \pi$ (due to GO part of the field), while $\mathcal{H}_-$ (oriented as $\mathcal{D}_-$) is its inverse below the real axis.

Besides, it is worth noticing that, in the explicitly causal expression in time domain for dispersive wedges (4.185), the contribution of GO poles of $\mathcal{F}_{a,A\to\infty}$ on real axis (the only poles which remain in time domain) can be expressed in closed form expressions from residue theorem.

NOTE 4.26.– Considering (4.173) for (4.185), we note that:

$$\overline{u}(t) = \text{Re}\Big(\frac{1}{\pi i}\int_{\mathcal{H}(\rho,t)} \mathcal{F}^{+}_{a,A\to\infty}(\alpha+\varphi,\tau_e(\alpha))d\alpha\Big). \tag{4.188}$$

4.9. References

[1] Jones, D.S. (1964). *The Theory of Electromagnetism*. Pergamon Press, London.

[2] Felsen, L.B. and Marcuvitz, N. (1973). *Radiation and Scattering of Waves*. Prentice Hall, Englewood Cliffs.

[3] Maliuzhinets, G.D. (1958). Inversion formula for the Sommerfeld integral. *Sov. Phys. Dokl.*, 3, 52–56.

[4] Maliuzhinets, G.D. (1958). Excitation, reflection and emission of surface waves from a wedge with given face impedances. *Sov. Phys. Dokl.*, 3, 752–755.

[5] Borovikov, V.A. (1962). Two-dimensional problem of diffraction on a polygon. *Sov. Phys. Dokl.*, 7(6), 499–501.

[6] Tuzhilin, A.A. (1973). The theory of Maliuzhinets inhomogeneous functional equations. *Diff. Urav.*, 9(11), 2058–2064.

[7] Bernard, J.M.L. (2005). Scattering by a three-part impedance plane: A new spectral approach. *Quart. J. of Mech. and Appl. Math.*, 58(3), 383–418.

[8] Bernard, J.M.L. (2006). A spectral approach for scattering by impedance polygons. *Quart. J. of Mech. and Appl. Math.*, 59(4), 518–548.

[9] Bernard, J.M.L. (1998). Diffraction at skew incidence by an anisotropic impedance wedge in electromagnetism theory: A new class of canonical cases. *J. of Phys. A: Math. Gen.*, 31, 595–613.

[10] Daniele, V.G. (2003). The Wiener-Hopf technique for impenetrable wedge having arbitrary aperture angle. *SIAM Journal of Applied Mathematics*, 63, 1442–1460.

[11] Lyalinov, M.A. (1999). Diffraction by a highly contrast transparent wedge. *J. Phys. A: Math. Gen.*, 32(11), 2183–2206.

[12] Sakharova, M.P. and Filippov, A.F. (1967). The solution of non-stationary problem of the diffraction at an impedance wedge using tabulated functions. *Zh. vychisl. Mat. Fiz.*, 7(3), 568–579.

[13] Bernard, J.M.L. (1993). Progresses on the diffraction by a wedge: Transient solution for line source illumination, single face contribution to scattered field, and new consequence of reciprocity on the spectral function. *Rev. Techn. Th.*, 25(4), 1209–1220.

[14] Bernard, J.M.L. (1994). On the time-domain scattering by a passive classical frequency dependent wedge-shaped region in a lossy dispersive medium. *Ann. Telecom.*, 49(11–12), 673–683.

[15] Bernard, J.M.L. (2014). *Advanced Theory of Diffraction by a Semi-infinite Impedance Cone*. Alpha Sciences International Ltd., Oxford.

[16] Tuzhilin, A.A. (1973). Diffraction of plane sound wave in an angular domain with absolutely hard and slippery face bounded by thin elastic plates. *Diff. Urav.*, 9(10), 1875–1888.

[17] Bernard, J.M.L. (1987). Diffraction by a metallic wedge covered with a dielectric material. *Wave Motion*, 9, 543–561.

[18] Osipov, A.V. (1996). Diffraction by a wedge with higher-order boundary conditions. *Radio Science*, 31(6), 1705–1720.

[19] Lyalinov, M.A. (1996). Diffraction of a plane wave by a wedge with thin material coatings. *Radio Science*, 31(6), 1721–1731.

[20] Michaeli, A. (1985). Contribution of a single face to the wedge diffracted field. *IEEE Trans. Antennas and Propagation*, 33(2), 221–223.

[21] Gradshteyn, I.S. and Ryzhik, I.M. (1980). *Table of Integrals, Series and Products.* Academic Press, London.

[22] Grolak, B. (2017). Analytic properties of the electromagnetic Green's function. *J. Math. Phys.*, 58, 071501.

[23] Bernard, J.M.L. (1998). A new class of canonical cases for the diffraction by an anisotropic impedance wedge. In *Proceeding Conf. IEEE AP-S*, Atlanta.

[24] Lyalinov, M.A. and Zhu, N.Y. (1999). Diffraction of a skewly incident plane wave by an anisotropic impedance wedge – A class of exactly solvable cases. *Wave Motion*, 30, 275–288.

[25] Bernard, J.M.L. (1989). Exact analytical solution for the diffraction at skew incidence by a class of wedge with absorbing material. *Rev. Techn. Th.*, 20–21, 523–527.

[26] Bernard, J.M.L. (1990). On the diffraction of an electromagnetic skew incident wave by a non perfectly conducting wedge. *Ann. Telecom.*, 45(1–2), 30–39.

[27] Osipov, A.V. (2014). Physical theory of diffraction for scatterers with low-reflection surface. *Radio Science*, 49, 1052–1064.

[28] Manara, G., Nepa, P., Pelosi, G., Vallechi, A. (2004). An approximate solution for skew incidence diffraction by an interior right-angled anisotropic impedance wedge. *Progr. Electrom. Res.*, 45, 45–75.

[29] Vallechi, A. (2012). Oblique incidence diffraction by edges in anisotropic impedance surfaces: A collection of exact solutions for specific non-axial tensor boundary conditions. *Electromagnetics*, 33(2), 73–98.

[30] Zavadskii, V.Y. and Sakharova, M.P. (1967). Application of the special function $\Psi_\Phi(z)$ in problems of wave diffraction in wedge-shaped regions. *Sov. Phys. Acoustics*, 13(1), 48–54.

[31] Bernard, J.M.L. (2013). On the scattering by a cavity in an impedance plane in 3D: Integral equations and new Green's function. *Quarterly J. Applied Math. Mech.*, 66(4), 609–640.

[32] Levine, L.M. (1964). A uniqueness theorem for the reduced wave equation. *Comm. Pure Appl. Math.*, 17, 147–176.

4.A. Appendix A: analysis of the solution of $s(\alpha \pm \Phi) - \varepsilon s(-\alpha \pm \Phi) = S^{\pm}(\alpha)$

We study here the solution of the functional equations for s:

$$s(\alpha \pm \Phi) - \varepsilon s(-\alpha \pm \Phi) = S^{\pm}(\alpha), \tag{4.189}$$

when s is regular as $|\text{Re}\alpha| \leq \Phi$, with ε being $+1$ or -1. We consider the case where the functions $S^{\pm}$ are analytic on the imaginary axis and $O(\exp(-a\ |\text{Im}\alpha\ |))$, $a > 0$, as $|\text{Im}\alpha| \rightarrow \infty$, and thus $s(i\infty) - \varepsilon s(-i\infty) = 0$. The solution s, regular as $|\text{Re}\alpha| \leq \Phi$ (even at infinity), can be written:

$$s(\alpha) = \frac{s(i\infty) + s(-i\infty)}{2} + \frac{s(i\infty) - s(-i\infty)}{2}\text{sign}(\text{Im}\alpha) + s_{\circ}(\alpha), \tag{4.190}$$

where $s_{\circ}$ is absolutely integrable on any line $\text{Re}\alpha = \alpha_{\circ}$ as $|\alpha_{\circ}| \leq \Phi$. This type of equation has been particularly solved in the works of Maliuzhinets, Tuzhilin and Bernard for the analytical determination of the solutions of wedge diffraction problems.

By use of Fourier-type transform $\mathcal{F}(\omega)$ of $s_{\circ}(\alpha) + \frac{s(i\infty) - s(-i\infty)}{2}\text{sign}(\text{Im}\alpha)$ along imaginary axis, in the sense of the finite part, we obtain a system of two equations:

$$\mathcal{F}(\omega)e^{\mp i\omega\Phi} - \varepsilon\mathcal{F}(-\omega)e^{\pm i\omega\Phi} = \mathcal{H}^{\pm}(\omega). \tag{4.191}$$

Multiplying them by $\exp(\mp i\omega\Phi)$ and making the difference, the function $\mathcal{F}$ satisfies:

$$2i\mathcal{F}(\omega)\sin(2\omega\Phi) = \mathcal{H}^{-}(\omega)e^{i\omega\Phi} - \mathcal{H}^{+}(\omega)e^{-i\omega\Phi}. \tag{4.192}$$

We notice that the analytic function obtained by dividing the right member of the previous equality with $2i\sin(2\omega\Phi)$ does verify (4.191), and then the inverse Fourier transform (integral taken in the sense of principal value) of this expression satisfies (4.189) for $|\text{Re}\alpha| < \Phi$. So, we have the developed form of s, for $|\text{Re}\alpha| < \Phi$:

$$\begin{aligned} s(\alpha) - \frac{s(i\infty) + s(-i\infty)}{2} &= \\ &= \frac{i}{\sqrt{2\pi}} pv \int_{-i\infty}^{+i\infty} \left(\frac{R_{+}(\omega)e^{-i\omega\Phi}}{i\sin(2\omega\Phi)} - \frac{R_{-}(\omega)e^{i\omega\Phi}}{i\sin(2\omega\Phi)} \right) e^{-i\omega\alpha} d\omega \\ &= \frac{i}{\sqrt{2\pi}} \sum_{\pm} \int_{0}^{i\infty} \frac{\pm R_{\pm}(\omega)(e^{-i\omega(\alpha\pm\Phi)} + \varepsilon e^{i\omega(\alpha\pm\Phi)})}{i\sin(2\omega\Phi)} d\omega \\ &= \frac{1}{\sqrt{2\pi}} \sum_{\pm} \int_{0}^{\infty} \frac{\pm R_{\pm}(i\nu)(e^{\nu(\alpha\pm\Phi)} + \varepsilon e^{-\nu(\alpha\pm\Phi)})}{\sinh(2\nu\Phi)} d\nu, \end{aligned} \tag{4.193}$$

with

$$R_{\pm}(\omega) = \frac{-i}{2\sqrt{2\pi}} \int_{-i\infty}^{+i\infty} S^{\pm}(\alpha')e^{i\omega\alpha'} d\alpha',$$

$$R_{\pm}(i\nu) = \frac{-i}{2\sqrt{2\pi}} \int_{-i\infty}^{+i\infty} S^{\pm}(\alpha')e^{-\nu\alpha'} d\alpha',$$

$$R_{\pm}(\omega) + \varepsilon R_{\pm}(-\omega) = 0, \tag{4.194}$$

where the quantity $s(i\infty) + s(-i\infty)$ is arbitrary when $\varepsilon = 1$ and equal to zero when $\varepsilon = -1$. It is worth noticing that after using pv (principal value) of the integral, it results in a well-defined analytical integral expression.

For the case where the function s is ∂v_{an} and $\partial \ln(\Psi_{an})$, we have $\varepsilon = -1$, while $R_{\pm}$ can be evaluated explicitly by the method of residues. So, we obtain a simple integral expression of v_{an} and Ψ_{an}, corresponding to (4.117) and (4.118).

In other respects, we notice that, since the functions $S^{\pm}$ are absolutely integrable, it is possible to change the order of integration, so that we obtain, for $|\mathrm{Re}\alpha| < \Phi$:

$$s(\alpha) = \frac{s(i\infty) + s(-i\infty)}{2} + \frac{-i}{8\Phi} \int_{-i\infty}^{+i\infty} d\alpha'(S^{+}(\alpha')\tan(\frac{\pi}{4\Phi}(\alpha + \Phi - \alpha'))$$
$$- S^{-}(\alpha')\tan(\frac{\pi}{4\Phi}(\alpha - \Phi - \alpha'))). \tag{4.195}$$

Considering $S^{\pm} = O(\exp(-a|\mathrm{Im}\alpha|))$ and the asymptotics of $\tan(.)$ terms, we have, when $\varepsilon = -1$:

$$s(\alpha) = \frac{\pm 1}{8\Phi} \int_{-i\infty}^{+i\infty} d\alpha'(S^{+}(\alpha') - S^{-}(\alpha')) + O(e^{-c|\mathrm{Im}\alpha|}), \tag{4.196}$$

as $\mathrm{Im}\alpha \to \pm\infty$, with $c > 0$ being some constants independent of α. Then, letting $s = \partial v_{an}$, we derive $v_{an}(\alpha) = A_1^{\pm} + O(e^{-c|\mathrm{Im}\alpha|})$, while, letting $s = \partial \ln(\Psi_{an})$, we obtain $\Psi_{an} = A_2^{\pm} \exp(\mu|\mathrm{Im}\alpha|)(1 + O(e^{-c|\mathrm{Im}\alpha|}))$ with $\mu = \pi/2\Phi$, $A_1^{\pm}$ and $A_2^{\pm}$ being some constants. This behavior, deduced from an expression valid for $|\mathrm{Re}\alpha| < \Phi$, remains valid in any band $|\mathrm{Re}\alpha| <$ constant, considering functional equations attached to these functions.

NOTE 4.27.– We notice, as Maliuzhinets, that:

$$pv\int_{-i\infty}^{+i\infty} \equiv \frac{1}{2}\Big(\int_{-i\infty-0^{+}}^{+i\infty-0^{+}} + \int_{-i\infty+0^{+}}^{+i\infty+0^{+}}\Big). \tag{4.197}$$

NOTE 4.28.– The expression (4.195) can be continued analytically for $|\mathrm{Re}\alpha| \geq \Phi$ by considering the residue due to the poles crossing the path of integration, so that we can easily verify that the continuation of this expression satisfies the initial functional equations. This latter expression continues to satisfy the functional equation when the analytic function $S^{\pm}$, regular on the imaginary axis, is only assumed to be absolutely integrable.

4.B. Appendix B: analysis of the reflection coefficient attached to a multilayered face of a wedge

Let us consider the reflection coefficient attached to each face $\varphi = \pm\Phi$, semi-planes composed of a passive surface with relative impedance $\sin\theta^{\pm}_{imp}$ covered by a passive material of total depth d with relative permittivity ϵ_r and permittivity μ_r varying in depth and frequency:

$$R^{\pm}(\beta) = \prod_{j=1}^{N_e} \frac{\cos\beta - \sin\theta^{\pm}_n}{\cos\beta + \sin\theta^{\pm}_n}, \tag{4.198}$$

where β is the angle with the outward normal at each face, and the complex constants $\sin\theta^{\pm}_n$ have real parts of arbitrary signs [16]–[18] with $|\mathrm{Re}(\theta^{\pm}_n)| \leq \pi/2$, $|\arg(i\omega)| < \pi/2$.

Let us consider to simplify the case with $|\arg(i\omega)| = |\arg(ik)| = \pi/2$, that is, ω and k are real. When the permittivity and the permeability of materials have positive real part, which is always the case at very large frequency, we can generally assume that the $\sin\theta^{\pm}_n$ with $\mathrm{Re}(\sin\theta^{\pm}_n) < 0$ (active modes) verify $\arg(ik)\mathrm{Im}(\sin\theta^{\pm}_n) < 0$ while the $\sin\theta^{\pm}_n$ with $\mathrm{Re}(\sin\theta^{\pm}_n) > 0$ (passive modes) have $\mathrm{Im}(\sin\theta^{\pm}_n)$ of arbitrary sign. In this case, the first poles of $f(\alpha;i\omega)$ from imaginary axis are, for each mode, at $\alpha = \pm(\pi + \theta^{\pm}_n + \Phi)$.

Besides, on limited domains of frequencies, permittivity and permeability can have negative real part (if, for example, metamaterials are present), and the term $\arg(ik)\mathrm{Im}(\sin\theta^{\pm}_n)$ for $\sin\theta^{\pm}_n$ with $\mathrm{Re}\sin\theta^{\pm}_n < 0$ (active modes) can have arbitrary sign. In this case, the first poles of $f(\alpha; i\omega)$ from imaginary axis are then, for each mode, at $\alpha = \pm(\pi + \theta^{\pm}_n + \Phi)$ when $\mathrm{Re}(\sin\theta^{\pm}_n) > 0$ (passive modes), while they are at $\alpha = \pm(-\theta^{\pm}_n + \Phi))$ (if $\arg(ik)\mathrm{Im}(\sin\theta^{\pm}_n) > 0$) or at $\alpha = \pm(\pi + \theta^{\pm}_n + \Phi)$ (if $\arg(ik)\mathrm{Im}(\sin\theta^{\pm}_n) < 0$) when $\mathrm{Re}(\sin\theta^{\pm}_n) < 0$ (active modes).

NOTE 4.29.– To analyze in more details f with complex poles, the reader can refer to the section where we give the solution for a wedge with arbitrary passive or active impedance faces, and we describe the suppression of nonphysical poles for f when active impedances are present.

NOTE 4.30.– We can generally assume that, as ω tends to infinity,

$$\epsilon_r \to 1, \mu_r \to 1, \text{and thus, } R^{\pm}(\beta) \to e^{-2i\omega d\cos\beta/c}\frac{\cos\beta - \sin\theta^{\pm}_{imp}}{\cos\beta + \sin\theta^{\pm}_{imp}}, \tag{4.199}$$

with $\mathrm{Re}(\theta^{\pm}_{imp}) > 0$, while, for the modes that are active with $\mathrm{Re}(\theta^{\pm}_{n}) < 0$:

$$\begin{aligned}&\mathrm{Re}(\theta^{\pm}_{n}) \to 0 \text{ as } |\mathrm{Re}(\omega)| \to \infty,\\ &|\mathrm{Im}(\theta^{\pm}_{n})| \to \infty \text{ as } \omega \to 0.\end{aligned} \tag{4.200}$$

NOTE 4.31.– Let us note that, in free space, we can write $k = \omega\sqrt{\epsilon_0\mu_0}$ $(= \omega/c)$ for the wavenumber, while for modelization of passive metamaterials (or plasmas), we have to write $k = \omega\sqrt{\epsilon}\sqrt{\mu}$ $(= \omega\sqrt{\epsilon_r}\sqrt{\mu_r}/c)$ to ensure passivity and $\mathrm{Re}(ik) > 0$.

4.C. Appendix C: some miscellaneous properties of spectral functions for an impedance wedge with passive impedance faces

4.C.1. *The spectral function* $f_0(\alpha)$ *for a wedge with straight passive faces*

Let us consider the spectral function $f_0(\alpha)$ associated with the diffraction of an unit plane wave of incidence angle $\varphi_\circ$ by a wedge with straight faces of relative impedances $\sin\theta^{\pm}$ with positive real part (passive impedances) [4]. It is given by:

$$\begin{aligned}&f_0(\alpha) = \Psi(\alpha)\sigma(\alpha)/\Psi(\varphi_\circ),\\ &\sigma(\alpha) = \mu\cos(\mu\varphi_\circ)/(\sin(\mu\alpha) - \sin(\mu\varphi_\circ)),\end{aligned} \tag{4.201}$$

where the function $\Psi(\alpha)$, without any zero or pole as $\mathrm{Re}\alpha < \Phi$, is the solution of:

$$\begin{aligned}&(\sin\alpha + \sin\theta^{+})\Psi(\alpha + \Phi) - (-\sin\alpha + \sin\theta^{+})\Psi(-\alpha + \Phi) = 0,\\ &(\sin\alpha - \sin\theta^{-})\Psi(\alpha - \Phi) - (-\sin\alpha - \sin\theta^{-})\Psi(-\alpha - \Phi) = 0,\end{aligned} \tag{4.202}$$

that is, when $\mathrm{Re}(\sin\theta^{\pm}) \neq 0$:

$$\begin{aligned}&\partial_\alpha \ln\frac{\Psi(\alpha \pm \Phi)}{\Psi(-\alpha \pm \Phi)} = \pm\partial_\alpha \ln\frac{(\sin\alpha - \sin\theta^{\pm})}{(\sin\alpha + \sin\theta^{\pm})},\\ &\frac{\Psi'}{\Psi}(\alpha \pm \Phi) + \frac{\Psi'}{\Psi}(-\alpha \pm \Phi) = S^{\pm}(\alpha),\\ &S^{\pm}(\alpha) = \pm\partial_\alpha \ln\frac{(\sin\alpha - \sin\theta^{\pm})}{(\sin\alpha + \sin\theta^{\pm})},\end{aligned} \tag{4.203}$$

(note: the case $\mathrm{Re}(\sin\theta^{\pm}) = 0$ can be considered by taking $\mathrm{Re}(\sin\theta^{\pm}) \to 0^{+}$). The expression of Ψ can be expressed from (4.115)–(4.117), considering Ψ in place of Ψ_{an} with $\delta = 0$. In passive case with $0 < \mathrm{Re}(\sin\theta^{\pm}) \leq \pi/2$, we can then write, as $|\mathrm{Re}\alpha| \leq \Phi + (N+1)\pi$:

$$
\begin{aligned}
\Psi(\alpha) = {} & B'_N \prod_{\pm}\Big(\prod_{m=0}^{N}[\Gamma(\frac{1}{2}+\frac{1}{4\Phi}(\alpha\pm\Phi+\theta^{\pm}+m\pi))\\
& \Gamma(\frac{1}{2}-\frac{1}{4\Phi}(\alpha\pm\Phi-(\theta^{\pm}+m\pi)))\Gamma(\frac{1}{2}+\frac{1}{4\Phi}(\alpha\pm\Phi+\pi-\theta^{\pm}+m\pi))\\
& \Gamma(\frac{1}{2}-\frac{1}{4\Phi}(\alpha\pm\Phi-(\pi-\theta^{\pm}+m\pi)))]^{(-1)^{m+1}}\\
& \times\exp\Big(\int_0^{\infty}(-e^{-\nu\pi})^{N+1}\frac{e^{-\nu\theta^{\pm}}+e^{-\nu(\pi-\theta^{\pm})}}{(1+e^{-\nu\pi})}\frac{(1-\cosh(\nu(\alpha\pm\Phi)))}{\nu\sinh(2\nu\Phi)}d\nu\Big)\Big),
\end{aligned}
\tag{4.204}
$$

where B'_N is an arbitrary constant. This expression is suitable for numerical calculus (with $N = 1$ or 2) or to derive the analytical properties of $\Psi(\alpha)$ (with $N \to \infty$) from those of Γ. We note that the function $\Psi(\alpha)$ in passive case, that is, when $0 < \mathrm{Re}(\sin\theta^{\pm}) \leq \pi/2$, has zeros at:

$$
\begin{aligned}
& \alpha = \mp\Phi + \epsilon(\theta^{\pm} + (4p+2)\Phi + 2m\pi)\\
& \text{and } \alpha = \mp\Phi + \epsilon(\pi - \theta^{\pm} + (4p+2)\Phi + 2m\pi),
\end{aligned}
\tag{4.205}
$$

and has poles at

$$
\begin{aligned}
& \alpha = \mp\Phi + \epsilon(\pi + \theta^{\pm} + (4p+2)\Phi + 2m\pi)\\
& \text{and } \alpha = \mp\Phi + \epsilon(2\pi - \theta^{\pm} + (4p+2)\Phi + 2m\pi),
\end{aligned}
\tag{4.206}
$$

for $m \geq 0$, $p \geq 0$, $\epsilon = 1$ and -1, while we have [12]:

$$
\Psi(\alpha) = A\cos(\frac{\pi\alpha}{2\Phi})(1 + O(\alpha^{\nu}e^{-\mu'|\mathrm{Im}\alpha|})),
\tag{4.207}
$$

for large $|\mathrm{Im}\alpha|$, with constants A, ν and $\mu' > 0$.

4.C.2. *Some miscellaneous results as* $\Phi = \pi/2$ *and Re*$(\sin\theta^{\pm}) \geq 0$

We note that, for $\Phi = \pi/2$, we have:

$$
\begin{aligned}
& \frac{(-\sin\alpha+\sin\theta^{-})}{\Psi(\alpha+\frac{\pi}{2})} - \frac{(\sin\alpha+\sin\theta^{-})}{\Psi(-\alpha+\frac{\pi}{2})}\\
& = (\sin\theta^{-} - \sin\theta^{+})(\frac{1}{\Psi(\alpha+\frac{\pi}{2})} - \frac{1}{\Psi(-\alpha+\frac{\pi}{2})}),\\
& \frac{(\sin\alpha+\sin\theta^{-})}{\Psi(\alpha-\frac{\pi}{2})} - \frac{(-\sin\alpha+\sin\theta^{-})}{\Psi(-\alpha-\frac{\pi}{2})} = 0.
\end{aligned}
\tag{4.208}
$$

Using that $\Psi(-\alpha)\Psi(\alpha) = A_0^2(\cos\alpha + \sin\theta^+)(\cos\alpha + \sin\theta^-)$, and $\Psi(\alpha)/\cos\alpha \to A_0$ as $\alpha \to i\infty$, we can let $s(\alpha) = \frac{A_0(\cos\alpha+\sin\theta^-)}{\Psi(\alpha)}$ and derive, from (4.195) in appendix A:

$$\frac{(\cos\alpha + \sin\theta^-)}{\Psi(\alpha)} - \frac{1}{A_0} =$$
$$= \frac{-i(\sin\theta^- - \sin\theta^+)}{4\pi} \int_{-i\infty}^{+i\infty} H(\alpha') \tan(\frac{1}{2}(\alpha + \frac{\pi}{2} - \alpha'))d\alpha',$$
$$\frac{i\Psi(-\alpha)/A_0^2}{(\cos\alpha + \sin\theta^+)} - \frac{i}{A_0} =$$
$$= \frac{(\sin\theta^- - \sin\theta^+)}{4\pi} \int_{-i\infty}^{+i\infty} H(\alpha') \tan(\frac{1}{2}(\alpha + \frac{\pi}{2} - \alpha'))d\alpha', \tag{4.209}$$

for $\frac{-3\pi}{2} < \text{Re}(\alpha) < \frac{\pi}{2}$, where $H(\alpha') = 2\sin\alpha'/(\Psi(\alpha' + \frac{\pi}{2})(\sin\alpha' + \sin\theta^+))$ and

$$H(\alpha) = \frac{1}{\Psi(\alpha + \frac{\pi}{2})} - \frac{1}{\Psi(-\alpha + \frac{\pi}{2})} = \frac{\sin\alpha/\sin\theta^+}{\Psi(\alpha + \frac{\pi}{2})} + \frac{\sin\alpha/\sin\theta^+}{\Psi(-\alpha + \frac{\pi}{2})}. \tag{4.210}$$

From oddness of $H(\alpha)$, $\tan(\frac{1}{2}(\alpha + \frac{\pi}{2} - \alpha'))$ can be replaced by $\frac{-\sin\alpha'}{(\cos(\alpha+\frac{\pi}{2})+\cos\alpha')}$, and we note that, after deformation of integration path, we have:

$$\int_{-i\infty}^{+i\infty} \frac{1}{\Psi(\varphi')} \frac{\cos\varphi'}{\cos\alpha + \sin\varphi'} d\varphi' = -2\pi i \frac{\frac{\sin\alpha+\sin\theta^-}{\Psi(\alpha-\frac{\pi}{2})} - \frac{1}{A_0}}{(\sin\theta^+ - \sin\theta^-)},$$
$$\int_{-i\infty}^{+i\infty} \frac{\cos\varphi'}{\Psi(\varphi')} \frac{\cos\varphi'}{\cos\alpha + \sin\varphi'} d\varphi' = 2\pi i \sin\theta^+ \frac{\frac{\sin\alpha+\sin\theta^-}{\Psi(\alpha-\frac{\pi}{2})} - \frac{1}{A_0}}{(\sin\theta^+ - \sin\theta^-)} + \frac{i\pi}{A_0}, \tag{4.211}$$

as $|\text{Re}\alpha| < \frac{\pi}{2}$. Similar expressions with $-\sin\varphi'$ in place of $\sin\varphi'$ can be obtained from $\Psi(-\alpha) \to \Psi(\alpha)$ when $\theta^\pm \to \theta^\mp$ (or by continuation with capture of poles).

4.D. Appendix D: functional equations of high order and behavior of fields at the edge

Let us consider high-order boundary conditions on each face, and associated functional equations on f, following:

$$\Big[\sin\alpha(\sum_{n=0}^{N} a_n^{\pm}\cos^{2n}\alpha)(f(\alpha\pm\Phi)+f(\alpha\pm\Phi))$$

$$+(\sum_{m=0}^{2N+1} b_m^{\pm}\cos^{m}\alpha)(f(\alpha\pm\Phi)-f(\alpha\pm\Phi))\Big]$$

$$\times(\sum_{p=0}^{P} d_p^{\pm}\cos^{p}\alpha)=\sum_{n=1}^{2N+P+1} c_n^{\pm}\sin\alpha\cos^{n-1}\alpha. \tag{4.212}$$

for f satisfying the conditions (a), (b) and (c).

4.D.1. *Combinations of derivatives at the edge for the* $c_n^{\pm}$

We write:

$$\sum_{p=0}^{P} d_p^{\pm}\cos^{p}\alpha\sum_{n=0}^{N} a_n^{\pm}\cos^{2n}\alpha=\sum_{n=0}^{M} g_n^{\pm}\cos^{n}\alpha,$$

$$\sum_{p=0}^{P} d_p^{\pm}\cos^{p}\alpha\sum_{m=0}^{2N+1} b_m^{\pm}\cos^{m}\alpha=\sum_{n=0}^{M+1} h_n^{\pm}\cos^{n}\alpha, \tag{4.213}$$

with $M=2N+P$. We can then reformulate the functional equations following:

$$\Big[\sin\alpha(\frac{\sum_{n=0}^{M} g_n^{\pm}\cos^{n}\alpha}{\cos^{m}\alpha})(f(\alpha\pm\Phi)+f(\alpha\pm\Phi))$$

$$+(\frac{\sum_{n=0}^{M+1} h_n^{\pm}\cos^{n}\alpha}{\cos^{m}\alpha})(f(\alpha\pm\Phi)-f(\alpha\pm\Phi))\Big]$$

$$=\sum_{n=1}^{m-1} c_n^{\pm}\frac{\sin\alpha}{\cos^{m-n+1}\alpha}+c_m^{\pm}\frac{\sin\alpha}{\cos\alpha}+\sum_{n=m+1}^{2N+P+1} c_n^{\pm}\sin\alpha\cos^{n-m-1}\alpha. \tag{4.214}$$

We then apply to this equation the operator:

$$\lim_{\rho\to 0}\frac{1}{4\pi i}\int_{\gamma}(.)e^{ik\rho\cos\alpha}d\alpha. \tag{4.215}$$

For that, we can note, that, deforming γ to take semi-infinite branches at $\mathrm{Re}\alpha=\pm\pi$, and taking account of periodicity of $(\sin^{p}\alpha)(\cos^{n}\alpha)$, we have:

$$\frac{1}{4\pi i}\int_{\gamma}(\sin^{p}\alpha)(\cos^{n}\alpha)e^{ik\rho\cos\alpha}d\alpha=0,\ \text{for } p\geq 0,\ n\geq 0, \tag{4.216}$$

while taking γ with large $|\mathrm{Im}\alpha| > d \gg 1$ and semi-infinite branches at $\mathrm{Re}\alpha = \pm\pi$, we can write:

$$\lim_{\rho\to 0}\frac{1}{4\pi i}\int_\gamma \frac{\sin\alpha}{\cos\alpha}e^{ik\rho\cos\alpha}d\alpha = -1, \lim_{\rho\to 0}\frac{1}{4\pi i}\int_\gamma \frac{\sin\alpha}{\cos^n\alpha}e^{ik\rho\cos\alpha}d\alpha = 0, n \geq 2. \tag{4.217}$$

and

$$\begin{aligned}
&\lim_{\rho\to 0}\frac{1}{4\pi i}\int_\gamma \frac{\sin\alpha}{\cos\alpha}(f(\alpha+\varphi)+f(-\alpha+\varphi))e^{ik\rho\cos\alpha}d\alpha \\
&= -(f(i\infty)+f(-i\infty)) = 0, \\
&\lim_{\rho\to 0}\frac{1}{4\pi i}\int_\gamma \frac{\sin\alpha}{\cos^n\alpha}(f(\alpha+\varphi)+f(-\alpha+\varphi))e^{ik\rho\cos\alpha}d\alpha = 0,\ n\geq 2,
\end{aligned} \tag{4.218}$$

then, to complete, we can notice, for $p \geq 0$, that we have:

$$\begin{aligned}
\frac{\partial^p}{\partial(ik\rho)^p}\frac{\partial u(\rho,\varphi)}{ik\rho\partial\varphi} &= \frac{1}{4\pi i}\int_\gamma \sin\alpha\cos^p\alpha(f(\alpha+\varphi)+f(-\alpha+\varphi))e^{ik\rho\cos\alpha}d\alpha, \\
\frac{\partial^p(u(\rho,\varphi))}{\partial(ik\rho)^p} &= \frac{1}{4\pi i}\int_\gamma \cos^p\alpha(f(\alpha+\varphi)-f(-\alpha+\varphi))e^{ik\rho\cos\alpha}d\alpha.
\end{aligned} \tag{4.219}$$

Therefore, we can write:

$$\lim_{\rho\to 0}\frac{1}{4\pi i}\Big(\sum_{n\geq m}^{M} g_n^{\pm}\frac{\partial^{n-m}}{\partial(ik\rho)^{n-m}}\frac{\partial}{ik\rho\partial\varphi} + \sum_{n\geq m}^{M+1} h_n^{\pm}\frac{\partial^{n-m}}{\partial(ik\rho)^{n-m}}\Big)u(\rho,\varphi)|_{\varphi=\pm\Phi} = -c_m^{\pm}, \tag{4.220}$$

for $m \leq M$, and

$$\frac{h_{M+1}^{\pm}}{4\pi i}u(\rho,\varphi)|_{\varphi=\pm\Phi} = \frac{h_{M+1}^{\pm}}{4\pi i}u(0,\varphi) = -c_{M+1}^{\pm}, \tag{4.221}$$

where we notice that, for $h_{M+1} = 0$, we have $c_{M+1}^{\pm} = 0$ and:

$$\frac{g_M^{\pm}}{h_M^{\pm}}\lim_{\rho\to 0}\Big(\frac{\partial u(\rho,\varphi)}{ik\rho\partial\varphi}\Big)|_{\varphi=\pm\Phi} = -\frac{4\pi i}{h_M^{\pm}}c_M^{\pm} - u(0,\varphi). \tag{4.222}$$

These results exhibit acceptable "combinations" of fields derivatives (each singular) that remains finite at the edge, keeping in mind that each derivatives can be singular as $\rho \to 0$, except the one of order 0, that is, $u(0,\varphi)$, which is finite [32]. These conditions on u can then contribute to define acceptable "contact" conditions between derivatives of u at $\varphi = \pm\Phi$ that will permit the unique determination of f, and thus u.

4.D.2. *Combinations of derivatives at the edge isolating* $\frac{\partial^m}{\partial(ik\rho)^m}\frac{\partial u}{ik\rho\partial\varphi}$

We now consider the expressions [26]:

$$\frac{\sum_{m=0}^{2N+1} b_m^{\pm} \cos^m \alpha}{\sum_{n=0}^{N} a_n^{\pm} \cos^{2n} \alpha} = \sum_{n=-1}^{M} \frac{e_n^{\pm}}{\cos^n \alpha} + E(\alpha),$$

$$\frac{\sum_{n=1}^{2N+P+1} c_n^{\pm} \cos^{n-1} \alpha}{\sum_{p=0}^{P} d_p^{\pm} \cos^p \alpha \sum_{n=0}^{N} a_n^{\pm} \cos^{2n} \alpha} = \sum_{n=0}^{M} \frac{l_n^{\pm}}{\cos^n \alpha} + L(\alpha), \tag{4.223}$$

where $E(\alpha)$ and $L(\alpha)$ are $O(1/\cos^{M+1}\alpha)$ when $|\mathrm{Im}\alpha| \to \infty$, $l_n^{\pm}$ being linear combination of $c_n^{\pm}$. This allows to reformulate the functional equations following:

$$\Big[(f(\alpha \pm \Phi) + f(-\alpha \pm \Phi))$$
$$+ \sum_{n=-1}^{M} \frac{e_n^{\pm}}{\cos^n \alpha} \frac{1}{\sin \alpha}(f(\alpha \pm \Phi) - f(-\alpha \pm \Phi))\Big]$$
$$= \sum_{n=0}^{M} \frac{l_n^{\pm}}{\cos^n \alpha} + [-\frac{E(\alpha)}{\sin \alpha}(f(\alpha \pm \Phi) - f(-\alpha \pm \Phi)) + L(\alpha)]. \tag{4.224}$$

Multiplying both sides of the functional equations by $\sin\alpha \cos^m \alpha$, and applying the operator:

$$\lim_{\rho \to 0} \frac{1}{4\pi i} \int (.) e^{ik\rho \cos \alpha} d\alpha, \tag{4.225}$$

we obtain [26]:

$$\lim_{\rho \to 0} \frac{1}{4\pi i} (\frac{\partial^m}{\partial(ik\rho)^m} \frac{\partial}{ik\rho\partial\varphi} + \sum_{n=-1}^{m} \frac{e_n^{\pm}}{\cos^n \alpha} \frac{\partial^{m-n}}{\partial(ik\rho)^{m-n}}) u(\rho, \varphi)|_{\varphi = \pm\Phi} = -l_{m+1}^{\pm}, \tag{4.226}$$

for $m \geq 0$, while, for $m = -1$, we have:

$$e_{-1}^{\pm} 2if(i\infty) = e_{-1}^{\pm} u(0^+, \varphi)|_{\varphi=\pm\Phi} = -l_0^{\pm},$$
$$\text{with } u(0^+, \varphi)|_{\varphi=+\Phi} = u(0^+, \varphi)|_{\varphi=-\Phi}. \tag{4.227}$$

We notice that, for $b_{2N+1}^{\pm} = 0$, we have $e_{-1}^{\pm} = 0$, thus $l_0^{\pm} = 0$, then $c_{2N+P+1}^{\pm} = 0$. As previously, these results also exhibit acceptable "combinations" of fields derivatives (with each derivative possibly singular) that remain finite at the edge, which are explicitly related to the coefficients $c^{\pm}$.

4.E. Appendix E: on the poles of $(\arctan\frac{n}{l})'$ for the evaluation of v_{an}

Refering to our solution for skew incidence [9],[23], we note that:

$$\arctan\frac{n}{l} = \frac{i}{2}\ln\Big(-\frac{n+il}{n-il}\Big),$$
$$(\arctan\frac{n}{l})' = \frac{i}{2}\Big(\frac{(n+il)'}{n+il} - \frac{(n-il)'}{n-il}\Big),$$
$$n(\alpha) = \cos\beta\sin 2\alpha,\ l(\alpha) = \cos^2\beta\sin^2\alpha - \cos^2\alpha - (\eta^2-1)\sin^2\beta, \qquad (4.228)$$

where $\eta = \sin\theta_1$ with 0 $<$Re$(\theta_1) \leq \pi/2$. We can write:

$$\begin{aligned}
n+il &= \cos\beta\sin 2\alpha + i(\cos^2\beta\sin^2\alpha - \cos^2\alpha - (\eta^2-1)\sin^2\beta)\\
&= i(\cos\beta\sin\alpha - i\cos\alpha)^2 - i(\eta^2-1)\sin^2\beta\\
&= i((\cos\beta\sin\alpha - i\cos\alpha)^2 + \cos^2\theta_1\sin^2\beta),\\
n-il &= \cos\beta\sin 2\alpha - i(\cos^2\beta\sin^2\alpha - \cos^2\alpha - (\eta^2-1)\sin^2\beta)\\
&= -i(\cos\beta\sin\alpha + i\cos\alpha)^2 + i(\eta^2-1)\sin^2\beta\\
&= -i((\cos\beta\sin\alpha + i\cos\alpha)^2 + \cos^2\theta_1\sin^2\beta),
\end{aligned} \qquad (4.229)$$

and, letting $\delta = \ln(\tan(\frac{\beta}{2}))$ with:

$$\begin{aligned}
\cos(i\delta) &= (\tan(\beta/2) + 1/\tan(\beta/2))/2 = 1/\sin\beta,\\
\sin(i\delta) &= i(\tan(\beta/2) - 1/\tan(\beta/2))/2 = -i\cot\beta,\\
\cos\beta &= i\sin i\delta/\cos i\delta,
\end{aligned} \qquad (4.230)$$

we have $n+il=0$ for $\alpha = \theta_1 - i\delta$ and $-\theta_1 - i\delta$, and $n-il=0$ for $\alpha = \theta_1 + i\delta$ and $-\theta_1 + i\delta$, which leads us, from the properties of $\sin\alpha$ and $\cos\alpha$, to derive all poles.

5

General Integral Identities for Bianisotropic Media and Related Equations, Properties and Coupling Expressions

5.1. Introduction

We consider advanced mathematical equations and properties on electromagnetic fields derived from integral identities in bianisotropic media: generalized reciprocity principle for adjoint systems, equivalence of surface sources, uniqueness and links between impedance surface for adjoint systems; then, we develop integral expressions of fields influenced by a perturbation of a system, as for an imperfectly conducting object inserted into a second one.

Therefore, we can consider the fields due to the coupling between two arbitrary scatterers A and A' perturbating a complex bianisotropic system, which generalizes our previous works [1]. As a particular example of application of our results on couplings for bianisotropic systems, we consider the case of the evaluation of the radiation of a cavity (noted A) inserted in an infinite plate; when, in practice, the ends of the plate (noted A') perturb the result, and then present an efficient method that permits the numerical suppression of the first-order couplings between A' and A by a simple and efficient postprocess. Thus, we can recover the scattering diagram of A when it is inserted in a plane (i.e. in the plate without ends) in a simple manner at any observation angle. Our method remains efficient for backscattering, from grazing to normal angles, in contrast of well-known difficulties at grazing incidence of common time domain filtering methods.

5.2. General integral identities for piecewise continuous media with finite losses

5.2.1. *Bianisotropic media and adjoint characteristics*

Let us consider the electromagnetic fields (E, H) and (E_0, H_0) in bianisotropic media, with characteristics (or constitutive parameters) $((\epsilon),(\xi),(\mu),(\varsigma))$ and $((\epsilon_0),(\xi_0),(\mu_0),(\varsigma_0))$ that are piecewise continuous in the domain Ω. For a time dependence $e^{i\omega t}$, the Maxwell equations attached to them [2]–[7] are given by:

$$\begin{aligned}
\text{curl}(H) &= i\omega((\epsilon)E + (\xi)H) + J,\\
\text{curl}(E) &= -i\omega((\mu)H + (\varsigma)E) - M,\\
\text{curl}(H_0) &= i\omega((\epsilon_0)E_0 + (\xi_0)H_0) + J_0,\\
\text{curl}(E_0) &= -i\omega((\mu_0)H_0 + (\varsigma_0)E_0) - M_0.
\end{aligned} \tag{5.1}$$

We can assume (5.1) as valid in whole Ω with finite characteristics, except at some isolated points (tips) or on some lines (edges) of discontinuity of parameters, where components of fields could be singular. In a similar manner, we define the fields (E_{0C}, H_{0C}) for media with characteristics $({}^t(\epsilon_0),-{}^t(\varsigma_0),{}^t(\mu_0),-{}^t(\xi_0))$, adjoint (or complementary) to $((\epsilon_0),(\xi_0),(\mu_0),(\varsigma_0))$, verifying:

$$\begin{aligned}
\text{curl}(H_{0C}) &= i\omega\,({}^t(\epsilon_0)E_{0C} - {}^t(\varsigma_0)H_{0C}) + J_{0C},\\
\text{curl}(E_{0C}) &= -i\omega({}^t(\mu_0)H_{0C} - {}^t(\xi_0)E_{0C}) - M_{0C},
\end{aligned} \tag{5.2}$$

where ${}^t(.)$ is the transpose operator.

The parameters $(\epsilon_{(0)})$ and $(\mu_{(0)})$ are the permittivity and the permeability tensors, while $(\xi_{(0)})$ and $(\varsigma_{(0)})$ are the bianisotropic coupling tensors [2]–[5]. The terms J_0, M_0, J_{0C}, M_{0C}, J and M are the primary sources, and we can consider them in the sense of distributions (or generalized functions) so that we can take account of Dirac (dipolar) sources (see pp. 35–37 of [6]). The fields in adjoint cases are strongly related, from some specific integral identities, in particular, from the generalized reciprocity theorem [2]–[4], that we will study and use in several configurations during our analysis.

NOTE 5.1.– The bianisotropic coupling tensors are related to the Tellegen (χ) and chirality (κ) tensors parameters, following $(\xi) = \frac{1}{c}((\chi) - i(\kappa))$ and $(\varsigma) = \frac{1}{c}((\chi) + i(\kappa))$, where c is the speed of light. In these expressions, (κ) is usually said to model reciprocal physical effects, while (χ) corresponds to nonreciprocal ones [2]–[5]. In the particular case where $(\chi) = 0$, ${}^t(\epsilon_0) = (\epsilon_0)$, ${}^t(\mu_0) = (\mu_0)$, and ${}^t(\kappa) = (\kappa)$, we note that the fields (E_{0C}, H_{0C}) and (E_0, H_0) satisfy identical Maxwell equations when $J_{0C} = J_0$ and $M_{0C} = M_0$.

5.2.2. *Basic general properties of fields*

We suppose afterwards the usual conditions on the fields satisfying (5.1) in Ω [6]:

a) The primary sources are bounded and the fields are composed of outward waves at infinity, along directions with no bounds of Ω. Considering, from now on, arbitrary fixed non-null small losses in unbounded part of media, the fields exponentially vanish if we let them tend the observation point to infinity (as for the limit absorption principle).

b) We assume that the fields remain bounded in an arbitrarily thin transition plane layer, without any primary source in it. This implies, with (5.1) and (5.2), that the curl of fields remain bounded. In the sense of the limit, for a piecewise continuous media with finite parameters, this leads us to the continuity of tangential components of fields at any smooth interfaces between two different media.

c) The fields can be singular at possible isolated points (tip of cone), or on some line of discontinuities (edge). Considering the primary sources apart, we suppose that the fields are integrable along any surface, including these discontinuities, which leads us to consider any further surface integrals as improper. Besides, in agreement with condition (b), we can assume that the fields component along any smooth line of discontinuities is continuous across it.

NOTE 5.2.– In media with large losses, the fields generally decrease exponentially as we move away from an interface (skin effect), and skin depth vanishes as losses tends to ∞.

NOTE 5.3.– To consider the equations (5.1) and (5.2) as valid at interfaces of piecewise continuous media with finite parameters is equivalent, in the sense of distributions, to let tangential surface sources vanish due to any jump in tangential fields expressions, which leads us to the continuity of tangential fields at smooth interfaces.

NOTE 5.4.– Let us note a point concerning mathematical definitions of parameters. In free space, we usually write $k = \omega\sqrt{\epsilon_0\mu_0}$ $(= \omega/c)$ for the wave number, while, for modelization of isotropic passive metamaterials, we need to take the precaution to write $k = \omega\sqrt{\epsilon}\sqrt{\mu}$ $(= \omega\sqrt{\epsilon_r}\sqrt{\mu_r}/c)$ to ensure passivity and $\mathrm{Re}(ik) > 0$, so that e^{-ikR} vanishes at infinity. For that, elementary expressions commonly used for usual materials often need to be changed for metamaterials. In this case, we note that:

$$\frac{k}{\omega\epsilon}\tan(kd) = \frac{\sqrt{\mu}}{\sqrt{\epsilon}}\tan(\omega\sqrt{\epsilon}\sqrt{\mu}\,d) \text{ is not equivalent to } \sqrt{\frac{\mu}{\epsilon}}\tan(\omega\sqrt{\epsilon\mu}\,d). \quad (5.3)$$

5.2.3. *Initial surface and volume integral identities for piecewise continuous media*

Let us consider in first step that the media parameters are continuous in Ω, and the bounded primary sources are disjoint from $\partial\Omega$. Using the basic vectorial equalities:

$$\begin{aligned} &\mathrm{div}(U \wedge W) = W.\mathrm{curl}(U) - U.\mathrm{curl}(W), \\ &U.{}^t(P)W = W.(P)U, \end{aligned} \tag{5.4}$$

with (5.1) and (5.2), and the divergence theorem [6], we can write:

$$\begin{aligned} &\int_\Omega \mathrm{div}(E \wedge H_{0C})dV = \int_{\partial\Omega} \widehat{n}_s.(E \wedge H_{0C})dS, \\ &\int_\Omega \mathrm{div}(E \wedge H_{0C})dV = \int_\Omega (H_{0C}.\mathrm{curl}E - E.\mathrm{curl}H_{0C})dV \\ &= \int_\Omega (-i\omega H_{0C}.(\mu)H - M.H_{0C} - i\omega E.{}^t(\epsilon_0)E_{0C} - J_{0C}.E)dV \\ &+ \int_\Omega (-i\omega H_{0C}.(\varsigma)E + i\omega E.{}^t(\varsigma_0)H_{0C})dV \\ &= \int_\Omega (-i\omega H_{0C}.(\mu)H - M.H_{0C} - i\omega E_{0C}.(\epsilon_0)E - J_{0C}.E)dV \\ &+ \int_\Omega (-i\omega H_{0C}.((\varsigma) - (\varsigma_0))E)dV, \end{aligned} \tag{5.5}$$

and,

$$\begin{aligned} &\int_\Omega \mathrm{div}(H \wedge E_{0C})dV = \int_{\partial\Omega} \widehat{n}_s.(H \wedge E_{0C})dS, \\ &\int_\Omega \mathrm{div}(H \wedge E_{0C})dV = \int_\Omega (E_{0C}.\mathrm{curl}(H) - H.\mathrm{curl}(E_{0C}))dV \\ &= \int_\Omega (i\omega E_{0C}.(\epsilon)E + J.E_{0C} + i\omega H.{}^t(\mu_0)H_{0C} + M_{0C}.H)dV \\ &+ \int_\Omega (i\omega E_{0C}.(\xi)H - i\omega H.{}^t(\xi_0)E_{0C})dV \\ &= \int_\Omega (i\omega E_{0C}.(\epsilon)E + J.E_{0C} + i\omega H_{0C}.(\mu_0)H + M_{0C}.H)dV \\ &+ \int_\Omega (i\omega E_{0C}.((\xi) - (\xi_0))H)dV, \end{aligned} \tag{5.6}$$

where $\widehat{n}_s$ is the unit normal to $\partial\Omega$ outward Ω.

After addition of these equations, we then obtain integral identities in the following form:

$$\begin{aligned}&\int_{\partial\Omega}(E_{0C}.(\widehat{n}_s\wedge H)+H_{0C}.(\widehat{n}_s\wedge E))dS\\&=\int_{\Omega}i\omega E_{0C}.((\epsilon)-(\epsilon_0))E\,dV-\int_{\Omega}i\omega H_{0C}.((\mu)-(\mu_0))H\,dV\\&-\int_{\Omega}i\omega H_{0C}.((\varsigma)-(\varsigma_0))EdV+\int_{\Omega}i\omega E_{0C}.((\xi)-(\xi_0))HdV\\&+\int_{\Omega}(M_{0C}.H-J_{0C}.E)\,dV-\int_{\Omega}(H_{0C}.M-E_{0C}.J)\,dV,\end{aligned} \quad (5.7)$$

and thus, with $\widehat{n}=-\widehat{n}_s$ the unit normal to $\partial\Omega$ inward Ω:

$$\begin{aligned}&\int_{\partial\Omega}(E_{0C}.(\widehat{n}\wedge H)+H_{0C}.(\widehat{n}\wedge E))dS\\&=-\int_{\Omega}i\omega E_{0C}.((\epsilon)-(\epsilon_0))E\,dV+\int_{\Omega}i\omega H_{0C}.((\mu)-(\mu_0))H\,dV\\&+\int_{\Omega}i\omega H_{0C}.((\varsigma)-(\varsigma_0))EdV-\int_{\Omega}i\omega E_{0C}.((\xi)-(\xi_0))HdV\\&+\int_{\Omega}(J_{0C}.E-M_{0C}.H)\,dV-\int_{\Omega}(E_{0C}.J-H_{0C}.M)\,dV.\end{aligned} \quad (5.8)$$

From continuity of tangential components of fields at boundaries when losses are finite, that is, at interface of media with finite parameters, we can consider, after concatenation of domains, that:

the equation (5.8) continues to be valid in general when Ω is filled
by piecewise continuous media with finite parameters, (5.9)

(note: infinite losses have to be taken in the sense of the limit).

5.3. Generalized reciprocity theorem and associated fields properties

Considering the fields (E, H) and (E_{0C}, H_{0C}) satisfying Maxwell equations (5.1) and (5.2) in media, respectively, adjoint (or complementary) each other in Ω, with characteristics:

$$\begin{aligned}&((\epsilon)=(\epsilon_0),(\xi)=(\xi_0),(\mu)=(\mu_0),(\varsigma)=(\varsigma_0))\\&\text{and }({}^t(\epsilon_0),-{}^t(\varsigma_0),{}^t(\mu_0),-{}^t(\xi_0))\text{ in }\Omega,\end{aligned} \quad (5.10)$$

piecewise continuous in Ω, we can apply (5.8) with:

$$(\epsilon) - (\epsilon_0) = (\xi) - (\xi_0) = (\mu) - (\mu_0) = (\varsigma) - (\varsigma_0) = 0, \tag{5.11}$$

and obtain the generalized reciprocity theorem in bianistropic media for adjoint systems, following:

$$\begin{aligned}&\int_{\partial\Omega} E_{0C}.(\widehat{n} \wedge H) + H_{0C}.(\widehat{n} \wedge E)dS \\ &= \int_{\Omega} (J_{0C}.E - M_{0C}.H)\, dV - \int_{\Omega} (E_{0C}.J - H_{0C}.M)\, dV,\end{aligned} \tag{5.12}$$

where $\widehat{n}$ is the unit normal to $\partial\Omega$, inward Ω. The surface integral is considered as improper, that is, with integration apart from the discontinuities on lines or at isolated points.

NOTE 5.5.– The case of media with infinite losses, or of sources in contact with discontinuities, can be considered in (5.12) in the sense of the limit, but also, of distributions theory.

NOTE 5.6.– We can also take the conjugate of fields and obtain some other generalized equalities, as those for the well-known Poynting vector (see (5.109) in appendix A).

5.3.1. *Property 1: principle of equivalence of surface sources enclosing the domain of a volume source in open space*

Let the domain Ω be simply connected (i.e. no holes in it), and Ω_1 a subdomain of it with $\partial\Omega \cap \partial\Omega_1 \equiv 0$ so that $\partial(\Omega\backslash\Omega_1) \equiv \partial\Omega \cup \partial\Omega_1$ (i.e. Ω_1 is a subset of Ω strictly disjoint from the frontier of Ω). We then consider that Ω_1 holds J and M, while $\Omega\backslash\Omega_1$ contains J_{0C} and M_{0C}. We have from (5.12):

$$\begin{aligned}&\int_{\Omega_1} E_{0C}.J - H_{0C}.M\, dV = -\int_{\partial\Omega_1} E_{0C}.(\widehat{n}_1 \wedge H) + H_{0C}.(\widehat{n}_1 \wedge E)dS, \\ &\int_{\Omega\backslash\Omega_1} J_{0C}.E - M_{0C}.H\, dV = -\int_{\partial\Omega_1} E_{0C}.(\widehat{n}_1 \wedge H) + H_{0C}.(\widehat{n}_1 \wedge E)dS \\ &+ \int_{\partial\Omega} (E_{0C}.(\widehat{n} \wedge H) + H_{0C}.(\widehat{n} \wedge E))dS,\end{aligned} \tag{5.13}$$

where $\widehat{n}$ is the inward unit normal to $\partial\Omega$, and $-\widehat{n}_1 = \widehat{n}_s$ is the unit normal to $\partial\Omega_1$ outward Ω_1 (note: $\Omega\backslash\Omega_1$ is the set of points of Ω not belonging to Ω_1). We then let $\Omega \rightarrow R^3$. For that, we need to consider the behavior of fields at infinity assumed in condition (a), with small losses at infinity. The fields exponentially vanish at infinity, and we can write in (5.13):

$$\lim_{\Omega \to R^3} \int_{\partial\Omega} (E_{0C}.(\widehat{n} \wedge H) + H_{0C}.(\widehat{n} \wedge E))dS = 0. \tag{5.14}$$

Therefore, considering that J_{0C} and M_{0C} in $\Omega \backslash \Omega_1$ can be chosen arbitrarily, we conclude from (5.13) when $\Omega \to R^3$, that:

for (E,H) in $R^3 \backslash \overline{\Omega}_1$, vanishing at infinity for small losses, the primary sources J and M enclosed by $\partial\Omega_1$ can be seen as strictly equivalent to the surface distributions: $J_e = (\widehat{n}_s \wedge H)\delta_{S_1}$ and $M_e = -(\widehat{n}_s \wedge E)\delta_{S_1}$. (5.15)

(note: this equivalence of sources can also be derived from the theory of distributions applied to Maxwell equations).

5.3.2. *Property 2: generalized reciprocity theorem in open space and generalized reciprocity principle*

For systems with media adjoint each other in $\Omega = R^3 \backslash \Omega'$, with Ω' a bounded subdomain of R^3, we can apply the generalized reciprocity theorem (5.12). We then rewrite it with (5.14), and obtain for the (infinite) open space $\Omega = R^3 \backslash \Omega'$:

$$\int_{\partial\Omega'} (E_{0C}.(\widehat{n} \wedge H) + H_{0C}.(\widehat{n} \wedge E))dS$$
$$= \int_{R^3 \backslash \Omega'} (J_{0C}.E - M_{0C}.H)\, dV - \int_{R^3 \backslash \Omega'} (E_{0C}.J - H_{0C}.M)\, dV, \tag{5.16}$$

where $\widehat{n}$ is the unit normal to $\partial\Omega'$ outward Ω'. In the case where $\Omega \equiv R^3$, this leads us to the generalized reciprocity principle in bianistropic media:

$$\int_{R^3} (J_{0C}.E - M_{0C}.H)\, dV = \int_{R^3} (E_{0C}.J - H_{0C}.M)\, dV. \tag{5.17}$$

5.3.3. *Uniqueness properties*

5.3.3.1. *Property 3: uniqueness property when primary sources J and M vanish*

Considering the fields (E, H) satisfying (5.1) with $J = M = 0$ in $\Omega \equiv R^3$ and the conditions (a), (b) and (c), we can apply (5.17) and write:

$$\int_{R^3} (J_{0C}.E - M_{0C}.H)\, dV = 0. \tag{5.18}$$

Since the choice of J_{OC} and M_{OC} is free, we then derive that $E = H = 0$, and thus:

$$\text{any field, verifying (5.1) in } \Omega \equiv R^3 \text{ and (a), (b) and (c),}$$
$$\text{with no primary sources, is null.} \tag{5.19}$$

In these circumstances:

$$\text{two fields } (E, H)\text{, satisfying (5.1) and the conditions (a), (b) and (c)}$$
$$\text{in whole } R^3 \text{ with the same primary sources, are necessarily identical.} \tag{5.20}$$

NOTE 5.7.– The vanishing of the real part of the Poynting vector is also often considered as one of conditions of uniqueness (see in particular section 9.1 of [6]), and concerning equalities for this vector in our case, the reader can refer to (5.109) of appendix A.

NOTE 5.8.– A surface wave that propagates on a plane with no primary sources is in contradiction with condition (a).

5.3.3.2. *Property 4: uniqueness when $\widehat{n}\wedge E$ and $\widehat{n}\wedge H$ both vanish on an open surface*

Let us consider Ω' an open connected subdomain of Ω, with the characteristics that are analytic with respect to coordinates in Ω. Taking $J = M = M_{OC} = 0$ in Ω' and $J_{OC} = J'_{0C}\delta(r - r_0)$ with $r_0 \in \Omega$, we have, from the application of (5.12):

$$J'_{0C}.E(r_0) = \int_{\partial\Omega'} (E_{OC}.(\widehat{n}\wedge H) + H_{OC}.(\widehat{n}\wedge E))dS \text{ for } r_0 \in \Omega',$$
$$\text{and } \int_{\partial\Omega'} (E_{OC}.(\widehat{n}\wedge H) + H_{OC}.(\widehat{n}\wedge E))dS = 0 \text{ for } r_0 \in \Omega\backslash\overline{\Omega'}. \tag{5.21}$$

We now take the case where:

$$(\widehat{n}\wedge H) = (\widehat{n}\wedge E) = 0 \text{ on a regular part } S_h \text{ of } \partial\Omega'. \tag{5.22}$$

making like a hole in $\partial\Omega'$, so that we can replace $\partial\Omega'$ by the open surface $\partial\Omega'\backslash S_h$ in the surface integral function of r_0 given in the first and second line of (5.21). Following analyticity of parameters, we can assume that (E_{OC}, H_{OC}), solution of (2) in Ω, is an analytic function with respect to coordinates of r_0 in Ω, and thus, that the surface integral function is continuous through the hole S_h. This function then continues to vanish from outside to inside Ω'. Therefore, we have $J'_{0C}.E(r_0) = 0$ in Ω for arbitrary vector J'_{0C}, and thus:

$$E(r_0) = 0 \text{ as } r_0 \in \Omega. \tag{5.23}$$

In the same manner, we can also consider $J = M = J_{0C} = 0$ in Ω' and $M_{0C} = M'_{0C}\delta(r - r_0)$ in (5.12), and then derive, when (5.22) applies, that $M'_{0C}.H(r_0) = 0$ in Ω for arbitrary vector M'_{0C}, and thus $H(r_0) = 0$ in Ω.

This implies the following uniqueness property:

$$\begin{aligned}&\text{if the characteristics of media are analytic in } \Omega\\ &\text{and } (\widehat{n} \wedge H) = (\widehat{n} \wedge E) = 0 \text{ on an open surface } S_h \subset \Omega,\\ &\text{then } E = H = 0 \text{ in } \Omega.\end{aligned} \tag{5.24}$$

This property also applies by continuity if S_h is on the frontier of Ω. Considering now the continuity of tangential fields at interface of media with finite losses in a connected domain Ω with piecewise analytic characteristics, we can apply (5.24) in whole generality if the open surface S_h belongs to $\overline{\Omega}$ (note: we avoid domains having some parts with strictly infinite losses, that is, strictly impenetrable).

5.3.4. *Property 5: vanishing surface integral equation*

For a domain $\overline{\Omega'}$ free of sources for (E, H) and (E_{0C}, H_{0C}) (i.e. J, M, J_{0C} and M_{0C} vanish in $\overline{\Omega'}$, with respective media adjoint each other in Ω'), we have (5.12):

$$\int_{\partial\Omega'} (E_{0C}.(\widehat{n} \wedge H) + H_{0C}.(\widehat{n} \wedge E))dS = 0. \tag{5.25}$$

We now consider elementary applications of this vanishing surface integral equation.

5.3.4.1. *Property 6: equivalence of two surface sources*

For the first case of application of (5.25), we use linearity of Maxwell equations. We consider the fields (E_{0C}, H_{0C}) associated with sources J_{0C} and M_{0C} vanishing in Ω', and two fields (E', H') and (E'', H'') associated with sources identical in Ω'. We then note that, when media considered for them are adjoint to media considered for (E_{0C}, H_{0C}) in Ω', we have, from (5.25):

$$\int_{\partial\Omega'} (E_{0C}.(\widehat{n} \wedge (H' - H'') + H_{0C}.(\widehat{n} \wedge (E' - E'')))dS = 0, \tag{5.26}$$

and thus the equivalence of contributions in this integral of $(\widehat{n} \wedge E', \widehat{n} \wedge H')|_{\partial\Omega'}$ and $(\widehat{n} \wedge E'', \widehat{n} \wedge H'')|_{\partial\Omega'}$, that can be written as:

$$\begin{aligned}&\int_{\partial\Omega'} (E_{0C}.(\widehat{n} \wedge H') + H_{0C}.(\widehat{n} \wedge E'))dS\\ &= \int_{\partial\Omega'} (E_{0C}.(\widehat{n} \wedge H'') + H_{0C}.(\widehat{n} \wedge E''))dS.\end{aligned} \tag{5.27}$$

5.3.4.2. *Property 7: links between surface impedance parameters for adjoint cases*

For the second case of application of (5.25), we can consider the case where Ω' encloses a part of the interface S of two distinct media 1 and 2 so that $\partial\Omega' \equiv S'_1 \cup S'_2$, where S'_1 is an open surface above S and S'_2 is an open one below S. We can write, from (5.25):

$$\int_{S'_1} (E_{0C}.(\widehat{n} \wedge H) + H_{0C}.(\widehat{n} \wedge E))dS$$
$$+ \int_{S'_2} (E_{0C}.(\widehat{n} \wedge H) + H_{0C}.(\widehat{n} \wedge E))dS = 0. \tag{5.28}$$

If we let losses tending to infinity in Ω' below, we have, at the limit, the fields that vanish below the interface:

$$E = E_{0C} = 0 \text{ and } H = H_{0C} = 0 \text{ on } S'_2, \tag{5.29}$$

which implies at the limit, from (5.28):

$$\int_{S'_1} (E_{0C}.(\widehat{n} \wedge H) + H_{0C}.(\widehat{n} \wedge E))dS = 0. \tag{5.30}$$

Considering the case where S'_1 tends to a smooth part of the interface S just above it, named S^+, we can let vary it, and then deduce that:

$$E_{0C}.(\widehat{n} \wedge H) + H_{0C}.(\widehat{n} \wedge E) = 0 \text{ on } S^+,$$
$$\text{i.e. } (\widehat{n} \wedge H_{0C}).(\widehat{n} \wedge (\widehat{n} \wedge E)) = (\widehat{n} \wedge H).(\widehat{n} \wedge (\widehat{n} \wedge E_{0C}) \text{ on } S^+. \tag{5.31}$$

Independently, in these circumstances, at interface of media with losses tending to infinity in the medium below the interface, we can assume surface impedance boundary conditions on tangential components of fields [6]:

$$(\widehat{n}_+ \wedge (\widehat{n}_+ \wedge E)) = -(Z_0)(\widehat{n}_+ \wedge H) \text{ on } S^+,$$
$$(\widehat{n}_+ \wedge (\widehat{n}_+ \wedge E_{0C})) = -(Z_{0C})(\widehat{n}_+ \wedge H_{0C}) \text{ on } S^+, \tag{5.32}$$

where $\widehat{n}_+$ is the unit normal to S^+, outward it. This implies, from (5.31):

$$(\widehat{n} \wedge H).((Z_{0C}) - {}^t(Z_0))(\widehat{n} \wedge H_{0C}) = 0 \text{ on } S^+, \tag{5.33}$$

where (Z_0) and (Z_{0C}) are tensors, assumed invertible and independent of fields (note: the cases with (Z_0) and (Z_{0C}) null or infinite can be considered in the limit sense).

From uniqueness, the tangential electric and magnetic field cannot be both null on a segment of S^+, except if the fields vanish everywhere above the interface. We then deduce, from (5.31), (5.32) and (5.33) and the free choice of primary sources exciting (E, H) and (E_{0C}, H_{0C}) above the interface, that we have the following link between respective surface impedances tensors attached to adjoint cases:

$$(Z_{0C}) = {}^t(Z_0). \tag{5.34}$$

5.3.4.3. *Property 8: impedance boundary conditions and holes in surface integrals*

We note that, for fields in adjoint cases, on any smooth part of a surface S^c holding impedance boundary conditions with the link (5.34) between impedances tensors, we can write:

$$(\widehat{n}_1 \wedge (\widehat{n}_1 \wedge E))|_{S^c} = -(Z)(\widehat{n}_1 \wedge H)|_{S^c},$$
$$(\widehat{n}_1 \wedge (\widehat{n}_1 \wedge E_{0C}))|_{S^c} = -\,{}^t(Z)(\widehat{n}_1 \wedge H_{0C})|_{S^c}, \tag{5.35}$$

where $\widehat{n}_1$ is the unit normal to S^c, outward the domain with infinite losses. We then have:

$$[E_{0C}.(\widehat{n} \wedge H) + H_{0C}.(\widehat{n} \wedge E))]_{S^c}$$
$$= [{}^t(Z)(\widehat{n}_1 \wedge H_{0C}).(\widehat{n} \wedge H) - (\widehat{n}_1 \wedge H_{0C}).(Z)(\widehat{n} \wedge H)]_{S^c} = 0, \tag{5.36}$$

and thus a hole in $\partial\Omega$, that is, $\partial\Omega$ replaced by $\partial\Omega\backslash S^c$, in all surface integrals previously considered for fields in adjoint cases (note: this result is intimately related to the limit case where we have a part of the surface included in a medium for which we let losses tend to infinity).

5.4. Generalized reciprocity principle and dyadic tensors fields

5.4.1. *Dirac sources in open space and reciprocity principle*

Let the primary sources $J_0 = J_0'\delta_{r_0}$ and $M_0 = M_0'\delta_{r_0}$ attached to (E_0, H_0), and the primary sources $J_{0C} = J_{0C}'\delta_{r_{0C}}$, $M_{0C} = M_{0C}'\delta_{r_{0C}}$ attached to (E_{0C}, H_{0C}), satisfying (5.1) and (5.2), with respective adjoint characteristics $((\epsilon_0),(\xi_0),(\mu_0),(\varsigma_0))$ and $({}^t(\epsilon_0),-{}^t(\varsigma_0),{}^t(\mu_0),-{}^t(\xi_0))$ in $\Omega \equiv R^3$ (note: by definition $\delta_{r_v} \equiv \delta(r - r_v)$). We then consider the system as composed of a bianisotropic object held by the domain Ω_0 and the open space $R^3\backslash\Omega_0$, which holds the primary sources. The generalized reciprocity principle (5.17) then applies, and the fields in adjoint cases are equivalent, in the sense that they simply derive each from other when the source and observation positions exchange, following:

$$\int_{R^3} (J_{0C}.E_0 - M_{0C}.H_0)\, dV = \int_{R^3} (E_{0C}.J_0 - H_{0C}.M_0)\, dV. \tag{5.37}$$

From now on, we will specify, case by case, the source terms in the arguments of fields.

5.4.2. *Generalized reciprocity for co-dipolar dyadic tensors*

We can define co-dipolar dyadic tensors, $[E_{0(C)}(r',r)]$ and $[H_{0(C)}(r',r)]$, in writing:

$$\int_{R^3} E_{0C}(J'_{0C}\delta_{r_{0C}}, M_{0C}=0; r).J_0 dV = J'_{0C}.[E_{0C}(r_{0C}, r_0)].J'_0,$$

$$\int_{R^3} E_0(J'_0\delta_{r_0}, M_0=0; r).J_{0C} dV = J'_0.[E_0(r_0, r_{0C})].J'_{0C},$$

$$\int_{R^3} H_{0C}(J_{0C}=0, M'_{0C}\delta_{r_{0C}}; r).M_0 dV = M'_{0C}.[H_{0C}(r_{0C}, r_0)].M'_0,$$

$$\int_{R^3} H_0(J_0=0, M'_0\delta_{r_0}; r).M_{0C} dV = M'_0.[H_0(r_0, r_{0C})].M'_{0C}, \tag{5.38}$$

and letting successively in (5.37), $M'_0 = M'_{0C} = 0$, then $J'_0 = J'_{0C} = 0$, we deduce that:

$$[E_{0C}(r',r)] = {}^t[E_0(r,r')],$$

$$[H_{0C}(r',r)] = {}^t[H_0(r,r')], \tag{5.39}$$

where ${}^t[\mathcal{H}]$ is the transpose of $[\mathcal{H}]$, such that $U.[\mathcal{H}].W = W.{}^t[\mathcal{H}].U$ (note: for details on dyadic tensors, the reader can refer to the sections on Cartesian (dyadic) tensors in [6]).

5.4.3. *Generalized reciprocity for cross-dipolar dyadic tensors*

In the same manner, we can define cross-dipolar dyadic tensors, $[E'_{0(C)}(r',r)]$ and $[H'_{0(C)}(r',r)]$, in writing:

$$\int_{R^3} E_{0C}(J_{0C}=0, M'_{0C}\delta_{r_{0C}}; r).J_0 dV = M'_{0C}.[E'_{0C}(r_{0C}, r_0)].J'_0,$$

$$\int_{R^3} H_0(J'_0\delta_{r_0}, M_0=0; r).M_{0C} dV = J'_0.[H'_0(r_0, r_{0C})].M'_{0C},$$

$$\int_{R^3} E_0(J_0=0, M'_0\delta_{r_0}; r).J_{0C} dV = M'_0.[E'_0(r_0, r_{0C})].J'_{0C},$$

$$\int_{R^3} H_{0C}(J'_{0C}\delta_{r_{0C}}, M_{0C}=0; r).M_0 dV = J'_{0C}.[H'_{0C}(r_{0C}, r_0)].M'_0, \tag{5.40}$$

and, letting successively $J'_{0C} = M'_0 = 0$, then $J'_0 = M'_{0C} = 0$, directly derive from (5.37):

$$[E'_{0C}(r',r)] = -{}^t[H'_0(r,r')],$$

$$[E'_0(r',r)] = -{}^t[H'_{0C}(r,r')]. \tag{5.41}$$

5.4.4. *Dyadic tensorial integral expressions of fields and reciprocity*

We can apply (5.38)–(5.39) and (5.40)–(5.41) to the dyadic tensorial expressions of fields. These formal expressions represent the fields influenced by arbitrary bounded primary sources J and M [6] following:

$$E_{0(C)}(J,M;r) = \int_{\Omega_1} Z_0 J(r').[\underline{\mathcal{F}}^0_{he,t(C)}(r',r)] - M(r').[\underline{\mathcal{D}}^0_{eh,t(C)}(r',r)]dV_{r'},$$

$$Z_0 H_{0(C)}(J,M;r) = \int_{\Omega_1} Z_0 J(r').[\underline{\mathcal{D}}^0_{he,t(C)}(r',r)] + M(r').[\underline{\mathcal{F}}^0_{eh,t(C)}(r',r)]dV_{r'}, \tag{5.42}$$

where Ω_1 is a domain holding J and M. From the reciprocity relations (5.38)–(5.39) and (5.40)–(5.41), we note that the dyadic tensors $[\underline{\mathcal{F}}^0_{he,t(C)}]$,$[\underline{\mathcal{D}}^0_{eh,t(C)}]$,$[\underline{\mathcal{D}}^0_{he,t(C)}]$ and $[\underline{\mathcal{F}}^0_{eh,t(C)}]$ verify:

$$[\underline{\mathcal{F}}^0_{he(,eh),t}(r_1,r'_1)] = {}^t[\underline{\mathcal{F}}^0_{he(,eh),tC}(r'_1,r_1)],$$

$$[\underline{\mathcal{D}}^0_{he(,eh),t}(r_1,r'_1)] = {}^t[\underline{\mathcal{D}}^0_{eh(,he),tC}(r'_1,r_1)]. \tag{5.43}$$

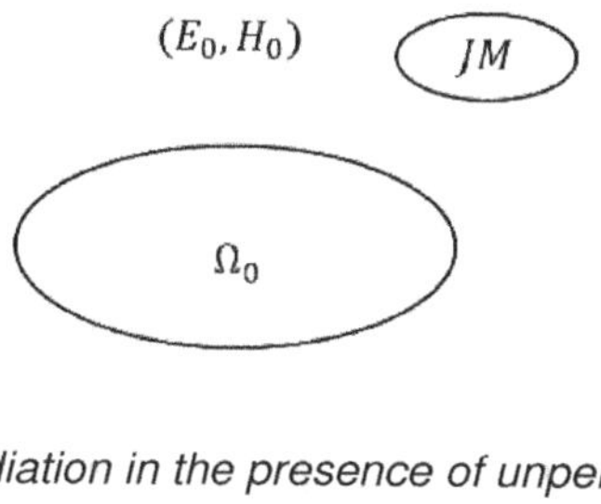

Figure 5.1. *Radiation in the presence of unperturbated object*

5.5. Fields influenced by a perturbation A of a bianisotropic object

5.5.1. *Definitions of fields*

As mentioned previously, (E_0, H_0) is the field radiated by primary sources in the presence of a bianisotropic object held by Ω_0 that we now call the unperturbated system. We then consider (E^0_{1S}, H^0_{1S}) the field in the presence of a scatterer A inserted in this initial system and define S as a closed surface enclosing A and limiting the simply connected domain Ω_S (no hole in it). We will note (E_{1S}, H_{1S}) the field identical to (E^0_{1S}, H^0_{1S}), except inside S, radiated by primary sources outside S, and by equivalent sources, enclosing A but inside S, in the presence of the unperturbated system attached to (E_0, H_0).

5.5.2. *Definition of primary sources*

The primary sources, for the fields $(E_0, H_0), (E^0_{1S}, H^0_{1S})$ and (E_{1S}, H_{1S}), are assumed to be the electric and magnetic dipolar sources positioned at $r_v \in R^3\backslash\overline{\Omega}_S$:

$$J_v\delta_{r_v} \text{ and } M_v\delta_{r_v} \text{ with } \delta_{r_v}(r) \equiv \delta(r - r_v), \tag{5.44}$$

where δ is the Dirac function [6]. For media with adjoint characteristics, the primary sources, for the fields $(E_{0C}, H_{0C}), (E^0_{1SC}, H^0_{1SC})$ and (E_{1SC}, H_{1SC}), are assumed to be the electric and magnetic dipolar sources positioned at $r' \in R^3\backslash\Omega_S$:

$$J'\delta_{r'} \text{ and } M'\delta_{r'} \text{ with } \delta_{r'} \equiv \delta(r - r'). \tag{5.45}$$

5.5.3. *Integral equalities for* (E_{1S}, H_{1S}) *and* (E^0_{1S}, H^0_{1S}) *on* S

Let us consider that the field (E^0_{1S}, H^0_{1S}) on S is known. We can write from the generalized integral reciprocity theorem in open space (5.16):

$$\begin{aligned}
&\int_{R^3\backslash\Omega_S} J'\delta_{r'}.E_{1S} - M'\delta_{r'}.H_{1S}\, dV_r - \int_{R^3\backslash\Omega_S} J_v\delta_{r_v}.E_{0C} - M_v\delta_{r_v}.H_{0C}\, dV_r\\
&= \int_S E_{0C}(r).(\widehat{n}\wedge H^0_{1S}(r)) + H_{0C}(r).(\widehat{n}\wedge E^0_{1S}(r))\, dS_r,\\
&\int_{R^3\backslash\Omega_S} J'\delta_{r'}.E_0 - M'\delta_{r'}.H_0\, dV_r - \int_{R^3\backslash\Omega_S} J_v\delta_{r_v}.E_{0C} - M_v\delta_{r_v}.H_{0C}\, dV_r = 0,
\end{aligned} \tag{5.46}$$

where $\widehat{n}$ is the outward unit normal to $\partial\Omega_S \equiv S$. By definition, we have:

$$(E^0_{1S}, H^0_{1S}) \equiv (E_{1S}, H_{1S}) \text{ on and outside } S, \tag{5.47}$$

with Ω_S containing A, while (E_0, H_0) and (E_{0C}, H_{0C}) are the unperturbated fields, in adjoint cases, in whole space, including $\overline{\Omega}_0$. The domain $\Omega = R^3\backslash\Omega_S$ holds the primary sources and a part of the unperturbated system.

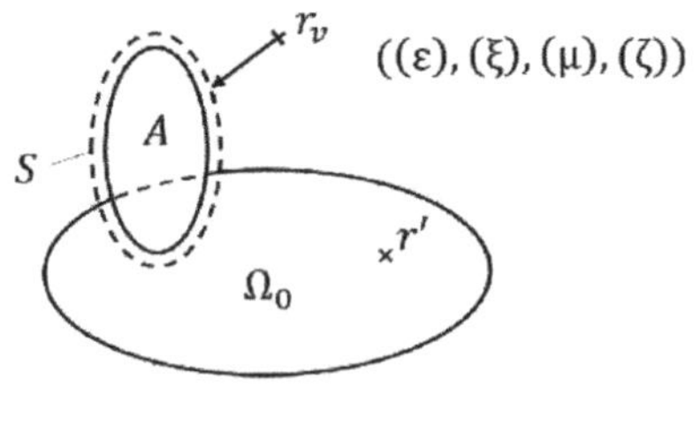

Figure 5.2. *Radiation in presence of the system perturbated by* A

5.5.4. *Integral expressions for* (E_{1S}, H_{1S}) *from fields known on* S

We now use (5.46) and present integral expressions for (E_{1S}, H_{1S}), outside and on S (up to S) in bianisotropic media, from (E_{1S}, H_{1S}) known on S.

5.5.4.1. *Definition of indicator function up to the frontier*

Let Ω_a be an open simply connected domain, Ω_S the open domain limited by S and $\Omega \equiv \Omega_a \backslash \Omega_S$. We can write, for $J'.E_{1S}$ continuous at point $r' \in \Omega$:

$$\int_\Omega J'\delta_{r'}.E_{1S}dV_r = 1_\Omega(r')J'.E_{1S} \text{ with } 1_\Omega(r') = \int_\Omega \delta(r'-r)dV_{r'}. \qquad (5.48)$$

The function $1_\Omega(r')$ is the indicator (also said characteristic) function of the domain $\Omega \equiv \Omega_a \backslash \Omega_S$, equal to 1 when $r' \in \Omega_a \backslash \overline{\Omega_S}$, equal to 0 when $r' \notin \overline{\Omega_a} \backslash \Omega_S$ and generally fractional when $r' \in S$. Letting Ω'_ϵ the ball of radius ϵ centered at $r' \in \Omega$, and this function verifies (see section 5.C.2 of appendix C):

$$1_\Omega(r') = \lim_{\epsilon\to 0}\int_\Omega \operatorname{div}(\operatorname{grad}(\frac{-1_{R^3\backslash\Omega_\epsilon}}{4\pi|r'-r|}))dV_{r'} = \lim_{\epsilon\to 0}\frac{1}{4\pi\epsilon^2}\int_{\partial\Omega'_\epsilon\cap\Omega} dS,$$

$$1_\Omega(r') = \frac{1}{4\pi}pv\int_{\partial\Omega}\widehat{n}.\operatorname{grad}_r(\frac{1}{|r-r'|})dS_r = \frac{1}{4\pi}pv\int_{\partial\Omega}\frac{\widehat{n}.(r'-r)}{|r'-r|^3}dS_r, \qquad (5.49)$$

where $\widehat{n}$ is the normal to $\partial\Omega$ directed inside Ω. We then note that $4\pi 1_{R^3\backslash\Omega_S}(r')$ is the solid angle, locally view from r', of the free space domain limited by S (note: on regular part of S, $1_\Omega(r') = \frac{1}{2}$).

5.5.4.2. *Integral expressions for* E_{1S} *and* H_{1S} *outside and on* S *for finite losses*

From the definitions of fields (E_{1S}, H_{1S}), we can assume that $J'.(E_{1S} - E_0)$ is continuous in the domain $R^3\backslash\Omega_S$ (up to S) for $J' = J_s$, where J_s is defined as an arbitrary vector, except on possible lines of discontinuities of E_{1S} on S, along which we define J_s as a vector locally tangent to them (note: we then take account of the property that, on any smooth part of lines of discontinuities, the fields component along it is continuous as we cross it).

Letting $\Omega \equiv R^3\backslash\Omega_S$, we can then rewrite (5.46) with the use of (5.48) and (5.37), when $J' = J_s$ and $M' = M_v = 0$, and obtain for $r' \in \Omega$ (outside and on S), as $r_v \in R^3\backslash\overline{\Omega_S}$:

$$1^c_\Omega(r')J'.E_{1S}(J_v\delta_{r_v}, M=0; r') - J'.E_0(J_v\delta_{r_v}, M=0; r') =$$

$$= \text{FP}\int_S E_{0C}(r).(\widehat{n}\wedge H_{1S}(r)) + H_{0C}(r).(\widehat{n}\wedge E_{1S}(r))dS_r$$

$$= \text{FP}\int_S (H_{0C}(r)\wedge\widehat{n}).E_{1S}(r)dS_r + \text{FP}\int_S (E_{0C}(r)\wedge\widehat{n}).H_{1S}(r)dS_r, \qquad (5.50)$$

where

$$1^c_\Omega(r') \equiv 1_\Omega(r') \text{ when the losses in the unperturbated system remains finite,} \quad (5.51)$$

the vector $\widehat{n}$ is the outward unit normal to $\partial\Omega_S \equiv S$, and FP is the finite part (FP) of integrals [7]. The FP is an operator regularizing integrals with singular kernels (see appendix B). The kernels are here $E_{0C} \wedge \widehat{n}$ and $H_{0C} \wedge \widehat{n}$ depending on r and r', which can be singular when $r' \to S$ (note: marginally, (5.50) also refines (5.25)).

In a similar manner, we can consider an integral expression for H_{1S} outside and on S. We then assume that $M'.(H_{1S} - H_0)$ is continuous in $R^3\backslash\Omega_S$ for $M' = M_s$, where M_s is an arbitrary vector, except on possible lines of discontinuities of H_{1S} on S, on which we define M_s as locally tangent to them. As previously for E_{1S}, we then rewrite (5.46) with the use of (5.48) and (5.37), and derive for H_{1S}, when $M' = M_s$ and $J' = J_v = 0$:

$$\begin{aligned} &1^c_\Omega(r')M'.Z_0H_{1S}(J=0, M_v\delta_{r_v}; r') - M'.Z_0H_0(J=0, M_v\delta_{r_v}; r') = \\ &= -\text{FP}\int_S E_{0C}(r).(\widehat{n} \wedge H_{1S}(r))dS_r + H_{0C}(r).(\widehat{n} \wedge E_{1S}(r))dS_r \\ &= -\text{FP}\int_S (E_{0C}(r) \wedge \widehat{n}).H_{1S}(r)dS_r - \text{FP}\int_S (H_{0C}(r) \wedge \widehat{n}).E_{1S}(r)dS_r. \end{aligned} \quad (5.52)$$

5.5.4.3. *Integral expressions outside and on S, taking account of infinite losses*

We modify (5.50) to extend its application to the case where S contains a subdomain of the unperturbated system with infinite losses whose surface partly coincides with S. This case is encountered, for example, if we want to consider an infinite plane with impedance boundary conditions, perturbated by a cavity inserted in it. For that, we begin by analyzing some particular properties of kernels.

We note that, taking $J' = J_s - (J_s.\widehat{n}_d(r'))\widehat{n}_d(r')$ (respectively, $J' = (J_s.\widehat{n}_d(r'))\widehat{n}_d(r')$) instead of $J' = J_s$, with $\widehat{n}_d(r')$ a function coinciding with $\widehat{n}$ for $r' \in S$, we can generally assume in (5.50) that the FP of the integral with $(E_{0C}(r) \wedge \widehat{n})$ kernel (respectively, $(H_{0C}(r) \wedge \widehat{n})$ kernel) is continuous at r', up to S included (note: FP generally implies boundedness but not necessarily the continuity up to S included).

Therefore, we can extend and slightly modify (5.50) to take account of a part S^c of S, piecewise smooth, coinciding with the boundary of a subdomain of the unperturbated system with infinite losses. In this case, we can assume impedance boundary conditions on S^c for the kernels following:

$$(\widehat{n} \wedge (\widehat{n} \wedge E_{0C})) = -(Z_{0C})(\widehat{n} \wedge H_{0C}) \text{ on } S^c \subset S, \quad (5.53)$$

as in (5.32). Therefore, in these circumstances, the FP of integral with kernel $(\widehat{n} \wedge H_{0C})$ (respectively, $(\widehat{n} \wedge E_{0C})$) is continuous as the one with $(\widehat{n} \wedge E_{0C})$ (respectively, $(\widehat{n} \wedge H_{0C})$) is, when we consider $J' = J_s - (J_s.\widehat{n}_d(r'))\widehat{n}_d(r')$ (respectively, $J' = (J_s.\widehat{n}_d(r'))\widehat{n}_d(r')$) on S^c, so that we can extend the domain of continuity up to S^c (note: this behavior is also related to the presence of an image of $J'\delta_{r'}$ relatively to S^c).

Therefore, the extension of (5.50) for this case amounts to taking account of this continuity up to S^c, and to completing the definition of $1^c_\Omega(r')$ on S^c, following:

$$\begin{aligned} &1^c_\Omega(r') \equiv 1_\Omega(r') \text{ as } r' \notin \overline{S^c}, \\ &1^c_\Omega(r') \equiv 1 \text{ as } r' \in S^c \backslash \partial S^c \text{ with } J' = J_s, \end{aligned} \tag{5.54}$$

and in (5.52),

$$\begin{aligned} &1^c_\Omega(r') \equiv 1_\Omega(r') \text{ as } r' \notin \overline{S^c}, \\ &1^c_\Omega(r') \equiv 1 \text{ as } r' \in S^c \backslash \partial S^c \text{ with } M' = M_s, \end{aligned} \tag{5.55}$$

where S^c is a piecewise smooth subdomain of S, on which the fields (E_{0C}, H_{0C}) (and thus also (E_0, H_0)), for the unperturbated system, satisfy impedance boundary conditions (note: $1^c_\Omega(r')$ on ∂S^c, which depends on the media configuration of the unperturbated system on this frontier with its consequences on the kernels, but, in practice, we often consider problems with $\partial S^c \equiv 0$ or $\partial S^c \to \infty$).

5.5.4.4. *General integral tensorial expressions of* (E_{1S}, H_{1S})

We can use the dyadic tensors expressions (5.42), with (5.43), to express (5.50) and (5.52) in a tensorial form. Considering (5.50), we have from (5.42):

$$E_{0C}(r) = Z_0 J'.[\underline{\mathcal{F}}^0_{he,tC}(r', r)], H_{0C}(r) = J'.[\underline{\mathcal{D}}^0_{he,t(C)}(r', r)], \tag{5.56}$$

and (5.50) can be rewritten as:

$$\begin{aligned} &1^c_\Omega(r')J'.E_{1S}(J_v\delta_{r_v}, M = 0; r') - J'.E_0(J_v\delta_{r_v}, M = 0; r') = \\ &= \text{FP}(J'. \int_S [\underline{\mathcal{F}}^0_{he,tC}(r', r)].Z_0 J_{es}(r) - [\underline{\mathcal{D}}^0_{he,tC}(r', r)].M_{es}(r) dS_r) \\ &= \text{FP}(J'. \int_S Z_0 J_{es}(r).([\underline{\mathcal{F}}^0_{he,t}(r, r')]) - M_{es}(r').([\underline{\mathcal{D}}^0_{eh,t}(r, r')]))dS_r), \end{aligned} \tag{5.57}$$

with

$$\begin{aligned} &J_{es}(r) = \widehat{n} \wedge H_{1S}(J_v\delta_{r_v}, M = 0; r), \\ &M_{es}(r) = -\widehat{n} \wedge E_{1S}(J_v\delta_{r_v}, M = 0; r), \end{aligned} \tag{5.58}$$

While, for (5.52), we have from (5.42):

$$E_{0C}(r) = -M'.[\underline{\mathcal{D}}^0_{eh,t(C)}(r', r)], H_{0C}(r) = M'.[\underline{\mathcal{F}}^0_{eh,t(C)}(r', r)]/Z_0, \tag{5.59}$$

and we obtain,

$$1^c_\Omega(r')M'.Z_0H_{1S}(J=0,M_v\delta_{r_v};r') - M'.Z_0H_0(J=0,M_v\delta_{r_v};r') =$$

$$= \text{FP}(M'.\int_S [\underline{\mathcal{D}}^0_{eh,tC}(r',r)].Z_0J_{ms}(r) + [\underline{\mathcal{F}}^0_{eh,tC}(r',r)].M_{ms}(r)dS_r)$$

$$= \text{FP}(M'.\int_S Z_0J_{ms}(r).[\underline{\mathcal{D}}^0_{he,t}(r,r')] + M_{ms}(r).[\underline{\mathcal{F}}^0_{eh,t}(r,r')]dS_r), \qquad (5.60)$$

with

$$J_{ms}(r) = \widehat{n}\wedge H_{1S}(J=0,M_v\delta_{r_v};r),$$
$$M_{ms}(r) = -\widehat{n}\wedge E_{1S}(J=0,M_v\delta_{r_v};r), \qquad (5.61)$$

where J' and M' are defined as previously for (5.50) and (5.52), and r_v is strictly exterior to S.

As a direct consequence, we note that we have:

$$1^c_\Omega(r')J'.Z_0H_{1S}(J_v\delta_{r_v},M=0;r') - J'.Z_0H_0(J_v\delta_{r_v},M=0;r')$$

$$= \text{FP}(J'.\int_S [\underline{\mathcal{D}}^0_{eh,tC}(r',r)].Z_0J_{es}(r) + [\underline{\mathcal{F}}^0_{eh,tC}(r',r)].M_{es}(r)dS_r)$$

$$= \text{FP}(J'.\int_S Z_0J_{es}(r).[\underline{\mathcal{D}}^0_{he,t}(r,r')] + M_{es}(r).[\underline{\mathcal{F}}^0_{eh,t}(r,r')]dS_r), \qquad (5.62)$$

and

$$1^c_\Omega(r')M'.E_{1S}(J=0,M_v\delta_{r_v};r') - M'.E_0(J=0,M_v\delta_{r_v};r')$$

$$= \text{FP}(M'.\int_S [\underline{\mathcal{F}}^0_{he,tC}(r',r)].Z_0J_{ms}(r) - [\underline{\mathcal{D}}^0_{he,tC}(r',r)].M_{ms}(r)dS_r)$$

$$= \text{FP}(M'.\int_S Z_0J_{ms}(r).([\underline{\mathcal{F}}^0_{he,t}(r,r')]) - M_{ms}(r).([\underline{\mathcal{D}}^0_{eh,t}(r,r')]))dS_r). \quad (5.63)$$

NOTE 5.9.– It is worth noticing that, we can take, for the definition of $J_{es(ms)}$ and $M_{es(ms)}$, (E^0_{1S},H^0_{1S}) in place of (E_{1S},H_{1S}) on S, that is,

$$J_{es(ms)} = \widehat{n}\wedge H^0_{1S},\ M_{es(ms)} = -\widehat{n}\wedge E^0_{1S}, \qquad (5.64)$$

but also, from (5.25), if r_v and $r'\notin S$ (we then exclude S):

$$J_{es(ms)} = \widehat{n}\wedge(H_{1S}-H_0),\ M_{es(ms)} = -\widehat{n}\wedge(E_{1S}-E_0). \qquad (5.65)$$

NOTE 5.10.– In our expressions of fields, we can let S be composed by piecewise continuous open parts S_i, $i=1,...n$, where E and H are continuous on S_i, except possibly on its contour $\mathcal{L}_i$ where only $\widehat{t}.E$ and $\widehat{t}.H$ are continuous across $\mathcal{L}_i$, with $\widehat{t}$ the unit tangential vector along $\mathcal{L}_i$.

NOTE 5.11.– Let us consider an impedance surface S^c of the unperturbated system outside the surface S enclosing A. If we deform S to be just in contact with S^c, we can then reduce S to an open surface $S\backslash S^c$ in the surface integral expressions.

5.5.5. *Definition of influence tensors from* $E_{1S} - E_0$ *and* $H_{1S} - H_0$

Let r_v and $r' \notin S$, so that we can take $1^c_\Omega(r') = 1$ and $FP \equiv$Identity in previous equations. We now define, for the perturbation of the field given by the equivalent sources on S, the influence tensors $\underline{[\mathcal{F}^1_{c,S}]}$, $\underline{[\mathcal{F}^1_{m,S}]}$, $\underline{[\mathcal{D}^1_{m,S}]}$, and $\underline{[\mathcal{D}^1_{c,S}]}$, as follows:

$$\begin{aligned}
&Z_0 J_v.\underline{[\mathcal{F}^1_{c,S}(r_v, r')]} = (E_{1S} - E_0)(J = J_v\delta_{r_v}, M = 0; r'),\\
&- M_v.\underline{[\mathcal{D}^1_{m,S}(r_v, r')]} = (E_{1S} - E_0)(J = 0, M = M_v\delta_{r_v}; r'),\\
&M_v.\underline{[\mathcal{F}^1_{m,S}(r_v, r')]} = (Z_0 H_{1S} - Z_0 H_0)(J = 0, M = M_v\delta_{r_v}; r'),\\
&Z_0 J_v.\underline{[\mathcal{D}^1_{c,S}(r_v, r')]} = (Z_0 H_{1S} - Z_0 H_0)(J = J_v\delta_{r_v}, M = 0; r'),
\end{aligned} \tag{5.66}$$

where J and M are electrical and magnetic sources.

In a similar manner, we can consider $\underline{[\mathcal{F}^1_{c(m),SC}]}$ and $\underline{[\mathcal{D}^1_{c(m),SC}]}$ for the adjoint system, and note that the application of the generalized principle of reciprocity (5.17) gives us:

$$\begin{aligned}
&J'.\underline{[\mathcal{F}^1_{c,SC}(r', r_v)]}.J_v = J_v.\underline{[\mathcal{F}^1_{c,S}(r_v, r')]}.J', \underline{[\mathcal{F}^1_{c,SC}(r', r)]} = {}^t\underline{[\mathcal{F}^1_{c,S}(r, r')]},\\
&M'.\underline{[\mathcal{F}^1_{m,SC}(r', r_v)]}.M_v = M_v.\underline{[\mathcal{F}^1_{m,S}(r_v, r')]}.M', \underline{[\mathcal{F}^1_{m,SC}(r', r)]} = {}^t\underline{[\mathcal{F}^1_{m,S}(r, r')]},\\
&M'.\underline{[\mathcal{D}^1_{m,SC}(r', r_v)]}.J_v = J_v.\underline{[\mathcal{D}^1_{c,S}(r_v, r')]}.M', \underline{[\mathcal{D}^1_{m,SC}(r', r)]} = {}^t\underline{[\mathcal{D}^1_{c,S}(r, r')]}
\end{aligned} \tag{5.67}$$

while, in other respects, concerning the system perturbated by A' bounded by S' in place of A bounded by S, we have equivalent properties and definitions with S replaced by S'.

5.6. Fields influenced by coupling between two scatterers A and A'

We consider the coupling of two scatterers A and A' limited by disjoint surfaces S and S' inserted in a complex system, which is composed of bianisotropic media and metamaterials. The scatterers A and A' can be considered as complex local perturbations of characteristics of the initial system. The field $(E_{1S(C)}, H_{1S(C)})$ is defined as previously for the system without A', and $(E_{1S'(C)}, H_{1S'(C)})$ in a similar manner for the system without A.

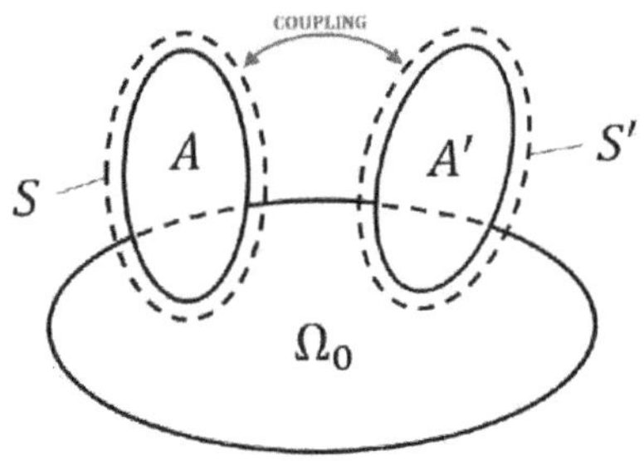

Figure 5.3. *Presence of perturbations A and A'. For a color version of this figure, see www.iste.co.uk/bernard/theoryofscattering.zip*

5.6.1. *Definitions of first-order coupling terms $E_{S,S'}$ and $H_{S,S'}$*

We now consider the system that consists of two scatterers A and A' that can be partly inserted in an imperfectly conducting object in Ω_0, illuminated by primary electric and magnetic dipoles at r_1:

$$J = J_1 \delta_{r_1} \text{ and } M = M_1 \delta_{r_1} \text{ with } r_1 \text{ strictly outside } S \text{ and } S', \tag{5.68}$$

while, for adjoint characteristics of media, we consider illumination by primary electric and magnetic dipoles at r'_1:

$$J = J'_1 \delta_{r'_1} \text{ and } M = M'_1 \delta_{r'_1} \text{ with } r'_1 \text{ strictly outside } S \text{ and } S'. \tag{5.69}$$

The sources $J_{1S'} = \widehat{n} \wedge H^0_{1S'}(r'_{S'})\delta_{S'}$ and $M_{1S'} = -\widehat{n} \wedge E^0_{1S'}(r'_{S'})\delta_{S'}$ influence A, and we define (E_2, H_2) as the field in presence of $J_{1S'}$ and $M_{1S'}$, of J and M, and of A partly inserted in Ω_0, but without A'.

We can say that the field (E_2, H_2) is partial, since it contains a term of coupling of A' on A, that we name $(E_{S,S'}, H_{S,S'})$, but no terms of coupling of A on A'. Thus, we can write:

$$\begin{aligned}(E_{S,S'}, H_{S,S'}) &= ((E_2, H_2) - (E_{1S}, H_{1S})) \\ &- ((E_{1S'}, H_{1S'}) - (E_0, H_0)),\end{aligned} \tag{5.70}$$

where the notation $(U_0, V_0) - (U_1, V_1) = (U_0 - U_1, V_0 - V_1)$ is used. The total term of coupling will be:

$$(E_{ctot}, H_{ctot}) = (E_{S,S'}, H_{S,S'}) + (E_{S',S}, H_{S',S}). \tag{5.71}$$

From now on, to simplify, a field U at r'_1 for primary sources $J = J_1\delta_{r_1}$ and $M = M_1\delta_{r_1}$ at r_1, will be denoted $U(r_1, r'_1)$, or $U(J_1\delta_{r_1}, M_1\delta_{r_1}; r'_1)$, or $U(J_1\delta_{r_1}; r'_1)$ when $M = 0$, or $U(M_1\delta_{r_1}; r'_1)$ when $J = 0$.

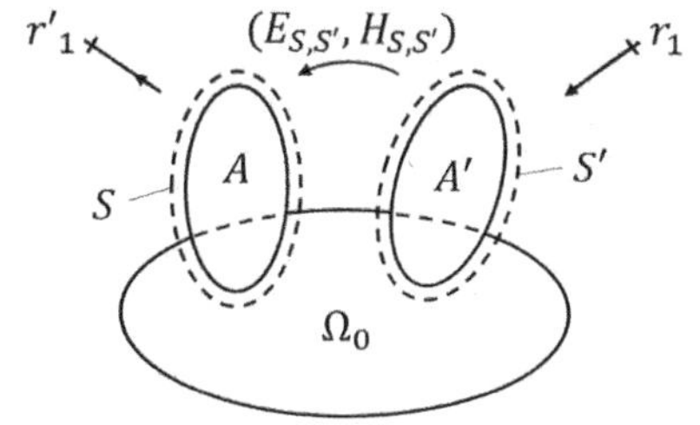

Figure 5.4. *Coupling influence of A' on A*

5.6.2. Expressions of $(E_{S,S'}, H_{S,S'})$ for $M'_1 = M_1 = 0$ and for $J'_1 = J_1 = 0$

To express $E_{S,S'}$ and $H_{S,S'}$, we use two times the generalized integral reciprocity equation (5.12) for dipoles sources at points r_1 and r'_1, strictly outside of S and S'. In what follows, we note Ω' the domain outside of S'. Coupling expressions and coupling tensors, for the fields due to couplings, will be, respectively, determined for $M'_1 = M_1 = 0$ and for $J'_1 = J_1 = 0$.

5.6.2.1. *General integral reduction for the fields due to coupling*

The systems for (E_2, H_2) and (E^0_{1S}, H^0_{1S}) are identical in the domain Ω', outside and on S', while the field (E_{1S}, H_{1S}) is identical to (E^0_{1S}, H^0_{1S}) outside and on S. We can then use (5.12); and, for r_1 and r'_1 strictly outside S, we can write:

$$\int_{\Omega'} (J'_1\delta(r'-r'_1)E_2(r_1,r') - M'_1\delta(r'-r'_1)H_2(r_1,r')$$
$$- J_1\delta(r'-r_1)E_{1SC}(r'_1,r') + M_1\delta(r'-r_1)H_{1SC}(r'_1,r'))dV_{r'}$$
$$= \int_{S'} (E_{1SC})(r'_1,r').(\widehat{n} \wedge H_2(r_1,r'))$$
$$+ (H_{1SC})(r'_1,r').(\widehat{n} \wedge E_2(r_1,r'))dS_{r'}, \tag{5.72}$$

In other respects, the systems for $(E_{1S'}, H_{1S'})$ and (E_0, H_0) are identical in the domain Ω', outside and on S', so that we have:

$$\int_{\Omega'} (J'_1\delta(r'-r'_1)E_{1S'}(r_1,r') - M'_1\delta(r'-r'_1)H_{1S'}(r_1,r')$$
$$- J_1\delta(r'-r_1)E_{0C}(r'_1,r') + M_1\delta(r'-r_1)H_{0C}(r'_1,r'))\, dV_{r'}$$
$$= \int_{S'} ((E_{0C})(r'_1,r').(\widehat{n} \wedge H_{1S'}(r_1,r'))$$
$$+ (H_{0C})(r'_1,r').(\widehat{n} \wedge E_{1S'}(r_1,r')))dS_{r'}, \tag{5.73}$$

where the subscript C means that we use adjoint parameters, and $\widehat{n}$ is the outward unit normal to S'. Since $(E_2 - E_{1S'}, H_2 - H_{1S'})$ and (E_{1SC}, H_{1SC}) correspond to systems, respectively, adjoint each other, and without sources inside and on S', we can use (5.26), following:

$$\int_{S'} E_{1SC}.(\widehat{n} \wedge (H_2 - H_{1S'})) + H_{1SC}.(\widehat{n} \wedge (E_2 - E_{1S'}))dS_{r'} = 0, \qquad (5.74)$$

and thus replace $(\widehat{n} \wedge H_2, -\widehat{n} \wedge E_2)$ by $(\widehat{n} \wedge H_{1S'}, -\widehat{n} \wedge E_{1S'})$ in (5.72) on S'. We then obtain after substraction:

$$\begin{aligned}&\int_{\Omega'} (J_1'\delta_{r_1'}(E_2 - E_{1S'})(r_1, r') - M_1'\delta_{r_1'}(H_2 - H_{1S'})(r_1, r') \\ &- J_1\delta_{r_1}(E_{1SC} - E_{0C})(r_1', r') + M_1\delta_{r_1}(H_{1SC} - H_{0C})(r_1', r'))\, dV_{r'} \\ &= \int_{S'} ((E_{1SC} - E_{0C})(r_1', r').(\widehat{n} \wedge H_{1S'}(r_1, r')) \\ &+ (H_{1SC} - H_{0C})(r_1', r').(\widehat{n} \wedge E_{1S'}(r_1, r')))dS_{r'}, \qquad (5.75)\end{aligned}$$

where we suppose known $(\widehat{n} \wedge (E_{1SC} - E_{0C}))$, $(\widehat{n} \wedge (H_{1SC} - H_{0C}))$, $(\widehat{n} \wedge H_{1S'})$ and $(\widehat{n} \wedge E_{1S'})$ on S' (note: $U.(\widehat{n} \wedge W) = -(\widehat{n} \wedge (\widehat{n} \wedge U)).(\widehat{n} \wedge W)$ for any U, W).

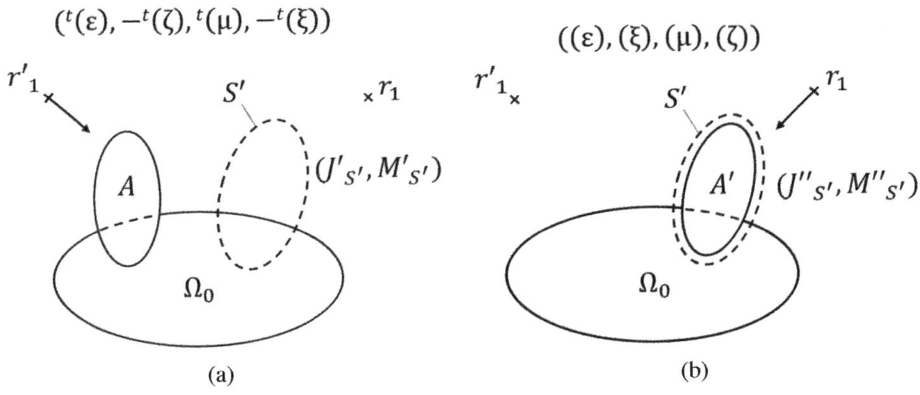

Figure 5.5. *(a)* $J'_{S'} = \widehat{n} \wedge (H_{1SC} - H_{0C})$ *and* $M'_{S'} = -\widehat{n} \wedge (E_{1SC} - E_{0C})$ *on* S' *for primary source at* r'_1, *(b)* $J''_{S'} = (\widehat{n} \wedge H_{1S'})$ *and* $M''_{S'} = -(\widehat{n} \wedge E_{1S'})$ *on* S' *for primary source at* r_1

5.6.2.2. *Expressions of* $(E_{S,S'}, H_{S,S'})$ *for* $M'_1 = M_1 = 0$

Taking $M'_1 = M_1 = 0$ in (5.75), we can then write:

$$J'_1.(E_2 - E_{1S'})(J_1\delta_{r_1}; r'_1) - J_1.(E_{1SC} - E_{0C})\,(J'_1\delta_{r'_1}; r_1)$$
$$= \int_{S'} (E_{1SC} - E_{0C})(J'_1\delta_{r'_1}; r').J'_{es',J_1\delta_{r_1}}(r')$$
$$- (H_{1SC} - H_{0C})(J'_1\delta_{r'_1}; r').M'_{es',J_1\delta_{r_1}}(r')dS_{r'} \tag{5.76}$$

for r_1 and r'_1 strictly outside S and S', where:

$$J'_{es'} = \widehat{n} \wedge H_{1S'}|_{S'} \text{ and } M'_{es'} = -\widehat{n} \wedge E_{1S'}|_{S'},$$
$$\text{or, } J'_{es'} = \widehat{n} \wedge (H_{1S'} - H_0)|_{S'} \text{ and } M'_{es'} = -\widehat{n} \wedge (E_{1S'} - E_0)|_{S'}. \tag{5.77}$$

Using reciprocity, we have:

$$J_1.(E_{1SC} - E_{0C})\,(J'_1\delta_{r'_1}; r_1) = J'_1.(E_{1S} - E_0)\,(J_1\delta_{r_1}; r'_1), \tag{5.78}$$

and thus, from (5.76) and (5.70), we finally obtain:

$$J'_1.E_{S,S'}(J_1\delta_{r_1}, r'_1)$$
$$= \int_{S'} (E_{1SC} - E_{0C})(J'_1\delta_{r'_1}; r').J'_{es',J_1\delta_{r_1}}(r')$$
$$- (H_{1SC} - H_{0C})(J'_1\delta_{r'_1}; r').M'_{es',J_1\delta_{r_1}}(r')dS_{r'}, \tag{5.79}$$

for r_1 and r'_1 strictly outside S and S'.

We can then write, using notations given in (5.66)–(5.67):

$$E_{S,S'}(J_1\delta_{r_1}, M_1\delta_{r_1} = 0; r'_1)$$
$$= \int_{S'} [\underline{\mathcal{F}^1_{c,SC}}(r'_1, r')].Z_0J'_{es',J_1\delta_{r_1}}(r') - [\underline{\mathcal{D}^1_{c,SC}}(r'_1, r')].M'_{es',J_1\delta_{r_1}}(r')dS_{r'}$$
$$= \int_{S'} Z_0J'_{es',J_1\delta_{r_1}}(r').[\underline{\mathcal{F}^1_{c,S}}(r', r'_1)] - M'_{es',J_1\delta_{r_1}}(r')[\underline{\mathcal{D}^1_{m,S}}(r', r'_1)]dS_{r'}, \tag{5.80}$$

and the associated magnetic field, following:

$$Z_0H_{S,S'}(J_1\delta_{r_1}, M_1\delta_{r_1} = 0; r'_1)$$
$$= \int_{S'} [\underline{\mathcal{D}^1_{m,SC}}(r'_1, r')].Z_0J'_{es',J_1\delta_{r_1}}(r') + [\underline{\mathcal{F}^1_{m,SC}}(r'_1, r')].M'_{es',J_1\delta_{r_1}}(r')dS_{r'}$$
$$= \int_{S'} Z_0J'_{es',J_1\delta_{r_1}}(r').[\underline{\mathcal{D}^1_{c,S}}(r', r'_1)] + M'_{es',J_1\delta_{r_1}}(r')[\underline{\mathcal{F}^1_{m,S}}(r', r'_1)]dS_{r'}. \tag{5.81}$$

5.6.2.3. *Expressions of* $(E_{S,S'}, H_{S,S'})$ *for* $J_1' = J_1 = 0$

In the same manner, taking $J_1' = J_1 = 0$ in (5.75), we have:

$$\begin{aligned}&- M_1'.(H_2 - H_{1S'})(M_1\delta_{r_1}; r_1') + M_1.(H_{1SC} - H_{0C})(M_1'\delta_{r_1'}; r_1)\\&= \int_{S'} (E_{1SC} - E_{0C})(M_1'\delta_{r_1'}; r').J'_{es',M_1\delta_{r_1}}(r')\\&- (H_{1SC} - H_{0C})(M_1'\delta_{r_1'}; r').M'_{es',M_1\delta_{r_1}}(r')dS_{r'},\end{aligned} \quad (5.82)$$

and thus, from generalized reciprocity principle,

$$\begin{aligned}&M_1'.(H_2(r_1, r_1') - H_{1S'}(r_1, r_1')) - M_1'(H_{1S}(r_1, r_1') - H_0(r_1, r_1'))\\&= - \int_{S'} (E_{1SC} - E_{0C})(M_1'\delta_{r_1'}; r').J'_{es',M_1\delta_{r_1}}(r')\\&- (H_{1SC} - H_{0C})(M_1'\delta_{r_1'}; r').M'_{es',M_1\delta_{r_1}}(r')dS_{r'},\end{aligned} \quad (5.83)$$

which implies

$$\begin{aligned}&M_1'.H_{S,S'}(J_1\delta_{r_1} = 0, M_1\delta_{r_1}; r_1')\\&= - \int_{S'} (E_{1SC} - E_{0C})(M_1'\delta_{r_1'}; r').J'_{es',M_1\delta_{r_1}}(r')\\&- (H_{1SC} - H_{0C})(M_1'\delta_{r_1'}; r').M'_{es',M_1\delta_{r_1}}(r')dS_{r'},\end{aligned} \quad (5.84)$$

for r_1 and r_1' strictly outside S and S'. We can then write, using the notations given in (5.66)–(5.67),

$$\begin{aligned}&Z_0 H_{S,S'}(J_1\delta_{r_1} = 0, M_1\delta_{r_1}; r_1')\\&= \int_{S'} [\underline{\mathcal{D}^1_{m,SC}}(r_1', r')].Z_0 J'_{es',M_1\delta_{r_1}}(r') + [\underline{\mathcal{F}^1_{m,SC}}(r_1', r')].M'_{es',M_1\delta_{r_1}}(r')dS_{r'}\\&= \int_{S'} Z_0 J'_{es',M_1\delta_{r_1}}(r').[\underline{\mathcal{D}^1_{c,S}}(r', r_1')] + M'_{es',M_1\delta_{r_1}}(r').[\underline{\mathcal{F}^1_{m,S}}(r', r_1')]dS_{r'},\end{aligned} \quad (5.85)$$

and the associated electric field, following,

$$\begin{aligned}&E_{S,S'}(J_1\delta_{r_1} = 0, M_1\delta_{r_1}; r_1')\\&= \int_{S'} [\underline{\mathcal{F}^1_{c,SC}}(r_1', r')].Z_0 J'_{es',M_1\delta_{r_1}}(r') - [\underline{\mathcal{D}^1_{c,SC}}(r_1', r')].M'_{es',M_1\delta_{r_1}}(r')dS_{r'}\\&= \int_{S'} Z_0 J'_{es',M_1\delta_{r_1}}(r').[\underline{\mathcal{F}^1_{c,S}}(r', r_1')] - M'_{es',M_1\delta_{r_1}}(r').[\underline{\mathcal{D}^1_{m,S}}(r', r_1')]dS_{r'},\end{aligned} \quad (5.86)$$

where $J'_{es'} = \widehat{n} \wedge H_{1S'}|_{S'}$ and $M'_{es'} = -\widehat{n} \wedge E_{1S'}|_{S'}$, or, $J'_{es'} = \widehat{n} \wedge (H_{1S'} - H_0)|_{S'}$ and $M'_{es'} = -\widehat{n} \wedge (E_{1S'} - E_0)|_{S'}$.

NOTE 5.12.– Considering a surface (S or S') partly included in a part of the unperturbated system where we let losses tend to infinity, the contribution of this part vanishes, making a hole in the surface expression of couplings.

5.6.3. *First-order coupling tensors* $\underline{[E^{c(,m)}_{S,S'}}(r_1, r'_1)]$ *and* $\underline{[H^{c(,m)}_{S,S'}}(r_1, r'_1)]$

5.6.3.1. *Coupling tensors* $\underline{[E^{c}_{S,S'}}(r_1, r'_1)]$ *and* $\underline{[H^{c}_{S,S'}}(r_1, r'_1)]$

We consider the dyadic tensors for the coupling of A' on A. After writing in (5.80)–(5.81):

$$E_{S,S'}(J_1\delta_{r_1}, M_1\delta_{r_1} = 0; r'_1) = J_1.\underline{[E^{c}_{S,S'}}(r_1, r'_1)],$$

$$Z_0H_{S,S'}(J_1\delta_{r_1}, M_1\delta_{r_1} = 0; r'_1) = J_1.\underline{[Z_0H^{c}_{S,S'}}(r_1, r'_1)],$$

$$Z_0J'_{es',J_1\delta_{r_1}} = J_1.\underline{[Z_0J'_{es',c}}(r_1, r')],\ M'_{es',J_1\delta_{r_1}} = J_1.\underline{[M'_{es',c}}(r_1, r')], \tag{5.87}$$

we obtain,

$$\begin{aligned}\underline{[E^{c}_{S,S'}}(r_1, r'_1)] &= (\int_{S'}(\underline{[Z_0J'_{es',c}}(r_1, r')].{}^t\underline{[\mathcal{F}^1_{c,SC}}(r'_1, r')]\\ &- \underline{[M'_{es',c}}(r_1, r')].{}^t\underline{[\mathcal{D}^1_{c,SC}}(r'_1, r')])dS_{r'}\\ &= (\int_{S'}(\underline{[Z_0J'_{es',c}}(r_1, r')].\underline{[\mathcal{F}^1_{c,S}}(r', r'_1)]\\ &- \underline{[M'_{es',c}}(r_1, r')].\underline{[\mathcal{D}^1_{m,S}}(r'_1, r')])dS_{r'},\end{aligned} \tag{5.88}$$

and

$$\begin{aligned}\underline{[Z_0H^{c}_{S,S'}}(r_1, r'_1)] &= \int_{S'}(\underline{[Z_0J'_{es',c}}(r_1, r')].{}^t\underline{[\mathcal{D}^1_{m,SC}}(r'_1, r')]\\ &+ \underline{[M'_{es',c}}(r_1, r')].{}^t\underline{[\mathcal{F}^1_{m,SC}}(r'_1, r')])dS_{r'}\\ &= \int_{S'}(\underline{[Z_0J'_{es',c}}(r_1, r')].\underline{[\mathcal{D}^1_{c,S}}(r', r'_1)]\\ &+ \underline{[M'_{es',c}}(r_1, r')].\underline{[\mathcal{F}^1_{m,S}}(r', r'_1)])dS_{r'}.\end{aligned} \tag{5.89}$$

5.6.3.2. *Coupling tensors* $\underline{[E^{m}_{S,S'}}(r_1, r'_1)]$ *and* $\underline{[H^{m}_{S,S'}}(r_1, r'_1)]$

In the same manner, noting in (5.85)–(5.86):

$$Z_0H_{S,S'}(J_1\delta_{r_1} = 0, M_1\delta_{r_1}; r'_1) = M_1.\underline{[Z_0H^{m}_{S,S'}}(r_1, r'_1)],$$

$$E_{S,S'}(J_1\delta_{r_1} = 0, M_1\delta_{r_1}; r'_1) = M_1.\underline{[E^{m}_{S,S'}}(r_1, r'_1)],$$

$$Z_0J'_{es',M_1\delta_{r_1}} = M_1.\underline{[Z_0J'_{es',m}}(r_1, r')],\ M'_{es',M_1\delta_{r_1}} = M_1.\underline{[M'_{es',m}}(r_1, r')], \tag{5.90}$$

we obtain

$$\begin{aligned}\underline{[Z_0 H^m_{S,S'}(r_1, r'_1)]} &= \int_{S'} \underline{[Z_0 J'_{es'}(r_1, r')]}.\,^t\underline{[\mathcal{D}^1_{m,SC}(r'_1, r')]} \\ &+ \underline{[M'_{es'}(r_1, r')]}.\,^t\underline{[\mathcal{F}^1_{m,SC}(r'_1, r')]} dS_{r'}, \\ &= \int_{S'} \underline{[Z_0 J'_{es'}(r_1, r')]}.\underline{[\mathcal{D}^1_{c,S}(r', r'_1)]} \\ &+ \underline{[M'_{es'}(r_1, r')]}.\underline{[\mathcal{F}^1_{m,S}(r', r'_1)]} dS_{r'},\end{aligned} \tag{5.91}$$

and

$$\begin{aligned}\underline{[E^m_{S,S'}(r_1, r'_1)]} &= \int_{S'} \underline{[Z_0 J'_{es'}(r_1, r')]}.\,^t\underline{[\mathcal{F}^1_{c,SC}(r'_1, r')]} \\ &- \underline{[M'_{es'}(r_1, r')]}.\,^t\underline{[\mathcal{D}^1_{c,SC}(r'_1, r')]} dS_{r'} \\ &= \int_{S'} \underline{[Z_0 J'_{es'}(r_1, r')]}.\underline{[\mathcal{F}^1_{c,S}(r', r'_1)]} \\ &- \underline{[M'_{es'}(r_1, r')]}.\underline{[\mathcal{D}^1_{m,S}(r', r'_1)]} dS_{r'}.\end{aligned} \tag{5.92}$$

NOTE 5.13.– In practice, we can consider that (E_0, H_0) is given either by closed form expressions (for r_1, r'_1 at large distance and/or simple geometries) or numerically, while $(E_{1S'}, H_{1S'})$ and (E_{1S}, H_{1S}) are generally calculated numerically.

5.6.4. *Reciprocity relations for coupling tensors*

Concerning the coupling of A on A', we can exchange A' and S' for A and S in previous equations. However, it is also possible to derive $E^c_{S',S}$ from $E^c_{S,S'}$ by some use of reciprocity. For this, we use (5.57)–(5.63) in (5.88) and notice that:

$$\begin{aligned}& J_1.\underline{[E^c_{S,S'}(r_1, r'_1)]}.J_2 = \\ &= C_0 \int_{S'} \int_{S} ((J_1.\underline{[Z_0 J'_{es',c}(r_1, r'_{S'})]}).(\underline{[\mathcal{F}^0_{he,t}(r'_{S'}, r'_S)]}) \\ &.(^t\underline{[Z_0 J'_{esC,c}(r'_1, r'_S)]}.J_2) dS_{r'_s} dS_{r'_{s'}} \\ &+ C_1 \int_{S'} \int_{S} (J_1.\underline{[Z_0 J'_{es',c}(r_1, r'_{S'})]}).(\underline{[\mathcal{D}^0_{he,t}(r'_{S'}, r'_S)]}) \\ &.(^t\underline{[M'_{esC,c}(r'_1, r'_S)]}.J_2) dS_{r'_s} dS_{r'_{s'}} \\ &+ C_1 \int_{S'} \int_{S} (J_1.\underline{[M'_{es',c}(r_1, r'_{S'})]}).(\underline{[\mathcal{D}^0_{eh,t}(r'_{S'}, r'_S)]})\end{aligned}$$

$$.({}^t[\underline{Z_0 J'_{esC,c}}(r'_1, r'_S)].J_2) dS_{r'_S} dS_{r'_{s'}}$$

$$+ C_2 \int_{S'} \int_S (J_1.[\underline{M'_{es',c}}(r_1, r'_{S'})]).([\underline{\mathcal{F}^0_{eh,t}}(r'_{S'}, r'_S)])$$

$$.({}^t[\underline{M'_{esC,c}}(r'_1, r'_S)].J_2) dS_{r'_S} dS_{r'_{s'}}, \tag{5.93}$$

where $[\underline{\mathcal{F}^0_{he(eh),t}}]$ and $[\underline{\mathcal{D}^0_{he(eh),t}}]$ are the Green's tensors corresponding to the unperturbated system, without A and A', and the C_j are constants. We can then transpose the scalar expression in double integrals without changing the results in each integral of (5.93). Using $[\underline{\mathcal{F}^0_{he(,eh),t}}(r, r')] = {}^t[\underline{\mathcal{F}^0_{he(,eh),tC}}(r', r)]$ and $[\underline{\mathcal{D}^0_{he(,eh),t}}(r, r')] = {}^t[\underline{\mathcal{D}^0_{eh(,he),tC}}(r', r)]$, we change the order of integration, and obtain $J_2.[\underline{E^c_{S',S}}(r'_1, r_1)].J_1$ for adjoint characteristics. Therefore, this implies:

$$[\underline{E^c_{S',S}}(r'_1, r_1)|_{((\epsilon),(\xi),(\mu),(\varsigma))}] = {}^t[\underline{E^c_{S,S'}}(r_1, r'_1)|_{({}^t(\epsilon),-{}^t(\varsigma),{}^t(\mu),-{}^t(\xi))}], \tag{5.94}$$

where $()|_{({}^t(\epsilon),-{}^t(\varsigma),{}^t(\mu),-{}^t(\xi))}$ means that the system is considered with adjoint characteristics. In the same manner, we can obtain the reciprocity relations:

$$[\underline{Z_0 H^m_{S',S}}(r'_1, r_1)|_{((\epsilon),(\xi),(\mu),(\varsigma))}] = {}^t[\underline{Z_0 H^m_{S,S'}}(r_1, r'_1)|_{({}^t(\epsilon),-{}^t(\varsigma),{}^t(\mu),-{}^t(\xi))}],$$

$$[\underline{H^c_{S',S}}(r'_1, r_1)|_{((\epsilon),(\xi),(\mu),(\varsigma))}] = -{}^t[\underline{E^m_{S,S'}}(r_1, r'_1)|_{({}^t(\epsilon),-{}^t(\varsigma),{}^t(\mu),-{}^t(\xi))}],$$

$$[\underline{E^m_{S',S}}(r'_1, r_1)|_{((\epsilon),(\xi),(\mu),(\varsigma))}] = -{}^t[\underline{H^c_{S,S'}}(r_1, r'_1)|_{({}^t(\epsilon),-{}^t(\varsigma),{}^t(\mu),-{}^t(\xi))}]. \tag{5.95}$$

We then note that the total coupling tensors of first order:

$$[\underline{E^{c(m)}_{ctot}}(r_1, r'_1)] = [\underline{E^{c(m)}_{S,S'}}(r_1, r'_1)] + [\underline{E^{c(m)}_{S',S}}(r_1, r'_1)],$$

$$[\underline{H^{m(c)}_{ctot}}(r_1, r'_1)] = [\underline{H^{m(c)}_{S,S'}}(r_1, r'_1)] + [\underline{H^{m(c)}_{S',S}}(r_1, r'_1)], \tag{5.96}$$

satisfy the reciprocity property:

$$[\underline{E^c_{ctot}}(r_1, r'_1)|_{((\epsilon),(\xi),(\mu),(\varsigma))}] = {}^t[\underline{E^c_{ctot}}(r'_1, r_1)|_{({}^t(\epsilon),-{}^t(\varsigma),{}^t(\mu),-{}^t(\xi))}],$$

$$[\underline{H^m_{ctot}}(r_1, r'_1)|_{((\epsilon),(\xi),(\mu),(\varsigma))}] = {}^t[\underline{H^m_{ctot}}(r'_1, r_1)|_{({}^t(\epsilon),-{}^t(\varsigma),{}^t(\mu),-{}^t(\xi))}],$$

$$[\underline{H^c_{ctot}}(r_1, r'_1)|_{((\epsilon),(\xi),(\mu),(\varsigma))}] = -{}^t[\underline{E^m_{ctot}}(r'_1, r_1)|_{({}^t(\epsilon),-{}^t(\varsigma),{}^t(\mu),-{}^t(\xi))}],$$

$$[\underline{E^m_{ctot}}(r_1, r'_1)|_{((\epsilon),(\xi),(\mu),(\varsigma))}] = -{}^t[\underline{H^c_{ctot}}(r'_1, r_1)|_{({}^t(\epsilon),-{}^t(\varsigma),{}^t(\mu),-{}^t(\xi))}], \tag{5.97}$$

which is in accordance with the generalized reciprocity principle.

NOTE 5.14.– The reciprocity relations derived in (5.94)–(5.97) are more general than the ones commonly used in hybrid methods [9].

5.7. Application to the numerical removal of couplings

In this section, we study the problem of suppressing numerically the coupling between A and A' to recover the contribution of A considered alone in the initial system, that is, without A', when we know:

– the fields for the total system, with A and A';

– the fields for the system with A' but without A.

That is, the case of the determination of the radiation of a scatterer A inserted in infinite plane when we know the results for a truncated plane with and without A, the scatterer A' being assumed as the ends of the plate. In what follows, we note:

$$V(r_1') \equiv J.[V(r_1, r_1')], \tag{5.98}$$

where $J\delta_{r_1}$ is the electric dipole source exciting the system, and $[V(r_1, r_1')]$ is the dyadic tensor attached to the field $V(r_1')$.

5.7.1. *Expression of the total field*

The dyadic tensor for the total field in presence of both scatterers A and A', respectively, limited by S and S', with the first-order coupling terms between them satisfying (5.88), is given by:

$$\begin{aligned}
[\underline{E_{tot}}(r_1, r_1')] &\simeq [\underline{E_{1S}}(r_1, r_1')] + [\underline{E_{1S'}}(r_1, r_1')] \\
&- [\underline{E_0}(r_1, r_1')] + [\underline{E^c_{S,S'}}(r_1, r_1')] + [\underline{E^c_{S',S}}(r_1, r_1')], \\
&= [\underline{E_{1S}}(r_1, r_1')] + [\underline{E_{1S'}}(r_1, r_1')] \\
&- [\underline{E_0}(r_1, r_1')] + [\underline{E^c_{S,S'}}(r_1, r_1')] + {}^t[\underline{E^c_{S,S'}}(r_1', r_1)|_C],
\end{aligned} \tag{5.99}$$

where E_{1S} (respectively, $E_{1S'}$) is the field without A' (respectively, without A), E_0 is the field of the system unperturbated (without A and A'), r_1 and r_1' are the respective positions of Dirac primary source and observation, while the subscript C specifies the adjoint system (note: in a similar manner, we can express H_{tot}, replacing E by H in (5.99), while, we remark that, for isotropic cases, the subscript C can be suppressed).

5.7.2. *A particular surface radiation integral, its properties and its relation with coupling*

We define E^f_S as the radiation of the sources $J_s = \widehat{n} \wedge H_{tot}$ and $M_s = -\widehat{n} \wedge E_{tot}$ on S, in the presence of the unperturbed system, with the associated tensor verifying:

$$[\underline{E_S^f}(r_1, r_1')] = \int_S Z_0[J_s].([\underline{\mathcal{F}}^0_{he,t}(r', r_1')]) - [M_s].([\underline{\mathcal{D}}^0_{eh,t}(r', r_1')])dS_{r'}$$

$$= \int_S [\underline{\mathcal{F}}^0_{he,tC}(r_1', r')].Z_0[J_s] - [\underline{\mathcal{D}}^0_{he,tC}(r_1', r')].[M_s]dS_{r'}, \tag{5.100}$$

for r_1 and r_1' outside S. Considering the total fields given by (5.99), and noticing that S' and S enclose disjoint domains, we can use (5.25) and then note that:

in the expressions of J_s and M_s with the total field,

$\widehat{n} \wedge (H_{1S'} + H^c_{S',S})$ and $-n \wedge (E_{1S'} + E^c_{S',S})$ on S

do not contribute to E^f_S from (5.25), (5.101)

so that we can write:

$$[\underline{E_S^f}(r_1, r_1')] \simeq ([\underline{E_{1S}}(r_1, r_1')] - [\underline{E_0}(r_1, r_1')]) + [\underline{E^c_{S,S'}}(r_1, r_1')]. \tag{5.102}$$

Exchanging r_1 for r_1', then using transpose, with the reciprocity relation (5.38)–(5.39) for E_0, (5.66)–(5.67) for E_{1S} and (5.94) for $E^c_{S,S'}$, we also obtain:

$$^t[\underline{E^f_{SC}}(r_1', r_1)] =$$

$$= ([\underline{E_{1S}}(r_1, r_1')] - [\underline{E_0}(r_1, r_1')]) + {}^t[\underline{E^c_{S,S'}}(r_1', r_1)|_C]$$

$$= ([\underline{E_{1S}}(r_1, r_1')] - [\underline{E_0}(r_1, r_1')]) + [\underline{E^c_{S',S}}(r_1, r_1')]. \tag{5.103}$$

NOTE 5.15.– In general, we note that, for bianisotropic system, $^t[\underline{E^f_{SC}}(r_1', r_1)] \neq [\underline{E^f_S}(r_1, r_1')]$.

5.7.3. *Efficient determination of the radiation of A without A'*

Adding (5.102) and (5.103), then substracting (5.99), we then note that we recover the field $(E_{1S} - E_0)$, scattered by A without A', following:

$$([\underline{E_{1S}}(r_1, r_1')] - [\underline{E_0}(r_1, r_1')]) =$$

$$= ([\underline{E^f_S}(r_1, r_1')] + {}^t[\underline{E^f_{SC}}(r_1', r_1)])$$

$$- ([\underline{E_{tot}}(r_1, r_1')] - [\underline{E_{1S'}}(r_1, r_1')]). \tag{5.104}$$

Therefore, we conclude that, if we know the fields:

– for the complete system with both scatterers A and A', for the calculus of $[\underline{E^f_S}(r_1, r_1')]$, $^t[\underline{E^f_{SC}}(r_1', r_1)]$ et $[\underline{E_{tot}}(r_1, r_1')]$;

– for the system with A' but without A, for the calculus of $[\underline{E_{1S'}}(r_1, r_1')]$.

Then it is possible, from (5.104), to know a second-order expression of the field diffracted by A without the presence of A'; that is, without the direct contribution of A' and the first-order coupling with it.

It is worth noticing that, in the literature, the terms $([\underline{E_{tot}}(r_1, r_1')] - [\underline{E_{1S'}}(r_1, r_1')])$ or $[\underline{E_S^f}(r_1, r_1')]$ are commonly used to evaluate the field diffracted by A without the presence of A', but, most often, it gives poor results (in particular, at grazing incidence for the case of the removal of the contribution of the ends of a plate).

It seems to us remarkable that, for the same cost, a simple combination of these terms given in (5.104) permits us to remove direct and first-order couplings for much better results.

NOTE 5.16.– A similar result can be obtained for the magnetic field, following:

$$
\begin{aligned}
&([\underline{H_{1S}}(r_1, r_1')] - [\underline{H_0}(r_1, r_1')]) = \\
&= ([\underline{H_S^f}(r_1, r_1')] + {}^t[\underline{H_{SC}^f}(r_1', r_1)]) \\
&- ([\underline{H_{tot}}(r_1, r_1')] - [\underline{H_{1S'}}(r_1, r_1')]),
\end{aligned}
\tag{5.105}
$$

where

$$
[Z_0 \underline{H_S^f}(r_1, r_1')] = \int_S Z_0[J_s].[\underline{\mathcal{D}}^0_{he,t}(r', r_1')] + [M_s].[\underline{\mathcal{E}}^0_{eh,t}(r', r_1')] dS_{r'}. \tag{5.106}
$$

NOTE 5.17.– Let notice that, for isotropic media, in monostatic case when $r_1 = r_1'$, we have:

$$
\begin{aligned}
&\widehat{w}.([\underline{E_S^f}(r_1, r_1)] + {}^t[\underline{E_S^f}(r_1, r_1)]).\widehat{w} \\
&= 2\widehat{w}.[\underline{E_S^f}(r_1, r_1)].\widehat{w},
\end{aligned}
\tag{5.107}
$$

and that, for isotropic A, grounded in a PEC plane [6], limited by a plane aperture S:

$$
E_S^f(r_1, r) = -2\text{curl} \int_S M_s(r') \frac{e^{-ik|r-r'|}}{4\pi|r-r'|} dS_{r'} \text{ with } M_s = -n \wedge E_{tot}. \tag{5.108}
$$

5.8. Numerical results

We discuss numerical removal of couplings from (5.104) in this section. For that, we consider the RCS (Radar Cross-Section) of a semicircular groove of radius $\lambda/10$ in a ground 2D plate of total length L, and the removal of couplings with the ends of the plate, as in [1]. The aim is then to recover the contribution of the groove on an infinite plane, and thus to suppress the perturbation due to the ends of the plate.

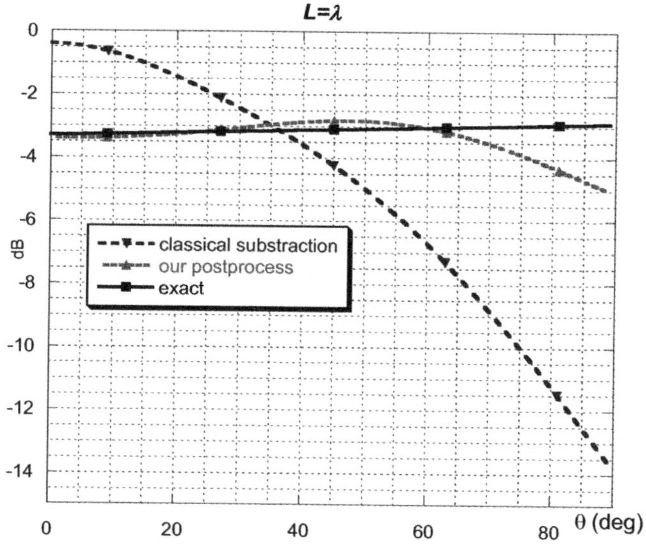

(a)

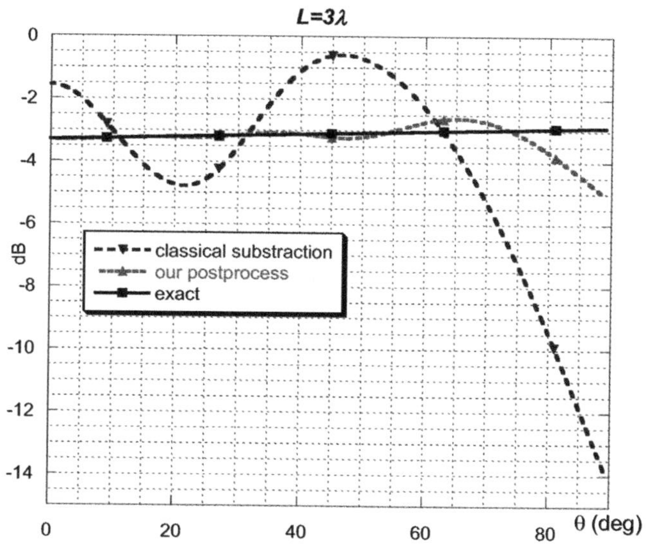

(b)

Figure 5.6. *Comparison of results for different truncation lengths L, when we classically substract the field of the plate without the groove, when we apply our postprocess (5.104), and when we apply the exact calculus for the groove in an infinite plane given in [10]. For a color version of this figure, see www.iste.co.uk/bernard/theoryofscattering.zip*

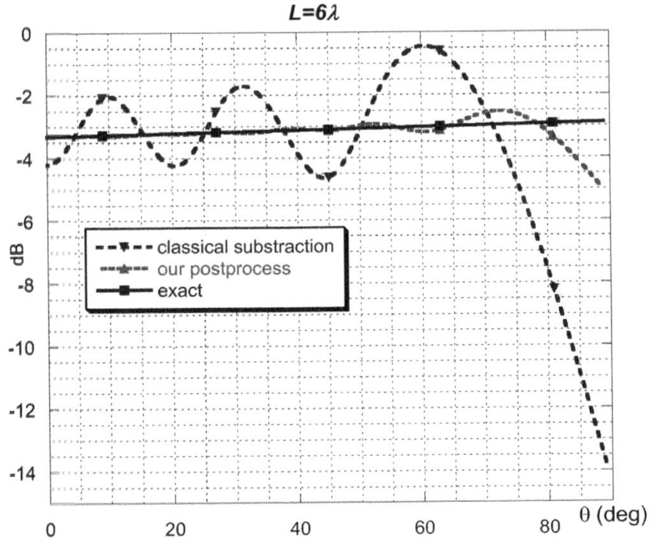

(c)

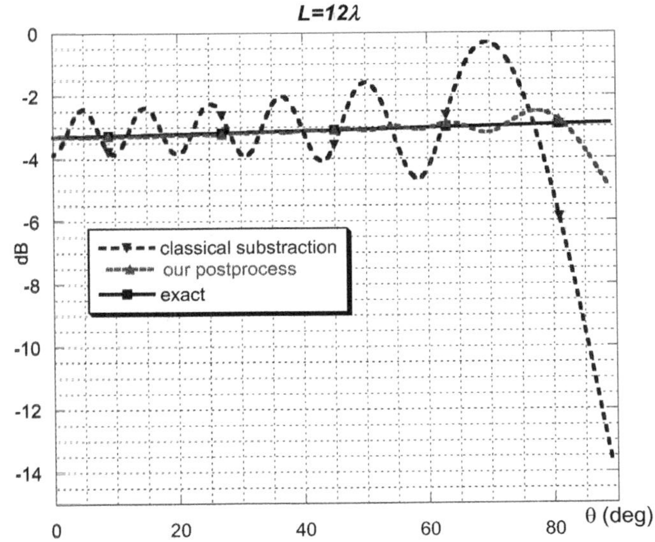

(d)

Figure 5.6. *Continued*

For this, we calculate the fields at infinity, with and without the groove, and on the aperture, for the calculus of E_S^f. We can then compare our results, when applying our postprocess (5.104), or when simply substracting the field for the plate without the groove, with the exact result given in [10], for different plate lengths $L = 1\lambda$, 3λ, 6λ, 12λ. The usual method of substraction gives poor results, while our postprocess gives excellent ones on a domain of angles that increases with L. The calculus of E_S^f and thus of (5.104) is very simple and fast, and the cost of our calculation is approximately the same as for the simple substraction method.

5.9. References

[1] Bernard, J.M.L. and Mallejac, N. (2016). Theory of couplings in electromagnetism: Novel results and numerical removal of couplings. In *ICEAA Proc., IEEE Xplore*, Piscataway.

[2] Kamenetskii, E.O. (2001). Nonreciprocal microwave bianisotropic materials: Reciprocity theorem and network reciprocity. *IEEE TAP*, 49(3), 361–366.

[3] Macher, W. (2014). Transfer operator theory and inter-reciprocity of non-reciprocal multiport antennas. *PIERS B*, 60, 169–193.

[4] Altman, C. and Suchy, K. (2011). *Reciprocity, Spatial Mappings and Time Reversal in Electromagnetics*. Springer, Dordrecht.

[5] Achouri, K. and Martin, O.J.F. (2021). Extension of Lorentz reciprocity and Poynting theorems for spatially dispersive media with quadrupolar response. *Phys. Rev. B*, 104, 165426.

[6] Jones, D.S. (1964). *The Theory of Electromagnetism*. Pergamon Press, Oxford.

[7] Jones, D.S. (1994). *Methods in Electromagnetic Wave Propagation*. Oxford University Press, Oxford.

[8] Bernard, J.M.L., Bouche, D., Andronov, I., Guyon, F. (2005). Expression du champ diffracté par une inclusion. *Ann. Telecom.*, 60(5–6), 630–648.

[9] Jin, J.M., Ni, S.M., Li, S.W. (1995). Hybridization of SBR and FEM for scattering by large bodies with cracks and cavities. *IEEE Trans. AP*, 43(10), 1130–1139.

[10] Hinders, M.K. and Yaghjian, A.D.Y. (1991). Dual-series solution to scattering from a semi-circular channel in a ground plane. *IEEE Micr. G. Wave Letters*, 1(9), 239–242.

[11] Vladimirov, V.S. (1979). *Generalized Functions in Mathematical Physics*. MIR Publishers, Moscow.

5.A. Appendix A: on some identities for conjugate fields

We can consider (5.1) and the conjugate of (5.2). In a manner similar to the one used for (5.8), we then obtain equalities, with, this time, the conjugate of (E_{0C}, H_{0C}) following:

$$\int_{\partial\Omega} H_{0C}^{*}.(\widehat{n}\wedge E)-E_{0C}^{*}.(\widehat{n}\wedge H)dS=\int_{\partial\Omega}(E_{0C}^{*}\wedge H+E\wedge H_{0C}^{*}).\widehat{n}dS$$

$$=\int_{\Omega} iE_{0C}^{*}.((\omega\epsilon)-(\omega\epsilon_0)^{*})EdV+\int_{\Omega} iH_{0C}^{*}.((\omega\mu)-(\omega\mu_0)^{*})HdV$$

$$+\int_{\Omega} iH_{0C}^{*}.((\omega\varsigma)+(\omega\varsigma_0)^{*})EdV+\int_{\Omega} iE_{0C}^{*}.((\omega\xi)+(\omega\xi_0)^{*})HdV$$

$$+\int_{\Omega}(J_{0C}^{*}.E+M_{0C}^{*}.H)\,dV+\int_{\Omega}(E_{0C}^{*}.J+H_{0C}^{*}.M)\,dV, \tag{5.109}$$

where (E,H) and (E_{0C},H_{0C}) correspond, respectively, to $((\epsilon),(\xi),(\mu),(\varsigma))$ and $({}^{t}(\epsilon_0),-{}^{t}(\varsigma_0),{}^{t}(\mu_0),-{}^{t}(\xi_0))$ characteristics, and $\widehat{n}$ is the inward normal to $\partial\Omega$. Considering this equation, we can obtain a generalization of the well-known integral equalities for the Poynting vector [6], in particular for parameters giving us $(E_{0C},H_{0C})=(E,H)$.

5.B. Appendix B: on some identities for surface integrals

5.B.1. *On the FP of a surface integral expression*

5.B.1.1. *FP by dropping some divergent terms*

The Hadamard FP (also called Hadamard regularization) is a method of regularizing divergent integral by dropping some divergent terms and keeping the FP. Let us keep in mind that, in our case, our integrals depend on a parameter r, and the FP part is not necessarily continuous relatively to it. For example, for integrals defined on a surface S, excluding a small domain belonging to the ball centered at r of radius ϵ, the continuity up to S is verified only if the limits when $r\to S$ and as $\epsilon\to 0$ can be interchanged without changing the result (note: for example, the integral expression of the indicator function 1_Ω of a domain Ω is bounded but not continuous up to its frontier included).

We now consider an example where S is a closed surface composed of piecewise continuous open part S_i, $i=1,...n$, and Ω the domain exterior to S. Letting:

$V=E$ or H,

$f(r')=\text{div}(u_0\dfrac{f_1(r')}{|r'-r|})=u_0.\text{grad}(\dfrac{f_1(r')}{|r'-r|})$ with $f_1(r')=O(1)$,

S_ϵ is the intersection of S with a ball Ω'_ϵ of radius ϵ centered at $r\in\overline{\Omega}$, (5.110)

we assume that $\widehat{n}(r').E(r')$ and $\widehat{n}(r').H(r')$ is continuous on S_i except possibly on its contour $\mathcal{L}_i$, while $\widehat{t}(r).E(r')$ and $\widehat{t}(r).H(r')$ is continuous across $\mathcal{L}_i$, $\widehat{t}(r')$ being the unit tangential vector along $\mathcal{L}_i$. Our aim is to extract the FP of the integral expression:

$$\mathcal{F}(r) = \int_{S_i \backslash S_\epsilon} (\widehat{n}(r') \wedge V(r')).\text{grad}_{r'}(f) dS_{r'}$$
$$= -\int_{S_i \backslash S_\epsilon} \widehat{n}(r').(\text{grad}_{r'}(f) \wedge V(r')) dS_{r'}, \qquad (5.111)$$

for $r \in \overline{\Omega}$. We begin by using:

$$\text{grad}_{r'}(f) \wedge V = \text{curl}_{r'}(fV) - f\text{curl}_{r'}(V),$$
$$\int_{S_i \backslash S_\epsilon} \widehat{n}(r').\text{curl}_{r'}(fV) dS_{r'} = (\int_{\mathcal{L}_i} - \int_{\mathcal{L}_\epsilon}) fV.dr', \qquad (5.112)$$

and obtain

$$\mathcal{F}(r) = -\int_{S_i \backslash S_\epsilon} \widehat{n}(r').\text{curl}_{r'}(fV) dS_{r'} + \int_{S_i \backslash S_\epsilon} \widehat{n}(r').\text{curl}_{r'}(V) f dS_{r'}$$
$$= -(\int_{\mathcal{L}_i} - \int_{\mathcal{L}_\epsilon}) fV.dr' + \int_{S_i \backslash S_\epsilon} \widehat{n}(r').\text{curl}_{r'}(V) f dS_{r'}, \qquad (5.113)$$

where $\widehat{n}$ is the outward normal to $S_{i(\epsilon)}$ and $\mathcal{L}_{i(\epsilon)}$ is positively oriented. The term $\int_{\mathcal{L}_\epsilon} fV.dr'$ vanishes as $\epsilon \to 0$ when $r \in \overline{\Omega} \backslash S_i$, and it is generally divergent as $\epsilon \to 0$ for any point of S_i. Dropping this term in $\mathcal{F}(r)$, the result unchanges when $r \in \overline{\Omega} \backslash S_i$ and becomes finite on $\overline{\Omega} \backslash \mathcal{L}_i$. We then obtain the FP of $\mathcal{F}(r)$, following:

$$\text{FP}(\mathcal{F}(r)) = \lim_{\epsilon \to 0} \Big(\int_{S_i \backslash S_\epsilon} (\widehat{n}(r') \wedge V(r')).\text{grad}_{r'}(f) dS_{r'} - \int_{\mathcal{L}_\epsilon} f(r')V(r').dr' \Big)$$
$$= pv \int_{S_i} \widehat{n}(r').\text{curl}_{r'}(V(r')) f(r') dS_{r'} - \int_{\mathcal{L}_i} f(r')V(r').dr', \qquad (5.114)$$

where pv denotes the principal value of Cauchy, and $\mathcal{L}_i$ is positively oriented (note: in applications, the integral along $\mathcal{L}_i$ can be named an integral of Kottler's type attached to S_i).

5.B.1.2. *FP by the theory of distributions*

It is worth noticing that the previous result is strongly connected with the theory of distributions. Indeed, noting Ω'_ϵ the ball of small radius ϵ centered at r, and letting:

$$1_{R^3 \backslash \Omega'_\epsilon} = 1 \text{ as } r' \in R^3 \backslash \overline{\Omega'_\epsilon}, \; = 0 \text{ as } r' \notin R^3 \backslash \Omega'_\epsilon,$$
$$\overline{f} \text{ with } \frac{1_{R^3 \backslash \Omega'_\epsilon}}{|r' - r|} \text{ instead of } f \text{ with } \frac{1}{|r' - r|},$$
$$\overline{f}(r') = 1_{R^3 \backslash \Omega'_\epsilon} u_0.\text{grad}(\frac{f_1(r')}{|r' - r|}) + \widehat{n}'_\epsilon.u_0 \frac{f_1(r')}{|r' - r|} \delta_{\partial \Omega'_\epsilon}, \qquad (5.115)$$

where $\widehat{n}'_\epsilon$ is the outward normal to Ω'_ϵ, it is sufficient to replace $f(r')$ by $\overline{f}(r')$ in the initial expression of $\mathcal{F}(r)$ to directly obtain the FP of $\mathcal{F}(r)$, following:

$$\begin{aligned} &\lim_{\epsilon\to 0}\int_{S_i}(\widehat{n}(r')\wedge V(r')).\text{grad}_{r'}(\overline{f})dS_{r'} \\ &= \lim_{\epsilon\to 0}(\mathcal{F}(r)+\int_{S_i} fV.(\widehat{n}'_\epsilon\wedge\widehat{n})\delta_{\partial\Omega'_\epsilon}dS_{r'}) \\ &= \lim_{\epsilon\to 0}(\mathcal{F}(r)-\int_{\mathcal{L}_\epsilon} fV.dr') = \text{FP}(\mathcal{F}(r)), \end{aligned} \tag{5.116}$$

where we have used that

$$\int_{S_i} fV.(\widehat{n}'_\epsilon\wedge\widehat{n})\delta_{\partial\Omega'_\epsilon}dS_{r'} = \int_{\mathcal{L}_\epsilon} fV.(\widehat{n}'_\epsilon\wedge\widehat{n})dl_{r'} \sim -\int_{\mathcal{L}_\epsilon} fV.dr' \text{ as } \epsilon\to 0. \tag{5.117}$$

In this case, the FP can be directly defined from the use of distributions theory, also named the theory of generalized functions [11].

NOTE 5.18.– We note that the components of $V(r')$ along the $\mathcal{L}_i$ are generally continuous across this frontier. Using $S\equiv\sum_i S_i$, we can then write, up to S (even on $\mathcal{L}_i$):

$$\begin{aligned} &\sum_i\int_{\mathcal{L}_i} f(r')V(r').dr' = 0, \\ &\sum_i \text{FP}\int_{S_i}(\widehat{n}(r')\wedge V(r')).\text{grad}_{r'}(f(r')))dS_{r'} = O(1). \end{aligned} \tag{5.118}$$

NOTE 5.19.– The fields component along the edge due to radiation of "one face" of a wedge (expression composed of simple and double layers contributions) is singular at the edge, while, considering radiations of "both faces", this component becomes regular at the edge.

5.B.2. *On the principal value (pv) of a surface integral*

In many cases, FP reduces to take principal value (pv) of the integral. This principal value remains finite, but can be discontinuous. An interesting example consists of the following:

$$\begin{aligned} &pv\int_{S_i} W_0(r').\text{grad}_{r'}(g(r',r))dS_{r'}, \\ &\text{with } W_0(r') = \widehat{n}(r')\wedge W(r') + w(r')\widehat{n}(r'), \end{aligned} \tag{5.119}$$

when we assume that

$$g(r', r)|r' - r| = h(r') \text{ is a smooth function on } S,$$
$$W(r') - W(r) = O(|r - r'|), w(r') - w(r) = O(|r - r'|). \tag{5.120}$$

5.B.2.1. *The part with* $\widehat{n}(r') \wedge W(r')$ *is finite and continuous*

For this case, we begin by writing:

$$pv \int_{S_i} (\widehat{n}(r') \wedge W(r')).\text{grad}_{r'}(g(r'))dS_{r'} =$$
$$\lim_{\epsilon \to 0} \int_{S_i \backslash S_\epsilon} (\widehat{n}(r') \wedge (W(r') - W(r))).\text{grad}_{r'}(g(r'))dS_{r'}$$
$$+ \lim_{\epsilon \to 0} \int_{S_i \backslash S_\epsilon} (\widehat{n}(r') \wedge W(r)).\text{grad}_{r'}(g(r'))dS_{r'}. \tag{5.121}$$

Using

$$\int_{S_i \backslash S_\epsilon} (\widehat{n}(r') \wedge W(r)).\text{grad}_{r'}(g(r'))dS_{r'}$$
$$= - \int_{S_i \backslash S_\epsilon} \widehat{n}(r').\text{curl}_{r'}(g(r')W(r))dS_{r'} = -(\int_{\mathcal{L}_i} - \int_{\mathcal{L}_\epsilon})g(r')W(r).dr',$$
$$\text{with } \lim_{\epsilon \to 0} \int_{\mathcal{L}_\epsilon} g(r')W(r).dr' = h(r) \lim_{\epsilon \to 0} \int_{\mathcal{L}_\epsilon} \frac{W(r)}{|r' - r|}.dr' = 0, \tag{5.122}$$

and noting that

$$\lim_{\epsilon \to 0} \int_{S_i \backslash S_\epsilon} (\widehat{n}(r') \wedge (W(r') - W(r))).\text{grad}_{r'}(g(r'))dS_{r'} = O(1), \tag{5.123}$$

we have

$$pv \int_{S_i} (\widehat{n}(r') \wedge W).\text{grad}_{r'}(f)dS_{r'} =$$
$$\lim_{\epsilon \to 0} \int_{S_i \backslash S_\epsilon} (\widehat{n}(r') \wedge (W(r') - W(r))).\text{grad}_{r'}(f)dS_{r'} - \int_{\mathcal{L}_i} g(r')W(r).dr', \tag{5.124}$$

which is continuous at any r' up to S_i (included), except possibly on $\mathcal{L}_i$ (note: the restriction on $\mathcal{L}_i$ does no more exist if we consider the closed surface $S \equiv \sum_i S_i$).

5.B.2.2. *The part with* $w(r')\widehat{n}(r')$ *is finite but discontinuous*

As for the part with $w(r')\widehat{n}(r')$, it is finite but discontinuous. Indeed, we can write:

$$pv\int_{S_i} w(r')\widehat{n}(r').\text{grad}_{r'}(g(r'))dS_{r'} = w(r)\ pv\int_{S_i}\widehat{n}(r').\text{grad}_{r'}(g(r'))dS_{r'}$$
$$+\int_{S_i}(w(r')-w(r))\widehat{n}(r').\text{grad}_{r'}(g(r'))dS_{r'}, \tag{5.125}$$

where we note the continuity of last integral; while, in the first line, the integral:

$$\lim_{\epsilon\to 0}\int_{S_i\backslash S_\epsilon}\widehat{n}(r').\text{grad}_{r'}(\frac{1}{|r'-r|})dS_{r'}, \tag{5.126}$$

is finite but discontinuous as r crosses S_i (see note 5.20 and appendix C for 1_Ω).

NOTE 5.20.– Let S'_i (respectively, $S'_{\epsilon'}$) be a disk of radius a (respectively, ϵ') at $z=0$, centered at the origin. We note that:

$$\mathcal{D}_{\epsilon'}(r) = -\int_{S_i\backslash S'_{\epsilon'}}\widehat{z}.\text{grad}_{r'}(\frac{-1/4\pi}{|r'-r|})dS_{r'} = \int_0^{2\pi}\int_{\epsilon'}^{a}\frac{\partial(\frac{1/4\pi}{\sqrt{\rho^2+(z-z')^2}})}{\partial z'}\rho d\rho d\varphi$$
$$=\int_{\epsilon'^2}^{a^2}\frac{z/4}{(\sqrt{v+z^2})^3}dv|_{z'=0} = \frac{z/2}{\sqrt{\epsilon'^2+z^2}} - \frac{z/2}{\sqrt{a^2+z^2}}, \tag{5.127}$$

and thus

$$\lim_{\epsilon'\to 0}\mathcal{D}_{\epsilon'}(r) = \text{sign}(z)/2 - \frac{z/2}{\sqrt{a^2+z^2}}, \tag{5.128}$$

is finite, but discontinuous at $z=0$.

5.B.3. *On properties of the surface divergence operator Div*

The term $\widehat{n}.\text{curl}_{r'}(V)$ on S_i only depends on tangential derivatives of tangential components. Besides, we note that:

$$\int_{S_i} -\widehat{n}.\text{curl}_{r'}(V)dS_{r'} = \int_{\mathcal{L}_i}\widehat{n}\wedge(\widehat{n}\wedge V).dr'$$
$$= -\int_{\mathcal{L}_i}(\widehat{n}\wedge V).\widehat{n}\wedge dr' = \int_{\mathcal{L}_i}(\widehat{n}\wedge V).\widehat{\nu}ds', \tag{5.129}$$

where $\widehat{\nu}$ is the outward normal to $\mathcal{L}_i$. Letting:

$$\text{Div}_{r'}(\widehat{n}\wedge V) = -(\widehat{n}.\text{curl}_{r'})(V)|_S, \tag{5.130}$$

in these integral equalities, we can then define the surface operator Div, as Jones in [7], following:

$$\mathrm{Div}(\widehat{n} \wedge V) = \lim_{A \to 0} \frac{1}{A} \int_{\mathcal{L}_i} (\widehat{n} \wedge V).\widehat{\nu} ds', \tag{5.131}$$

where A is the area of S_i. These equalities are particularly useful, in particular when $V = E$ or H, and $\mathrm{curl}_{r'}(V)$, are given by Maxwell equations.

5.C. Appendix C: note on distributions and some integrals expressions

5.C.1. *Elementary equalities for distributions*

Let us consider $\Omega' \backslash \Omega'_\epsilon$, with Ω'_ϵ a ball of small radius ϵ centered at r, and Ω' a domain with a piecewise smooth frontier. We note that, in the sense of distributions (or generalized functions) [6]:

$$\begin{aligned}
&\mathrm{curl}_{r'}(1_{\Omega'\backslash\Omega'_\epsilon} V) = 1_{\Omega'\backslash\Omega'_\epsilon} \mathrm{curl}_{r'}(V) - (\widehat{n}' \wedge V)\delta_{\partial\Omega'} + (\widehat{n}'_\epsilon \wedge V)\delta_{\partial\Omega'_\epsilon},\\
&\mathrm{grad}_{r'}(1_{\Omega'\backslash\Omega'_\epsilon} g) = 1_{\Omega'\backslash\Omega'_\epsilon} \mathrm{grad}_{r'}(g) - (\widehat{n}' g)\delta_{\partial\Omega'} + (\widehat{n}'_\epsilon g)\delta_{\partial\Omega'_\epsilon},\\
&\mathrm{div}_{r'}(1_{\Omega'\backslash\Omega'_\epsilon} V) = 1_{\Omega'\backslash\Omega'_\epsilon} \mathrm{div}_{r'}(V) - (\widehat{n}'.V)\delta_{\partial\Omega'} + (\widehat{n}'_\epsilon.V)\delta_{\partial\Omega'_\epsilon},\\
&(\Delta_{r'} + k_0^2)(1_{R^3\backslash\Omega'_\epsilon}(r')\frac{-e^{-ik|r'-r|}}{4\pi|r'-r|}) \to \delta(r'-r) \text{ as } \epsilon \to 0,
\end{aligned} \tag{5.132}$$

where $1_{\Omega'\backslash\Omega'_\epsilon}$ is the indicator function of $\Omega'\backslash\Omega'_\epsilon$, with $1_{\Omega'\backslash\Omega'_\epsilon}(r') = 1$ as $r' \in \Omega'\backslash\Omega'_\epsilon$, $= 0$ as $r' \notin \overline{\Omega'\backslash\Omega'}_\epsilon$. The unit vector $\widehat{n}'$ (respectively, $\widehat{n}'_\epsilon$) is the outward normal to $\partial\Omega'$ (respectively, $\partial\Omega'_\epsilon$), while $1_{R^3\backslash\Omega'_\epsilon}$ is $1_{\Omega'\backslash\Omega'_\epsilon}$ as $\Omega' \to R^3$, and $\Delta \equiv$div(grad(.)) (note: $1_{R^3\backslash\Omega'_\epsilon}\frac{-e^{-ik|r'-r|}}{4\pi|r'-r|}$ is also called $fp(\frac{-e^{-ik|r'-r|}}{4\pi|r'-r|})$ as $\epsilon \to 0$).

5.C.2. *Some integrals concerning the indicator function* $1_{\Omega'}(r)$ *up to* $\partial\Omega'$

We analyze the properties of integrals on Ω' in relation with:

$$(\mathrm{div}(\mathrm{grad}\ (.)) + k^2)\frac{-1_{R^3\backslash\Omega'_\epsilon} e^{-ik|r'-r|}}{4\pi|r'-r|} = \delta(r'-r), \tag{5.133}$$

for $k = 0$ and for $\mathrm{Re}(ik) > 0$ in (5.133). In our analysis, $\partial\Omega'$ is piecewise smooth, and we consider integrals related to the definition of $1_{\Omega'}(r)$ for r up to $\partial\Omega'$ included, where this function can be fractional and discontinuous.

5.C.2.1. *The case* $k = 0$

Let us consider $g(r')$ an analytic function with $\text{grad}(g(r')) = O(1)$ in R^3. We can write:

$$I_\epsilon(r) = \int_{R^3} 1_{\Omega'} g(r')\text{div}(\text{grad}(\frac{-1_{R^3\setminus\Omega'_\epsilon}}{4\pi|r'-r|}))dV_{r'}$$

$$= \int_{R^3} 1_{\Omega'} g(r')\text{div}(1_{R^3\setminus\Omega'_\epsilon}\text{grad}_{r'}(\frac{-1}{4\pi|r'-r|}) - \frac{\widehat{n}'_\epsilon \delta_{\partial\Omega'_\epsilon}}{4\pi|r'-r|})dV$$

$$= \int_{\partial\Omega'_\epsilon} \frac{\text{grad}(1_{\Omega'} g(r'))\widehat{n}'_\epsilon}{4\pi|r'-r|}dS + \int_{\Omega'} g(r')\text{div}(1_{R^3\setminus\Omega'_\epsilon}\text{grad}_{r'}(\frac{-1}{4\pi|r'-r|}))dV, \tag{5.134}$$

and thus

$$I_\epsilon(r) = \int_{\partial\Omega'_\epsilon} \frac{\text{grad}(1_{\Omega'} g(r'))\widehat{n}'_\epsilon}{4\pi|r'-r|}dS + \int_{\Omega'\setminus\Omega'_\epsilon} \frac{-g(r')}{4\pi}\text{div}(\text{grad}_{r'}(\frac{1}{|r'-r|}))dV$$

$$+ \int_{\Omega'} g(r')\delta_{\partial\Omega'_\epsilon}(\widehat{n}'_\epsilon.\text{grad}_{r'}(\frac{-1}{4\pi|r'-r|}))dV$$

$$= \int_{\partial\Omega'_\epsilon} \frac{\text{grad}(1_{\Omega'} g(r'))\widehat{n}'_\epsilon}{4\pi|r'-r|}dS + 0 + \int_{\partial\Omega'_\epsilon} \frac{1_{\Omega'} g(r')\widehat{n}'_\epsilon.\text{grad}_{r'}(\frac{-1}{|r'-r|})}{4\pi}dS$$

$$= \int_{\partial\Omega'_\epsilon\cap\partial\Omega'} \frac{-\widehat{n}'.\widehat{n}'_\epsilon\frac{g(r')}{|r'-r|}}{4\pi}dl + \int_{\partial\Omega'_\epsilon} \frac{1_{\Omega'}\frac{\text{grad}(g(r'))}{|r'-r|}\widehat{n}'_\epsilon dS}{4\pi} + \int_{\partial\Omega'_\epsilon} \frac{1_{\Omega'}\frac{g(r')}{|r'-r|^2}dS}{4\pi}. \tag{5.135}$$

Since $\widehat{n}'.\widehat{n}'_\epsilon = 0$ on points (if any) of $\partial\Omega'_{\epsilon\to 0} \cap \partial\Omega'$, we then obtain:

$$\lim_{\epsilon\to 0} I_\epsilon(r) = g(r)\lim_{\epsilon\to 0}\frac{1}{4\pi\epsilon^2}\int_{\Omega'\cap\partial\Omega'_\epsilon} dS = g(r)1_{\Omega'}(r),$$

$$\text{and thus, } \text{div}(\text{grad}(\frac{-1_{R^3\setminus\Omega'_\epsilon}(r')}{4\pi|r'-r|})) = \delta(r'-r). \tag{5.136}$$

So defined, we have:

$$1_{\Omega'}(r) = \lim_{\epsilon\to 0} I_\epsilon(r) \text{ for } g(r) \equiv 1, \text{ and thus, } 1_{\Omega'}(r) = \lim_{\epsilon\to 0}\frac{1}{4\pi\epsilon^2}\int_{\Omega'\cap\partial\Omega'_\epsilon} dS. \tag{5.137}$$

The function $1_{\Omega'}(r)$ is then $= 1$ in $\Omega'\setminus\partial\Omega'$, $= 0$ outside $\overline{\Omega'}$, and is fractional on $\partial\Omega'$. We note in particular that, on the smooth part of $\partial\Omega'$, $1_{\Omega'} = \frac{1}{2}$, while, at the tip of a circular cone of exterior half angle θ_c, $1_{\Omega'} = \frac{1-\cos\theta_c}{2}$, and at the edge of a wedge of exterior half angle Φ, $1_{\Omega'} = \frac{\Phi}{\pi}$.

In other respects, we can express it in a different manner. For that, we can write:

$$\lim_{\epsilon\to 0} I_\epsilon(r') = \lim_{\epsilon\to 0}\int_{R^3} \text{grad}(1_{\Omega'} g(r')).\text{grad}(\frac{1_{R^3\backslash\Omega'_\epsilon}}{4\pi|r'-r|}))dV$$

$$= \lim_{\epsilon\to 0}\int_{\Omega'\backslash\Omega'_\epsilon} \text{grad}(g(r')).\text{grad}(\frac{1}{4\pi|r'-r|}))dV$$

$$+ \lim_{\epsilon\to 0}\int_{\partial\Omega'_\epsilon} 1_{\Omega'}\text{grad}(g(r')).(\frac{\widehat{n}'_\epsilon}{4\pi|r'-r|}))dS$$

$$- \lim_{\epsilon\to 0}\int_{\partial\Omega'} 1_{R^3\backslash\Omega'_\epsilon}\widehat{n}' g(r').\text{grad}(\frac{1}{4\pi|r'-r|})dS, \qquad (5.138)$$

then

$$\lim_{\epsilon\to 0} I_\epsilon(r') = \lim_{\epsilon\to 0}\int_{\Omega'\backslash\Omega'_\epsilon} \text{grad}(g(r'))\text{grad}(\frac{1}{4\pi|r'-r|}))dV$$

$$+ pv\int_{\partial\Omega'} g(r')(-\widehat{n}').\text{grad}(\frac{1}{4\pi|r'-r|})dS, \qquad (5.139)$$

which leads us to

$$1_{\Omega'}(r) = pv\int_{\partial\Omega'} \widehat{n}'.\text{grad}(\frac{-1}{4\pi|r'-r|})dS. \qquad (5.140)$$

5.C.2.2. *The case* $Re(ik) > 0$ *in comparison to the case* $k = 0$

Let us consider:

$$J_\epsilon(r') = \int_{R^3} 1_{\Omega'} g(r')(\text{div}(\text{grad}(.)) + k^2)(\frac{-1_{R^3\backslash\Omega'_\epsilon}}{4\pi|r'-r|})dV_{r'}, \qquad (5.141)$$

If we develop the integral in a manner similar to (5.135), the influence of k is negligible, and we obtain, like in (5.136):

$$\lim_{\epsilon\to 0} J_\epsilon(r) = g(r)\lim_{\epsilon\to 0}\frac{1}{4\pi\epsilon^2}\int_{\Omega'\cap\partial\Omega'_\epsilon} dS, \qquad (5.142)$$

in perfect agreement with $(\text{div}(\text{grad}(.)) + k^2)\frac{-1_{R^3\backslash\Omega'_\epsilon} e^{-ik|r'-r|}}{4\pi|r'-r|} = \delta(r'-r)$.

Now, developing, this time, the integral in a manner similar to (5.139), the parameter k clearly influences the form of the expression, although its value, by definition, should remain the same for any k. So, writing:

$$\lim_{\epsilon \to 0} J_\epsilon(r') = \int_{R^3} 1_{\Omega'} g(r')(\mathrm{div}(\mathrm{grad}(.)) + k^2) \frac{-1_{R^3 \backslash \Omega'_\epsilon}}{4\pi |r' - r|} dV$$

$$= \lim_{\epsilon \to 0} \int_{\Omega' \backslash \Omega'_\epsilon} \mathrm{grad}(g(r')).\mathrm{grad}(\frac{e^{-ik|r'-r|}}{4\pi |r' - r|}) dV$$

$$+ pv \int_{\partial \Omega'} g(r')(-\widehat{n}').\mathrm{grad}(\frac{e^{-ik|r'-r|}}{4\pi |r' - r|}) dS$$

$$+ pv \int_{\Omega'} k^2 g(r') \frac{-1_{R^3 \backslash \Omega'_\epsilon} e^{-ik|r'-r|}}{4\pi |r' - r|} dV, \qquad (5.143)$$

we remark that the integrals on $\partial\Omega'$ vanish as $\Omega' \to R^3$ when $\mathrm{Re}(ik) > 0$, which is not the case for $k = 0$, while the one on Ω' vanish for $k = 0$, which is not the case for $k \neq 0$. However, the sum of integrals should remain the same for $k = 0$ and for $\mathrm{Re}(ik) > 0$. To show it, let us consider the example of $g(r') = 1$ and $\Omega' \to R^3$. After noticing that:

$$-\int_0^\infty k^2 e^{-ik|r|} |r| d|r| = 1 \text{ and } \lim_{|r| \to \infty} e^{-ik|r|} = 0 \text{ when } \mathrm{Re}(ik) > 0, \qquad (5.144)$$

we note that:

$$\lim_{\Omega' \to R^3} \int_{\partial \Omega'} (-\widehat{n}').\mathrm{grad}(\frac{e^{-ik|r'-r|}}{4\pi |r' - r|}) dS = \left| \begin{matrix} =0 \text{ if } \mathrm{Re}(ik)>0 \\ =1 \text{ if } k=0 \end{matrix} \right. ,$$

$$\lim_{\Omega' \to R^3} pv \int_{\Omega'} k^2 \frac{-e^{-ik|r'-r|}}{4\pi |r' - r|} dV = \left| \begin{matrix} =1 \text{ if } \mathrm{Re}(ik)>0 \\ =0 \text{ if } k=0 \end{matrix} \right. , \qquad (5.145)$$

and, as expected, the sum is the same (=1), for both cases $k = 0$ and $\mathrm{Re}(ik) > 0$.

5.D. Appendix D: integral expressions and equations when S is piecewise smooth

5.D.1. *General integral expressions for piecewise smooth* S

Let us consider an initial system with bianisotropic media, perturbated by a scatterer bounded by a closed surface $S = \sum_i S_i$, with geometric (or materials) discontinuities along the frontiers ∂S_i. The ∂S_i, when they are positively oriented, are denoted $\mathcal{L}_i$. Considering the regularity of the field, we now assume the continuity across $\mathcal{L}_i$ of fields components along it (note: this condition is related to canonical wedge cases and the uniqueness theorem of Levine). Applying (5.50), we have:

$$u_0.(1^c_\Omega E - E_0)(r) = \sum_i FP \int_{S_i} (\widehat{n} \wedge H).E_{0C}(J_{0C}; r')$$

$$+ (\widehat{n} \wedge E).H_{0C}(J_{0C}; r') dS_{r'}, \qquad (5.146)$$

where $\widehat{n}$ is the outward normal to S, and $u_0.E$ is continuous at $r \in \Omega \equiv R^3 \backslash \Omega_S$, outside and on S. The integrals are considered for their FP (see details in appendix B). The fields E and E_0 are, respectively, the field radiated in the presence of the perturbated and the unperturbated system, while (E_{0C}, H_{0C}) are the fields attached to radiation of primary source $J_{0C} = u_0\delta(r' - r)$ in the presence of the unperturbated system with adjoint (complementary) characteristics.

Concerning the particular case of points along $\mathcal{L}_i$, we can write:

$u_0 = \widehat{u}_e(r)$ tangent to the discontinuities along $\mathcal{L}_i$,

$$\widehat{u}_e(r).(1^c_\Omega E - E_{inc})(r) = \sum_i FP \int_{S_i} (\widehat{n} \wedge H).E_{0C}(J_{0C}; r')$$

$$+ (\widehat{n} \wedge E).H_{0C}(J_{0C}; r')dS_{r'}, \tag{5.147}$$

where we consider the continuity across $\mathcal{L}_i$ of the component of E along them. In a similar manner, for $J_0 = J = 0$, $M_{0C} = -u_0\delta(r' - r)$, we have, from (5.52):

$$u_0.(1^c_\Omega H - H_0)(r) = \sum_i FP \int_{S_i} (\widehat{n} \wedge H).E_{0C}(M_{0C}; r')$$

$$+ (\widehat{n} \wedge E).H_{0C}(M_{0C}; r')dS_{r'}, \tag{5.148}$$

where $u_0.H$ is continuous at $r \in \Omega \equiv R^3 \backslash \Omega_S$, while H and H_0 are, respectively, the field radiated in the presence of the perturbated and the unperturbated system. For points along $\mathcal{L}_i$, we can then take account of the continuity across the $\mathcal{L}_i$ of the component of H along them, as previously done for E.

5.D.2. Generalities when the domain $\Omega \equiv R^3 \backslash \Omega_S$ holds the isotropic free space

Let us consider the case where the domain $\Omega \equiv R^3 \backslash \Omega_S$, exterior to S, is filled by isotropic homogeneous medium. Different choices of (E_{0C}, H_{0C}) are possible with this constraint, in particular the one where the unperturbated system in R^3 is the free space. In this case, considering previous equations, we can write:

$$u_0.(1_\Omega H - H_{inc})(r) = \sum_i FP \int_{S_i} ((\widehat{n} \wedge H).E_{0C}(M_{0C}; r')$$

$$+ (\widehat{n} \wedge E).H_{0C}(M_{0C}; r'))dS_{r'},$$

$$\text{with } M_{0C} = -u_0\delta(r' - r),\ E_{0C} = \text{curl}_{r'}(u_0 \frac{e^{-ikR}}{4\pi R}) = -u_0 \wedge \text{grad}_{r'}(\frac{e^{-ikR}}{4\pi R})$$

$$\text{and } H_{0C} = i\omega\varepsilon u_0 \frac{e^{-ikR}}{4\pi R} - \text{grad div } \frac{u_0}{i\omega\mu} \frac{e^{-ikR}}{4\pi R} \text{ for } R \neq 0, \tag{5.149}$$

and

$$u_0.(1_\Omega E - E_{inc})(r) = \sum_i FP \int_{S_i} ((\widehat{n} \wedge H).E_{0C}(J_{0C}; r')$$

$$+ (\widehat{n} \wedge E).H_{0C}(J_{0C}; r'))dS_{r'},$$

with $J_{0C} = u_0 \delta(r' - r)$, $H_{0C} = \text{curl}_{r'}(u_0 \dfrac{e^{-ikR}}{4\pi R}) = -u_0 \wedge \text{grad}_{r'}(\dfrac{e^{-ikR}}{4\pi R})$

$$\text{and } E_{0C} = -i\omega\mu u_0 \frac{e^{-ikR}}{4\pi R} + \text{grad div } \frac{u_0}{i\omega\varepsilon} \frac{e^{-ikR}}{4\pi R} \text{ for } R \neq 0, \tag{5.150}$$

with $R = |r' - r|$ and $\text{div}(\frac{u_0 e^{-ikR}}{4\pi R}) = u_0.\text{grad}(\frac{e^{-ikR}}{4\pi R})$. We note that, considering appendix B, we have for $r \in \Omega \backslash \mathcal{L}_i$:

$$- FP \int_{S_i} (\widehat{n} \wedge E).\text{grad div } \frac{u_0}{i\omega\mu} \frac{e^{-ikR}}{4\pi R} dS_{r'}$$

$$= FP \int_{S_i} \widehat{n}(r').((\text{grad div } \frac{u_0}{i\omega\mu} \frac{e^{-ikR}}{4\pi R}) \wedge E)dS_{r'}$$

$$= -pv \int_{S_i} \widehat{n}(r').\text{curl}_{r'}(E)\text{div}_{r'}(\frac{\frac{u_0}{i\omega\mu} e^{-ikR}}{4\pi R})dS_{r'} + \int_{\mathcal{L}_i} \text{div}_{r'}(\frac{\frac{u_0}{i\omega\mu} e^{-ikR}}{4\pi R})E.dr'$$

$$= pv \int_{S_i} (\widehat{n}(r').H(r'))\, u_0.\text{grad}_{r'}(\frac{e^{-ikR}}{4\pi R})dS_{r'} + \int_{\mathcal{L}_i} \frac{u_0.\text{grad}(\frac{e^{-ikR}}{4\pi R})}{i\omega\mu} E(r').dr', \tag{5.151}$$

and

$$FP \int_{S_i} (\widehat{n} \wedge H).\text{grad div } \frac{u_0}{i\omega\varepsilon} \frac{e^{-ikR}}{4\pi R} dS_{r'}$$

$$= -FP \int_{S_i} \widehat{n}.((\text{grad div } \frac{u_0}{i\omega\varepsilon} \frac{e^{-ikR}}{4\pi R}) \wedge H)dS_{r'}$$

$$= pv \int_{S_i} \widehat{n}(r').\text{curl}_{r'}(H)\text{div}_{r'}(\frac{u_0}{i\omega\varepsilon} \frac{e^{-ikR}}{4\pi R})dS_{r'} - \int_{\mathcal{L}_i} (\text{div}(\frac{u_0}{i\omega\varepsilon} \frac{e^{-ikR}}{4\pi R}))H.dr'$$

$$= pv \int_{S_i} (\widehat{n}(r').E)\, u_0.\text{grad}(\frac{e^{-ikR}}{4\pi R})dS_{r'} - \int_{\mathcal{L}_i} (\frac{u_0}{i\omega\varepsilon}.\text{grad}(\frac{e^{-ikR}}{4\pi R}))H.dr', \tag{5.152}$$

where $-\widehat{n}.\text{curl}_{r'}(E)|_S = \text{Div}(\widehat{n} \wedge E)$, $-\widehat{n}.\text{curl}_{r'}(H)|_S = \text{Div}(\widehat{n} \wedge H)$, and the contour integral along $\mathcal{L}_i$ (positively oriented) is commonly named the Kottler integral attached to S_i.

NOTE 5.21.– As noticed previously, we can also let $1_{R^3 \backslash \Omega'_\epsilon} \frac{e^{-ikR}}{4\pi R}$ instead of $\frac{e^{-ikR}}{4\pi R}$ (in the sense of distributions) in the expressions of E_{0C} and H_{0C} to directly obtain the FP expressions.

5.D.3. *Integrals expressions in free space and discontinuities*

Assuming that all lines of discontinuities are along $\mathcal{L}_i$, we can use the continuity, across the oriented contours $\mathcal{L}_i$, of the components parallel to them. This will imply that, when $S = \sum_i S_i$ is a closed surface:

$$\sum_i \int_{\mathcal{L}_i} (\frac{u_0}{i\omega\mu}.\text{grad}(\frac{e^{-ikR}}{4\pi R}))E.dr' = 0,$$

$$\sum_i \int_{\mathcal{L}_i} (\frac{u_0}{i\omega\varepsilon}.\text{grad}(\frac{e^{-ikR}}{4\pi R}))H.dr' = 0. \tag{5.153}$$

Using the notations,

$$-(\widehat{n}.\text{curl}_{r'})(E)|_S = \text{Div}_{r'}(\widehat{n} \wedge E),$$

$$-(\widehat{n}.\text{curl}_{r'})(H)|_S = \text{Div}_{r'}(\widehat{n} \wedge H), \tag{5.154}$$

with $\text{Div}_{r'}$ as defined in appendix B, we can develop in whole generality the expressions of fields, outside and on $S \equiv \sum_i S_i$, following:

$$u_0.(1_\Omega H - H_{inc})(r) = \sum_i \Bigg(pv \int_{S_i} -(\widehat{n} \wedge H).(u_0 \wedge \text{grad}_{r'}(\frac{e^{-ikR}}{4\pi R}))$$

$$+ (\widehat{n} \wedge E).i\omega\varepsilon u_0 \frac{e^{-ikR}}{4\pi R} + \text{Div}(\widehat{n} \wedge E)(\frac{u_0}{i\omega\mu}.\text{grad}_{r'}(\frac{e^{-ikR}}{4\pi R}))dS_{r'}$$

$$+ \int_{\mathcal{L}_i} (\frac{u_0}{i\omega\mu}.\text{grad}_{r'}(\frac{e^{-ikR}}{4\pi R}))E.dr' \Bigg), \tag{5.155}$$

and

$$u_0.(1_\Omega E - E_{inc})(r) = -\sum_i \Bigg(pv \int_{S_i} \text{Div}(\widehat{n} \wedge H)(\frac{u_0}{i\omega\varepsilon}.\text{grad}_{r'}(\frac{e^{-ikR}}{4\pi R}))$$

$$+ (\widehat{n} \wedge H).i\omega\mu u_0 \frac{e^{-ikR}}{4\pi R} + (\widehat{n} \wedge E).(u_0 \wedge \text{grad}_{r'}(\frac{e^{-ikR}}{4\pi R}))dS_{r'}$$

$$+ \int_{\mathcal{L}_i} (\frac{u_0}{i\omega\varepsilon}.\text{grad}_{r'}(\frac{e^{-ikR}}{4\pi R}))H.dr' \Bigg), \tag{5.156}$$

where we take care that u_0 on discontinuities is directed along them, while noting that:

$$(\widehat{n} \wedge V).(u_0 \wedge \text{grad}_{r'}(\frac{e^{-ikR}}{4\pi R})) = ((u_0.\widehat{n})V - (u_0.V)\widehat{n})).\text{grad}_{r'}(\frac{e^{-ikR}}{4\pi R}). \tag{5.157}$$

The contour integrals along $\mathcal{L}_i$ (positively oriented) is commonly named the Kottler integrals attached to S_i, and their sum, as previously noticed, is null on a closed surface, because of continuity of fields components along $\mathcal{L}_i$.

5.D.4. *Integral equations for impedance boundary conditions on* S

5.D.4.1. *Impedance, continuity and discontinuities*

We now consider impedance boundary conditions on each S_i composing S, following:

$$(\widehat{n} \wedge E)|_{S_i} = (Z)_i(\widehat{n} \wedge (\widehat{n} \wedge H))|_{S_i} \tag{5.158}$$

for $i = 1, ..., n$. All lines of discontinuities are along the oriented contours $\mathcal{L}_i$. We now assume that the surface S is closed, which implies, from continuity of tangential components across $\mathcal{L}_i$, that we have:

$$\begin{aligned}&\sum_i \int_{\mathcal{L}_i} (\frac{u_0}{i\omega\mu}.\text{grad}_{r'}(\frac{e^{-ikR}}{4\pi R}))E.dr' = 0,\\&\sum_i \int_{\mathcal{L}_i} (\frac{u_0}{i\omega\varepsilon}.\text{grad}_{r'}(\frac{e^{-ikR}}{4\pi R}))H.dr' = 0,\end{aligned} \tag{5.159}$$

From impedance conditions, we can remark:

concerning the tangential components of fields on S,

the boundness of parallel fields components along $\mathcal{L}_i$ implies

also the ones of perpendicular ones when $(Z)_i$ is invertible,

but in contrary to components parallel to $\mathcal{L}_i$ that are continuous across it,

the perpendicular components can be discontinuous across it. (5.160)

In what follows, we note:

$\widehat{u}_e^i(r)$ is the unit tangent to $\mathcal{L}_i$ (positively oriented) as $r \in \mathcal{L}_i$,

$\widehat{\nu}_e^i(r) = -\widehat{n} \wedge \widehat{u}_e^i(r)$ is the outward normal to $\mathcal{L}_i$, with $\widehat{n}$ defined on S_i. (5.161)

5.D.4.2. *For* $r \in \Omega\backslash\mathcal{L}_i$*, that is, on and outside* S*, except at discontinuities*

Considering $(Z)_i$ smooth on each S_i, we can write, for $r \in \Omega\backslash\mathcal{L}_i$:

$$\begin{aligned}(1_\Omega H - H_{inc})(r) = \sum_i \Bigg(pv \int_{S_i} &((\widehat{n} \wedge H) \wedge \text{grad}_{r'}(\frac{e^{-ikR}}{4\pi R}))+\\&+ \text{Div}((Z)_i(\widehat{n} \wedge (\widehat{n} \wedge H)))\frac{1}{i\omega\mu}\text{grad}_{r'}(\frac{e^{-ikR}}{4\pi R}))\\&+ (Z)_i(\widehat{n} \wedge (\widehat{n} \wedge H))i\omega\varepsilon\frac{e^{-ikR}}{4\pi R})dS_{r'}\Bigg),\end{aligned} \tag{5.162}$$

and

$$(1_\Omega E - E_{inc})(r) = -\sum_i \left(pv \int_{S_i} \text{Div}(\widehat{n} \wedge H) \frac{1}{i\omega\varepsilon} \text{grad}_{r'} (\frac{e^{-ikR}}{4\pi R}) \right.$$

$$\left. + (\widehat{n} \wedge H) i\omega\mu \frac{e^{-ikR}}{4\pi R} + (Z)_i (\text{grad}_{r'} (\frac{e^{-ikR}}{4\pi R}) \wedge (\widehat{n} \wedge (\widehat{n} \wedge H)) dS_{r'} \right). \quad (5.163)$$

We then note that we can obtain integral equations on tangential fields components for $r \in S_i \backslash \mathcal{L}_i$, $i = 1, 2, ...n$, by considering, in first choice:

$$(\widehat{n} \wedge H)|_{S_i} \text{ from (5.162) as } r \in S_i \backslash \mathcal{L}_i,$$

$$\text{(which reduces to surface MFIE for } (Z)_i = 0\text{)}, \quad (5.164)$$

and, in second choice, from a combination of (5.162) and (5.163):

$$(\widehat{n} \wedge E)|_{S_i} - (Z)_i (\widehat{n} \wedge (\widehat{n} \wedge H))|_{S_i} = 0 \text{ as } r \in S_i \backslash \mathcal{L}_i$$

$$\text{(which reduces to surface EFIE for } (Z)_i = 0\text{)}, \quad (5.165)$$

with $J_s = (\widehat{n} \wedge H)|_{S_i}$, $(H.\widehat{u}_e^i)$ and $(H.\widehat{\nu}_e^i)$ on $\mathcal{L}_i$, $i = 1, 2, ..., n$, as unknowns.

5.D.4.3. *For $r \in \mathcal{L}_i$ along lines of discontinuities on S*

For $r \in \mathcal{L}_i$, we take $u_0 \equiv u_0(r)$ as $r \in \mathcal{L}_i$, following:

$$u_0(r) \equiv \widehat{u}_e^i(r) \text{ is the unit tangent to } \mathcal{L}_i \text{ positively oriented as } r \in \mathcal{L}_i. \quad (5.166)$$

The components of E and H along $\widehat{u}_e^i(r)$ are components that are continuous across discontinuities on $\mathcal{L}_i$, and we can write for $r \in \mathcal{L}_i$:

$$u_0(r).1_\Omega H(r) = u_0(r).H_{inc}(r)$$

$$\sum_i \left(pv \int_{S_i} (-(\widehat{n} \wedge H(r')).(u_0(r) \wedge \text{grad}_{r'} (\frac{e^{-ikR}}{4\pi R})) + \right.$$

$$+ \text{Div}_{r'} ((Z)_i (\widehat{n} \wedge (\widehat{n} \wedge H(r')))) (\frac{u_0(r)}{i\omega\mu} .\text{grad}_{r'} (\frac{e^{-ikR}}{4\pi R}))$$

$$\left. + (Z)_i (\widehat{n} \wedge (\widehat{n} \wedge H(r'))).i\omega\varepsilon u_0(r) \frac{e^{-ikR}}{4\pi R}) dS_{r'} \right), \quad (5.167)$$

and

$$u_0(r).1_\Omega E(r) = u_0(r).E_{inc}(r)$$

$$-\sum_i \left(pv \int_{S_i} \text{Div}(\widehat{n} \wedge H(r')) (\frac{u_0(r)}{i\omega\varepsilon} .\text{grad}_{r'} (\frac{e^{-ikR}}{4\pi R})) \right.$$

$$+ (\widehat{n} \wedge H(r')).i\omega\mu u_0(r)\frac{e^{-ikR}}{4\pi R}$$

$$+ (Z)_i(\widehat{n} \wedge (\widehat{n} \wedge H(r'))).(u_0(r) \wedge \text{grad}(\frac{e^{-ikR}}{4\pi R}))dS_{r'}\Big). \tag{5.168}$$

Therefore, those equations on $\mathcal{L}_i$ complete the system of integral equations given in (5.164) and (5.165) on $S\backslash\mathcal{L}_i$.

5.D.5. *Integral expressions when S is a perfectly conducting surface in free space*

Let us consider the example of a perfectly conducting surface S in free space, with a discontinuity line l. The electric tangential field E vanishes on S, and we can now write:

$$u_0.(1_\Omega H - H_{inc})(r) = pv \int_S (\widehat{n} \wedge H(r')).\text{curl}_{r'}(u_0\frac{e^{-ik|r'-r|}}{4\pi|r'-r|})dS_{r'}, r \in \Omega\backslash l$$

$$\widehat{u}_e.(1_\Omega H - H_{inc})(r) = pv \int_S (\widehat{n} \wedge H(r')).\text{curl}_{r'}(\widehat{u}_e(r)\frac{e^{-ik|r'-r|}}{4\pi|r'-r|})dS_{r'}, r \in l, \tag{5.169}$$

where $\widehat{u}_e(r)$ is the unit vector tangent to l. In these equations, we note that:

$$(n \wedge H).\text{curl}_{r'}(u_0\frac{e^{-ik|r'-r|}}{4\pi|r'-r|}) = u_0.((\widehat{n} \wedge H) \wedge \text{grad}_{r'}(\frac{e^{-ik|r'-r|}}{4\pi|r'-r|})),$$

$$(\widehat{n} \wedge H).\text{curl}_{r'}(\widehat{u}_e(r)\frac{e^{-ik|r'-r|}}{4\pi|r'-r|}) = \text{grad}_{r'}(\frac{e^{-ik|r'-r|}}{4\pi|r'-r|}).(\widehat{u}_e(r) \wedge (\widehat{n} \wedge H))$$

$$= \text{grad}_{r'}(\frac{e^{-ik|r'-r|}}{4\pi|r'-r|}).((\widehat{u}_e(r).H(r'))\widehat{n}(r') - (\widehat{u}_e(r).\widehat{n}(r'))H(r')), \tag{5.170}$$

where the component $\widehat{u}_e(r).H(r')$ is continuous across l, and $(\widehat{u}_e(r).\widehat{n}) \to 0$, when $r' \to r \in l$. Therefore, we obtain:

$$(1_\Omega H - H_{inc})(r) = pv \int_S (\widehat{n} \wedge H) \wedge \text{grad}_{r'}(\frac{e^{-ik|r'-r|}}{4\pi|r'-r|})dS_{r'}, r \in \Omega\backslash l,$$

$$1_\Omega\widehat{u}_e(r).H(r) = \widehat{u}_e(r).H_{inc}(r) + pv \int_S \text{grad}_{r'}(\frac{e^{-ik|r'-r|}}{4\pi|r'-r|}).((\widehat{u}_e(r).H(r'))\widehat{n}(r')$$

$$- (\widehat{u}_e(r).\widehat{n}(r'))H(r'))dS_{r'},\ r \in l. \tag{5.171}$$

NOTE 5.22.– If we want to consider the system of a perfectly conducting plane, perturbated by a scatterer above it with impedance boundary conditions, we can choose to change the free space Green functions for the one of the perfectly conducting plane. For that, we have to add the contributions of the image sources on each side of equations (5.162), (5.163), (5.167) and (5.168), which corresponds to making modifications following:

$$M_0 = [-u_0\delta(r' - r)] \rightarrow M_0 = [-u_0\delta(r' - r)] + [-(u_0 - 2(u_0.\widehat{z})\widehat{z})\delta(r' - r_{im})],$$

$$u_0\frac{e^{-ikR}}{4\pi R} \rightarrow u_0\frac{e^{-ikR}}{4\pi R} + (u_0 - 2(u_0.\widehat{z})\widehat{z})\frac{e^{-ikR_{im}}}{4\pi R_{im}}, \qquad (5.172)$$

and

$$J_0 = [u_0\delta(r' - r)] \rightarrow J_0 = [u_0\delta(r' - r)] + [-(u_0 - 2(u_0.\widehat{z})\widehat{z})\delta(r' - r_{im})],$$

$$u_0\frac{e^{-ikR}}{4\pi R} \rightarrow u_0\frac{e^{-ikR}}{4\pi R} - (u_0 - 2(u_0.\widehat{z})\widehat{z})\frac{e^{-ikR_{im}}}{4\pi R_{im}}, \qquad (5.173)$$

with $r_{im} = r - 2(r.\widehat{z})\widehat{z}$, $R = |r' - r|$ and $R_{im} = |r' - r_{im}|$.

6

Exact and Asymptotic Reductions of Surface Radiation Integrals with Complex Exponential Arguments to Efficient Contour Integrals

6.1. Introduction

We present here an efficient method for the reduction of surface radiation integrals to non-singular contour integrals, when we assume to know, as in physical optics (PO) [1]–[5], the analytic expression of surface currents whose electromagnetic radiation is calculated. This method applies to a large class of oscillatory surface integrals with complex exponential arguments, allowing us to save computer time as the surface dimensions increase. In what follows, we look for near-field solutions, which signifies that we seek for expressions which give a correct result from infinity to a short distance to the surface, when the illumination is not necessarily plane but only locally plane on the surface.

Although many authors have studied these type of problems [6]–[22], we present a novel practical solution, leading to simple non-singular expressions [23], in the case of large planar or moderately curved surfaces:

a) In monostatic near-field case, for the radiation of PO currents carried by a perfectly or imperfectly reflective surface, excited by a point source whose radiation can be of arbitrary pattern, without any approximation in the complex exponential argument of the integrand.

b) In bistatic near-field case, when we consider integrands that have complex exponential arguments that are quadratic functions of the coordinates, as for Gaussian beam propagation and scattering problems, or when the exponential argument of the Green function for the radiation of plane or curved plates is approached by second-order approximation.

c) In monostatic near-field case for the radiation of PO currents supported by curved plates whose principal curvatures have arbitrary signs for convex, concave or saddle geometries.

It is worth noticing that the functions involved in case (a) and case (b) are clearly different, even if case (b) can also be considered for a reduction in monostatic case. The case (c) profits from complex scaling transformations and exact results developed in case (b) to extend the results of case (a) to curved surfaces.

The already known methods are not general enough to solve our problems, either because they depend too closely on Helmholtz or Maxwell equations [6]–[12], they assume specific geometries [9]–[11], [13]–[15] or specific illuminations [16]–[20], or they do not allow us to consider complex exponential arguments [17] or complex surfaces due to a complex change of variables [6]–[22]; besides, they often do not propose non-singular general expressions [6], [8], [12], [19], [22], which leads to severe drawbacks, either for an exact calculus (see section 1 of [12] on the use of Asvestas works [9]–[11]) or for asymptotic derivations (see section 4 of [6]). In contrast, our global reduction approach does not assume that integrands depend on Helmholtz or Maxwell equations; it gives explicit non-singular reduction and avoids the use of special function, and it allows us to consider changes of variables with complex coordinates.

This chapter is organized as follows. After presenting the problem in section 6.2, we develop our method for an analytic contour reduction of surface integrals in section 6.3. Then we detail its application to case (a) (sections 6.4 and 6.5) and to case (b) with complex scaling (section 6.6). The use of complex scaling is then generalized in sections 6.7 and 6.8, and the results, developed initially in case (a) for plane plates, are extended to surfaces with low curvatures (case (c)) in section 6.9. The numerical comparison of classic surface integration results with our contour integral reduction is then presented in section 6.11, showing very good agreements for (a), (b) and (c) cases. Some developments are given in the appendices, concerning in particular the reductions for imperfectly reflective surfaces.

6.2. Formulation

Let S be a regular open surface of low curvature, defined, in the (x', y', z') Cartesian coordinate system of origin O', as a smooth perturbation of a surface S' of the plane $z' = 0$ with unit normal $\widehat{z}$. We search to develop a method that allows the reduction to a contour integral along $\partial S'$ of surface integral expressions of electric $\boldsymbol{E}$ and magnetic $\boldsymbol{H}$ fields in free space, given by the radiation of surface currents $\boldsymbol{J}$ and

$\boldsymbol{M}$ on S [1, section 1.26] following:

$$\boldsymbol{E}(\boldsymbol{r}) = \frac{Z_0}{4\pi ik}\int_S (\text{grad}_{\boldsymbol{r}}\text{div}_{\boldsymbol{r}} + k^2)(\psi \boldsymbol{J})dS - \frac{1}{4\pi}\int_S \text{curl}_{\boldsymbol{r}}(\psi \boldsymbol{M})dS,$$

$$Z_0\boldsymbol{H}(\boldsymbol{r}) = \frac{Z_0}{4\pi}\int_S \text{curl}_{\boldsymbol{r}}(\psi \boldsymbol{J})dS + \frac{1}{4\pi ik}\int_S (\text{grad}_{\boldsymbol{r}}\text{div}_{\boldsymbol{r}} + k^2)(\psi \boldsymbol{M})dS,$$

$$\psi(\widetilde{R}) = \frac{e^{-ik\widetilde{R}}}{\widetilde{R}}, \widetilde{R} = \sqrt{\boldsymbol{R}.\boldsymbol{R}},\ \boldsymbol{R} = \boldsymbol{r} - (\boldsymbol{r}' + z'\widehat{\boldsymbol{z}}) = \widetilde{R}\boldsymbol{u}, \tag{6.1}$$

where $\boldsymbol{r} = (\boldsymbol{a} + z\widehat{\boldsymbol{z}}) \notin S$ is the observation point,and $(\boldsymbol{r}' + z'\widehat{\boldsymbol{z}})$ is the current point of integration on S, with $z' = \zeta_s(\boldsymbol{r}')$, $\boldsymbol{r}' = x'\widehat{\boldsymbol{x}} + y'\widehat{\boldsymbol{y}} \in S'$, for a free space impedance Z_0, a wavenumber $k = \frac{2\pi}{\lambda}$ and a time dependence $e^{i\omega t}$. We then assume that $\boldsymbol{J}$ and $\boldsymbol{M}$ are analytically defined. For example, we can consider that the currents, when excited by a point source at $\boldsymbol{r}_0$, are of the form:

$$\sum_n \boldsymbol{f}_n(\boldsymbol{u}_0)\frac{e^{-ik\widetilde{R}_0}}{(k\widetilde{R}_0)^n}, \tag{6.2}$$

where $\boldsymbol{R}_0 = (\boldsymbol{r}' + z'\widehat{\boldsymbol{z}}) - \boldsymbol{r}_0, \widetilde{R}_0 = \sqrt{\boldsymbol{R}_0.\boldsymbol{R}_0}$, $\boldsymbol{u}_0 = \frac{\boldsymbol{R}_0}{\widetilde{R}_0}$ and the $\boldsymbol{f}_n$ are regular functions.

In what follows, we study the reduction of different classes of integrals derived from (6.1):

– in monostatic case, when $\boldsymbol{r} = \boldsymbol{r}_0$, and thus $\boldsymbol{R} = -\boldsymbol{R}_0$ in (6.1), for a planar (case (a)) or moderately curved (case (c)) surface without any approximation of the exponential arguments in the integrand;

– in bistatic case, when $\boldsymbol{r}$ and $\boldsymbol{r}_0$ can be distinct, for S with low curvature and complex illuminations of Gaussian type (case (b)), when the integrals can be approximated in the form:

$$\frac{e^{-ik(\widetilde{r}+\widetilde{r}_0)}}{\widetilde{r}\widetilde{r}_0}\int_{S'} e^{(\alpha_1 x'+\alpha_2 y')+(\beta_1 x'^2+\beta_2 y'^2+\beta_3 x'y')}h(\boldsymbol{r}')dS', \tag{6.3}$$

with some smooth function $h(\boldsymbol{r}')$, $\widetilde{r} = \sqrt{\boldsymbol{r}.\boldsymbol{r}}$, $\widetilde{r}_0 = \sqrt{\boldsymbol{r}_0.\boldsymbol{r}_0}$ and some complex constants α_i, β_j. In this situation, exact results, in particular when $h(\boldsymbol{r}')$ is a polynomial, are obtained.

It is worth noticing that we will not need to suppose that $\boldsymbol{J}$ and $\boldsymbol{M}$ derive from fields satisfying Helmholtz or Maxwell equations as in [6]–[12], so that we can consider a large class of functions or asymptotic approximations for the integrands.

We develop two fundamental ways of investigation for the reduction of a surface integral to a contour integral, then some basic examples illustrating their complementarity, which subsequently leads us to efficient solutions for our problems. For that, complex changes of variables, and thus domains of integration with complex coordinates will be considered for the reduction.

6.3. A combination of two fundamental ways for an efficient reduction to a contour integral

The reduction that we propose is carried out by a correct combination of two fundamental ways of investigation and associated general basic results that are perfectly adapted to solve the class of problems that we study in a simple manner.

DEFINITION 6.1.– *Let the plane* $z = 0$ *be defined with unit normal* $\widehat{\boldsymbol{z}}$, *and a unit standard basis composed of* $\widehat{\boldsymbol{x}}$ *and* $\widehat{\boldsymbol{y}} = \widehat{\boldsymbol{z}} \times \widehat{\boldsymbol{x}}$. *Each point is denoted as* $\boldsymbol{r}' = x'\widehat{\boldsymbol{x}} + y'\widehat{\boldsymbol{y}}$ *in* (x', y') *coordinates. In the applications, analytical changes of variables can transform the initial domain of integration with real coordinates into a domain with complex coordinates. The dot and cross products of vectors with complex coordinates are here considered as in real space (without complex conjugation), following:*

$$\boldsymbol{u}.\boldsymbol{v} = u_x v_x + u_y v_y,\ (\boldsymbol{u})^2 = \boldsymbol{u}\,.\boldsymbol{u} = u_x u_x + u_y u_y,$$
$$\boldsymbol{u} \times \boldsymbol{v} = (u_x v_y - u_y v_x)\widehat{\boldsymbol{z}},\ \widehat{\boldsymbol{z}} \times \boldsymbol{u} = -u_y\widehat{\boldsymbol{x}} + u_x\widehat{\boldsymbol{y}}, \tag{6.4}$$

where $\boldsymbol{u} = u_x\widehat{\boldsymbol{x}} + u_y\widehat{\boldsymbol{y}}$ *and* $\boldsymbol{v} = v_x\widehat{\boldsymbol{x}} + v_y\widehat{\boldsymbol{y}}$ *express with complex numbers* $u_{x(y)}$ *and* $v_{x(y)}$. *In this situation, we assume that* S' *is a subset of a complex manifold* S'_0 *of the plane* $z = 0$, *both simply connected and oriented. The domains* S'_0 *and* S' *both result from the transformation of simply connected (two-dimensional) real manifolds of the plane* $z = 0$ *with contours positively oriented, by an invertible function* P, *that we suppose thereafter linear to simplify, with oriented contours both resulting from the transform by* P *of the real contours. Letting* $P(\boldsymbol{u}) = P_x(\boldsymbol{u})\widehat{\boldsymbol{x}} + P_y(\boldsymbol{u})\widehat{\boldsymbol{y}}$, *we notice that* $P(\boldsymbol{u}).P(\boldsymbol{u}) = \prod_{\pm}(P_x(\boldsymbol{u}) \pm iP_y(\boldsymbol{u}))$.

NOTE 6.1.– On this occasion, it is worth noticing that a linear invertible transform of a (closed) simply connected domain is also a (closed) simply connected domain (note: for an injective continuous transform, see the Browner's domain invariance theorem).

NOTE 6.2.– If we remove a simply connected subset from a simply connected domain, we create a hole, but this new domain remains the union of two simply connected domains. Consequently, it can be assumed that our results on reduction of integrals on simply connected surfaces, described in what follows, can be extended to multiply connected domains of integration.

6.3.1. *First way*

6.3.1.1. *Some elements of analysis in real space*

Let S' be first a real domain in cylindrical coordinates (ρ, φ) and $F(\rho^2, \varphi)$ be a function which is regular on any straight line from the origin to any point of real S'; we can write:

$$\int_{S'} \frac{\partial F(\rho^2, \varphi)}{\partial(\rho^2)} dS' = \int_{\mathcal{L}} \int_{0+}^{\rho(\varphi)} \frac{\partial F(\rho^2, \varphi)}{\partial(\rho^2)} \rho d\rho d\varphi = \frac{1}{2} \int_{\mathcal{L}} (F(\rho^2, \varphi) - F(0, \varphi)) d\varphi, \tag{6.5}$$

where $\mathcal{L}$ is positively oriented. In this expression, $d\varphi = (\widehat{\boldsymbol{z}} \times \frac{\overline{\rho}}{\rho^2}).d\overline{\rho}$ with $\overline{\rho}$ the current point of integration. In another manner, we can recover it from the Stokes theorem [1, equ. (34) chap.2.9] using:

$$\begin{aligned} &\text{curl}(\widehat{\boldsymbol{z}} \times \frac{\overline{\rho}}{\rho^2}(F(\rho^2,\varphi) - F(0,\varphi))) = \widehat{\boldsymbol{z}}\, \text{div}(\frac{\overline{\rho}}{\rho^2}(F(\rho^2,\varphi) - F(0,\varphi))) \\ &= \widehat{\boldsymbol{z}}\, \frac{2\partial((\overline{\rho}.\frac{\overline{\rho}}{\rho^2})(F(\rho^2,\varphi) - F(0,\varphi)))}{\partial(\rho^2)} = \widehat{\boldsymbol{z}}\, \frac{2\partial(F(\rho^2,\varphi))}{\partial(\rho^2)}, \end{aligned} \tag{6.6}$$

when $F(\rho^2,\varphi) - F(0,\varphi)$ is analytic on S', so that we have:

$$\int_{S'} \frac{\partial(F(\rho^2,\varphi))}{\partial(\rho^2)} dS' = \frac{1}{2} \int_{\mathcal{L}} (F(\rho^2,\varphi) - F(0,\varphi))(\widehat{\boldsymbol{z}} \times \frac{\overline{\rho}}{\rho^2}).d\overline{\rho}. \tag{6.7}$$

We note that, shifting the origin to a real point $\boldsymbol{a}$ (Figure 6.1), with $\overline{\rho} = \boldsymbol{r'} - \boldsymbol{a}$, $\frac{\boldsymbol{r'}-\boldsymbol{a}}{\sqrt{(\boldsymbol{r'}-\boldsymbol{a})^2}} = \widehat{\boldsymbol{x}} \cos\varphi + \widehat{\boldsymbol{y}} \sin\varphi$, we can write, when $F((\boldsymbol{r'} - \boldsymbol{a})^2,\varphi) - F(0,\varphi)$ remains analytic on S':

$$\int_{S'} \frac{\partial(F((\boldsymbol{r'} - \boldsymbol{a})^2,\varphi))}{\partial((\boldsymbol{r'} - \boldsymbol{a})^2)} dS' = \frac{1}{2} \int_{\mathcal{L}} (F((\boldsymbol{r'} - \boldsymbol{a})^2,\varphi) - F(0,\varphi)) \frac{\widehat{\boldsymbol{z}} \times (\boldsymbol{r'} - \boldsymbol{a})}{(\boldsymbol{r'} - \boldsymbol{a})^2} .d\boldsymbol{r'}, \tag{6.8}$$

with $\int_{\mathcal{L}} d\varphi = 2\pi$ if $\boldsymbol{a} \in S'$ and $\int_{\mathcal{L}} d\varphi = 0$ if $\boldsymbol{a} \notin S'$, for real S' and $\boldsymbol{a}$.

6.3.1.2. *Reduction in complex coordinates*

The Stokes theorem, usually used in a domain with real coordinates, can be extended to domains with complex coordinates [24, p.213] by the formalism of exterior differential forms on complex manifolds, letting $dS' = (dx' \wedge dy')$ [25], [26, chap.7], [27, chap.3]. Considering the definition 6.1, S' is now a subset of a complex manifold S'_0 of the plane $z = 0$, both simply connected, resulting from a linear invertible transform of real connected domains with positively oriented contours.

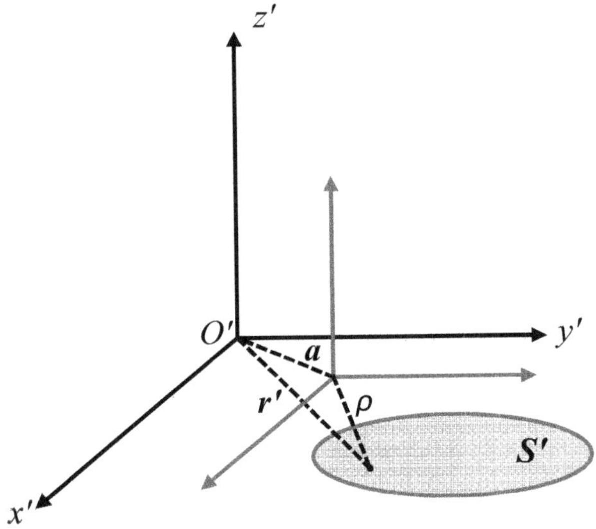

Figure 6.1. *Geometry and shift of the origin*

DEFINITION 6.2.– *Let* $\boldsymbol{r}' = x'\widehat{\boldsymbol{x}} + y'\widehat{\boldsymbol{y}}$ *denote a point of the four-dimensional plane with complex coordinates* x' *and* y', *and* $\boldsymbol{a} = a_x\widehat{\boldsymbol{x}} + a_y\widehat{\boldsymbol{y}}$ *denote a fixed complex point, with* $\widehat{\boldsymbol{z}} \times \widehat{\boldsymbol{x}} = \widehat{\boldsymbol{y}}$ *and* $\widehat{\boldsymbol{y}} \times \widehat{\boldsymbol{z}} = \widehat{\boldsymbol{x}}$, *we can write:*

$$z'_\pm = x' \pm iy' - (a_x \pm ia_y) = (x' - a_x) \pm i(y' - a_y) = (\boldsymbol{r}' - \boldsymbol{a}).(\widehat{\boldsymbol{x}} \pm i\widehat{\boldsymbol{y}}),$$

$$e^{\pm i\varphi} = \frac{z'_\pm}{(z'_+ z'_-)^{\frac{1}{2}}}, \varphi = \mp i \ln(\frac{z_\pm}{(z'_+ z'_-)^{\frac{1}{2}}}), (\boldsymbol{r}' - \boldsymbol{a})^2 = z'_+ z'_- = (z'_\pm e^{\mp i\varphi})^2,$$

$$\cos\varphi = \frac{1}{2}(\frac{z_\pm}{(z'_+ z'_-)^{\frac{1}{2}}} + (\frac{z_\pm}{(z'_+ z'_-)^{\frac{1}{2}}})^{-1}), \sin\varphi = \frac{1}{2i}(\frac{z_\pm}{(z'_+ z'_-)^{\frac{1}{2}}} - (\frac{z_\pm}{(z'_+ z'_-)^{\frac{1}{2}}})^{-1}),$$

$$\boldsymbol{r}' - \boldsymbol{a} = \frac{(z'_+ + z'_-)}{2}\widehat{\boldsymbol{x}} + \frac{(z'_+ - z'_-)}{2i}\widehat{\boldsymbol{y}} = z'_\pm e^{\mp i\varphi}(\cos\varphi\, \widehat{\boldsymbol{x}} + \sin\varphi\, \widehat{\boldsymbol{y}}), \tag{6.9}$$

where the functions $(.)^{1/2}$ *and* $\ln(.)$ *are analytical, apart from branch cuts, and the numbers* $z'_+ z'_-$ *and* φ *can be complex. We notice that* $z_+ = z_- = 0$ *only if* $\boldsymbol{r}' = \boldsymbol{a}$, *but that, in complex coordinates, the set of points satisfying* $(\boldsymbol{r}' - \boldsymbol{a})^2 = 0$ *are the two 2-dimensional straight lines verifying:*

$$\boldsymbol{r}' = \boldsymbol{a} + t(\widehat{\boldsymbol{x}} \pm i\widehat{\boldsymbol{y}}) \text{ where } z_\pm = 0, \text{ with arbitrary complex number } t,$$

$$\text{while } \widehat{\boldsymbol{x}} \pm i\widehat{\boldsymbol{y}} = \frac{\mathbf{c} \pm i\widehat{\boldsymbol{z}} \times \mathbf{c}}{c_x \mp ic_y} \text{ for any complex vector } \mathbf{c} \text{ with } \mathbf{c}^2 \neq 0. \tag{6.10}$$

(note: if $(\boldsymbol{r'} - \boldsymbol{a})^2 \to 0$ while $\boldsymbol{r'}$ does not tend to $\boldsymbol{a}$, $\cos\varphi$ and $\sin\varphi$ tend to infinity). We also note that we have:

$$d\varphi = \frac{-i}{2} d\ln(\frac{z'_+}{z'_-}) = \frac{-i}{2}(\frac{dz'_+}{z'_+} - \frac{dz'_-}{z'_-}) = (\widehat{\boldsymbol{z}} \times \frac{\boldsymbol{r'} - \boldsymbol{a}}{(\boldsymbol{r'} - \boldsymbol{a})^2}).d\boldsymbol{r'}, \tag{6.11}$$

where the logarithm function has to be considered apart from its branch cut, or, more generally, by analytic continuation across it. The term $d\varphi$ is independent of the choice of $\widehat{\boldsymbol{x}}$ and $\widehat{\boldsymbol{y}} = \widehat{\boldsymbol{z}} \times \widehat{\boldsymbol{x}}$, and it vanishes along any line $\boldsymbol{r'} = \boldsymbol{a} + t\boldsymbol{v}$, with variable t and constant $\boldsymbol{v}$, so that φ remains identical for $\boldsymbol{r'} - \boldsymbol{a} = t\boldsymbol{v}$ as t varies. On this occasion, we also remark that $\frac{z'_+}{z'_-}$ describes a closed contour along any closed path in the (x', y')-plane.

DEFINITION 6.3.– *Considering $\boldsymbol{a} = a_x\widehat{\boldsymbol{x}} + a_y\widehat{\boldsymbol{y}}$ a complex point of the plane, a function $F((\boldsymbol{r'} - \boldsymbol{a})^2, \varphi)$ is said to be c-regular relatively to $\boldsymbol{r'} = x'\widehat{\boldsymbol{x}} + y'\widehat{\boldsymbol{y}}$ on S'_0 ('c' for "centered"), if there exists a four-dimensional complex vicinity of S'_0, where $F((\boldsymbol{r'} - \boldsymbol{a})^2, \varphi) - F(0, \varphi)$ is analytic relatively to $((\boldsymbol{r'} - \boldsymbol{a})^2)^{\frac{1}{2}} \cos\varphi = (x' - a_x)$ and $((\boldsymbol{r'} - \boldsymbol{a})^2)^{\frac{1}{2}} \sin\varphi = (y' - a_y)$, and vanishes when $(\boldsymbol{r'} - \boldsymbol{a})^2 \to 0$, while $\frac{\partial(F((\boldsymbol{r'}-\boldsymbol{a})^2,\varphi))}{\partial((\boldsymbol{r'}-\boldsymbol{a})^2)}$ is bounded. Noticing that the set of points verifying $(\boldsymbol{r'} - \boldsymbol{a})^2 = 0$ is only a two-dimensional manifold (see definition 6.2), $\frac{\partial(F((\boldsymbol{r'}-\boldsymbol{a})^2,\varphi))}{\partial((\boldsymbol{r'}-\boldsymbol{a})^2)}$ is also analytic relatively to $(x' - a_x)$ and $(y' - a_y)$ on S'_0 [28, sect. 23.2].*

Let us assume that $F((\boldsymbol{r'} - \boldsymbol{a})^2, \varphi)$ is c-regular on S'_0 for any complex point $\boldsymbol{a}$ belonging to it. We define the condition of continuation for $F((\boldsymbol{r'} - \boldsymbol{a})^2, \varphi)$ relatively to $\boldsymbol{a}$, from $\boldsymbol{a} = \boldsymbol{a'}$ to S'_0, as follows: it exists an open connected finite domain $\mathcal{H}$ of complex points $\boldsymbol{a}$ (four-dimensional set) including $\boldsymbol{a'}$ and a point of S'_0, where F remains c-regular on S'_0.

NOTE 6.3.– The functions $(\boldsymbol{r'} - \boldsymbol{a})^2(x' - a)^n$ and $(\boldsymbol{r'} - \boldsymbol{a})^2(y' - a)^n$, $n \geq 0$, are c-regular, and their derivatives relatively to $(\boldsymbol{r'} - \boldsymbol{a})^2$ are $(1 + n/2)(x' - a)^n$ and $(1 + n/2)(y' - a)^n$.

NOTE 6.4.– If $F(t, s)$ is an entire function of t independent of s, then the c-regularity on S'_0 is verified for any $\boldsymbol{a}$. If $F((\boldsymbol{r'} - \boldsymbol{a})^2, \varphi)$ is c-regular on S'_0, then it is also the case for $(\boldsymbol{r'} - \boldsymbol{a})^2(F((\boldsymbol{r'} - \boldsymbol{a})^2, \varphi) - F(0, \varphi))$, and for any primitive $\int^{(\boldsymbol{r'}-\boldsymbol{a})^2}(F(t, \varphi) - F(0, \varphi))dt$, definite on S'_0 and at points with $(\boldsymbol{r'} - \boldsymbol{a})^2 = 0$.

THEOREM 6.1.– *Let S' be a subdomain of a complex manifold S'_0, both simply connected oriented complex manifolds, where S'_0 and thus S' result from a linear invertible transform of a real simply connected domain with positively oriented contours, as indicated in definition 6.1, with $\mathcal{L}$ the oriented contour of S' resulting from this transformation. We consider $F((\boldsymbol{r'} - \boldsymbol{a})^2, \varphi)$ as a complex function satisfying* *the c-regularity on S'_0, described in definition 6.3, for any point $\boldsymbol{a}$ belonging to it. The equality:*

$$\int_{S'} \frac{\partial(F((\boldsymbol{r'}-\boldsymbol{a})^2,\varphi))}{\partial((\boldsymbol{r'}-\boldsymbol{a})^2)} dS' = \frac{1}{2}\int_{\mathcal{L}} (F((\boldsymbol{r'}-\boldsymbol{a})^2,\varphi) - F(0,\varphi)) \frac{\widehat{\boldsymbol{z}} \times (\boldsymbol{r'}-\boldsymbol{a})}{(\boldsymbol{r'}-\boldsymbol{a})^2} .d\boldsymbol{r'}, \tag{6.12}$$

applies for $\boldsymbol{a} \in S'_0$, and for any $\boldsymbol{a}$ verifying the condition of continuation in $\mathcal{H}$ to S'_0 given in definition 6.3 (note: $\boldsymbol{r'}$ denotes $x'\widehat{\boldsymbol{x}} + y'\widehat{\boldsymbol{y}}$ with $\widehat{\boldsymbol{y}} = \widehat{\boldsymbol{z}} \times \widehat{\boldsymbol{x}}$ and $\widehat{\boldsymbol{x}} = \widehat{\boldsymbol{y}} \times \widehat{\boldsymbol{z}}$). This equality is based on analytical continuation from real to complex coordinates, where it is worth noticing that $(\boldsymbol{r'}-\boldsymbol{a})^2 = 0$ does no more imply $\boldsymbol{r'} = \boldsymbol{a}$.

PROOF.– The equality (6.12) corresponds to a continuation of (6.8) for points with complex coordinates in the domain of regularity of F. We first state (6.12) for $\boldsymbol{a} \in S'_0$. From the properties of the exterior product and the definition of z'_+ and z'_- in (6.9), we can write:

$$dS' = (dx' \wedge dy') = \frac{i}{2}(dz'_+ \wedge dz'_-) = \frac{-i}{4}(z'_- dz'_+ + z'_+ dz'_-) \wedge (\frac{dz'_+}{z'_+} - \frac{dz'_-}{z'_-})$$

$$= \frac{1}{2} d(z'_+ z'_-) \wedge \frac{-i}{2} d\ln(\frac{z'_+}{z'_-}) = \frac{1}{2} d(\boldsymbol{r'}-\boldsymbol{a})^2 \wedge d\varphi, \tag{6.13}$$

From definition 6.1, S'_0 results from a linear transformation of a simply connected real manifold relatively to (x', y') coordinates, and when $\boldsymbol{a} \in S'_0$, it can be mapped from $\boldsymbol{a}$ to ∞ at constant φ with straight semi-lines, except on some cut(s) where φ is constant on both sides. In this case, we can integrate in the integral on S' relatively to $(\boldsymbol{r'}-\boldsymbol{a})^2$ variable from (6.13) and derive the contour integral along $\mathcal{L}$ given in (6.12). Let us notice that when $\boldsymbol{a} \in S'$, we integrate from $\boldsymbol{r'} = \boldsymbol{a}$ to a point of $\mathcal{L}$ at constant φ, while, when $\boldsymbol{a} \in S'_0$ but $\boldsymbol{a} \notin S'$, the limits of integration are two points of $\mathcal{L}$ at constant φ.

From the condition of continuation given in definition 6.3, we can extend the validity of this equality to any point $\boldsymbol{a} = \boldsymbol{a}'$ belonging to the domain $\mathcal{H}$, which comprises a subset $S'_{\boldsymbol{a}}$ of S'_0. By definition of the c-regularity, the cuts in φ continue to have no influence; the left-hand side of (6.12) is analytic relatively to complex coordinates a_x and a_y; the right-hand side, continuous at any point of $\mathcal{H}$, is analytic when $(\boldsymbol{r'}-\boldsymbol{a})^2 \neq 0$ on $\mathcal{L}$. That restriction disappears from the generalized principle of continuation [28, sect. 23.2], and the analyticity extends at any point $\boldsymbol{a}$ of $\mathcal{H}$ by noticing that $(\boldsymbol{r'}-\boldsymbol{a})^2|_{\boldsymbol{r'}\in\mathcal{L}} = 0$ describes a piecewise analytic p-dimensional manifold with $p \leq 3$. Therefore, considering that we have the analyticity in $\mathcal{H}$ and that (6.12) is stated on $S'_{\boldsymbol{a}}$, we can use the identity theorem [28, sect. 6.1] and conclude the validity of (6.12) for any point of $\mathcal{H}$.

Since the integration in (6.12) is on φ variable, the part of the contour integral, restricted to the factor $F(0, \varphi)$, could express in a simple form, and we now analyze it when $F(0, \varphi) \equiv 1$.

PROPOSITION 6.1.– *Let $\mathcal{D}_1$ be a smooth semi-line of the complex plane from the origin to ∞, and $\mathcal{L}$ is a closed oriented contour of points $\boldsymbol{r'} = x'\widehat{\boldsymbol{x}} + y'\widehat{\boldsymbol{y}}$ with complex coordinates x' and y'. We have, when $(\boldsymbol{r'} - \boldsymbol{a})^2 \neq 0$ on $\mathcal{L}$:*

$$\int_{\mathcal{L}} (\widehat{\boldsymbol{z}} \times \frac{\boldsymbol{r'} - \boldsymbol{a}}{(\boldsymbol{r'} - \boldsymbol{a})^2}).d\boldsymbol{r'} = \frac{-i}{2} \int_{\mathcal{L}} d\ln(\frac{z'_+}{z'_-}) = \pi(N_+ - N_-), \tag{6.14}$$

where N_+ (respectively, N_-) is the number of times that the complex $\frac{z'_+}{z'_-}$ crosses $\mathcal{D}_1$ in the positive sense (respectively, negative). For a complex surface S' with complex contour $\mathcal{L}$ verifying definition 6.1, that is, resulting from an invertible linear transformation of a real simply connected surface with a positively oriented contour, we can write, when $(\boldsymbol{r'} - \boldsymbol{a})^2 \neq 0$ on $\mathcal{L}$:

$$\int_{\mathcal{L}} (\widehat{\boldsymbol{z}} \times \frac{\boldsymbol{r'} - \boldsymbol{a}}{(\boldsymbol{r'} - \boldsymbol{a})^2}).d\boldsymbol{r'} = \pi(\epsilon^+ 1_{S'_+}(a_x + ia_y) + \epsilon^- 1_{S'_-}(a_x - ia_y)). \tag{6.15}$$

Letting $x'_\pm$ =Re$(x' \pm iy')$, $y'_\pm$ =Im$(x' \pm iy')$, and noting that $x'_\pm + iy'_\pm = x' \pm iy'$, the functions $1_{S'_\pm}$ are the indicator functions of the domains $S'_\pm$ of the complex plane, defined following:

$$\begin{aligned} &1_{S'_\pm}(z) = \frac{1}{2\pi i(\pm\epsilon^\pm)} \int_{\mathcal{L}_\pm} \frac{d(x'_\pm + iy'_\pm)}{x'_\pm + iy'_\pm - z} = 1 \textit{ if } z \in S'_\pm, = 0 \textit{ if } z \notin S'_\pm, \\ &S'_\pm = \{z, 1_{S'_\pm}(z) = 1\} \equiv \{z = x' \pm iy', M(x', y') \in S'\}, \end{aligned} \tag{6.16}$$

where $\mathcal{L}_\pm$ and $\epsilon^\pm$ verify:

$$\begin{aligned} &\mathcal{L}_\pm = \{z = x'_\pm + iy'_\pm (= x' \pm iy'), \textit{ as } M(x', y') \textit{ follows the contour } \mathcal{L}\}, \\ &\pm\epsilon^\pm = 1 \textit{ (resp. } \pm\epsilon^\pm = -1) \textit{ for } \mathcal{L}_\pm \textit{ positively (respectively, negatively) oriented.} \end{aligned} \tag{6.17}$$

Let us note that, for a real positively oriented contour $\mathcal{L}$ and a point $\boldsymbol{a}$ with real coordinates, we have $1_{S'_+}(a_x + ia_y) = 1_{S'_-}(a_x - ia_y)$, while the contours $\mathcal{L}_+$ and $\mathcal{L}_-$ are, respectively, positively and negatively oriented, so that $\epsilon^\pm = 1$.

PROOF.– From (6.11), we have, when $(\boldsymbol{r'} - \boldsymbol{a})^2 \neq 0$ on $\mathcal{L}$:

$$\int_{\mathcal{L}} (\widehat{\boldsymbol{z}} \times \frac{\boldsymbol{r'} - \boldsymbol{a}}{(\boldsymbol{r'} - \boldsymbol{a})^2}).d\boldsymbol{r'} = \frac{-i}{2} \int_{\mathcal{L}} d\ln(\frac{z'_+}{z'_-}), \tag{6.18}$$

where the logarithm function is considered by analytic continuation along the closed path covered by the uniform function $\frac{z'_+}{z'_-}$ on $\mathcal{L}$. We then obtain (6.14), since

$(N_+ - N_-)$ is the number of turns done by $\frac{z'_+}{z'_-}$ around the origin, along the closed contour $\mathcal{L}$. For (6.15), (6.16) and (6.17), we write:

$$\int_{\mathcal{L}} (\widehat{\boldsymbol{z}} \times \frac{\boldsymbol{r}' - \boldsymbol{a}}{(\boldsymbol{r}' - \boldsymbol{a})^2}) d\boldsymbol{r}' = \frac{-i}{2} \int_{\mathcal{L}} (\frac{dz'_+}{z'_+} - \frac{dz'_-}{z'_-})$$
$$= \frac{1}{2i} \int_{\mathcal{L}_+} \frac{d(x' + iy')}{x' + iy' - (a_x + ia_y)} - \frac{1}{2i} \int_{\mathcal{L}_-} \frac{d(x' - iy')}{x' - iy' - (a_x - ia_y)}$$
$$= \epsilon^+ \pi 1_{S'_+}(a_x + ia_y) + \epsilon^- \pi 1_{S'_-}(a_x - ia_y), \tag{6.19}$$

where the residue theorem is used, on the oriented complex contours $\mathcal{L}_\pm = \{x' \pm iy'$, as $M(x', y')$ follows the oriented contour $\mathcal{L}\}$, which limit the complex connected domains $S'_\pm = \{x' \pm iy',\ M(x', y') \in S'\}$, noticing that $x' \pm iy'$ =Re$(x' \pm iy')$ + iIm$(x' \pm iy')$.

NOTE 6.5.– A linear invertible transform of a (closed) simply connected domain is a (closed) simply connected domain (note: more generally, for an injective continuous transform, see the Browner's domain invariance theorem).

PROPOSITION 6.2.– *Let $\mathcal{W}$ be an open simply connected subset of the domain of analyticity of:*

$$\frac{\partial F((\boldsymbol{r}' - \boldsymbol{a})^2, \varphi)}{\partial (\boldsymbol{r}' - \boldsymbol{a})^2} = s(\boldsymbol{r}'), \tag{6.20}$$

relatively to the complex coordinates of $\boldsymbol{r}' = x'\widehat{\boldsymbol{x}} + y'\widehat{\boldsymbol{y}}$ that includes $\boldsymbol{r}' = \boldsymbol{a}$ and S'. Considering two points $\boldsymbol{r}'_{1,2}$ such that $(\boldsymbol{r}'_1 - \boldsymbol{a}) = t_1(\boldsymbol{r}'_2 - \boldsymbol{a})$ with $(\boldsymbol{r}'_2 - \boldsymbol{a})^2 \neq 0$, we suppose that it exists a complex path $\mathcal{T}_{\mathcal{L}}$ from $t = t_1$ to $t = 1$, defined such that $\boldsymbol{r}' = t(\boldsymbol{r}'_2 - \boldsymbol{a}) + \boldsymbol{a}$ remains in $\mathcal{W}$, for which, by definition, the complex angle φ of $(\boldsymbol{r}' - \boldsymbol{a})$ remains the same as t varies (see definition 6.2). Therefore, by analytic continuation in an open simply connected subset, we can integrate $s(\boldsymbol{r}')$ relatively to $(\boldsymbol{r}' - \boldsymbol{a})^2$ at constant φ along $\mathcal{T}_{\mathcal{L}}$ and write:

$$F((\boldsymbol{r}'_2 - \boldsymbol{a})^2, \varphi) - F((\boldsymbol{r}'_1 - \boldsymbol{a})^2, \varphi) = 2(\boldsymbol{r}'_2 - \boldsymbol{a})^2 \int_{\mathcal{T}_{\mathcal{L}}} s(t(\boldsymbol{r}'_2 - \boldsymbol{a}) + \boldsymbol{a})\, t dt \tag{6.21}$$

which could be used in (6.12) of theorem 6.1, when $\boldsymbol{r}'_1 = \boldsymbol{a}$.

PROOF.– We integrate s along $\mathcal{T}_{\mathcal{L}}$, relatively to $(\boldsymbol{r}' - \boldsymbol{a})^2$ variable at constant φ, by letting $\boldsymbol{r}' - \boldsymbol{a} = t(\boldsymbol{r}'_2 - \boldsymbol{a})$ and $d(\boldsymbol{r}' - \boldsymbol{a})^2 = 2(\boldsymbol{r}'_2 - \boldsymbol{a})^2 t dt$.

NOTE 6.6.– Let us notice that a complex anisotropic scaling can lead us to work on a complex manifold. If we dilate a real surface S' by a factor $p \neq 0$ along the direction $\boldsymbol{c}_p$ with:

$$\mathcal{P}: \boldsymbol{r}' \rightarrow \boldsymbol{r}'_p = x'_p \widehat{\boldsymbol{c}}_p + y'_p(\widehat{\boldsymbol{z}} \times \widehat{\boldsymbol{c}}_p),\; x'_p = p\,(\boldsymbol{r}'.\widehat{\boldsymbol{c}}_p),\; y'_p = \boldsymbol{r}'.(\widehat{\boldsymbol{z}} \times \widehat{\boldsymbol{c}}_p), \quad (6.22)$$

where $\widehat{\boldsymbol{c}}_p = \frac{\boldsymbol{c}_p}{\sqrt{\boldsymbol{c}_p.\boldsymbol{c}_p}}$ (with $\boldsymbol{c}_p^2 \neq 0$ and $p \neq 0$), we obtain a complex surface S'_p, with:

$$dS'_p = dx'_p \wedge dy'_p = p\, dx' \wedge dy' = p\, dS',$$
$$\boldsymbol{v}.\mathcal{P}(\boldsymbol{r}') = \mathcal{P}(\boldsymbol{v}).\boldsymbol{r}',\; (\widehat{\boldsymbol{z}} \times \mathcal{P}(\boldsymbol{v})).\mathcal{P}(\boldsymbol{r}') = (\widehat{\boldsymbol{z}} \times p\boldsymbol{v}).\boldsymbol{r}'. \quad (6.23)$$

NOTE 6.7.– The choice to orient (or not) a domain stands out from the following equalities:

$$\int_{\chi([a,b])} f(x)dx = \int_{[a,b]} f(\chi(t))|\chi'(t)|dt,\; \int_{\chi(a)}^{\chi(b)} f(x)dx = \int_a^b f(\chi(t))\chi'(t)dt, \quad (6.24)$$

by noticing that the first one, contrarily to the second, cannot be applied to complex a and b.

NOTE 6.8.– Let us notice that on each straight part of $\mathcal{L}$ along $\widehat{\boldsymbol{v}}$, when $d\boldsymbol{r}' = \widehat{\boldsymbol{v}}ds$ and $\boldsymbol{r}' - \boldsymbol{a} = \boldsymbol{r}'_0 + \widehat{\boldsymbol{v}}s$ with constant $\boldsymbol{r}'_0$, we have $(\widehat{\boldsymbol{z}} \times (\boldsymbol{r}' - \boldsymbol{a})).d\boldsymbol{r}' = \boldsymbol{g}_0 ds$, with $\boldsymbol{g}_0 = \widehat{\boldsymbol{z}} \times \boldsymbol{r}'_0$.

6.3.2. *Second way*

We consider here, as in previous section, the formalism of exterior differential forms, with $dS' = (dx' \wedge dy')$ [25]–[26, chap.7].

THEOREM 6.2.– *Let S' be a simply connected oriented complex manifold of complex points $\boldsymbol{r}' = x'\widehat{\boldsymbol{x}} + y'\widehat{\boldsymbol{y}}$, satisfying the definition 6.1 and limited by the oriented contour $\mathcal{L}$, and a complex constant point $\boldsymbol{a} = a_x\widehat{\boldsymbol{x}} + a_y\widehat{\boldsymbol{y}}$. We consider a function $F((\boldsymbol{r}' - \boldsymbol{a})^2,\varphi)$, analytic relatively to $((\boldsymbol{r}' - \boldsymbol{a})^2)^{\frac{1}{2}} \cos\varphi = (x' - a_x)$ and $((\boldsymbol{r}' - \boldsymbol{a})^2)^{\frac{1}{2}} \sin\varphi = (y' - a_y)$ on S'. We can write:*

$$\int_{S'} \boldsymbol{b}.grad_{\boldsymbol{r}'}(F((\boldsymbol{r}' - \boldsymbol{a})^2,\varphi) - A_0)dS' = \int_{\mathcal{L}} (F((\boldsymbol{r}' - \boldsymbol{a})^2,\varphi) - A_0)(\widehat{\boldsymbol{z}} \times \boldsymbol{b}).d\boldsymbol{r}', \quad (6.25)$$

where $\boldsymbol{b}$ and A_0 are arbitrary constants, $grad_{\boldsymbol{r}'}$ is the gradient restricted to complex (x', y') coordinates of the plane $z = 0$ and the cut in φ can be ignored. From:

$$grad_{\boldsymbol{r}'}(.) = 2(\boldsymbol{r}' - \boldsymbol{a})\frac{\partial(.)}{\partial((\boldsymbol{r}' - \boldsymbol{a})^2)} + \frac{(\widehat{\boldsymbol{z}} \times (\boldsymbol{r}' - \boldsymbol{a}))}{(\boldsymbol{r}' - \boldsymbol{a})^2}\frac{\partial(.)}{\partial\varphi}, \quad (6.26)$$

we then have

$$\int_{S'}(\boldsymbol{r'}-\boldsymbol{a}).(\boldsymbol{b}\frac{\partial(F((\boldsymbol{r'}-\boldsymbol{a})^2,\varphi))}{\partial((\boldsymbol{r'}-\boldsymbol{a})^2)}-\frac{1}{2}\frac{(\widehat{\boldsymbol{z}}\times\boldsymbol{b})}{(\boldsymbol{r'}-\boldsymbol{a})^2}\frac{\partial(F((\boldsymbol{r'}-\boldsymbol{a})^2,\varphi))}{\partial\varphi})dS'$$

$$=\frac{1}{2}\int_{\mathcal{L}}(F((\boldsymbol{r'}-\boldsymbol{a})^2,\varphi)-A_0)(\widehat{\boldsymbol{z}}\times\boldsymbol{b}).d\boldsymbol{r'}. \qquad (6.27)$$

PROOF.– In real coordinates, we could use the Stokes theorem [1, chap.2.9], in combination with $\text{grad}(F((\boldsymbol{r'}-\boldsymbol{a})^2,\varphi))\times\widehat{\boldsymbol{d}}$ $=\text{curl}\,(\widehat{\boldsymbol{d}}F((\boldsymbol{r'}-\boldsymbol{a})^2,\varphi))$, where $\widehat{\boldsymbol{d}}$ is a constant vector. Alternatively, in the formalism of exterior forms for complex manifolds, we can choose to use gradient in Cartesian coordinates with $dS'=dx'\wedge dy'=-dy'\wedge dx'$ following:

$$\int_{S'}b.\text{grad}_{\boldsymbol{r'}}(F)dS'=\int_{S'}b.\widehat{\boldsymbol{x}}\frac{\partial F}{\partial x'}dx'\wedge dy'-\int_{S'}b.\widehat{\boldsymbol{y}}\frac{\partial F}{\partial y'}dy'\wedge dx'. \qquad (6.28)$$

From the analyticity of F relatively to x' and y', the cut in φ is omitted. We integrate $\frac{\partial F}{\partial x'}dx'$ and $\frac{\partial F}{\partial y'}dy'$ parts, while noticing $(\widehat{\boldsymbol{z}}\times b).d\boldsymbol{r'}=b.\widehat{\boldsymbol{x}}dy'-b.\widehat{\boldsymbol{y}}dx'$ and obtain (6.25). Then, from:

$$dF=\frac{\partial F}{\partial(\boldsymbol{r'}-\boldsymbol{a})^2}d(\boldsymbol{r'}-\boldsymbol{a})^2+\frac{\partial F}{\partial\varphi}d\varphi=\text{grad}_{\boldsymbol{r'}}(F).d(\boldsymbol{r'}-\boldsymbol{a}),$$

$$d(\boldsymbol{r'}-\boldsymbol{a})=\frac{1}{2}\frac{\boldsymbol{r'}-\boldsymbol{a}}{(\boldsymbol{r'}-\boldsymbol{a})^2}d(\boldsymbol{r'}-\boldsymbol{a})^2+(\widehat{\boldsymbol{z}}\times(\boldsymbol{r'}-\boldsymbol{a}))d\varphi, \qquad (6.29)$$

we derive (6.26), which, combined with (6.25), gives (6.27).

NOTE 6.9.– We can simply recover the Cauchy integral theorem from (6.25). Letting $\epsilon=+1$ or -1, we take $\boldsymbol{b}=\frac{1}{2}(\widehat{\boldsymbol{x}}+i\epsilon\widehat{\boldsymbol{y}})$ and $A_0=0$ in (6.25), with $F((\boldsymbol{r'}-\boldsymbol{a})^2,\varphi)=u(z_+,z_-)$ analytical on S'. We can use that $\boldsymbol{b}.\text{grad}_{\boldsymbol{r'}}\equiv\frac{1}{2}(\frac{\partial}{\partial x'}+i\epsilon\frac{\partial}{\partial y'})$, $(\widehat{\boldsymbol{z}}\times\boldsymbol{b})=-\frac{i\epsilon}{2}(\widehat{\boldsymbol{x}}+i\epsilon\widehat{\boldsymbol{y}})$, $(\widehat{\boldsymbol{x}}+i\epsilon\widehat{\boldsymbol{y}}).d\boldsymbol{r'}=dz_\epsilon$ and write from (6.25):

$$\int_{S'}(\frac{\partial}{\partial x'}+i\epsilon\frac{\partial}{\partial y'})u(z_+,z_-)dS'=-i\epsilon\int_{\mathcal{L}}u(z_+,z_-)dz_\epsilon. \qquad (6.30)$$

In the particular case where $u(z_+,z_-)=v(z_\epsilon)$, considering the regularity on S', we have $\frac{1}{2}(\frac{\partial}{\partial x'}\pm i\frac{\partial}{\partial y'})v(z_\pm)=\frac{\partial}{\partial z_\mp}v(z_\pm)=[\frac{\partial}{\partial z_\mp}(z_\pm)]v'(z_\pm)=0$ on S', and we recover the Cauchy integral theorem:

$$\int_{\mathcal{L}}v(z_\pm)dz_\pm=0 \text{ when } v'(z_\pm) \text{ exists on whole } S'. \qquad (6.31)$$

NOTE 6.10.– Using (6.25) with $F((\boldsymbol{r'}-\boldsymbol{a})^2,\varphi)=g(\boldsymbol{r'})F_1((\boldsymbol{r'}-\boldsymbol{a})^2)$, we have:

$$\int_{S'}\boldsymbol{b}.\text{grad}_{\boldsymbol{r'}}(g(\boldsymbol{r'})F_1((\boldsymbol{r'}-\boldsymbol{a})^2))dS'=\int_{\mathcal{L}}(g(\boldsymbol{r'})F_1((\boldsymbol{r'}-\boldsymbol{a})^2)-A_0)(\widehat{\boldsymbol{z}}\times\boldsymbol{b}).d\boldsymbol{r'}, \qquad (6.32)$$

and it is worth noticing that, in asymptotics, when $F_1((\boldsymbol{r'}-\boldsymbol{a})^2)$ is highly oscillatory and $g(\boldsymbol{r'})$ remains smooth as $(\boldsymbol{r'}-\boldsymbol{a})^2$ varies, we have:

$$\begin{aligned}
&\text{grad}_{\boldsymbol{r'}}(g(\boldsymbol{r'})F((\boldsymbol{r'}-\boldsymbol{a})^2)) = \text{grad}_{\boldsymbol{r'}}(g(\boldsymbol{r'}))F((\boldsymbol{r'}-\boldsymbol{a})^2) \\
&+ 2(\boldsymbol{r'}-\boldsymbol{a})g(\boldsymbol{r'})\frac{\partial F((\boldsymbol{r'}-\boldsymbol{a})^2)}{\partial((\boldsymbol{r'}-\boldsymbol{a})^2)} \sim 2(\boldsymbol{r'}-\boldsymbol{a})g(\boldsymbol{r'})\frac{\partial F((\boldsymbol{r'}-\boldsymbol{a})^2)}{\partial((\boldsymbol{r'}-\boldsymbol{a})^2)}. \qquad (6.33)
\end{aligned}$$

PROPOSITION 6.3.– *Let $F_0((\boldsymbol{r'}-\boldsymbol{a})^2)$ be a scalar function of complex points $\boldsymbol{r'} = x'\widehat{\boldsymbol{x}} + y'\widehat{\boldsymbol{y}}$ analytic on S', and $\boldsymbol{F}_{0v}$ a vectorial function, whose each component is analytic on S', we have:*

$$\begin{aligned}
&\int_{S'} b.(\boldsymbol{r'}-\boldsymbol{a})\frac{\partial(F_0((\boldsymbol{r'}-\boldsymbol{a})^2))}{\partial((\boldsymbol{r'}-\boldsymbol{a})^2)}dS' = \frac{1}{2}\int_{\mathcal{L}} F_0((\boldsymbol{r'}-\boldsymbol{a})^2)(\widehat{\boldsymbol{z}} \times b).d\boldsymbol{r'} \\
&\int_{S'} (\boldsymbol{r'}-\boldsymbol{a}).\frac{\partial(\boldsymbol{F}_{0v}((\boldsymbol{r'}-\boldsymbol{a})^2))}{\partial((\boldsymbol{r'}-\boldsymbol{a})^2)}dS' = \frac{1}{2}\int_{\mathcal{L}} (\widehat{\boldsymbol{z}} \times \boldsymbol{F}_{0v}((\boldsymbol{r'}-\boldsymbol{a})^2).d\boldsymbol{r'}. \qquad (6.34)
\end{aligned}$$

PROOF.– Using theorem 6.2, with $F((\boldsymbol{r'}-\boldsymbol{a})^2,\varphi) = F_0((\boldsymbol{r'}-\boldsymbol{a})^2)$ independent of φ, we obtain the first line, then the second line by considering each component of $\boldsymbol{F}_{0v}$ as F_0.

On this occasion, let us remark that:

$$\begin{aligned}
&\text{if } \frac{\partial F_0((\boldsymbol{r'}-\boldsymbol{a})^2)}{\partial((\boldsymbol{r'}-\boldsymbol{a})^2)} = \frac{e^{-v\widetilde{R}}}{\widetilde{R}} \text{ then } F_0((\boldsymbol{r'}-\boldsymbol{a})^2) = 2\int^{\widetilde{R}} e^{-vt}dt = \frac{-2e^{-v\widetilde{R}}}{v} + C, \\
&\text{if } \frac{\partial F_0((\boldsymbol{r'}-\boldsymbol{a})^2)}{\partial((\boldsymbol{r'}-\boldsymbol{a})^2)} = e^{-v(\boldsymbol{r'}-\boldsymbol{a})^2} \text{ then } F_0((\boldsymbol{r'}-\boldsymbol{a})^2) = \frac{-e^{-v(\boldsymbol{r'}-\boldsymbol{a})^2}}{v} + C, \quad (6.35)
\end{aligned}$$

where $\widetilde{R} = \sqrt{\boldsymbol{R}.\boldsymbol{R}} = \sqrt{(\boldsymbol{r'}-\boldsymbol{a})^2 + z^2}$, and C, v are some arbitrary constants.

THEOREM 6.3.– *Let $B_0((\boldsymbol{r'}-\boldsymbol{a})^2,\varphi)$ be a function, analytic relatively to $((\boldsymbol{r'}-\boldsymbol{a})^2)^{\frac{1}{2}}\cos\varphi = (x'-a_x)$ and $((\boldsymbol{r'}-\boldsymbol{a})^2)^{\frac{1}{2}}\sin\varphi = (y'-a_y)$ on S', we have, for $n \geq 0$:*

$$\begin{aligned}
&\boldsymbol{d}_{n+1}.\int_{S'}\prod_{j=1}^{n}(\boldsymbol{d}_j.(\boldsymbol{r'}-\boldsymbol{a}))\, grad(B_0((\boldsymbol{r'}-\boldsymbol{a})^2,\varphi))dS' \\
&= \int_{\mathcal{L}}\prod_{j=1}^{n}(\boldsymbol{d}_j.(\boldsymbol{r'}-\boldsymbol{a}))B_0((\boldsymbol{r'}-\boldsymbol{a})^2,\varphi)(\widehat{\boldsymbol{z}} \times \boldsymbol{d}_{n+1}).d\boldsymbol{r'} \\
&- \boldsymbol{d}_{n+1}.\sum_{m=1}^{n}(\boldsymbol{d}_m\int_{S'}\prod_{j\neq m}(\boldsymbol{d}_j.(\boldsymbol{r'}-\boldsymbol{a}))B_0((\boldsymbol{r'}-\boldsymbol{a})^2,\varphi)dS'), \qquad (6.36)
\end{aligned}$$

where the $\boldsymbol{d}_j$ are constants vectors, and we can let $\prod_j^0(..) = 1$, $\prod_{j\neq m}^1(..) = 1$, $\prod_{j\neq m}^0(..) = 0$.

PROOF.– Since $f\text{grad}(h) = \text{grad}(fh) - h\,\text{grad}(f)$ and $\text{grad}(\boldsymbol{d}_m.\boldsymbol{r}') = \boldsymbol{d}_m$, we have:

$$\prod_{j=1}^{n}(\boldsymbol{d}_j.(\boldsymbol{r}'-\boldsymbol{a}))\text{grad}(B_0((\boldsymbol{r}'-\boldsymbol{a})^2,\varphi)) =$$

$$= \text{grad}(\prod_{j=1}^{n}(\boldsymbol{d}_j.(\boldsymbol{r}'-\boldsymbol{a}))B_0((\boldsymbol{r}'-\boldsymbol{a})^2,\varphi))$$

$$-\sum_{m=1}^{n}\boldsymbol{d}_m\prod_{j\neq m}^{n}(\boldsymbol{d}_j.(\boldsymbol{r}'-\boldsymbol{a}))B_0((\boldsymbol{r}'-\boldsymbol{a})^2,\varphi). \tag{6.37}$$

Multiplying this equation by $\boldsymbol{d}_{n+1}$ and integrating, then using (6.25), we obtain the proposition. This result will be useful, in particular when $B_0((\boldsymbol{r}'-\boldsymbol{a})^2,\varphi)$ is independent of the variable φ, as we will see in theorem 6.4.

6.3.3. *Why these distinct ways are complementary: a practical example*

The first and second theorems permit us to consider respective reductions of surface integrals for functions of types $F((\boldsymbol{r}'-\boldsymbol{a})^2,\varphi)$ and $(\boldsymbol{r}'-\boldsymbol{a})F((\boldsymbol{r}'-\boldsymbol{a})^2,\varphi)$, which, correctly combined, avoid the use of fractional powers of $(\boldsymbol{r}'-\boldsymbol{a})^2$ and special functions. Let us illustrate it by some basic results.

For that, let us us show how to reduce, for arbitrary integer $n+1 \geq 0$:

$$I_{n+1} = \int_{S'}\prod_{j=1}^{n+1}(\boldsymbol{d}_j.(\boldsymbol{r}'-\boldsymbol{a}))F_0((\boldsymbol{r}'-\boldsymbol{a})^2)dS',$$

$$\text{with } I_0 = \int_{S'}F_0((\boldsymbol{r}'-\boldsymbol{a})^2)dS', \tag{6.38}$$

to a contour integral of the plane $z = 0$, where the polynomial $\prod_{j=1}^{n+1}(\boldsymbol{d}_j.(\boldsymbol{r}'-\boldsymbol{a}))$ is defined with arbitrary complex constant vectors $\boldsymbol{d}_j$ of the plane $z = 0$, and $\prod_{j=1}^{0}(..) = 1$.

We now state, from theorem 6.1, a specific proposition for the even case $n+1 = 2q$, then a general proposition for arbitrary integer n defined from the application of theorem 6.3 (derived from 6.2). In this example, the results of theorems 6.1 and 6.2 will then appear clearly complementary to finally solve the problem in whole generality. We consider in what follows, that we are in an open connected domain of the variable $\boldsymbol{a}$ including a point of S', where $F_0((\boldsymbol{r}'-\boldsymbol{a})^2)$ remains analytic on S', which allows the application of theorems 6.1 and 6.2.

6.3.3.1. *Exact reduction for an even polynomial factors*

PROPOSITION 6.4.– *We have, from the theorem 6.1, for* $n+1=2q\geq 0$,

$$I_{2q}=\int_{S'}\prod_{j=1}^{2q}(\boldsymbol{d}_j.(\boldsymbol{r'}-\boldsymbol{a}))F_0((\boldsymbol{r'}-\boldsymbol{a})^2)dS'$$

$$=\frac{1}{2}\int_{\mathcal{L}}G_{2q}(\varphi)(B_{2q}((\boldsymbol{r'}-\boldsymbol{a})^2)-B_{2q}(0))[(\widehat{\boldsymbol{z}}\times\frac{\boldsymbol{r'}-\boldsymbol{a}}{(\boldsymbol{r'}-\boldsymbol{a})^2}).d\boldsymbol{r'}],\qquad(6.39)$$

where,

$$B_{2q}((\boldsymbol{r'}-\boldsymbol{a})^2)=\int^{(\boldsymbol{r'}-\boldsymbol{a})^2}t^qF_0(t)dt,\ \prod_{j=1}^{2q}\boldsymbol{d}_j.(\boldsymbol{r'}-\boldsymbol{a})=G_{2q}(\varphi)((\boldsymbol{r'}-\boldsymbol{a})^2)^q,$$

$$B_0((\boldsymbol{r'}-\boldsymbol{a})^2)=\int^{(\boldsymbol{r'}-\boldsymbol{a})^2}F_0(t)dt,\ G_0(\varphi)=1,\qquad(6.40)$$

with constant $\boldsymbol{d}_j$, *and the case* $q=0$ *is considered with* $\prod_{j=1}^{0}(..)\equiv 1$.

PROOF.– Letting $\frac{\partial(F((\boldsymbol{r'}-\boldsymbol{a})^2),\varphi)}{\partial((\boldsymbol{r'}-\boldsymbol{a})^2)}=G_{2q}(\varphi)(\boldsymbol{r'}-\boldsymbol{a})^{2q}F_0((\boldsymbol{r'}-\boldsymbol{a})^2)$ in theorem 6.1, we obtain the equality (6.39), where we notice that B_{2q} depends on $q-1$ successive primitives of F_0.

It is worth noticing that we have:

$$B_{2q}(t')=\int_0^{(\boldsymbol{r'}-\boldsymbol{a})^2}t^qe^{-vt}dt+C\text{ if }F_0(t')=e^{-vt'},$$

$$B_{2q}(t')=2\int_{\sqrt{z^2}}^{\widetilde{R}}(t^2-z^2)^q\frac{e^{-vt}}{t^{m-1}}dt+C\text{ if }F_0(t')=\frac{e^{-v\widetilde{R}}}{(\widetilde{R})^m},\qquad(6.41)$$

where we let $\widetilde{R}=\sqrt{\boldsymbol{R}.\boldsymbol{R}}=\sqrt{t'+z^2}$ with $t'=(\boldsymbol{r'}-\boldsymbol{a})^2$ and $\boldsymbol{r'}=x'\widehat{\boldsymbol{x}}+y'\widehat{\boldsymbol{y}}$, while $\boldsymbol{a}$, v, C and z are some constants. In these cases, the function $B_{2q}((\boldsymbol{r'}-\boldsymbol{a})^2)-B_{2q}(0)$ can be expressed exactly or asymptotically with elementary functions.

6.3.3.2. *Exact reduction for polynomial factors of arbitrary order*

A direct use of theorem 6.1 for the determination of I_{n+1}, when $n+1$ is odd, would lead to a contour integral with fractional powers of $(\boldsymbol{r'}-\boldsymbol{a})^2$. To avoid it, we now state a relation, directly derived from theorem 6.3 (derived from 6.2), that lowers the degree of the polynomial in whole generality. The repeated use of it, in combination of results for I_1 (proposition 6.3) and I_0 (proposition 6.4), allows a reduction without fractional powers for polynomial factors of arbitrary order.

THEOREM 6.4.– *We have:*

$$I_0 = \int_{S'} F_0((\boldsymbol{r'} - \boldsymbol{a})^2)dS'$$
$$= \frac{1}{2}\int_{\mathcal{L}} (B_0((\boldsymbol{r'} - \boldsymbol{a})^2) - B_0(0))[(\widehat{\boldsymbol{z}} \times \frac{\boldsymbol{r'} - \boldsymbol{a}}{(\boldsymbol{r'} - \boldsymbol{a})^2}).d\boldsymbol{r'}], \tag{6.42}$$

and we can write, for arbitrary integer $n \geq 0$*:*

$$I_{n+1} = \int_{S'} \boldsymbol{d}_{n+1}.(\boldsymbol{r'} - \boldsymbol{a}) \prod_{j=1}^{n} (\boldsymbol{d}_j.(\boldsymbol{r'} - \boldsymbol{a}))F_0((\boldsymbol{r'} - \boldsymbol{a})^2)dS'$$
$$= \frac{1}{2}\int_{\mathcal{L}} \prod_{j=1}^{n} (\boldsymbol{d}_j.(\boldsymbol{r'} - \boldsymbol{a}))B_0((\boldsymbol{r'} - \boldsymbol{a})^2)(\widehat{\boldsymbol{z}} \times \boldsymbol{d}_{n+1}).d\boldsymbol{r'}$$
$$- \frac{1}{2}\sum_{m=1}^{n} ((\boldsymbol{d}_{n+1}.\boldsymbol{d}_m) \int_{S'} \prod_{j \neq m}^{n} (\boldsymbol{d}_j.(\boldsymbol{r'} - \boldsymbol{a}))B_0((\boldsymbol{r'} - \boldsymbol{a})^2)dS'), \tag{6.43}$$

where B_0 *is a primitive of* F_0*, the* $\boldsymbol{d}_j$*, numbered from* 1 *to* $n+1$*, are arbitrary constant vectors of the plane* $z = 0$*,* $\boldsymbol{a} = a_x\widehat{\boldsymbol{x}} + a_y\widehat{\boldsymbol{y}}$ *is a constant vector,* $\boldsymbol{r'} = x'\widehat{\boldsymbol{x}} + y'\widehat{\boldsymbol{y}}$ *is the current point of integration on the plane, and* $\prod_j^0(..) = 1$*,* $\prod_{j\neq m}^1(..) = 1$ *and* $\prod_{j\neq m}^0(..) = 0$*.*

Whatever the parity of $n + 1$*, this equation allows us, by repeated applications, to reduce* I_{n+1} *to a combination of contour integrals, depending on* $M = \lfloor (n + 3)/2 \rfloor$ *successive primitives of* F_0*, denoted* $F_0^{-(q+1)}(x) = \frac{1}{q!}\int^x (x - t)^q F_0(t)dt$*,* $q = 0, ..., M - 1$*.*

PROOF.– Using theorem 6.3 with the expression (6.26) of the gradient, we derive (6.43) of the proposition. The case $n = 0$ corresponds to proposition 6.3. The equation (6.43), after repeated applications, allows the reduction of the polynomial rank in surface integral term, until a complete reduction of I_{n+1} to a combination of contour integrals. The result then depends on a set of $\lfloor (n + 1)/2 \rfloor + 1$ (i.e. $\lfloor (n + 3)/2 \rfloor$) successive primitives F_0^{-q} of F_0, which can be expressed by the use of integration by parts following $F_0^{-q}(x) = \frac{1}{q!}\int^x (x - t)^q F_0(t)dt$.

A generalization of the proposition will be derived in theorem 6.13, by considering $\mathcal{R}^2$ instead of $(\boldsymbol{r'} - \boldsymbol{a})^2$, where $\mathcal{R}^2$ is an arbitrary polynomial of x' and y' of rank 2.

NOTE 6.11.– The reduction to a contour integral is not unique. When we use theorem 6.4, the order of the $\boldsymbol{d}_j$ could have an influence on the form of final reduction.

NOTE 6.12.– Letting $\mathrm{I}_{n+1} = \boldsymbol{d}_{n+1}.\boldsymbol{\mathcal{I}}_n$, we note that:

$$\begin{aligned}\boldsymbol{\mathcal{I}}_n &= \int_{S'} (\boldsymbol{r'} - \boldsymbol{a}) \prod_{j=1}^{n} (\boldsymbol{d}_j.(\boldsymbol{r'} - \boldsymbol{a}))F_0((\boldsymbol{r'} - \boldsymbol{a})^2)dS' \\ &= \frac{-1}{2}((\widehat{\boldsymbol{z}} \times \int_{\mathcal{L}} \prod_{j=1}^{n} (\boldsymbol{d}_j.(\boldsymbol{r'} - \boldsymbol{a}))B_0((\boldsymbol{r'} - \boldsymbol{a})^2)d\boldsymbol{r'}) \\ &+ \sum_{m=1}^{n} (\boldsymbol{d}_m \int_{S'} \prod_{j \neq m}^{n} (\boldsymbol{d}_j.(\boldsymbol{r'} - \boldsymbol{a}))B_0((\boldsymbol{r'} - \boldsymbol{a})^2)dS')).\end{aligned} \tag{6.44}$$

NOTE 6.13.– The reduction to a contour integral generally supposes the knowledge of a primitive relatively to $(\boldsymbol{r'} - \boldsymbol{a})^2$. When it cannot be expressed in closed form, its expression can often be derived asymptotically as in lemma 6.2 (see appendix D).

6.4. Reduction to contour integrals in case (a) for a pure dipolar illumination of a large perfectly conducting flat plate

We study here the reduction of PO radiation integrals belonging to case (a) for pure electric and magnetic dipolar illumination from results of previous sections. A rewriting of the PO integrals, and the expressions of primitives involved in the reduction, are given for flat surfaces. A generalization for non-dipolar illuminations and for the case of S with low curvature will be respectively given in sections 6.5 and 6.9.

Our method allows us to study the PO diffraction for perfectly and imperfectly conducting surfaces. To simplify the reading, we analyze here the first case and choose to describe the latter case (surface with impedance boundary conditions) in appendix C.

6.4.1. *Basic expansion of kernels in general radiation surface integral expressions*

Let us consider the expressions of the electric and magnetic fields, $\boldsymbol{E}$ and $\boldsymbol{H}$, radiated by electric and magnetic currents $\boldsymbol{J}$ and $\boldsymbol{M}$ impressed on a real arbitrary surface S, in free space with impedance Z_0 and wavenumber $k = \frac{2\pi}{\lambda}$, at the observation point $\boldsymbol{r} = (\boldsymbol{a} + z\,\widehat{\boldsymbol{z}}) \notin S$ [1], given in (6.1), where $\psi(\widetilde{R}) = \frac{e^{-ik\widetilde{R}}}{\widetilde{R}}$, $\widetilde{R} = \sqrt{\boldsymbol{R}.\boldsymbol{R}}$, $\boldsymbol{R} = \boldsymbol{a} + z\widehat{\boldsymbol{z}} - (\boldsymbol{r'} + z'\widehat{\boldsymbol{z}})$, $\boldsymbol{R} = \widetilde{R}\boldsymbol{u}$, and $(\boldsymbol{r'} + z'\widehat{\boldsymbol{z}})$ is the point of integration on S. We can write in this expression:

$$(\text{grad}_{\boldsymbol{r}}\text{div}_{\boldsymbol{r}} + k^2)(\psi \boldsymbol{J}) = \widetilde{R}(\frac{\psi'}{\widetilde{R}})'(\boldsymbol{J}.\boldsymbol{u})\boldsymbol{u} + \frac{\psi'}{\widetilde{R}}\boldsymbol{J} + k^2\psi\boldsymbol{J},$$

$$\text{curl}_{\boldsymbol{r}}(\psi\boldsymbol{J}) = -\boldsymbol{J} \times \text{grad}_{\boldsymbol{r}}\psi = \frac{\psi'}{\widetilde{R}}\boldsymbol{R} \times \boldsymbol{J}, \tag{6.45}$$

where $\psi' = \partial_{\widetilde{R}}\psi$ and $(\frac{\psi'}{\widetilde{R}})' = \partial_{\widetilde{R}}(\frac{\psi'}{\widetilde{R}})$.

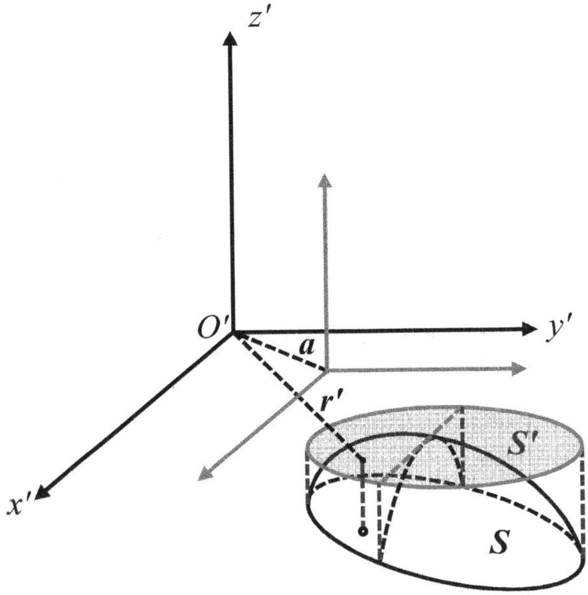

Figure 6.2. *Geometry of a curved plate S and its plane projection S'*

To simplify, using $\Delta\psi = \frac{1}{R^2}((\widetilde{R})^3\frac{\psi'}{\widetilde{R}})' = -k^2\psi$, we note that $\widetilde{R}(\frac{\psi'}{\widetilde{R}})' + k^2\psi = -3\frac{\psi'}{\widetilde{R}}$, and with $(\boldsymbol{J}.\boldsymbol{u})\boldsymbol{u} = \boldsymbol{u} \times (\boldsymbol{u} \times \boldsymbol{J}) + \boldsymbol{J}$, we derive:

$$(\text{grad}_{\boldsymbol{r}}\text{div}_{\boldsymbol{r}} + k^2)(\psi\boldsymbol{J}) = \widetilde{R}(\frac{\psi'}{\widetilde{R}})'\boldsymbol{u} \times (\boldsymbol{u} \times \boldsymbol{J}) - \frac{2\psi'}{\widetilde{R}}\boldsymbol{J}. \tag{6.46}$$

This gives us, at $\boldsymbol{r} = (\boldsymbol{a} + z\,\widehat{\boldsymbol{z}}) \notin S$,

$$\boldsymbol{E} = \frac{Z_0}{4\pi i k}\int_S (\widetilde{R}(\frac{\psi'}{\widetilde{R}})'\boldsymbol{u} \times (\boldsymbol{u} \times \boldsymbol{J}) - \frac{2\psi'}{\widetilde{R}}\boldsymbol{J})dS - \frac{1}{4\pi}\int_S \frac{\psi'}{\widetilde{R}}\boldsymbol{R} \times \boldsymbol{M}dS,$$

$$Z_0\boldsymbol{H} = \frac{Z_0}{4\pi}\int_S \frac{\psi'}{\widetilde{R}}\boldsymbol{R} \times \boldsymbol{J}dS + \frac{1}{4\pi i k}\int_S (\widetilde{R}(\frac{\psi'}{\widetilde{R}})'\boldsymbol{u} \times (\boldsymbol{u} \times \boldsymbol{M}) - \frac{2\psi'}{\widetilde{R}}\boldsymbol{M})dS. \tag{6.47}$$

We now study the case where S is equal to its projection S' on the plane $z = 0$. A generalization for the case of S with low curvature, which will be given in sections 6.8 and 6.9.

NOTE 6.14.– Letting $\boldsymbol{E} = \int_S \mathcal{G}_e dS$ and $\boldsymbol{H} = \int_S \mathcal{G}_h dS$, we have $\mathcal{G}_e \sim Z_0(\mathcal{G}_h \times \boldsymbol{u})$ for large $k\widetilde{R}$.

6.4.2. *Exact reduction to contour integrals when S is a flat plate S' perfectly conducting*

In the monostatic case, the primary point source and the observation point have the same position. The monostatic PO radiation integrals in perfectly conducting case ($\boldsymbol{M}|_S = 0$), for electric and magnetic dipolar illuminations, can be then reworked in such a form that all surface integrals can be exactly reduced to contour integrals, with all required primitives in closed forms.

LEMMA 6.1.– *Let us consider a surface integral on a real flat plate S' of the plane $z = 0$, which is a combination of integrals of the type:*

$$\int_{S'} \mathcal{F}(\boldsymbol{R}^2)dS', \int_{S'} \frac{\mathcal{F}(\boldsymbol{R}^2)(\boldsymbol{R}^2 - z^2)}{\boldsymbol{R}^2}G(\varphi)dS' \text{ and } \int_{S'} (\boldsymbol{a} - \boldsymbol{r}')\mathcal{F}(\boldsymbol{R}^2)dS', \tag{6.48}$$

where $\boldsymbol{R}^2 = (\boldsymbol{r}' - \boldsymbol{a})^2 + z^2$, $G(\varphi) = \frac{((\boldsymbol{r}'-\boldsymbol{a}).\widehat{\mathbf{d}})(\boldsymbol{r}'-\boldsymbol{a})}{(\boldsymbol{r}'-\boldsymbol{a})^2}$, with $z \neq 0$ and $\widehat{\mathbf{d}}$ some constants. If the point $\boldsymbol{a}$ belongs to an open connected domain $\mathcal{H}$ of analyticity which includes a point of S' (see definition 6.3), where $\mathcal{F}(\boldsymbol{R}^2)$ and $\mathcal{F}(\boldsymbol{R}^2)/\boldsymbol{R}^2$ are analytic on S', we can, with expressions of primitives of $\mathcal{F}(t)$ and $\mathcal{F}(t)/t$, reduce the surface integrals to non-singular contour integrals positively oriented, from proposition 6.4 (derived from theorem 6.1) and proposition 6.3 (derived from theorem 6.2).

PROOF.– To reduce the integrals to contour ones, we let successively $F_0((\boldsymbol{r}' - \boldsymbol{a})^2)$ as $\mathcal{F}(\boldsymbol{R}^2)$ and $\frac{\mathcal{F}(\boldsymbol{R}^2)(\boldsymbol{R}^2-z^2)}{\boldsymbol{R}^2}$ in proposition 6.4, involving the primitives of $\mathcal{F}(t)$ and $\mathcal{F}(t)/t$, and $\frac{\partial(F((\boldsymbol{r}'-\boldsymbol{a})^2)}{\partial((\boldsymbol{r}'-\boldsymbol{a})^2)}$ as $\mathcal{F}(\boldsymbol{R}^2)$ in proposition 6.3, involving the primitive of $\mathcal{F}(t)$.

6.4.2.1. *Physical optics fields and reduction for an electric dipolar source*

THEOREM 6.5.– *The PO diffracted field, radiated by a perfectly conducting flat surface S' of the plane $z' = 0$, when illuminated by an electric dipole $\boldsymbol{J}_0(\boldsymbol{r}_1) = \widehat{\boldsymbol{d}}\delta(\boldsymbol{r}_1 - \boldsymbol{r}_0)$, can be written at $\boldsymbol{r} = z\widehat{\boldsymbol{z}} + \boldsymbol{a} \notin S'$, when $\boldsymbol{r}_0 = \boldsymbol{r}$ (i.e. in monostatic case), following:*

$$\boldsymbol{E} = \frac{-Z_0\widehat{\boldsymbol{d}}}{16\pi^2 ik}\int_{S'} z\widetilde{R}((\frac{\psi'}{\widetilde{R}})^2)'dS' + \frac{Z_0}{16\pi^2 ik}\int_{S'} z\widetilde{R}((\frac{\psi'}{\widetilde{R}})^2)'(\boldsymbol{u}.\widehat{\boldsymbol{d}})\boldsymbol{u}dS'$$
$$+\frac{Z_0(\widehat{\boldsymbol{z}}.\widehat{\boldsymbol{d}})}{4\pi^2 ik}\int_{S'}(\frac{\psi'}{\widetilde{R}})^2(\boldsymbol{a}-\boldsymbol{r}')dS' + \frac{Z_0 z((\widehat{\boldsymbol{z}}.\widehat{\boldsymbol{d}})\widehat{\boldsymbol{z}}-\widehat{\boldsymbol{d}})}{4\pi^2 ik}\int_{S'}(\frac{\psi'}{\widetilde{R}})^2 dS',$$
$$\boldsymbol{H} = \frac{-\widehat{\boldsymbol{d}}}{8\pi^2}\times\int_{S'} z\boldsymbol{R}(\frac{\psi'}{\widetilde{R}})^2 dS', \qquad (6.49)$$

where $(.)' \equiv \partial_{\widetilde{R}}(.)$, $\psi(\widetilde{R}) = \frac{e^{-ik\widetilde{R}}}{\widetilde{R}}$, $\widetilde{R} = \sqrt{\boldsymbol{R}.\boldsymbol{R}}$, *while noting that, on S' (i.e. at $z' = 0$), we have,* $\boldsymbol{R} = (\boldsymbol{a}-\boldsymbol{r}') + z\,\widehat{\boldsymbol{z}}, \boldsymbol{R}^2 = (\boldsymbol{r}'-\boldsymbol{a})^2 + z^2$ *and:*

$$(\boldsymbol{u}.\widehat{\boldsymbol{d}})\boldsymbol{u} = \frac{(\widehat{\boldsymbol{d}}.\widehat{\boldsymbol{z}})\widehat{\boldsymbol{z}}z^2}{\boldsymbol{R}^2} + \frac{(\widehat{\boldsymbol{d}}.\widehat{\boldsymbol{z}} + \widehat{\boldsymbol{z}}(\widehat{\boldsymbol{d}}.(.)))z(\boldsymbol{a}-\boldsymbol{r}')}{\boldsymbol{R}^2}$$
$$+\frac{\boldsymbol{R}^2 - z^2}{\boldsymbol{R}^2}\frac{((\boldsymbol{r}'-\boldsymbol{a}).\widehat{\boldsymbol{d}})(\boldsymbol{r}'-\boldsymbol{a})}{(\boldsymbol{r}'-\boldsymbol{a})^2}. \qquad (6.50)$$

In this case, using the following exact primitives:

$$\int\frac{((\frac{\psi'}{\widetilde{R}})^2)'}{\widetilde{R}}d\boldsymbol{R}^2 = \frac{2(ik+\frac{1}{\widetilde{R}})^2 e^{-i2k\widetilde{R}}}{\boldsymbol{R}^4}, \int(\frac{\psi'}{\widetilde{R}})^2 d\boldsymbol{R}^2 = -\frac{(1+2ik\widetilde{R})e^{-i2k\widetilde{R}}}{2\boldsymbol{R}^4},$$
$$\int\widetilde{R}((\frac{\psi'}{\widetilde{R}})^2)'d\boldsymbol{R}^2 = 2(-k^2 + \frac{3ik}{\widetilde{R}} + \frac{3}{2\boldsymbol{R}^2})\frac{e^{-i2k\widetilde{R}}}{\boldsymbol{R}^2}, \qquad (6.51)$$

(see details in appendix A), the lemma 6.1 applies, and we can exactly reduce every integral of (6.49) to a non-singular contour integral, whose integrand involves no special function.

PROOF.– We can write the PO field with (6.47), considering that $\boldsymbol{J} = 2\widehat{\boldsymbol{z}} \times \boldsymbol{H}_i$ and $\boldsymbol{M} = 0$ on S', where $\boldsymbol{H}_i$ is radiated by the source $\boldsymbol{J}_0(\boldsymbol{r}_1) = \widehat{\boldsymbol{d}}\delta(\boldsymbol{r}_1 - \boldsymbol{r}_0)$, and thus:

$$2\pi\boldsymbol{J} = \widehat{\boldsymbol{z}}\times(\boldsymbol{R}_0\times\frac{\psi_0'}{\widetilde{R}_0}\widehat{\boldsymbol{d}}) = (\widehat{\boldsymbol{z}}.\widehat{\boldsymbol{d}})\frac{\psi_0'}{\widetilde{R}_0}\boldsymbol{R}_0 + z_0\frac{\psi_0'}{\widetilde{R}_0}\widehat{\boldsymbol{d}},\ \boldsymbol{R}_0 = (\boldsymbol{r}'-\boldsymbol{r}_0). \qquad (6.52)$$

In the monostatic case, when $\boldsymbol{r} = \boldsymbol{r}_0 = z\widehat{\boldsymbol{z}} + \boldsymbol{a} \notin S'$ and $\boldsymbol{R} = -\boldsymbol{R}_0 = \widetilde{R}\boldsymbol{u}$, we can write:

$$\boldsymbol{E} = \frac{Z_0}{8\pi^2 ik}\int_{S'} z\widetilde{R}(\frac{\psi'}{\widetilde{R}})'\boldsymbol{u}\times(\boldsymbol{u}\times\frac{\psi'}{\widetilde{R}}\widehat{\boldsymbol{d}}) + 2(\frac{\psi'}{\widetilde{R}})^2((\widehat{\boldsymbol{z}}.\widehat{\boldsymbol{d}})\boldsymbol{R} - z\widehat{\boldsymbol{d}})dS'$$

$$= \frac{-Z_0\widehat{\boldsymbol{d}}}{16\pi^2 ik}\int_{S'} z\widetilde{R}((\frac{\psi'}{\widetilde{R}})^2)'dS' + \frac{Z_0}{16\pi^2 ik}\int_{S'} z\widetilde{R}((\frac{\psi'}{\widetilde{R}})^2)'(\boldsymbol{u}.\widehat{\boldsymbol{d}})\boldsymbol{u}dS' +$$

$$+ \frac{Z_0(\widehat{\boldsymbol{z}}.\widehat{\boldsymbol{d}})}{4\pi^2 ik}\int_{S'}(\frac{\psi'}{\widetilde{R}})^2(\boldsymbol{a}-\boldsymbol{r}')dS' + \frac{Z_0((\widehat{\boldsymbol{z}}.\widehat{\boldsymbol{d}})\widehat{\boldsymbol{z}} - \widehat{\boldsymbol{d}})}{4\pi^2 ik}\int_{S'} z(\frac{\psi'}{\widetilde{R}})^2 dS', \quad (6.53)$$

and

$$\boldsymbol{H} = \frac{-1}{8\pi^2}\int_{S'}(\frac{\psi'}{\widetilde{R}})^2\boldsymbol{R}\times(\widehat{\boldsymbol{z}}\times(\boldsymbol{R}\times\widehat{\boldsymbol{d}}))dS' = \frac{-\widehat{\boldsymbol{d}}}{8\pi^2}\times\int_{S'} z\boldsymbol{R}(\frac{\psi'}{\widetilde{R}})^2 dS', \quad (6.54)$$

and thus (6.49). Lemma 6.1 then applies, and we can reduce the expressions of $\boldsymbol{E}$ and $\boldsymbol{H}$ to contour integrals, by using the exact primitives given in (6.51) (derived in appendix A). Let us notice that, for large $k\widetilde{R}$ in the integrals:

$$\boldsymbol{E} \sim \frac{Z_0}{8\pi^2 ik}\int_{S'}\frac{z\widetilde{R}((\frac{\psi'}{\widetilde{R}})^2)'}{2}((\boldsymbol{u}.\widehat{\boldsymbol{d}})\boldsymbol{u} - \widehat{\boldsymbol{d}})dS', \boldsymbol{H} \sim \frac{-\widehat{\boldsymbol{d}}}{8\pi^2}\times\int_{S'} z\boldsymbol{R}(\frac{\psi'}{\widetilde{R}})^2 dS'. \quad (6.55)$$

6.4.2.2. *Physical optics fields and reduction for a magnetic dipolar source*

THEOREM 6.6.– *The PO diffracted field, radiated by a perfectly conducting surface S' of the plane $z' = 0$, when illuminated by a magnetic dipole $\boldsymbol{M}_0(\boldsymbol{r}_1) = \widehat{\boldsymbol{d}}\delta(\boldsymbol{r}_1 - \boldsymbol{r}_0)$, can be written at $\boldsymbol{r} = z\widehat{\boldsymbol{z}} + \boldsymbol{a} \notin S'$, when $\boldsymbol{r}_0 = \boldsymbol{r}$ (monostatic case), following:*

$$\boldsymbol{E} = \frac{1}{8(\pi ik)^2}\int_{S'} z(\frac{\psi'}{\widetilde{R}})'(\frac{\psi'}{\widetilde{R}})'(\boldsymbol{R}\times\widehat{\boldsymbol{d}})dS' -$$

$$-\frac{1}{4(\pi ik)^2}\int_{S'}\widehat{\boldsymbol{z}}\times(\widetilde{R}((\frac{\psi'}{\widetilde{R}})^2)'(\boldsymbol{u}(\boldsymbol{u}.\widehat{\boldsymbol{d}}) - \widehat{\boldsymbol{d}}) - 2(\frac{\psi'}{\widetilde{R}})^2\widehat{\boldsymbol{d}})dS',$$

$$\boldsymbol{H} = \frac{1}{8\pi^2 ikZ_0}\int_{S'}\frac{z}{2}\widetilde{R}((\frac{\psi'}{\widetilde{R}})^2)'(\widehat{\boldsymbol{d}} - \boldsymbol{u}(\boldsymbol{u}.\widehat{\boldsymbol{d}})) - 2(\frac{\psi'}{\widetilde{R}})^2\boldsymbol{R}\times(\widehat{\boldsymbol{z}}\times\widehat{\boldsymbol{d}})dS', \quad (6.56)$$

with $(.)' \equiv \partial_{\widetilde{R}}(.)$, $\psi(\widetilde{R}) = \frac{e^{-ik\widetilde{R}}}{\widetilde{R}}$ and (6.50). Using the exact primitives given in (6.51) and:

$$\int(\frac{\psi'}{\widetilde{R}})'(\frac{\psi'}{\widetilde{R}})'d\boldsymbol{R}^2 = ik^3\frac{\psi^2}{\widetilde{R}} + 3(\frac{\psi}{\widetilde{R}})^2(\frac{k^2}{2} - (ik + \frac{1}{\widetilde{R}})^2), \quad (6.57)$$

lemma 6.1 applies, and we can exactly reduce every integral of (6.56) to a non-singular contour integral, whose integrand involves no special function.

PROOF.– We can write the PO field with (6.47), considering that the sources $\boldsymbol{J} = 2\widehat{\boldsymbol{z}} \times \boldsymbol{H}_i$ and $\boldsymbol{M} = 0$ on S', where $\boldsymbol{H}_i$ is radiated by $\boldsymbol{M}_0(\boldsymbol{r}_1) = \widehat{\boldsymbol{d}}\delta(\boldsymbol{r}_1 - \boldsymbol{r}_0)$, and thus:

$$\boldsymbol{J} = \frac{\widehat{\boldsymbol{z}} \times (\tilde{R}_0(\frac{\psi_0'}{\tilde{R}_0})'\boldsymbol{u}_0 \times (\boldsymbol{u}_0 \times \widehat{\boldsymbol{d}}) - \frac{2\psi'}{\tilde{R}}\widehat{\boldsymbol{d}})}{2\pi ikZ_0}, \; \boldsymbol{R}_0 = (\boldsymbol{r}' - \boldsymbol{r}_0), \; \boldsymbol{u}_0 = \frac{\boldsymbol{R}_0}{\tilde{R}_0}. \quad (6.58)$$

In the monostatic case $\boldsymbol{R} = -\boldsymbol{R}_0 = z\widehat{\boldsymbol{z}} + \boldsymbol{a} - \boldsymbol{r}'$, we then obtain:

$$\begin{aligned}
\boldsymbol{E} &= \frac{1}{8(\pi ik)^2}\int_{S'} z\tilde{R}(\frac{\psi'}{\tilde{R}})'(\boldsymbol{u} \times (\frac{\psi'}{\tilde{R}})'\widehat{\boldsymbol{d}}) - 2\psi'(\frac{\psi'}{\tilde{R}})'\widehat{\boldsymbol{z}} \times (\boldsymbol{u} \times (\boldsymbol{u} \times \widehat{\boldsymbol{d}}))dS' \\
&- \frac{1}{8(\pi ik)^2}\int_{S'} \frac{2\psi'}{\tilde{R}}\widehat{\boldsymbol{z}} \times (\tilde{R}(\frac{\psi'}{\tilde{R}})'\boldsymbol{u} \times (\boldsymbol{u} \times \widehat{\boldsymbol{d}}) - \frac{2\psi'}{\tilde{R}}\widehat{\boldsymbol{d}})dS' \\
&= \frac{1}{8(\pi ik)^2}\int_{S'} z(\frac{\psi'}{\tilde{R}})'(\frac{\psi'}{\tilde{R}})'(\boldsymbol{R} \times \widehat{\boldsymbol{d}})dS' - \\
&- \frac{1}{4(\pi ik)^2}\int_{S'} \widehat{\boldsymbol{z}} \times (\tilde{R}((\frac{\psi'}{\tilde{R}})^2)'(\boldsymbol{u}(\boldsymbol{u}.\widehat{\boldsymbol{d}}) - \widehat{\boldsymbol{d}}) - 2(\frac{\psi'}{\tilde{R}})^2\widehat{\boldsymbol{d}})dS',
\end{aligned} \quad (6.59)$$

and

$$\begin{aligned}
\boldsymbol{H} &= \frac{1}{8\pi^2 ikZ_0}\int_{S'} \frac{\psi'}{\tilde{R}}\boldsymbol{R} \times (\widehat{\boldsymbol{z}} \times (\tilde{R}(\frac{\psi'}{\tilde{R}})'\boldsymbol{u} \times (\boldsymbol{u} \times \widehat{\boldsymbol{d}}) - \frac{2\psi'}{\tilde{R}}\widehat{\boldsymbol{d}}))dS' \\
&= \frac{1}{8\pi^2 ikZ_0}\int_{S'} \frac{z}{2}\tilde{R}((\frac{\psi'}{\tilde{R}})^2)'(\widehat{\boldsymbol{d}} - \boldsymbol{u}(\boldsymbol{u}.\widehat{\boldsymbol{d}})) - 2(\frac{\psi'}{\tilde{R}})^2\boldsymbol{R} \times (\widehat{\boldsymbol{z}} \times \widehat{\boldsymbol{d}})\,dS',
\end{aligned} \quad (6.60)$$

and thus (6.56). Considering these expressions with (6.50), lemma 6.1 applies, and we can completely reduce them to contour integrals, by using the primitives given in (6.51) and (6.57) (derived in appendix A). Let us notice that, for large $k\tilde{R}$:

$$\boldsymbol{E} \sim \frac{-\widehat{\boldsymbol{d}}\times}{8\pi^2(ik)^2}\int_{S'} z\boldsymbol{R}((\frac{\psi'}{\tilde{R}})')^2 dS', \boldsymbol{H} \sim \frac{1}{Z_0}\int_{S'} \frac{z(\tilde{R}((\frac{\psi'}{\tilde{R}})^2)'(\widehat{\boldsymbol{d}} - \boldsymbol{u}(\boldsymbol{u}.\widehat{\boldsymbol{d}}))}{2ik8\pi^2}dS'. \quad (6.61)$$

6.5. Reduction to contour integrals in case (a) for a weighted dipolar illumination of a large perfectly conducting flat plate

6.5.1. *Integral expressions for a weighted dipolar illumination*

We now consider the incident field, as the combination of the radiations of electric and magnetic dipoles $\boldsymbol{J}_0(\boldsymbol{r}_1) = \widehat{\boldsymbol{d}}_e\delta(\boldsymbol{r}_1 - \boldsymbol{r}_0)$ and $\boldsymbol{M}_0(\boldsymbol{r}_1) = \widehat{\boldsymbol{d}}_h\delta(\boldsymbol{r}_1 - \boldsymbol{r}_0)$ weighted by $f_{e,h}(\boldsymbol{u}_0)$, following:

$$\boldsymbol{E}_i(\boldsymbol{r}) \sim \frac{1}{4\pi ik}(f_e(\boldsymbol{u}_0)\tilde{R}_0(\frac{\psi_0'}{\tilde{R}_0})'(\boldsymbol{u}_0 \times (\boldsymbol{u}_0 \times Z_0\widehat{\boldsymbol{d}}_e)) - ikf_h(\boldsymbol{u}_0)\frac{\psi_0'}{\tilde{R}_0}\boldsymbol{R}_0 \times \widehat{\boldsymbol{d}}_h)$$

$$\sim \frac{ik}{4\pi}\frac{e^{-ik\tilde{R}_0}}{\tilde{R}_0}(f_e(\boldsymbol{u}_0)(\boldsymbol{u}_0 \times (\boldsymbol{u}_0 \times Z_0\widehat{\boldsymbol{d}}_e)) + f_h(\boldsymbol{u}_0)\,(\boldsymbol{u}_0 \times \widehat{\boldsymbol{d}}_h)),$$

$$\boldsymbol{H}_i(\boldsymbol{r}) \sim \frac{1}{4\pi ikZ_0}(ikf_e(\boldsymbol{u}_0)\frac{\psi_0'}{\tilde{R}_0}(\boldsymbol{R}_0 \times Z_0\widehat{\boldsymbol{d}}_e) + f_h(\boldsymbol{u}_0)(\frac{\psi_0'}{\tilde{R}_0})'\boldsymbol{R}_0 \times (\boldsymbol{u}_0 \times \widehat{\boldsymbol{d}}_h))$$

$$\sim \frac{ik}{4\pi Z_0}\frac{e^{-ik\tilde{R}_0}}{\tilde{R}_0}(-f_e(\boldsymbol{u}_0)(\boldsymbol{u}_0 \times Z_0\widehat{\boldsymbol{d}}_e) + f_h(\boldsymbol{u}_0)(\boldsymbol{u}_0 \times (\boldsymbol{u}_0 \times \widehat{\boldsymbol{d}}_h))), \qquad (6.62)$$

for large $k\tilde{R}_0$, where $\boldsymbol{R}_0 = \boldsymbol{r} - \boldsymbol{r}_0 = \tilde{R}_0\boldsymbol{u}_0$, $\tilde{R}_0 = \sqrt{\boldsymbol{R}_0.\boldsymbol{R}_0}$, and the functions $f_{e,h}(\boldsymbol{u}_0)(\boldsymbol{u}_0 \times \widehat{\boldsymbol{d}}_h)$ are analytic, with $\boldsymbol{E}_i(\boldsymbol{r}) \sim -Z_0\boldsymbol{u}_0 \times \boldsymbol{H}_i(\boldsymbol{r})$.

In this case, the near-field monostatic diffracted field expressions at $\boldsymbol{r} = \boldsymbol{r}_0 \notin S'$, for a perfectly conducting plate S' illuminated by a source of arbitrary radiation pattern, can be derived in PO from (6.47) with $\boldsymbol{J} = 2\widehat{\boldsymbol{z}} \times \boldsymbol{H}_i$ and $\boldsymbol{M} = 0$ on S'. These PO expressions then remain valid from infinity to a distance qualified to be large relatively to the wavelength λ but that can be small relatively to the dimensions of S'.

Therefore, we obtain, in the monostatic case, from a combination of (6.55) and (6.61) weighted by $f_{e,h}$, the expressions:

$$\boldsymbol{E}(\boldsymbol{r}) \sim \frac{1}{8\pi^2(ik)^2}(\int_{S'} zf_e(-\boldsymbol{u})\frac{ik}{2}((\frac{\psi'}{\tilde{R}})^2)'(\boldsymbol{R} \times (\boldsymbol{u} \times Z_0\widehat{\boldsymbol{d}}_e))dS'$$

$$+\int_{S'} zf_h(-\boldsymbol{u})((\frac{\psi'}{\tilde{R}})')^2(\boldsymbol{R} \times \widehat{\boldsymbol{d}}_h)dS'),$$

$$\boldsymbol{H}(\boldsymbol{r}) \sim \frac{1}{8\pi^2(ik)^2Z_0}(\int_{S'} zf_e(-\boldsymbol{u})(ik)^2(\frac{\psi'}{\tilde{R}})^2(\boldsymbol{R} \times Z_0\widehat{\boldsymbol{d}}_e)dS'$$

$$-\int_{S'} zf_h(-\boldsymbol{u})\frac{ik}{2}((\frac{\psi'}{\tilde{R}})^2)'(\boldsymbol{R} \times (\boldsymbol{u} \times \widehat{\boldsymbol{d}}_h))dS'), \qquad (6.63)$$

for large $k\widetilde{R}$, with $\boldsymbol{r} = \boldsymbol{r}_0 = \boldsymbol{a} + z\widehat{\boldsymbol{z}}$, $\widetilde{R} = \sqrt{\boldsymbol{R}.\boldsymbol{R}}$, $\boldsymbol{R} = \widetilde{R}\boldsymbol{u} = \boldsymbol{a} + z\widehat{\boldsymbol{z}} - \boldsymbol{r}'$, $\boldsymbol{u} = -\boldsymbol{u}_0$. In this frame, let us notice that if we want to consider a source with a spherical pattern, instead of a dipole with toric pattern, we can apply the weighting factors $f_{e,h}(\boldsymbol{u}_0) = \beta_{e,h}/\sqrt{(\boldsymbol{u}_0 \times \widehat{\boldsymbol{d}}_{e,h})^2}$, with $(\boldsymbol{u}_0 \times \widehat{\boldsymbol{d}}_{e,h})^2 = 1 - (\boldsymbol{u}_0.\widehat{\boldsymbol{d}}_{e,h})^2$.

6.5.2. *Reduction to contour integrals for even/odd weight functions* $f_{e,h}$

Thereafter, we reduce $\boldsymbol{E}$ in (6.63) to a contour integral, for $f_h \equiv 0$ (respectively, $f_e \equiv 0$), when f_e (respectively, f_h) is even or odd relatively to coordinates of $\frac{\boldsymbol{r}'-\boldsymbol{a}}{R}$ in the integral expressions, for the cases with $\widehat{\boldsymbol{d}}_{e,h}.\widehat{\boldsymbol{z}} = 0$.

6.5.2.1. *Electric case with* $f_h \equiv 0$, $\widehat{\boldsymbol{d}} = \widehat{\boldsymbol{d}}_e$

Let us consider the case with $f_h \equiv 0$, $\widehat{\boldsymbol{d}} = \widehat{\boldsymbol{d}}_e$. The plate S' is then illuminated by:

$$\boldsymbol{E}_i \sim \frac{ikZ_0}{4\pi}\frac{e^{-ik\widetilde{R}_0}}{\widetilde{R}_0} f_e(\boldsymbol{u}_0)(\boldsymbol{u}_0(\boldsymbol{u}_0.\widehat{\boldsymbol{d}}) - \widehat{\boldsymbol{d}}), \tag{6.64}$$

and the PO diffracted field at $\boldsymbol{r} = \boldsymbol{a} + z\widehat{\boldsymbol{z}} \notin S'$, when $\boldsymbol{r} = \boldsymbol{r}_0$ (monostatic case) is given by:

$$\boldsymbol{E} \sim -\frac{Z_0}{8\pi k^2}\frac{i}{\lambda}\int_{S'} z\widetilde{R}((\frac{\psi'}{\widetilde{R}})^2)' f_e(-\boldsymbol{u})((\boldsymbol{u}.\widehat{\boldsymbol{d}})\boldsymbol{u} - \widehat{\boldsymbol{d}})dS'. \tag{6.65}$$

THEOREM 6.7.– *When* $f_e(-\boldsymbol{u}) = \frac{f(\frac{-z}{\widetilde{R}}, \frac{\boldsymbol{r}'-\boldsymbol{a}}{\widetilde{R}}) + f(\frac{-z}{\widetilde{R}}, \frac{\boldsymbol{a}-\boldsymbol{r}'}{\widetilde{R}})}{2} = \frac{g(\frac{z}{\widetilde{R}}, \frac{\boldsymbol{a}-\boldsymbol{r}'}{\widetilde{R}})}{\sqrt{1-(\boldsymbol{u}.\widehat{\boldsymbol{d}})^2}}$ *and* $\widehat{\boldsymbol{d}}.\widehat{\boldsymbol{z}} = 0$, *the integral reduces to:*

$$\begin{aligned}\boldsymbol{E} \sim \frac{-i}{\lambda}\Big(\frac{1}{2}\int_{\mathcal{L}} &(F_1((\boldsymbol{r}' - \boldsymbol{a})^2, \varphi) f_e(-\boldsymbol{u}) \\ &- F_1(0, \varphi) g(1{,}0))[(\widehat{\boldsymbol{z}} \times \frac{\boldsymbol{r}' - \boldsymbol{a}}{(\boldsymbol{r}' - \boldsymbol{a})^2}).d\boldsymbol{r}'] \\ &- \frac{1}{2}\int_{\mathcal{L}}(F_2((\boldsymbol{r}' - \boldsymbol{a})^2) f_e(-\boldsymbol{u}) - F_2(0) g(1{,}0))[(\widehat{\boldsymbol{z}} \times \widehat{\boldsymbol{d}}).d\boldsymbol{r}']\Big),\end{aligned} \tag{6.66}$$

where $\mathcal{L} \equiv \partial S'$ *is oriented counterclockwise, and the primitives* F_1 *and* F_2 *satisfy:*

$$F_1 = \frac{Z_0 z}{8\pi k^2} \int \widetilde{R}((\frac{\psi'}{\widetilde{R}})^2)'(\frac{(\boldsymbol{r'}-\boldsymbol{a})^2 G(\varphi)}{\boldsymbol{R}^2} - \widehat{\boldsymbol{d}}) d\boldsymbol{R}^2,$$

$$F_2 = \frac{Z_0 z^2 \widehat{\boldsymbol{z}}}{8\pi k^2} \int \frac{\widetilde{R}((\frac{\psi'}{\widetilde{R}})^2)'}{\boldsymbol{R}^2} d\boldsymbol{R}^2,$$

$$F_1 \sim -\frac{Z_0 z e^{-2ik\widetilde{R}}}{4\pi \boldsymbol{R}^2}(\frac{(\boldsymbol{r'}-\boldsymbol{a})^2 G(\varphi)}{\boldsymbol{R}^2} - \widehat{\boldsymbol{d}}),\ F_2 \sim -\frac{Z_0 z^2 \widehat{\boldsymbol{z}} e^{-2ik\widetilde{R}}}{4\pi \boldsymbol{R}^4}, \tag{6.67}$$

for $z \neq 0$, *with* $G(\varphi) = \frac{((\boldsymbol{r'}-\boldsymbol{a}).\widehat{\boldsymbol{d}})(\boldsymbol{r'}-\boldsymbol{a})}{(\boldsymbol{r'}-\boldsymbol{a})^2}$, $F_1(0,\varphi) = \frac{Z_0}{4\pi}\frac{\widehat{\boldsymbol{d}} e^{-2ikz}}{z}$ *and* $F_2(0) = -\frac{Z_0}{4\pi}\frac{\widehat{\boldsymbol{z}} e^{-2ikz}}{z^2}$, $f_e(-\boldsymbol{u})$ *regular on* S'. *The VV polarization term is given by* $\boldsymbol{E}.\widehat{\boldsymbol{d}}$ *when* $\boldsymbol{r}_0.\widehat{\boldsymbol{d}} = \widehat{\boldsymbol{z}}.\widehat{\boldsymbol{d}} = 0$,*that is,* $\widehat{\boldsymbol{d}} = \frac{\boldsymbol{r}_0 \times \widehat{\boldsymbol{z}}}{\sqrt{(\boldsymbol{r}_0 \times \widehat{\boldsymbol{z}})^2}}$ *as* $\boldsymbol{r}_0 \times \widehat{\boldsymbol{z}} \neq 0$.

PROOF.– We can detail (6.65), noticing that, when $\widehat{\boldsymbol{d}}.\widehat{\boldsymbol{z}} = 0$, we have:

$$(\boldsymbol{u}.\widehat{\boldsymbol{d}})\boldsymbol{u} = \frac{(\boldsymbol{r'}-\boldsymbol{a})^2 G(\varphi)}{\boldsymbol{R}^2} + \frac{(\boldsymbol{a}-\boldsymbol{r'}).\widehat{\boldsymbol{d}}}{\boldsymbol{R}^2} z\widehat{\boldsymbol{z}},\ G(\varphi) = \frac{((\boldsymbol{r'}-\boldsymbol{a}).\widehat{\boldsymbol{d}})(\boldsymbol{r'}-\boldsymbol{a})}{(\boldsymbol{r'}-\boldsymbol{a})^2}. \tag{6.68}$$

Considering that $\frac{g(\frac{z}{\widetilde{R}}, \frac{\boldsymbol{a}-\boldsymbol{r'}}{\widetilde{R}})}{\sqrt{1-(\boldsymbol{u}.\widehat{\boldsymbol{d}})^2}}$ is a regular even function of $\frac{\boldsymbol{a}-\boldsymbol{r'}}{\widetilde{R}}$ for $z \neq 0$, we use theorem 6.1 with the proposition 6.13 (see appendix D) for the part of the integrand with $(\frac{(\boldsymbol{r'}-\boldsymbol{a})^2 G(\varphi)}{\boldsymbol{R}^2} - \widehat{\boldsymbol{d}})$ factor, then proposition 6.14 (see appendix D) for the part of the integrand with $(\boldsymbol{a}-\boldsymbol{r'}).\widehat{\boldsymbol{d}}$ factor, and thus derive (6.66).

For an illumination with an isotropic pattern, we can choose $g \equiv 1$, and write:

$$\boldsymbol{E}.\widehat{\boldsymbol{d}} \sim -\frac{Z_0}{4\pi}\frac{i}{\lambda}\Big(\frac{1}{2}\int_{\mathcal{L}} \frac{(\frac{z e^{-2ik\widetilde{R}}}{\boldsymbol{R}^2}\sqrt{1-(\frac{(\boldsymbol{r'}-\boldsymbol{a})}{\widetilde{R}}.\widehat{\boldsymbol{d}})^2} - \frac{e^{-2ikz}}{z})}{(\boldsymbol{r'}-\boldsymbol{a})^2}[(\widehat{\boldsymbol{z}} \times (\boldsymbol{r'}-\boldsymbol{a})).d\boldsymbol{r'}]\Big), \tag{6.69}$$

which is the VV polarization term when $\boldsymbol{r}_0.\widehat{\boldsymbol{d}} = 0$, that is, $\widehat{\boldsymbol{d}} = \frac{\boldsymbol{r}_0 \times \widehat{\boldsymbol{z}}}{\sqrt{(\boldsymbol{r}_0 \times \widehat{\boldsymbol{z}})^2}}$ as $\boldsymbol{r}_0 \times \widehat{\boldsymbol{z}} \neq 0$.

Let us notice that $(\widehat{\boldsymbol{z}} \times (\boldsymbol{r'}-\boldsymbol{a})).\widehat{\boldsymbol{v}}$ is constant on each straight part of $\mathcal{L}$ along $\widehat{\boldsymbol{v}}$, and that, for real $\boldsymbol{a} \notin S'$, the parts of the integrals with $\frac{e^{-2ikz}}{z}$ factor vanish.

THEOREM 6.8.– *When* $f_e(-\boldsymbol{u}) = \frac{f(\frac{-z}{\widetilde{R}}, \frac{\boldsymbol{r'}-\boldsymbol{a}}{\widetilde{R}}) - f(\frac{-z}{\widetilde{R}}, \frac{\boldsymbol{a}-\boldsymbol{r'}}{\widetilde{R}})}{2} = \frac{(\boldsymbol{r'}-\boldsymbol{a})}{\widetilde{R}}.\frac{\boldsymbol{h}(\frac{z}{\widetilde{R}}, \frac{\boldsymbol{a}-\boldsymbol{r'}}{\widetilde{R}})}{\sqrt{1-(\boldsymbol{u}.\widehat{\boldsymbol{d}})^2}}$ *and* $\widehat{\boldsymbol{d}}.\widehat{\boldsymbol{z}} = 0$, *it reduces to:*

$$\boldsymbol{E} \sim -\frac{i}{\lambda}\Big(\frac{1}{2}\int_{\mathcal{L}} F_1((\boldsymbol{r}'-\boldsymbol{a})^2,\varphi)[(\widehat{\boldsymbol{z}}\times \boldsymbol{h}(\frac{z}{\underset{\sim}{R}},\frac{\boldsymbol{a}-\boldsymbol{r}'}{\underset{\sim}{R}})).d\boldsymbol{r}']+$$

$$+\frac{1}{2}\int_{\mathcal{L}} F_2((\boldsymbol{r}'-\boldsymbol{a})^2)'(\frac{\widehat{\boldsymbol{d}}.(\boldsymbol{a}-\boldsymbol{r}')f_e(-\boldsymbol{u})}{\underset{\sim}{R}})[(\widehat{\boldsymbol{z}}\times\frac{\boldsymbol{r}'-\boldsymbol{a}}{(\boldsymbol{r}'-\boldsymbol{a})^2}).d\boldsymbol{r}']\Big), \tag{6.70}$$

where, for $z \neq 0$, the primitives F_1 and F_2 are written as follows:

$$F_1 = \frac{Z_0 z}{8\pi k^2}\int ((\frac{\psi'}{\underset{\sim}{R}})^2)'(\frac{(\boldsymbol{r}'-\boldsymbol{a})^2 G(\varphi)}{R^2} - \widehat{\boldsymbol{d}})dR^2,$$

$$F_2 = \frac{Z_0 z^2\widehat{\boldsymbol{z}}}{8\pi k^2}\int ((\frac{\psi'}{\underset{\sim}{R}})^2)'dR^2,$$

$$F_1 \sim -\frac{Z_0}{4\pi}\frac{z e^{-2ik\underset{\sim}{R}}}{\underset{\sim}{R}\boldsymbol{R}^2}(\frac{(\boldsymbol{r}'-\boldsymbol{a})^2 G(\varphi)}{R^2} - \widehat{\boldsymbol{d}}),\ F_2 \sim -\frac{Z_0}{4\pi}\frac{z^2\widehat{\boldsymbol{z}}e^{-2ik\underset{\sim}{R}}}{\underset{\sim}{R}\boldsymbol{R}^2}, \tag{6.71}$$

with $G(\varphi) = \frac{((\boldsymbol{r}'-\boldsymbol{a}).\widehat{\boldsymbol{d}})(\boldsymbol{r}'-\boldsymbol{a})}{(\boldsymbol{r}'-\boldsymbol{a})^2}$, $f_e(-\boldsymbol{u})$ regular on S', $\mathcal{L} \equiv \partial S'$ oriented counterclockwise. The VV polarization term is given by $\boldsymbol{E}.\widehat{\boldsymbol{d}}$, when $\widehat{\boldsymbol{d}} = \frac{\boldsymbol{r}_0\times\widehat{\boldsymbol{z}}}{\sqrt{(\boldsymbol{r}_0\times\widehat{\boldsymbol{z}})^2}}$ as $\boldsymbol{r}_0 \times \widehat{\boldsymbol{z}} \neq 0$.

PROOF.– Considering that $\frac{\boldsymbol{h}(\frac{z}{\underset{\sim}{R}},\frac{\boldsymbol{a}-\boldsymbol{r}'}{\underset{\sim}{R}})}{\sqrt{1-(\boldsymbol{u}.\widehat{\boldsymbol{d}})^2}}$ is a regular even function of the coordinates of $\frac{\boldsymbol{r}'-\boldsymbol{a}}{\underset{\sim}{R}}$, $\widehat{\boldsymbol{d}}.\widehat{\boldsymbol{z}} = 0$, we use proposition 6.14 and theorem 6.1 with proposition 6.13 (see appendix D), and derive (6.70) with (6.71).

6.5.2.2. *Magnetic case with $f_e \equiv 0$, $\widehat{\boldsymbol{d}} = \widehat{\boldsymbol{d}}_h$*

Let us consider the case with $f_e \equiv 0$, $\widehat{\boldsymbol{d}} = \widehat{\boldsymbol{d}}_h$. The plate S' is then illuminated by:

$$\boldsymbol{E}_i \sim \frac{ik}{4\pi}(\boldsymbol{u}_0\times\widehat{\boldsymbol{d}})f_h(\boldsymbol{u}_0)\frac{e^{-ik\underset{\sim}{R_0}}}{\underset{\sim}{R_0}}, \tag{6.72}$$

and the PO diffracted field at $\boldsymbol{r} = \boldsymbol{a} + z\widehat{\boldsymbol{z}} \notin S'$, when $\boldsymbol{r} = \boldsymbol{r}_0$ (monostatic case), satisfies:

$$\boldsymbol{E}.\boldsymbol{v}_0 \sim \frac{-z}{8(\pi ik)^2}\int_{S'}((\frac{\psi'}{\underset{\sim}{R}})')^2\boldsymbol{w}_0.\boldsymbol{R}\,f_h(-\boldsymbol{u})dS' \text{ with } \boldsymbol{w}_0 = \boldsymbol{v}_0\times\widehat{\boldsymbol{d}}. \tag{6.73}$$

THEOREM 6.9.– *When $f_h(-\boldsymbol{u}) = \frac{f(\frac{-z}{\underset{\sim}{R}},\frac{\boldsymbol{r}'-\boldsymbol{a}}{\underset{\sim}{R}})+f(\frac{-z}{\underset{\sim}{R}},\frac{\boldsymbol{a}-\boldsymbol{r}'}{\underset{\sim}{R}})}{2} = \frac{g(\frac{z}{\underset{\sim}{R}},\frac{\boldsymbol{a}-\boldsymbol{r}'}{\underset{\sim}{R}})}{\sqrt{1-(\boldsymbol{u}.\widehat{\boldsymbol{d}})^2}}$ and $\widehat{\boldsymbol{d}}.\widehat{\boldsymbol{z}} = 0$, the integral reduces to:*

$$\boldsymbol{E}.\boldsymbol{v}_0 \sim \int_{\mathcal{L}} \frac{(\frac{ze^{-2ik\widetilde{R}}f_h(-\boldsymbol{u})}{\boldsymbol{R}^2} - \frac{\widetilde{R}}{z}\frac{e^{-2ikz}g(1,0)}{z})(\widehat{\boldsymbol{z}} \times \frac{-\boldsymbol{w}_0(\boldsymbol{r}'-\boldsymbol{a})^2+(\boldsymbol{r}'-\boldsymbol{a})(zw_{0z})}{\widetilde{R}})}{(4\pi)^2(\boldsymbol{r}'-\boldsymbol{a})^2/ik}.d\boldsymbol{r}' \tag{6.74}$$

where $\boldsymbol{w}_0 = \boldsymbol{v}_0 \times \widehat{\boldsymbol{d}}$. *For* $\boldsymbol{v}_0 = (\widehat{\boldsymbol{d}} \times \frac{\boldsymbol{r}_0}{\widetilde{r}_0})$, *with* $\widehat{\boldsymbol{d}} = \frac{\boldsymbol{r}_0 \times \widehat{\boldsymbol{z}}}{\sqrt{(\boldsymbol{r}_0 \times \widehat{\boldsymbol{z}})^2}}$ *as* $\boldsymbol{r}_0 \times \widehat{\boldsymbol{z}} \neq 0$, *we have:*

$$\boldsymbol{E}.\boldsymbol{v}_0 \sim \frac{ik}{4\pi}\Big(\int_{\mathcal{L}} \frac{\frac{ze^{-2ik\widetilde{R}}f_h(-\boldsymbol{u})}{\boldsymbol{R}^2} - \frac{\widetilde{R}}{z}\frac{e^{-2ikz}g(1,0)}{z}}{4\pi(\boldsymbol{r}'-\boldsymbol{a})^2}$$
$$[\widehat{\boldsymbol{z}} \times (\frac{(\boldsymbol{r}'-\boldsymbol{a})\widetilde{R}}{\widetilde{r}_0} - \frac{\boldsymbol{r}'(\boldsymbol{r}'-\boldsymbol{a})^2}{\widetilde{R}\widetilde{r}_0})].d\boldsymbol{r}'\Big), \tag{6.75}$$

denoted the HH polarization term. The contour $\mathcal{L} \equiv \partial S'$ *is oriented counterclockwise.*

PROOF.– Noticing that $w_0.\boldsymbol{R} = -\boldsymbol{w}_0.(\boldsymbol{r}' - \boldsymbol{a}) + zw_{0z}$, and the function $f_h(-\boldsymbol{u})$ is even relatively to the coordinates of $\frac{\boldsymbol{r}'-\boldsymbol{a}}{\widetilde{R}}$, we derive, from theorem 6.1 with proposition 6.13 and proposition 6.14 (see appendix D):

$$\boldsymbol{E}.\boldsymbol{v}_0 \sim \frac{i}{\lambda}\frac{1}{2}\int_{\mathcal{L}}(F((\boldsymbol{r}'-\boldsymbol{a})^2) - F(0))(\widehat{\boldsymbol{z}} \times (-\boldsymbol{w}_0 + \frac{\boldsymbol{r}'-\boldsymbol{a}}{(\boldsymbol{r}'-\boldsymbol{a})^2}(zw_{0z}))).d\boldsymbol{r}',$$
$$F = \frac{z}{4\pi k^2 ik}\int((\frac{\psi'}{\widetilde{R}})')^2 f_h(-\boldsymbol{u})d\boldsymbol{R}^2. \tag{6.76}$$

From primitives given in appendix A, we derive $F \sim \frac{1}{4\pi}\frac{ze^{-2ik\widetilde{R}}}{\widetilde{R}^3}f_h(-\boldsymbol{u})$ and thus (6.74). For polarization HH, $\boldsymbol{v}_0 = (\widehat{\boldsymbol{d}} \times \frac{\boldsymbol{r}_0}{\widetilde{r}_0})$ with $\widehat{\boldsymbol{d}} = \frac{\boldsymbol{r}_0 \times \widehat{\boldsymbol{z}}}{\sqrt{(\boldsymbol{r}_0 \times \widehat{\boldsymbol{z}})^2}}$ as $\boldsymbol{r}_0 \times \widehat{\boldsymbol{z}} \neq 0$ and $\boldsymbol{w}_0 = \frac{\boldsymbol{r}_0}{\widetilde{r}_0} = \frac{\boldsymbol{a}+z\widehat{\boldsymbol{z}}}{\widetilde{r}_0}$, we then derive (6.78) from:

$$-\boldsymbol{w}_0(\boldsymbol{r}'-\boldsymbol{a})^2 + (\boldsymbol{r}'-\boldsymbol{a})(zw_{0z}) = \frac{(-\boldsymbol{r}' + (\boldsymbol{r}'-\boldsymbol{a}))(\boldsymbol{r}'-\boldsymbol{a})^2 + (\boldsymbol{r}'-\boldsymbol{a})z^2}{\widetilde{r}_0}, \tag{6.77}$$

For an illumination with isotropic pattern, that is, $g \equiv 1$ and $f_h(-\boldsymbol{u}) = \frac{1}{\sqrt{1-(\boldsymbol{u}.\widehat{\boldsymbol{d}})^2}}$, we have:

$$\boldsymbol{E}.\boldsymbol{v}_0 \sim \frac{ik}{4\pi}\int_{\mathcal{L}} \frac{(\frac{ze^{-2ik\widetilde{R}}}{\boldsymbol{R}^2\sqrt{1-(\boldsymbol{u}.\widehat{\boldsymbol{d}})^2}} - \frac{\widetilde{R}}{z}\frac{e^{-2ikz}}{z})}{4\pi(\boldsymbol{r}'-\boldsymbol{a})^2}[\widehat{\boldsymbol{z}} \times (\frac{(\boldsymbol{r}'-\boldsymbol{a})\widetilde{R}}{\widetilde{r}_0} - \frac{\boldsymbol{r}'(\boldsymbol{r}'-\boldsymbol{a})^2}{\widetilde{R}\widetilde{r}_0})].d\boldsymbol{r}'. \tag{6.78}$$

Let us notice that $(\widehat{z} \times (\boldsymbol{r'} - \boldsymbol{a})).\widehat{\boldsymbol{v}}$ is constant on each straight part of $\mathcal{L}$ along $\widehat{\boldsymbol{v}}$, and that, for real $\boldsymbol{a} \notin S'$, the parts of the integrals with $\frac{\widetilde{R}}{z}\frac{e^{-2ikz}}{z}$ factor vanish.

THEOREM 6.10.– *When* $f_h(-\boldsymbol{u}) = \frac{f(\frac{-z}{\widetilde{R}},\frac{\boldsymbol{r'}-\boldsymbol{a}}{\widetilde{R}})-f(\frac{-z}{\widetilde{R}},\frac{\boldsymbol{a}-\boldsymbol{r'}}{\widetilde{R}})}{2} = \frac{(\boldsymbol{r'}-\boldsymbol{a})}{\widetilde{R}}.\frac{\boldsymbol{h}(\frac{z}{\widetilde{R}},\frac{\boldsymbol{a}-\boldsymbol{r'}}{\widetilde{R}})}{\sqrt{1-(\boldsymbol{u}.\widehat{\boldsymbol{d}})^2}}$ *and* $\widehat{\boldsymbol{d}}.\widehat{z} = 0$, *the integral reduces to:*

$$\boldsymbol{E}.\boldsymbol{v}_0 \sim \frac{ik}{4\pi}\Big(\int_{\mathcal{L}} \frac{ze^{-2ik\widetilde{R}}\big(\frac{(\boldsymbol{w}_0.(\boldsymbol{a}-\boldsymbol{r'}))f_h(-\boldsymbol{u})}{\widetilde{R}}\big)}{4\pi \boldsymbol{R}^2}(\widehat{z} \times \frac{(\boldsymbol{r'}-\boldsymbol{a})}{(\boldsymbol{r'}-\boldsymbol{a})^2}).d\boldsymbol{r'}$$

$$+ \int_{\mathcal{L}} \frac{ze^{-2ik\widetilde{R}}(zw_{0z})}{4\pi \boldsymbol{R}^4}(\widehat{z} \times \frac{\boldsymbol{h}(\frac{z}{\widetilde{R}},\frac{\boldsymbol{a}-\boldsymbol{r'}}{\widetilde{R}})}{\sqrt{1-(\boldsymbol{u}.\widehat{\boldsymbol{d}})^2}}).d\boldsymbol{r'}\Big), \qquad (6.79)$$

with $\boldsymbol{w}_0 = \boldsymbol{v}_0 \times \widehat{\boldsymbol{d}}$. *For* $\boldsymbol{v}_0 = (\widehat{\boldsymbol{d}} \times \frac{\boldsymbol{r}_0}{\widetilde{r}_0})$ *with* $\widehat{\boldsymbol{d}} = \frac{\boldsymbol{r}_0 \times \widehat{z}}{\sqrt{(\boldsymbol{r}_0 \times \widehat{z})^2}}$ *as* $\boldsymbol{r}_0 \times \widehat{z} \neq 0$, *and* $\boldsymbol{w}_0 = \frac{\boldsymbol{r}_0}{\widetilde{r}_0} = \frac{\boldsymbol{a}+z\widehat{z}}{\widetilde{r}_0}$, $\boldsymbol{E}.\boldsymbol{v}_0$ *is the HH polarization term.*

PROOF.– $\frac{\boldsymbol{h}(\frac{z}{\widetilde{R}},\frac{\boldsymbol{a}-\boldsymbol{r'}}{\widetilde{R}})}{\sqrt{1-(\boldsymbol{u}.\widehat{\boldsymbol{d}})^2}}$ is a regular even function of the coordinates of $\frac{\boldsymbol{r'}-\boldsymbol{a}}{\widetilde{R}}$ with $\widehat{\boldsymbol{d}}.\widehat{z} = 0$. We can then use proposition 6.14 (see appendix D) and theorem 6.1 with proposition 6.13 (see appendix D) and derive (6.79).

6.6. Definition of complex scaling in case (b) and elementary exact surface integral reduction

As defined in section 6.2, we consider the case (b) and study the reduction of a plane surface radiation integral, whose integrand has an exponential term with an argument composed of a complex polynomial of rank 2 relatively to plane coordinates. This form is generally encountered in bistatic case, when the surface of the scatterer is a flat or a curved plate, if it is illuminated by a Gaussian beam [13]–[15], or if its largest dimension D is such that $\sqrt{kD^3}$ is much smaller than the source and observation distances from S' [2]–[4] (see (6.208)–(6.209) in appendix B). It is worth noticing that, for a flat plate, the importance of the nonlinear part of the polynomial vanishes in the far-field region (i.e. at large distance d relatively to λ, with $d > 2D^2/\lambda$ so that phase error is within $\pi/8$ [1]), while it remains significant if the scatterer is a curved plate.

We begin by analyzing the exact reduction of the elementary surface integral in this section:

$$\int_{S'} e^{\mathcal{R}^2} dS' \text{ with } \mathcal{R}^2 = (\alpha_1 x' + \alpha_2 y') - ik(h_1 x'^2 + h_2 y'^2 + h_3 x'y'), \qquad (6.80)$$

for arbitrary complex α_i and h_j, to a contour integral without any use of special functions. For that, we define a general complex scaling transformation to cover a domain of applications complementary to the one of results given in previous sections. More complex integrands will be considered in sections 6.7 and 6.8.

6.6.1. *Transformation of the polynomial argument* $\mathcal{R}^2$

We begin by developing a particular rewriting of polynomial $\mathcal{R}^2$. This will lead us to an efficient complex scaling transform that will permit the extension of our contour reduction method.

PROPOSITION 6.5.– *We can write, for any h_j coefficients with $h_3^2 \neq 4h_1h_2$:*

$$h_1x^2 + h_2y^2 + h_3xy = \xi((1 - c_{p,x}^2)x^2 + (1 - c_{p,y}^2)y^2 - 2c_{p,x}c_{p,y}xy), \tag{6.81}$$

where $\mathbf{c}_{px}$ and $\mathbf{c}_{py}$ are the (complex) coordinates of a vector $\mathbf{c}_p$. Subsequently, we have:

$$2c_{p,x}c_{p,y} = -h_3/\xi,\ c_{p,x}^2 = 1 - h_1/\xi,\ c_{p,y}^2 = 1 - h_2/\xi,$$
$$\xi = \frac{(h_1 + h_2) + \epsilon\sqrt{(h_1 - h_2)^2 + h_3^2}}{2}, \tag{6.82}$$

where the choice of $\epsilon = +1$ or -1 is free. We can then let:

$$c_{p,x} = \epsilon'_x\sqrt{1 - h_1/\xi},\ c_{p,y} = \epsilon'_y\sqrt{1 - h_2/\xi},$$
$$\epsilon_x'^2 = \epsilon_y'^2 = 1,\ \epsilon'_x\epsilon'_y = \frac{-h_3/\xi}{2\sqrt{1 - h_1/\xi}\sqrt{1 - h_2/\xi}}\ \textit{for}\ h_3 \neq 0, \tag{6.83}$$

or,

$$c_{p,x} = \epsilon_x\sqrt{\xi - h_1}/\sqrt{\xi},\ c_{p,y} = \epsilon_y\sqrt{\xi - h_2}/\sqrt{\xi},$$
$$\epsilon_x^2 = \epsilon_y^2 = 1,\ \epsilon_x\epsilon_y = \frac{-h_3}{2\sqrt{\xi - h_1}\sqrt{\xi - h_2}}\ \textit{for}\ h_3 \neq 0, \tag{6.84}$$

where we can choose arbitrarily ϵ_x (or ϵ'_x) equal to $+1$ or -1, and take indifferently $-\mathbf{c}_p$ in place of $\mathbf{c}_p$ (note: in the case $h_3 = 0$, every $\epsilon_{x(,y)}$ and $\epsilon'_{x(,y)}$ can be chosen either $+1$ or -1 freely). Letting $(\mathbf{c}_p)_\epsilon$ for $\mathbf{c}_p$ when $\epsilon = \pm 1$, it is worth noticing that:

$$\frac{(c_{px})_\epsilon}{(c_{py})_\epsilon} = -\frac{(c_{py})_{-\epsilon}}{(c_{px})_{-\epsilon}},\ (\mathbf{c}_p)_\epsilon^2 - 1 = \frac{1}{(\mathbf{c}_p)_{-\epsilon}^2 - 1}, \tag{6.85}$$

and thus,

$$(\mathbf{c}_p)_\epsilon = \frac{(c_{py})_\epsilon}{(c_{px})_{-\epsilon}}(-(c_{py})_{-\epsilon}\widehat{\boldsymbol{x}} + (c_{px})_{-\epsilon}\widehat{\boldsymbol{y}}) = \frac{(c_{py})_\epsilon}{(c_{px})_{-\epsilon}}(\widehat{\boldsymbol{z}} \times (\mathbf{c}_p)_{-\epsilon}),$$
$$\frac{(\mathbf{c}_p)_\epsilon^2}{(\mathbf{c}_p)_{-\epsilon}^2} = \frac{(c_{py})_\epsilon^2}{(c_{px})_{-\epsilon}^2} = \frac{(c_{px})_\epsilon^2}{(c_{py})_{-\epsilon}^2} = (\mathbf{c}_p)_\epsilon^2 - 1. \tag{6.86}$$

For the particular case where $h_3 = 0$, we obtain, after choosing $\epsilon = +1$ or $\epsilon = -1$ in the definition of ξ :

$$\xi = h_1, c_{p,x} = 0, c_{p,y} = \sqrt{h_1 - h_2}/\sqrt{h_1}, p^2 = 1 - (\mathbf{c}_p)^2 = h_2/h_1,$$
$$\xi = h_2, c_{p,x} = \sqrt{h_2 - h_1}/\sqrt{h_2}, c_{p,y} = 0, p^2 = 1 - (\mathbf{c}_p)^2 = h_1/h_2. \quad (6.87)$$

We also note that, when $h_3^2 \to 4h_1h_2$, we can take $\xi = h_1 + h_2$ and write:

$$(h_1x'^2 + h_2y'^2 + h_3x'y') \to (\sqrt{h_1}x' - \epsilon'\sqrt{h_2}y')^2,\ \epsilon' = -h_3/(2\sqrt{h_1}\sqrt{h_2}),$$

$$\xi = h_1 + h_2,\ c_{p,x} = \frac{\sqrt{h_2}}{\sqrt{h_1 + h_2}},\ c_{p,y} = \epsilon'\frac{\sqrt{h_1}}{\sqrt{h_1 + h_2}},\ p^2 = 1 - (\mathbf{c}_p)^2 = 0. \quad (6.88)$$

PROOF.– Letting

$$(h_1x'^2 + h_2y'^2 + h_3x'y') = \xi((\mathbf{r}')^2 - (\mathbf{c}_p.\mathbf{r}')^2), \quad (6.89)$$

we obtain (6.81) with (6.82), then $2c_{p,x}c_{p,y} = -h_3/\xi$, $c_{p,x}^2 = 1 - h_1/\xi$, $c_{p,y}^2 = 1 - h_2/\xi$. Thus, $4(1 - h_1/\xi)(1 - h_2/\xi) = (h_3/\xi)^2$ and we have:

$$\xi = \frac{(h_1 + h_2) + \epsilon\sqrt{(h_1 - h_2)^2 + 4(h_3/2)^2}}{2},$$

$$c_{p,x} = \epsilon'_x\sqrt{1 - h_1/\xi} = \frac{\epsilon_x\sqrt{\xi - h_1}}{\sqrt{\xi}},\ c_{p,y} = \epsilon'_y\sqrt{1 - h_2/\xi} = \frac{\epsilon_y\sqrt{\xi - h_2}}{\sqrt{\xi}},$$

$$\epsilon'_x\epsilon'_y = \frac{-h_3/(2\xi)}{\sqrt{1 - \frac{h_1}{\xi}}\sqrt{1 - \frac{h_2}{\xi}}}, \epsilon_x\epsilon_y = \frac{-h_3/2}{\sqrt{\xi - h_1}\sqrt{\xi - h_2}}, \epsilon^2_{x(,y)} = \epsilon'^2_{x(,y)} = 1, \quad (6.90)$$

where the choice of ϵ, ϵ_x and ϵ'_x, either $+1$ or -1, is free (note: we can take $-\mathbf{c}_p$ in place of $\mathbf{c}_p$). Letting ξ_ϵ and $(\mathbf{c}_p)_\epsilon$, specifying the choice of ϵ, in place of ξ and $\mathbf{c}_p$, we can write:

$$\frac{(c_{px})_\epsilon}{(c_{py})_\epsilon} = \frac{(c_{px})_\epsilon(c_{py})_\epsilon}{(c_{py})_\epsilon(c_{py})_\epsilon} = \frac{-h_3}{2(\xi_\epsilon - h_2)} = \frac{-h_3}{(h_1 - h_2) + \epsilon\sqrt{(h_1 - h_2)^2 + h_3^2}}$$
$$= \frac{(h_1 - h_2) - \epsilon\sqrt{(h_1 - h_2)^2 + h_3^2}}{h_3} = -\frac{(c_{py})_{-\epsilon}}{(c_{px})_{-\epsilon}},$$
$$(\mathbf{c}_p)^2_\epsilon = 2 - \frac{h_1 + h_2}{\xi_\epsilon} = 1 - 2\frac{-\xi_\epsilon + (h_1 + h_2)}{(h_1 + h_2) + \epsilon\sqrt{(h_1 - h_2)^2 + h_3^2}}$$
$$= 1 - \frac{(h_1 + h_2) - \epsilon\sqrt{(h_1 - h_2)^2 + h_3^2}}{(h_1 + h_2) + \epsilon\sqrt{(h_1 - h_2)^2 + h_3^2}} = 1 - \frac{1}{1 - (\mathbf{c}_p)^2_{-\epsilon}} = \frac{(\mathbf{c}_p)^2_{-\epsilon}}{(\mathbf{c}_p)^2_{-\epsilon} - 1}.$$
(6.91)

PROPOSITION 6.6.– *We can write:*

$$(\alpha_1 x' + \alpha_2 y') - ik(h_1 x'^2 + h_2 y'^2 + h_3 x'y') = \boldsymbol{w}'.\boldsymbol{r}' - ik\xi((\boldsymbol{r}')^2 - (\mathbf{c}_p.\boldsymbol{r}')^2)$$
$$= \boldsymbol{w}'_m.\boldsymbol{r}'_{\boldsymbol{p}} - ik\xi(\boldsymbol{r}'_{\boldsymbol{p}})^2 = v(\boldsymbol{r}'_p - \boldsymbol{a}_p)^2 + v_1, \tag{6.92}$$

with $\boldsymbol{w}'.\boldsymbol{r}' = \alpha_1 x' + \alpha_2 y'$, $\boldsymbol{w}'.\boldsymbol{r}' = \boldsymbol{w}'_m.\boldsymbol{r}'_{\boldsymbol{p}}$, $(\boldsymbol{r}'_{\boldsymbol{p}}) = (\boldsymbol{r}')^2 - (\mathbf{c}_p.\boldsymbol{r}')^2$, $v = -ik\xi$ *and* $\mathbf{c}_p$ *defined in proposition 6.5. Letting* $\mathbf{c}_p = \sqrt{\mathbf{c}_p^2}\,\widehat{\mathbf{c}}_p$ *and* $p = \sqrt{1-\mathbf{c}_p^2}$, *the complex points* $\boldsymbol{r}'_{\boldsymbol{p}}$, $\boldsymbol{w}'_m$ *and* $\boldsymbol{a}_p$ *verify:*

$$\boldsymbol{r}'_{\boldsymbol{p}} = (p(\widehat{\mathbf{c}}_p.\boldsymbol{r}')\widehat{\mathbf{c}}_p + (\boldsymbol{r}' - \widehat{\mathbf{c}}_p(\widehat{\mathbf{c}}_p.\boldsymbol{r}')) = \boldsymbol{r}' - \mathbf{c}_p(\mathbf{c}_p.\boldsymbol{r}')/(1+p),$$
$$\boldsymbol{w}'_m = \frac{(\widehat{\mathbf{c}}_p \boldsymbol{w}')}{p}\widehat{\mathbf{c}}_p + (\boldsymbol{w}' - \widehat{\mathbf{c}}_p(\widehat{\mathbf{c}}_p \boldsymbol{w}')) = \boldsymbol{w}' + \mathbf{c}_p(\mathbf{c}_p.\boldsymbol{w}')/(p(1+p)),$$
$$\boldsymbol{a}_p = \frac{\boldsymbol{w}'_m}{2ik\xi},\ v = -ik\xi,\ v_1 = ik\xi(\boldsymbol{a}_p)^2 = \frac{\boldsymbol{w}'^2_m}{4ik\xi}, \tag{6.93}$$

where ξ *and* p *satisfy:*

$$\xi = \frac{(h_1+h_2) + \epsilon\sqrt{(h_1-h_2)^2 + h_3^2}}{2} = \frac{2(h_1 h_2 - (h_3/2)^2)}{(h_1+h_2) - \epsilon\sqrt{(h_1-h_2)^2 + h_3^2}},$$
$$p = \sqrt{1-\mathbf{c}_p^2} = \sqrt{-1 + \frac{h_1+h_2}{\xi}} = \sqrt{\frac{(h_1+h_2) - \epsilon\sqrt{(h_1-h_2)^2+h_3^2}}{(h_1+h_2) + \epsilon\sqrt{(h_1-h_2)^2+h_3^2}}} \tag{6.94}$$

with $\epsilon = +1$ *or* -1, $(p\xi)^2 = h_1 h_2 - (\frac{h_3}{2})^2$.

This can be considered in general, and for some particular cases, in the sense of the limit. So, if $h_3^2/(4h_1h_2) \to 1$, *following the choice of* ϵ *for* ξ, *we have* $\xi \to 0$ *with* $p \to \infty$ *or* $\xi \to h_1 + h_2$ *with* $p \to 0$, *and we can take:*

$$\widehat{c}_{p,x} = \frac{\sqrt{h_1}}{\sqrt{h_1+h_2}},\ \widehat{c}_{p,y} = \frac{-\epsilon'\sqrt{h_2}}{\sqrt{h_1+h_2}},\ \text{with } p \sim \sqrt{\frac{4(h_1+h_2)^2}{4h_1h_2 - h_3^2}} \to \infty,$$
$$\widehat{c}_{p,x} = \frac{\sqrt{h_2}}{\sqrt{h_1+h_2}},\ \widehat{c}_{p,y} = \frac{\epsilon'\sqrt{h_1}}{\sqrt{h_1+h_2}},\ \text{with } p \sim \sqrt{\frac{4h_1h_2 - h_3^2}{4(h_1+h_2)^2}} \to 0, \tag{6.95}$$

with $\epsilon' = \frac{-h_3/2}{\sqrt{h_1}\sqrt{h_2}}$. *Besides, if* $-h_3^2/(h_1-h_2)^2 \to 1$, *we have* $\xi \to \frac{h_1+h_2}{2}$ *with* $p \to 1$, *and thus* $\mathbf{c}_p^2 \to 0$ *(note:* $\mathbf{c}_p^2 = 0$ *does not imply* $\mathbf{c}_p = 0$ *in complex coordinates), and we can take:*

$$\boldsymbol{r}'_{\boldsymbol{p}} \to \boldsymbol{r}' - \frac{(\mathbf{c}_p.\boldsymbol{r}')\mathbf{c}_p}{2} \text{ and } \boldsymbol{w}'_m \to \boldsymbol{w}' + \frac{(\mathbf{c}_p.\boldsymbol{w}')\mathbf{c}_p}{2},$$
$$\text{with } c_{p,x} = \frac{\sqrt{h_2-h_1}}{\sqrt{h_1+h_2}},\ c_{p,y} = \epsilon_y\frac{\sqrt{h_1-h_2}}{\sqrt{h_1+h_2}},\ \epsilon_y = \frac{-h_3}{\sqrt{h_2-h_1}\sqrt{h_1-h_2}}, \tag{6.96}$$

which implies, for $(h_1 - h_2)$ *and* $h_3 \to 0$*, that we have* $\boldsymbol{c}_p \to 0$ *and:*

$$\boldsymbol{r'_p} \to \boldsymbol{r'},\ \boldsymbol{w'_m} \to \boldsymbol{w'},\ \boldsymbol{a}_p \to \frac{\boldsymbol{w'}}{2ik\xi} \text{ with } \xi \to h_{1(,2)}. \tag{6.97}$$

PROOF.– Considering (6.92), we choose $\boldsymbol{w'_m}$ and $\boldsymbol{r'_p}$ verifying:

$$\boldsymbol{w'_m}.\boldsymbol{r'_p} = \boldsymbol{w'}.\boldsymbol{r'} = \alpha_1 x' + \alpha_2 y',\ (\boldsymbol{r'_p})^2 = ((\boldsymbol{r'})^2 - (\boldsymbol{c}_p.\boldsymbol{r'})^2), \tag{6.98}$$

as in (6.93), with $\widehat{\boldsymbol{c}}_p = \frac{\boldsymbol{c}_p}{\sqrt{\boldsymbol{c}_p.\boldsymbol{c}_p}}$ and $\boldsymbol{c}_p$ defined in proposition 6.5. We then have:

$$(\alpha_1 x' + \alpha_2 y') - ik(h_1 x'^2 + h_2 y'^2 + h_3 x'y') =$$

$$= \boldsymbol{w'_m}\boldsymbol{r'_p} - ik\xi(\boldsymbol{r'_p})^2 = -ik\xi((\boldsymbol{r'_p} - \frac{\boldsymbol{w'_m}}{2ik\xi})^2 - \frac{\boldsymbol{w'}^2_m}{(2ik\xi)^2}) = v(\boldsymbol{r'_p} - \boldsymbol{a}_p)^2 + v_1, \tag{6.99}$$

with $\boldsymbol{a}_p = \frac{\boldsymbol{w'_m}}{2ik\xi}$, $v = -ik\xi$, $v_1 = \frac{\boldsymbol{w'}^2_m}{4ik\xi}$, which completes (6.93). Using $p \sim 1 - \boldsymbol{c}_p^2/2$ when $\boldsymbol{c}_p^2 \to 0$, we derive (6.96). On this occasion, let us notice that we could have chosen to take $p = -\sqrt{1 - \boldsymbol{c}_p^2}$ in place of $p = \sqrt{1 - \boldsymbol{c}_p^2}$ in the definitions.

6.6.2. *Reduction of the surface integral to a contour integral*

Considering the expression of complex points $\boldsymbol{r'_p}$ given in proposition 6.6, we define a complex change of variables (complex scaling) and an efficient reduction to contours integrals.

6.6.2.1. *Complex scaling transform* $\mathcal{P}$ *and definition of* S'_p

DEFINITION 6.4.– *Using* $\boldsymbol{r'_p}$ *given in (6.93) of proposition 6.6, we consider* $\mathcal{P}$*, the complex anisotropic dilation (or scaling) by a factor* $p = \sqrt{1 - \boldsymbol{c}_p^2}$ *along the vector* $\widehat{\boldsymbol{c}}_p = \boldsymbol{c}_p/\sqrt{\boldsymbol{c}_p^2}$ *(with* $\boldsymbol{c}_p^2 \neq 0$*), defined by:*

$$\mathcal{P}\colon \boldsymbol{r'} = x'\widehat{\boldsymbol{x}} + y'(\widehat{\boldsymbol{z}} \times \widehat{\boldsymbol{x}}) \to \boldsymbol{r'_p} = x'_p\widehat{\boldsymbol{c}}_p + y'_p(\widehat{\boldsymbol{z}} \times \widehat{\boldsymbol{c}}_p),$$

$$\text{with } x'_p = p\,(\boldsymbol{r'}.\widehat{\boldsymbol{c}}_p), y'_p = \boldsymbol{r'}.(\widehat{\boldsymbol{z}} \times \widehat{\boldsymbol{c}}_p), \tag{6.100}$$

when $h_3^2 \neq 4h_1h_2$*, that is,* $p \neq 0$*. This function is linear invertible and satisfies:*

$$\mathcal{P}(\boldsymbol{r'}) = (p(\widehat{\boldsymbol{c}}_p.\boldsymbol{r'})\widehat{\boldsymbol{c}}_p + (\boldsymbol{r'} - \widehat{\boldsymbol{c}}_p(\widehat{\boldsymbol{c}}_p.\boldsymbol{r'})) = \boldsymbol{r'} - \frac{\boldsymbol{c}_p(\boldsymbol{c}_p.\boldsymbol{r'})}{1+p} = \boldsymbol{r'_p},$$

$$\mathcal{P}^{-1}(\boldsymbol{r'_p}) = \frac{(\widehat{\boldsymbol{c}}_p\boldsymbol{r'_p})}{p}\widehat{\boldsymbol{c}}_p + (\boldsymbol{r'_p} - \widehat{\boldsymbol{c}}_p(\widehat{\boldsymbol{c}}_p\boldsymbol{r'_p})) = \boldsymbol{r'_p} + \frac{\boldsymbol{c}_p(\boldsymbol{c}_p.\boldsymbol{r'_p})}{p(1+p)} = \boldsymbol{r'},$$

$$\mathcal{P}(\boldsymbol{c}_p) = p\boldsymbol{c}_p, \mathcal{P}((\widehat{\boldsymbol{z}} \times \boldsymbol{c}_p)) = (\widehat{\boldsymbol{z}} \times \boldsymbol{c}_p), \tag{6.101}$$

As in definition 6.1, we then define a complex simply connected manifold S'_p and its oriented contour $\mathcal{L}_p$ as the transformation by $\mathcal{P}$ of a real, simply connected manifold of the plane $z = 0$, now named S', and its real positively oriented contour $\mathcal{L}$. Letting $dS' = dx' \wedge dy'$ and $dS'_p = dx'_p \wedge dy'_p$, we have $p\,dS' = dS'_p$.
Besides, we remark that:

$$grad_{\boldsymbol{r}'}(g(\boldsymbol{r}')) = \mathcal{P}(grad_{\boldsymbol{r}'_p}(g(\boldsymbol{r}'))),$$
$$\mathcal{P}^2(\boldsymbol{r}') = \boldsymbol{r}' - (\boldsymbol{c}_p.\boldsymbol{r}')\boldsymbol{c}_p,\ p^2\mathcal{P}^{-2}(\boldsymbol{r}') = p^2\boldsymbol{r}' + (\boldsymbol{c}_p.\boldsymbol{r}')\boldsymbol{c}_p, \tag{6.102}$$

and notice, in the limit case $-h_3^2/(h_1 - h_2)^2 \to 1$, that is, when $p \to 1$ and thus $(\boldsymbol{c}_p)^2 \to 0$:

$$\boldsymbol{r}'_p = \mathcal{P}(\boldsymbol{r}') \to \boldsymbol{r}' - \frac{(\boldsymbol{c}_p.\boldsymbol{r}')\boldsymbol{c}_p}{2} \text{ and } \boldsymbol{w}'_m = \mathcal{P}^{-1}(\boldsymbol{w}') \to \boldsymbol{w}' + \frac{(\boldsymbol{c}_p.\boldsymbol{w}')\boldsymbol{c}_p}{2}, \tag{6.103}$$

(note: $(\boldsymbol{c}_p)^2 \to 0$ implies $\boldsymbol{c}_p \to A_1(\widehat{\boldsymbol{x}} \pm i\widehat{\boldsymbol{y}})$, where $\widehat{\boldsymbol{x}} \pm i\widehat{\boldsymbol{y}} = \frac{\mathbf{c} \pm i\widehat{\mathbf{z}} \times \mathbf{c}}{c_x \mp ic_y}$ for any complex vector $\mathbf{c}$ with $c_x \mp ic_y \neq 0$). Therefore, for $(h_1 - h_2) \to 0$ and $h_3 \to 0$, we have $\boldsymbol{c}_p \to 0$ and:

$$\mathcal{P}(\boldsymbol{r}') \to \boldsymbol{r}' \text{ and } \boldsymbol{w}'_m = \mathcal{P}^{-1}(\boldsymbol{w}') \to \boldsymbol{w}' \text{ with } \xi \to h_{1(,2)}. \tag{6.104}$$

NOTE 6.15.– For $h_3 = 0$, we note that we have the two following choices:

$$\xi = h_1, \widehat{\boldsymbol{c}}_p = \widehat{\boldsymbol{y}}, p^2 = h_2/h_1, \mathcal{P}(\boldsymbol{r}') = (\widehat{\boldsymbol{x}}.\boldsymbol{r}')\widehat{\boldsymbol{x}} + p\,(\widehat{\boldsymbol{y}}.\boldsymbol{r}')\widehat{\boldsymbol{y}},$$
$$\xi = h_2, \widehat{\boldsymbol{c}}_p = \widehat{\boldsymbol{x}}, p^2 = h_1/h_2,\ \mathcal{P}(\boldsymbol{r}') = p(\widehat{\boldsymbol{x}}.\boldsymbol{r}')\widehat{\boldsymbol{x}} + (\widehat{\boldsymbol{y}}.\boldsymbol{r}')\widehat{\boldsymbol{y}}. \tag{6.105}$$

NOTE 6.16.– We note that, for $n \geq 1$:

$$\mathcal{P}^{2n}(\boldsymbol{r}') = \boldsymbol{r}' - (\sum_{k=0}^{n-1} p^{2k})(\boldsymbol{c}_p.\boldsymbol{r}')\boldsymbol{c}_p,\ \mathcal{P}^{2n+1}(\boldsymbol{r}') = \mathcal{P}(\boldsymbol{r}') - (p\sum_{k=0}^{n-1} p^{2k})(\boldsymbol{c}_p.\boldsymbol{r}')\boldsymbol{c}_p, \tag{6.106}$$

with $\sum_{k=0}^{n-1} p^{2k} = \frac{1-p^{2n}}{1-p^2}$ for $n \geq 1$, from which we derive $\mathcal{P}^{-m}(\boldsymbol{r}')$ from $\mathcal{P}^{-1}(\boldsymbol{c}_p) = \boldsymbol{c}_p/p$.

NOTE 6.17.– From (6.26), we have $\partial_{(\boldsymbol{r}'_p - \boldsymbol{a}_p)^2} \equiv \frac{(\boldsymbol{r}'_p - \boldsymbol{a}_p)}{2(\boldsymbol{r}'_p - \boldsymbol{a}_p)^2}.\mathrm{grad}_{\boldsymbol{r}'_p}$ and using $\mathrm{grad}_{\boldsymbol{r}'_p} \equiv \mathcal{P}^{-1}(\mathrm{grad}_{\boldsymbol{r}'})$, we note that $\partial_{(\boldsymbol{r}'_p - \boldsymbol{a}_p)^2} \equiv \frac{(\boldsymbol{r}' - \mathcal{P}^{-1}(\boldsymbol{a}_p))}{2(\boldsymbol{r}'_p - \boldsymbol{a}_p)^2}.\mathrm{grad}_{\boldsymbol{r}'}$.

6.6.2.2. *First step reduction of integral along the complex transformed contour $\mathcal{L}_p$*

PROPOSITION 6.7.– *Considering $\boldsymbol{r}'_p = \mathcal{P}(\boldsymbol{r}')$, with $\boldsymbol{a}_p = \frac{\boldsymbol{w}'_m}{2ik\xi}$, v and p given in (6.93), and an arbitrary constant A_0, we can write:*

$$\int_{S'} e^{(\alpha_1 x' + \alpha_2 y') - ik(h_1 x'^2 + h_2 y'^2 + h_3 x' y')} \cdot dS' =$$

$$= \frac{1}{2p} \int_{\mathcal{L}_p} \left(\frac{\widehat{\boldsymbol{z}} \times (\boldsymbol{r}'_p - \boldsymbol{a}_p)(e^{\boldsymbol{w}'_m \cdot \boldsymbol{r}'_p - ik\xi(\boldsymbol{r}'_p)^2} - A_0)}{\boldsymbol{w}'_m \cdot \boldsymbol{r}'_p - ik\xi(\boldsymbol{r}'_p)^2 - \frac{\boldsymbol{w}'^2_m}{4ik\xi}} \right) . d\boldsymbol{r}'_p$$

$$+ \frac{(A_0 - e^{\frac{\boldsymbol{w}'^2_m}{4ik\xi}})}{-2ikp\xi} \int_{\mathcal{L}_p} (\widehat{\boldsymbol{z}} \times \frac{\boldsymbol{r}'_p - \boldsymbol{a}_p}{(\boldsymbol{r}'_p - \boldsymbol{a}_p)^2}) . d\boldsymbol{r}'_p, \tag{6.107}$$

where $\boldsymbol{w}'_m \cdot \boldsymbol{r}'_p - ik\xi(\boldsymbol{r}'_p)^2 - \frac{\boldsymbol{w}'^2_m}{4ik\xi} = v(\boldsymbol{r}'_p - \boldsymbol{a}_p)^2$, $v = -ik\xi$. Let $\mathcal{D}_1$ be a smooth semi-line of the complex plane from the origin to ∞; we note that, when $(\boldsymbol{r}'_p - \boldsymbol{a}_p)^2 \neq 0$ on $\mathcal{L}_p$:

$$\int_{\mathcal{L}_p} (\widehat{\boldsymbol{z}} \times \frac{\boldsymbol{r}'_p - \boldsymbol{a}_p}{(\boldsymbol{r}'_p - \boldsymbol{a}_p)^2}) . d\boldsymbol{r}'_p = \frac{-i}{2} \int_{\mathcal{L}_p} d \ln(\chi(\boldsymbol{r}'_p)) = \pi(N_+ - N_-), \tag{6.108}$$

where N_+ (respectively, N_-) is the number of times $\chi(\boldsymbol{r}'_p)$ crosses $\mathcal{D}_1$ in positive sense (respectively, negative) along the closed contour $\mathcal{L}_p$, and:

$$\chi(\boldsymbol{r}'_p) = \frac{z'_{\boldsymbol{a}+}}{z'_{\boldsymbol{a}-}}, \; z'_{\boldsymbol{a}\pm} = x'_a \pm i y'_a \text{ as } \boldsymbol{r}'_p - \boldsymbol{a}_p = x'_a \widehat{\boldsymbol{c}}_p + y'_a(\widehat{\boldsymbol{z}} \times \widehat{\boldsymbol{c}}_p). \tag{6.109}$$

The first contour integral in (6.107) is in whole generality non-singular as we take $A_0 = e^{\frac{\boldsymbol{w}'^2_m}{4ik\xi}}$, but, when $(\boldsymbol{r}'_p - \boldsymbol{a}_p)^2 \neq 0$ on $\mathcal{L}_p$, the choice $A_0 = 0$ could be suitable for asymptotic evaluations if v is large.

PROOF.– From the proposition 6.6 and $p\, dS' = dS'_p$, we have:

$$\int_{S'} e^{(\alpha_1 x' + \alpha_2 y') - ik(h_1 x'^2 + h_2 y'^2 + h_3 x' y')} dS' = \frac{1}{p} \int_{S'_p} e^{v(\boldsymbol{r}'_p - \boldsymbol{a}_p)^2 + v_1} dS'_p. \tag{6.110}$$

We can then apply theorem 6.1 to the surface integral following:

$$\int_{S'_p} e^{v(\boldsymbol{r}'_p - \boldsymbol{a}_p)^2 + v_1} dS_p = \frac{1}{2} \Big(\int_{\mathcal{L}_p} \left(\frac{\widehat{\boldsymbol{z}} \times (\boldsymbol{r}'_p - \boldsymbol{a}_p)(e^{-ik\xi(\boldsymbol{r}'_p - \boldsymbol{a}_p)^2 + v_1} - A_0)}{v(\boldsymbol{r}'_p - \boldsymbol{a}_p)^2} \right) . d\boldsymbol{r}'_p +$$

$$+ (A_0 - e^{v_1}) \int_{\mathcal{L}_p} (\widehat{\boldsymbol{z}} \times \frac{\boldsymbol{r}'_p - \boldsymbol{a}_p}{v(\boldsymbol{r}'_p - \boldsymbol{a}_p)^2}) . d\boldsymbol{r}'_p \Big), \tag{6.111}$$

and derive (6.107), definite for $(\boldsymbol{r}'_{\boldsymbol{p}} - \boldsymbol{a}_p)^2 \neq 0$ on $\mathcal{L}_p$, or, in whole generality, for $A_0 = e^{v_1}$. We obtain (6.108) from proposition 6.1.

NOTE 6.18.– When the terms $\frac{(\boldsymbol{r}'_{\boldsymbol{p}} \pm i(\boldsymbol{r}'_{\boldsymbol{p}} \times \widehat{\boldsymbol{z}})).\widehat{\boldsymbol{c}}_p}{(\boldsymbol{a}_p \pm i(\boldsymbol{a}_p \times \widehat{\boldsymbol{z}})).\widehat{\boldsymbol{c}}_p}$ are small as $\boldsymbol{r}'_{\boldsymbol{p}} \in \mathcal{L}_p$, we note that:

$$\max |\mathrm{Im}(\ln(\frac{1 - \frac{(\boldsymbol{r}'_{\boldsymbol{p}} + i(\boldsymbol{r}'_{\boldsymbol{p}} \times \widehat{\boldsymbol{z}})).\widehat{\boldsymbol{c}}_p}{(\boldsymbol{a}_p + i(\boldsymbol{a}_p \times \widehat{\boldsymbol{z}})).\widehat{\boldsymbol{c}}_p}}{1 - \frac{(\boldsymbol{r}'_{\boldsymbol{p}} - i(\boldsymbol{r}'_{\boldsymbol{p}} \times \widehat{\boldsymbol{z}})).\widehat{\boldsymbol{c}}_p}{(\boldsymbol{a}_p - i(\boldsymbol{a}_p \times \widehat{\boldsymbol{z}})).\widehat{\boldsymbol{c}}_p}}))|_{\boldsymbol{r}'_{\boldsymbol{p}} \in \mathcal{L}_p} \ll 1, \tag{6.112}$$

and thus, the term $\mathrm{Im}(\ln(\frac{z'_{a+}}{z'_{a-}}))$ varies little, $\frac{z'_{a+}}{z'_{a-}}$ does not run round the origin, $N_+ = N_-$ and the winding number $(N_+ - N_-)$ is null in (6.108).

6.6.2.3. *Some properties of $\mathcal{P}(\boldsymbol{r}') = \boldsymbol{r}'_{\boldsymbol{p}}$ to analyze and transform the integral on $\mathcal{L}_p$*

PROPOSITION 6.8.– *From the definitions of $\mathcal{P}$ and $\mathcal{P}^{-1}$ (see definition 6.4), we have:*

$$\mathcal{P}(\boldsymbol{u}).\boldsymbol{v} = \boldsymbol{u}.\mathcal{P}(\boldsymbol{v}),\ \mathcal{P}(\widehat{\boldsymbol{z}} \times \boldsymbol{u}) = p\,\widehat{\boldsymbol{z}} \times \mathcal{P}^{-1}(\boldsymbol{u}),\ \mathcal{P}(\widehat{\boldsymbol{z}} \times \mathcal{P}(\boldsymbol{v})) = p(\widehat{\boldsymbol{z}} \times \boldsymbol{v}), \tag{6.113}$$

for any vectors $\boldsymbol{u}$ and $\boldsymbol{v}$ of the plane. Thus, we can write:

$$\boldsymbol{u}.d\boldsymbol{r}'_{\boldsymbol{p}} = \mathcal{P}(\boldsymbol{u}).d\boldsymbol{r}',\ (\widehat{\boldsymbol{z}} \times \boldsymbol{r}'_{\boldsymbol{p}}).d\boldsymbol{r}'_{\boldsymbol{p}} = p\,(\widehat{\boldsymbol{z}} \times \boldsymbol{r}').d\boldsymbol{r}', \tag{6.114}$$

and note that $\boldsymbol{w}'_m.\boldsymbol{r}'_{\boldsymbol{p}} = \boldsymbol{w}'.\boldsymbol{r}'$ with $\boldsymbol{r}'_{\boldsymbol{p}} = \mathcal{P}(\boldsymbol{r}')$ implies:

$$\mathcal{P}(\boldsymbol{w}'_m) = \boldsymbol{w}',\ \textit{and}\ \boldsymbol{w}'_m = \mathcal{P}^{-1}(\boldsymbol{w}'). \tag{6.115}$$

Concerning the properties relatively to the choice of $\epsilon = \pm 1$ (factor used in (6.90) for ξ), we can let $(\mathbf{c}_p)_\pm = (\widehat{\mathbf{c}}_p)_\pm\sqrt{(\mathbf{c}_p^2)_\pm}$, $p_\pm$, $\xi_\pm$, $(\mathcal{P})_\pm$ and $(\boldsymbol{r}'_{\boldsymbol{p}})_\pm$, be the respective terms $(\mathbf{c}_p) = (\widehat{\mathbf{c}}_p)\sqrt{(\mathbf{c}_p^2)}$, ξ, p, $\mathcal{P}$ and $(\boldsymbol{r}'_{\boldsymbol{p}})$ for $\epsilon = \pm 1$, and notice that:

$$p_\pm^2 = \frac{\xi_{\mp 1}}{\xi_{\pm 1}} = \frac{1}{p_\mp^2},\ (\xi_{\pm 1}p_\pm)^2 = (\xi_{\mp 1}p_\mp)^2 = \xi_+\xi_- = h_1h_2 - (h_3/2)^2,$$

$$p_\pm^2 = 1 - (\mathbf{c}_p^2)_\pm = -1 + \frac{h_1 + h_2}{\xi_\pm},\ (\boldsymbol{r}'_{\boldsymbol{p}})_\pm = p_\pm(\boldsymbol{r}'_{\boldsymbol{p}})_\mp,\ (\mathcal{P})_\pm = p_\pm(\mathcal{P})_\mp,$$

$$(\widehat{\mathbf{c}}_p)_\pm = \epsilon_{c\pm}(\widehat{\boldsymbol{z}} \times (\widehat{\mathbf{c}}_p)_\mp),\ \epsilon_{c\pm} = \frac{(c_{py})_\pm\sqrt{(\mathbf{c}_p)^2_\mp}}{(c_{px})_\mp\sqrt{(\mathbf{c}_p)^2_\pm}} = -\frac{(c_{px})_\pm\sqrt{(\mathbf{c}_p)^2_\mp}}{(c_{py})_\mp\sqrt{(\mathbf{c}_p)^2_\pm}}, \tag{6.116}$$

where $(\epsilon_c)^2 = 1$, and $-\epsilon \equiv \mp$ as $\epsilon \equiv \pm$. We remark that $\xi_{-\epsilon} = 0$ implies $h_3^2 = 4h_1h_2$, $\xi_\epsilon = h_1 + h_2$, and $(\mathbf{c}_p^2)_\epsilon = 1$ if $h_1 + h_2 \neq 0$, while $(\mathbf{c}_p^2)_\epsilon = 1$ implies $\xi_{-\epsilon} = 0$ and

$\xi_\epsilon = h_1 + h_2$, *and notice that* $(\mathbf{c}_p^2)_\epsilon = 0$ *implies* $(\mathbf{c}_p^2)_{-\epsilon} = 0$, $2\xi_\pm = h_1 + h_2$ *and* $p_\pm = 1$. *(note: even if* $(\xi p)^2 = h_1 h_2 - (h_3/2)^2$, *we do not have to calculate* $p\xi$ *as* $\sqrt{(p\xi)^2}$, *but with each term* p *and* ξ *determined independently to avoid the cut of square root).*

PROOF.– The properties in (6.113) derives from linearity. For (6.116), we consider (6.82) and notice that:

$$1 - (\mathbf{c}_p^2)_\pm = -1 + \frac{h_1 + h_2}{\xi_\pm} = \frac{\xi_\mp}{\xi_\pm} = \frac{1}{1 - (\mathbf{c}_p^2)_\mp},$$

$$\frac{(\widehat{c}_{px})_\epsilon}{(\widehat{c}_{py})_\epsilon} = \frac{(c_{px})_\epsilon (c_{py})_\epsilon}{(c_{py})_\epsilon (c_{py})_\epsilon} = \frac{-h_3}{2(\xi_\epsilon - h_2)} = \frac{-h_3}{(h_1 - h_2) + \epsilon\sqrt{(h_1 - h_2)^2 + h_3^2}}$$

$$= \frac{(h_1 - h_2) - \epsilon\sqrt{(h_1 - h_2)^2 + h_3^2}}{h_3} = -\frac{(c_{py})_{-\epsilon}}{(c_{px})_{-\epsilon}} = -\frac{(\widehat{c}_{py})_{-\epsilon}}{(\widehat{c}_{px})_{-\epsilon}}. \quad (6.117)$$

PROPOSITION 6.9.– *Let from now on:*

$$\mathcal{M}(\boldsymbol{w}') = p^2 \mathcal{P}^{-1}(\boldsymbol{w}'_m), \text{ with } p\mathcal{P}(\widehat{\boldsymbol{z}} \times \boldsymbol{w}'_m) = \widehat{\boldsymbol{z}} \times \mathcal{M}(\boldsymbol{w}'),$$

$$\mathcal{N}(\boldsymbol{w}') = p\,\boldsymbol{w}'_m, \text{ with } v_1 = ik\xi(\boldsymbol{a}_p)^2 = \frac{\boldsymbol{w}'^2_m}{4ik\xi} = \frac{(\mathcal{N}(\boldsymbol{w}'))^2}{4ik\xi p^2}, \quad (6.118)$$

where $\boldsymbol{a}_p = \frac{\boldsymbol{w}'_m}{2ik\xi}$, $\boldsymbol{w}'_m = \mathcal{P}^{-1}(\boldsymbol{w}')$, *and* $p = \sqrt{1 - \mathbf{c}_p^2}$. *We can write, for* $\boldsymbol{r}'_{\boldsymbol{p}} = \mathcal{P}(\boldsymbol{r}')$:

$$\frac{(\widehat{\boldsymbol{z}} \times \frac{(\boldsymbol{r}'_p - \boldsymbol{a}_p)}{(\boldsymbol{r}'_p - \boldsymbol{a}_p)^2}).d\boldsymbol{r}'_{\boldsymbol{p}}}{-2ikp\xi} = \frac{\frac{1}{2}(\widehat{\boldsymbol{z}} \times (\boldsymbol{r}' - \frac{\mathcal{M}(\boldsymbol{w}')}{2ik\xi p^2})).d\boldsymbol{r}'}{\boldsymbol{w}'.\boldsymbol{r}' - ik(h_1 x'^2 + h_2 y'^2 + h_3 x'y') - \frac{(\mathcal{N}(\boldsymbol{w}'))^2}{4ik\xi p^2}}, \quad (6.119)$$

where

$$\frac{\mathcal{M}(\boldsymbol{w}')}{2ik\xi p^2} = \mathcal{P}^{-1}(\boldsymbol{a}_p) = \frac{(1 - \mathbf{c}_p^2)\boldsymbol{w}' + (\boldsymbol{c}_p.\boldsymbol{w}')\boldsymbol{c}_p}{2ik\xi p^2}$$

$$= \frac{1}{2ik(\xi p)^2}((h_2 w'_x - \frac{h_3 w'_y}{2})\widehat{\boldsymbol{x}} + (h_1 w'_y - \frac{h_3 w'_x}{2})\widehat{\boldsymbol{y}}), \quad (6.120)$$

and

$$\frac{(\mathcal{N}(\boldsymbol{w}'))^2}{4ik\xi p^2} = \frac{\boldsymbol{w}'^2_m}{4ik\xi} = \frac{(1 - \mathbf{c}_p^2)\boldsymbol{w}'^2 + (\boldsymbol{c}_p.\boldsymbol{w}')^2}{4ik\xi p^2}$$

$$= \frac{1}{4ik(\xi p)^2}(h_2 w'^2_x + h_1 w'^2_y - h_3 w'_x w'_y), \quad (6.121)$$

with $(\xi p)^2 = \xi_+\xi_- = h_1h_2 - (h_3/2)^2$. *Let us notice that when* $h_3 = 0$ *and* $h_1 = h_2 \to 0$,

$$\frac{(\widehat{\boldsymbol{z}} \times \frac{(\boldsymbol{r'_p}-\boldsymbol{a}_p)}{(\boldsymbol{r'_p}-\boldsymbol{a}_p)^2}).d\boldsymbol{r'_p}}{-2ikp\xi} = \frac{(\widehat{\boldsymbol{z}} \times \boldsymbol{w'})}{(\boldsymbol{w'})^2}.d\boldsymbol{r'}. \tag{6.122}$$

PROOF.– Considering (6.92) and (6.114)–(6.118), we obtain (6.119), then we derive (6.120) from:

$$\begin{aligned}
\mathcal{M}(\boldsymbol{w'}) &= p^2\mathcal{P}^{-2}(\boldsymbol{w'})\\
&= (\widehat{\boldsymbol{c}}_p.\boldsymbol{w'})\widehat{\boldsymbol{c}}_p + (1-\boldsymbol{c}_p^2)(\boldsymbol{w'} - \widehat{\boldsymbol{c}}_p(\widehat{\boldsymbol{c}}_p.\boldsymbol{w'})) = (1-\boldsymbol{c}_p^2)\boldsymbol{w'} + (\boldsymbol{c}_p.\boldsymbol{w'})\boldsymbol{c}_p\\
&= (-1+\frac{h_1+h_2}{\xi})\boldsymbol{w'} + (1-\frac{h_1}{\xi})w'_x\widehat{\boldsymbol{x}} + (1-\frac{h_2}{\xi})w'_y\widehat{\boldsymbol{y}} - \frac{h_3}{2\xi}(w'_y\widehat{\boldsymbol{x}} + w'_x\widehat{\boldsymbol{y}})\\
&= \frac{1}{\xi}(h_2w'_x\widehat{\boldsymbol{x}} + h_1w'_y\widehat{\boldsymbol{y}} - \frac{h_3}{2}(w'_y\widehat{\boldsymbol{x}} + w'_x\widehat{\boldsymbol{y}}))\\
&= \frac{1}{\xi}((h_2w'_x - \frac{h_3w'_y}{2})\widehat{\boldsymbol{x}} + (h_1w'_y - \frac{h_3w'_x}{2})\widehat{\boldsymbol{y}}),
\end{aligned} \tag{6.123}$$

and

$$\begin{aligned}
(\mathcal{N}(w'))^2 &= \frac{\boldsymbol{w'}^2_m}{4ik\xi} = (p\mathcal{P}^{-1}(\boldsymbol{w'}))^2 = (\widehat{\boldsymbol{c}}_p.\boldsymbol{w'})^2 + (1-\boldsymbol{c}_p^2)(\boldsymbol{w'} - \widehat{\boldsymbol{c}}_p(\widehat{\boldsymbol{c}}_p.\boldsymbol{w'}))^2\\
&= (1-\boldsymbol{c}_p^2)\boldsymbol{w'}^2 + (\boldsymbol{c}_p.\boldsymbol{w'})^2\\
&= (-1+\frac{h_1+h_2}{\xi})(w'^2_x + w'^2_y) + (1-\frac{h_1}{\xi})w'^2_x + (1-\frac{h_2}{\xi})w'^2_y - \frac{h_3}{\xi}w'_xw'_y\\
&= \frac{1}{\xi}(h_2w'^2_x + h_1w'^2_y - h_3w'_xw'_y).
\end{aligned} \tag{6.124}$$

PROPOSITION 6.10.– *Let* $z'_{\boldsymbol{a}\pm} = (\boldsymbol{r'_p} - \boldsymbol{a}_p).(\widehat{\boldsymbol{x}} \pm i\widehat{\boldsymbol{y}}) = x'_a \pm iy'_a$, *where* $\boldsymbol{r'_p} - \boldsymbol{a}_p = x'_a\widehat{\boldsymbol{c}}_p + y'_a(\widehat{\boldsymbol{z}} \times \widehat{\boldsymbol{c}}_p)$ *with* $\boldsymbol{r'_p} = \mathcal{P}(\boldsymbol{r'})$, *then* $z'_{\boldsymbol{a}\pm} = 0$ *on lines* $\boldsymbol{r'} = \mathcal{P}^{-1}(\boldsymbol{a}_p) + t\boldsymbol{u}_c$, $z\boldsymbol{u}_c = \widehat{\boldsymbol{c}}_p \pm ip(\widehat{\boldsymbol{z}} \times \widehat{\boldsymbol{c}}_p)$. *For real* $\boldsymbol{r'}$, *each equation* $z'_{\boldsymbol{a}\pm} = 0$ *has no or an infinity of solutions if* $\frac{\mp ip\xi + \frac{h_3}{2}}{h_2} (= \frac{h_1}{\pm ip\xi + \frac{h_3}{2}})$ *is real, and a single one* $\boldsymbol{r}_\pm$ *else. In the latter case, we notice that* $\boldsymbol{r}_+ = \boldsymbol{r}_- = \mathcal{P}^{-1}(\boldsymbol{a}_p)$ *if* $\mathcal{P}^{-1}(\boldsymbol{a}_p)$ *is a point with real coordinates, for example, when* $\boldsymbol{a}_p$ *and* $\mathcal{P}(\boldsymbol{r'})$ *are real.*

PROOF.– From $z'_{\boldsymbol{a}\pm} = (\boldsymbol{r'_p} - \boldsymbol{a}_p).(\widehat{\boldsymbol{x}} \pm i\widehat{\boldsymbol{y}})$, $(\widehat{\boldsymbol{x}} \pm i\widehat{\boldsymbol{y}})^2 = 0$ and $\widehat{\boldsymbol{x}} \pm i\widehat{\boldsymbol{y}} = \frac{\mathbf{c} \pm i\widehat{\boldsymbol{z}} \times \mathbf{c}}{c_x \mp ic_y}$ as $\boldsymbol{c}^2 \neq 0$, we have $z'_{\boldsymbol{a}\pm} = 0$ as $\boldsymbol{r'} = \mathcal{P}^{-1}(\boldsymbol{a}_p) + t\boldsymbol{u}_c$, $\boldsymbol{u}_c = p\mathcal{P}^{-1}(\widehat{\boldsymbol{c}}_p \pm i(\widehat{\boldsymbol{z}} \times \widehat{\boldsymbol{c}}_p))$ and $z'_{\boldsymbol{a}\pm} = \frac{(\boldsymbol{r'} - \mathcal{P}^{-1}(\boldsymbol{a}_p)).\boldsymbol{v}_c}{c_x \mp ic_y}$ with $\boldsymbol{v}_c = \mathcal{P}(\boldsymbol{c}_p \pm i\widehat{\boldsymbol{z}} \times \boldsymbol{c}_p) = p\boldsymbol{c}_p \pm i\widehat{\boldsymbol{z}} \times \boldsymbol{c}_p$; we can consider solutions of $\boldsymbol{v}_c.\boldsymbol{r}' = 0$ with:

$$\frac{v_{cx}}{v_{cy}} = \frac{pc_x \mp ic_y}{pc_y \pm ic_x} = \frac{\mp ip(c_x^2+c_y^2) + c_xc_y(p^2-1)}{p^2c_y^2+c_x^2}$$

$$= \frac{\mp ip - c_xc_y}{1-c_y^2} = \frac{\mp ip\xi + \frac{h_3}{2}}{h_2}, \tag{6.125}$$

and deduce the last statement. Let us remark that $\boldsymbol{u}_c.\boldsymbol{v}_c = 0$ and $(p\xi)^2 = h_1h_2 - (\frac{h_3}{2})^2$.

6.6.2.4. *Final reduction to an integral on contour $\mathcal{L}$ of S' for arbitrary quadratic arguments*

THEOREM 6.11.– *We can reduce the integral of an exponential with arbitrary quadratic argument on S', to an integral along the positively oriented contour $\mathcal{L}$ of S', following:*

$$\int_{S'} e^{\boldsymbol{w}'.\boldsymbol{r}'-ik(h_1x'^2+h_2y'^2+h_3x'y')}dS' =$$

$$= \frac{1}{2}\int_{\mathcal{L}}\Big[\frac{(e^{\boldsymbol{w}'.\boldsymbol{r}'-ik(h_1x'^2+h_2y'^2+h_3x'y')} - A_0)}{\boldsymbol{w}'.\boldsymbol{r}' - ik(h_1x'^2+h_2y'^2+h_3x'y') - \frac{\boldsymbol{w}'^2_m}{4ik\xi}}\Big][(\widehat{\boldsymbol{z}}\times(\boldsymbol{r}' - \frac{\mathcal{M}(\boldsymbol{w}')}{2ik\xi p^2})).d\boldsymbol{r}']$$

$$+\frac{(A_0 - e^{\frac{\boldsymbol{w}'^2_m}{4ik\xi}})}{-2ikp\xi}\int_{\mathcal{L}_p}(\widehat{\boldsymbol{z}}\times\frac{\boldsymbol{r}'_{\boldsymbol{p}}-\boldsymbol{a}_p}{(\boldsymbol{r}'_{\boldsymbol{p}}-\boldsymbol{a}_p)^2}).d\boldsymbol{r}'_{\boldsymbol{p}}, \tag{6.126}$$

with $\frac{\mathcal{M}(\boldsymbol{w}')}{2ik\xi p^2}$ and $\frac{\boldsymbol{w}'^2_m}{4ik\xi}$ detailed in proposition 6.9, arbitrary A_0 and $\widehat{\boldsymbol{y}} = \widehat{\boldsymbol{z}}\times\widehat{\boldsymbol{x}}$. The integral on $\mathcal{L}$ is non-singular when we take $A_0 = e^{\frac{\boldsymbol{w}'^2_m}{4ik\xi}}$. However, for $A_0 \neq e^{\frac{\boldsymbol{w}'^2_m}{4ik\xi}}$, we can notice, from proposition 6.1, that, when $(\boldsymbol{r}'_{\boldsymbol{p}}-\boldsymbol{a}_p)^2 \neq 0$ on $\mathcal{L}$:

$$\int_{\mathcal{L}_p}(\widehat{\boldsymbol{z}}\times\frac{\boldsymbol{r}'_p-\boldsymbol{a}_p}{(\boldsymbol{r}'_{\boldsymbol{p}}-\boldsymbol{a}_p)^2}).d\boldsymbol{r}'_{\boldsymbol{p}} = \frac{-i}{2}\int_{\mathcal{L}} d\ln(\chi(\mathcal{P}(\boldsymbol{r}'))) = \pi(N_+ - N_-), \tag{6.127}$$

where N_+ (respectively, N_-) is the number of times the complex $\chi(\mathcal{P}(\boldsymbol{r}'))$ crosses any smooth semi-line $\mathcal{D}_1$ from the origin to ∞, in the positive sense (respectively, negative), with $\chi(\mathcal{P}(\boldsymbol{r}')) = \frac{x'_a+iy'_a}{x'_a-iy'_a}$, $\mathcal{P}(\boldsymbol{r}') - \boldsymbol{a}_p = x'_a\widehat{\boldsymbol{c}}_{\boldsymbol{p}} + y'_a(\widehat{\boldsymbol{z}}\times\widehat{\boldsymbol{c}}_{\boldsymbol{p}})$, $\mathcal{P}(\boldsymbol{r}') = \boldsymbol{r}'_{\boldsymbol{p}}$.

Although this expression is in whole generality non-singular for $A_0 = e^{\frac{\boldsymbol{w}'^2_m}{4ik\xi}}$, it is worth noticing that, when $(\boldsymbol{r}'_{\boldsymbol{p}}-\boldsymbol{a}_p)^2 \neq 0$ on $\mathcal{L}$, the choice $A_0 = 0$ could be suitable for an asymptotic evaluation as k is large.

PROOF.– For that, we rewrite proposition 6.7 from a direct use of propositions 6.8 and 6.9.

NOTE 6.19.– Concerning $\boldsymbol{r}'$ deduced from complex $\boldsymbol{r}'_{\boldsymbol{p}} - \boldsymbol{a}_p$, we note that, after a use of (6.9), we have:

$$\boldsymbol{r}' = \mathcal{P}^{-1}(\boldsymbol{a}_p) + h'_p\,\boldsymbol{u}_{\varphi_p}, \boldsymbol{u}_{\varphi_p} = \mathcal{P}^{-1}(\cos\varphi_p\,\widehat{\boldsymbol{x}} + \sin\varphi_p\,\widehat{\boldsymbol{y}}),\ h'^2_p = (\boldsymbol{r}'_{\boldsymbol{p}} - \boldsymbol{a}_p)^2, \tag{6.128}$$

except on lines $\boldsymbol{r}' = \mathcal{P}^{-1}(\boldsymbol{a}_p) + t\boldsymbol{u}_c$ with $\boldsymbol{u}_c = \widehat{\boldsymbol{c}}_p \pm ip(\widehat{\boldsymbol{z}} \times \widehat{\boldsymbol{c}}_p)$, where $h'^2_p = (\boldsymbol{r}'_{\boldsymbol{p}} - \boldsymbol{a}_p)^2 = 0$ and $\boldsymbol{u}_{\varphi_p}$ is then indefinite (see (6.10)). To avoid this possible situation on $\mathcal{L}$ (and thus $\mathcal{L}_p$), we can choose to apply infinitely small perturbations in the complex plane (permitted in particular in proposition 6.1).

NOTE 6.20.– For the case where $\boldsymbol{a}_p$ tends to infinity, in particular when $h_1h_2 \to (h_3/2)^2$, we can use that, when (6.112) is satisfied, we have $\int_{\mathcal{L}} d\ln(\chi(\mathcal{P}(\boldsymbol{r}'))) = 0$.

6.7. Complex scaling and extended cases of applications for exact reductions

We now consider the surface integral of a function $g(\boldsymbol{r}')f(\mathcal{R}^2)$, completing and generalizing results of the previous section, when $\mathcal{R}^2$ is a polynomial of coordinates:

$$\begin{aligned}\mathcal{R}^2 &= d_0 + (\alpha_1 x' + \alpha_2 y') - ik(h_1 x'^2 + h_2 y'^2 + h_3 x'y') \\ &= d_0 + v(\boldsymbol{r}'_{\boldsymbol{p}} - \boldsymbol{a}_p)^2 + v_1,\end{aligned} \tag{6.129}$$

with $\boldsymbol{r}'_{\boldsymbol{p}} = \mathcal{P}(\boldsymbol{r}')$, and $\mathcal{P}$ a complex scaling verifying the definition 6.4. So, we determine the exact reduction of the general surface integral:

$$\int_{S'} g(r')f(\mathcal{R}^2)dS', \tag{6.130}$$

when the primitive of f satisfies the general conditions of analyticity given in definition 6.3, and $g(r')$ is a polynomial of coordinates. We will define, as previously, the complex surface S'_p, of contour $\mathcal{L}_p$, as the transform by $\mathcal{P}$ of the real surface S' of arbitrary contour $\mathcal{L}$, with $dS' = dx' \wedge dy'$, $dS'_p = dx'_p \wedge dy'_p$, and $dS' = dS'_p/p$.

6.7.1. *Exact reduction when* $g \equiv 1$

THEOREM 6.12.– *Let* F_v *be a primitive of* f*, and we can write:*

$$\int_{S'} f(\mathcal{R}^2)dS' = \frac{1}{2}\int_{\mathcal{L}}[\frac{F_v(\mathcal{R}^2) - A_0}{\mathcal{R}^2 - (\frac{\boldsymbol{w}'^2_m}{4ik\xi} + d_0)}][(\widehat{\boldsymbol{z}} \times (\boldsymbol{r}' - \mathcal{P}^{-1}(\boldsymbol{a}_p))).d\boldsymbol{r}']$$

$$+\frac{A_0 - F_v(\frac{\boldsymbol{w}'^2_m}{4ik\xi} + d_0)}{-2ikp\xi}\int_{\mathcal{L}_p}(\widehat{\boldsymbol{z}} \times \frac{\boldsymbol{r}'_p - \boldsymbol{a}_p}{(\boldsymbol{r}'_{\boldsymbol{p}} - \boldsymbol{a}_p)^2}).d\boldsymbol{r}'_{\boldsymbol{p}}, \tag{6.131}$$

where A_0 is an arbitrary constant, $p = \sqrt{1 - c_p^2}$, and:

$$\mathcal{P}^{-1}(\boldsymbol{a}_p) = \frac{1}{2ik(h_1h_2 - (h_3/2)^2)}((h_2w'_x - \frac{h_3w'_y}{2})\widehat{\boldsymbol{x}} + (h_1w'_y - \frac{h_3w'_x}{2})\widehat{\boldsymbol{y}}),$$

$$\frac{\boldsymbol{w'}^2_m}{4ik\xi} = \frac{1}{4ik(h_1h_2 - (h_3/2)^2)}(h_2w'^2_x + h_1w'^2_y - h_3w'_xw'_y), \tag{6.132}$$

while $\int_{\mathcal{L}_p}(\widehat{\boldsymbol{z}} \times \frac{\boldsymbol{r}'_p - \boldsymbol{a}_p}{(\boldsymbol{r}'_p - \boldsymbol{a}_p)^2}).d\boldsymbol{r}'_p$ can be determined with (6.127) (note: for most generality, instead of calculating $p\xi$ on the second line of (6.131) as $\sqrt{(p\xi)^2}$, it will be preferable to determine each term p and ξ separately to avoid problems with the cut of square root).

PROOF.– In proposition 6.7 and theorem 6.11, we change the exponential function for f, and let:

$$\frac{F_v(\mathcal{R}^2)}{v} = \int f(d_0 + v(\boldsymbol{r}'_p - \boldsymbol{a}_p)^2 + v_1)d(\boldsymbol{r}'_p - \boldsymbol{a}_p)^2 = \frac{1}{v}\int f(\mathcal{R}^2)d\mathcal{R}^2. \tag{6.133}$$

We can then write:

$$\begin{aligned}\int_{S'} f(\mathcal{R}^2)dS' &= \frac{1}{p}\int_{S'_p} f(d_0 + v(\boldsymbol{r}'_p - \boldsymbol{a}_p)^2 + v_1)dS'_p \\ &= \int_{\mathcal{L}_p} \frac{F_v(\mathcal{R}^2) - A_0}{2p}\frac{(\widehat{\boldsymbol{z}} \times (\boldsymbol{r}'_p - \boldsymbol{a}_p)).d\boldsymbol{r}'_p}{\mathcal{R}^2 - (v_1 + d_0)} \\ &+ \frac{(A_0 - F(v_1 + d_0))}{2vp}\int_{\mathcal{L}_p}(\widehat{\boldsymbol{z}} \times \frac{\boldsymbol{r}'_p - \boldsymbol{a}_p}{(\boldsymbol{r}'_p - \boldsymbol{a}_p)^2}).d\boldsymbol{r}'_p,\end{aligned} \tag{6.134}$$

and (6.131) from $(\widehat{\boldsymbol{z}} \times (\boldsymbol{r}'_p - \boldsymbol{a}_p)).d\boldsymbol{r}'_p = p(\widehat{\boldsymbol{z}} \times (\boldsymbol{r}' - \mathcal{P}^{-1}(\boldsymbol{a}_p))).d\boldsymbol{r}'$.

6.7.2. Exact reduction when $g(\boldsymbol{r}')$ is a polynomial of coordinates

Considering that $\boldsymbol{r}' = \mathcal{P}^{-1}(\boldsymbol{r}'_p)$, and:

$$\begin{aligned}\widehat{\boldsymbol{x}}.(\boldsymbol{r}' - \mathcal{P}^{-1}(\boldsymbol{a}_p)) &= \widehat{\boldsymbol{x}}.\mathcal{P}^{-1}(\boldsymbol{r}'_p - \boldsymbol{a}_p) = \mathcal{P}^{-1}(\widehat{\boldsymbol{x}}).(\boldsymbol{r}'_p - \boldsymbol{a}_p), \\ \widehat{\boldsymbol{y}}.(\boldsymbol{r}' - \mathcal{P}^{-1}(\boldsymbol{a}_p)) &= \widehat{\boldsymbol{y}}.\mathcal{P}^{-1}(\boldsymbol{r}'_p - \boldsymbol{a}_p) = \mathcal{P}^{-1}(\widehat{\boldsymbol{y}}).(\boldsymbol{r}'_p - \boldsymbol{a}_p),\end{aligned} \tag{6.135}$$

we notice, from Taylor expansion, that we have, for any polynomial $g(\boldsymbol{r}')$ of order m:

$$\begin{aligned}g(\boldsymbol{r}') = \sum_{j=0}^{m}\frac{1}{j!}\sum_{k=0}^{j}\binom{j}{k}\frac{\partial^j g|_{\boldsymbol{r}'=\mathcal{P}^{-1}(\boldsymbol{a}_p)}}{\partial x'^k \partial y'^{j-k}} \\ [(\mathcal{P}^{-1}(\widehat{\boldsymbol{x}}).(\boldsymbol{r}'_p - \boldsymbol{a}_p))^k(\mathcal{P}^{-1}(\widehat{\boldsymbol{y}}).(\boldsymbol{r}'_p - \boldsymbol{a}_p))^{j-k}].\end{aligned} \tag{6.136}$$

THEOREM 6.13.– *Considering a polynomial $g(\boldsymbol{r'})$ of order m, we can develop the exact reduction of the integral $\int_{S'} g(\boldsymbol{r'})f(\mathcal{R}^2)dS'$ on the surface S' of the plane. For that, we need the reduction of integral:*

$$I_0 = \int_{S'} f(\mathcal{R}^2)dS', \tag{6.137}$$

given by theorem 6.12 in (6.131), and of the elementary integrals:

$$I_{n+1} = \int_{S'} \boldsymbol{d}_{n+1}.(\boldsymbol{r}'_p - \boldsymbol{a}_p)\prod_{j=1}^{n}(\boldsymbol{d}_j.(\boldsymbol{r}'_p - \boldsymbol{a}_p))f(\mathcal{R}^2)dS', \tag{6.138}$$

with $1 \leq n+1 \leq m$, constants $\boldsymbol{d}_j$, and complex scaling $\mathcal{P}$ for (6.129). For that, we generalize theorem 6.4 and write:

$$I_{n+1} = \frac{-1}{2ik\xi}\Big(\frac{1}{p}\int_{\mathcal{L}}\prod_{j=1}^{n}(\boldsymbol{d}_j.(\boldsymbol{r}'_p - \boldsymbol{a}_p))F_v(\mathcal{R}^2)\mathcal{P}(\widehat{\boldsymbol{z}}\times\boldsymbol{d}_{n+1}).d\boldsymbol{r'}$$
$$-\sum_{m=1}^{n}((\boldsymbol{d}_{n+1}.\boldsymbol{d}_m)\int_{S'}\prod_{j\neq m}^{n}(\boldsymbol{d}_j.(\boldsymbol{r}'_p - \boldsymbol{a}_p))F_v(\mathcal{R}^2)dS')\Big), \tag{6.139}$$

where F_v is a primitive of f, $\mathcal{P}(\widehat{\boldsymbol{z}}\times\boldsymbol{d}_{n+1}) = p\,\widehat{\boldsymbol{z}}\times\mathcal{P}^{-1}(\boldsymbol{d}_{n+1})$, and $\prod_{j=1}^{0}(..) = 1$, $\prod_{j\neq i}^{1}(..) = 1$, and $\prod_{j\neq i}^{0}(..) = 0$. Subsequently, we have decreased the rank of polynomial in the surface integral term, and it becomes possible, by a repeated application of (6.139), to reduce I_{n+1} to a combination of contour integrals depending on M= $\lfloor(n+3)/2\rfloor$ successive primitives of f, denoted $\frac{1}{q!}\int^x(x-t)^q f(t)dt$, $q = 0, ..., M-1$.

PROOF.– Taking up theorem 6.4, $(\boldsymbol{r}'_p - \boldsymbol{a}_p)$ and dS'_p instead of $(\boldsymbol{r'} - \boldsymbol{a})$ and dS', with $v = -ik\xi$ and $dS'_p \equiv pdS'$, we derive (6.139). By repeated applications, we can decrease the rank, and using theorem 6.12, reduce I_{n+1} to a combination of contour integrals for any $n \geq 0$, using repeated integrals of f (see the results given in appendix A for successive primitives and their asymptotics).

NOTE 6.21.– Letting $\mathbf{I}_{n+1} = \mathcal{P}^{-1}(\boldsymbol{d}_{n+1}).\boldsymbol{\mathcal{I}}_n$ in (6.139), we use $\mathcal{P}(\widehat{\boldsymbol{z}}\times\boldsymbol{d}_{n+1}) = p\widehat{\boldsymbol{z}}\times\mathcal{P}^{-1}(\boldsymbol{d}_{n+1})$ and $\boldsymbol{d}_{n+1}.\boldsymbol{d}_m = \mathcal{P}^{-1}(\boldsymbol{d}_{n+1}).\mathcal{P}(\boldsymbol{d}_m)$, and then note that:

$$\boldsymbol{\mathcal{I}}_n = \frac{1}{2ik\xi}\Big(\widehat{\boldsymbol{z}}\times\int_{\mathcal{L}}\prod_{j=1}^{n}(\boldsymbol{d}_j.(\boldsymbol{r}'_p - \boldsymbol{a}_p))F_v(\mathcal{R}^2)d\boldsymbol{r'}$$
$$+\sum_{m=1}^{n}(\mathcal{P}(\boldsymbol{d}_m)\int_{S'}\prod_{j\neq m}^{n}(\boldsymbol{d}_j.(\boldsymbol{r}'_p - \boldsymbol{a}_p))F_v(\mathcal{R}^2)dS')\Big). \tag{6.140}$$

NOTE 6.22.– When $(h_1 - h_2) \to 0$ and $h_3 \to 0$, we have $\boldsymbol{r}'_{\boldsymbol{p}} \to \boldsymbol{r}'$, $p \to 1$, $\boldsymbol{a}_p \to \frac{\boldsymbol{w}'}{2ik\xi}$, $\xi \to h_{1(,2)}$, and using $\frac{F_v(\mathcal{R}^2)}{v} = \int f(\mathcal{R}^2) d(\boldsymbol{r}'_{\boldsymbol{p}} - \boldsymbol{a}_p)^2$ with $v = -ik\xi$, we recover theorem 6.4.

NOTE 6.23.– From (6.139), we note that:

$$\int_{S'} \boldsymbol{d}.(\boldsymbol{r}' - \mathcal{P}^{-1}(\boldsymbol{a}_{\boldsymbol{p}})) f(\mathcal{R}^2) dS' = \frac{1}{p} \int_{S'_p} \mathcal{P}^{-1}(\boldsymbol{d}).(\boldsymbol{r}'_{\boldsymbol{p}} - \boldsymbol{a}_p) f(\mathcal{R}^2) dS'_p$$

$$= \frac{1}{2p} \int_{\mathcal{C}} \frac{[F_v(\mathcal{R}^2) - A_0]}{v} [\mathcal{P}\ (\widehat{\boldsymbol{z}} \times \mathcal{P}^{-1}(\boldsymbol{d})).d\boldsymbol{r}'], \tag{6.141}$$

where $\frac{\mathcal{P}\ (\widehat{\boldsymbol{z}} \times \mathcal{P}^{-1}(\boldsymbol{d}))}{\xi p} = \frac{(\widehat{\boldsymbol{z}} \times \mathcal{P}^{-2}(\boldsymbol{d}))}{\xi}$ with $\frac{\mathcal{P}^{-2}(\boldsymbol{d})}{\xi} = \frac{\boldsymbol{d}.\widehat{\boldsymbol{c}}_{\boldsymbol{p}}}{p^2 \xi} \widehat{\boldsymbol{c}}_{\boldsymbol{p}} + \frac{\boldsymbol{d}.(\widehat{\boldsymbol{z}} \times \widehat{\boldsymbol{c}}_p)}{\xi} \widehat{\boldsymbol{z}} \times \widehat{\boldsymbol{c}}_p$. For $h_3 = 0$, we have, using (6.87), $\frac{\mathcal{P}^{-2}(\boldsymbol{d})}{\xi} = \frac{\boldsymbol{d}.\widehat{x}}{h_1} \widehat{x} + \frac{\boldsymbol{d}.\widehat{y}}{h_2} \widehat{y}$.

6.8. Complex scaling and applications to asymptotic reductions

We now consider asymptotic reduction of $\int_{S'} g(r') f(\mathcal{R}^2) dS'$ when $f(\mathcal{R}^2) = e^{\mathcal{R}^2}$ or $f(\mathcal{R}^2) = \frac{e^{2\sqrt{\mathcal{R}^2}}}{\sqrt{\mathcal{R}^2}}$ are highly oscillating functions, while $g(\boldsymbol{r}')$ is a smooth analytical function on S' and $\mathcal{R}^2$ is a polynomial.

6.8.1. *The case with $f(\mathcal{R}^2) = e^{\mathcal{R}^2}$ and semi-asymptotic reduction*

To simplify the reading, we choose to detail in appendix the analysis of the case with $f(\mathcal{R}^2) = e^{\mathcal{R}^2}$, for $\mathcal{R}^2$ given by a polynomial of order 2, like in (6.129), when $g(\boldsymbol{r}')$ is a general smooth function, with proposition 6.15 and semi-asymptotic reduction given in theorem 6.19 of appendix E.

6.8.2. *The case with $f(\mathcal{R}^2) = \frac{e^{2\sqrt{\mathcal{R}^2}}}{\sqrt{\mathcal{R}^2}}$ and asymptotic reduction for curved plate*

If we want to generalize the results of section 6.5 and take account of a moderate curvature of the surface of integration in the radiation of surface currents, we have to consider $(z - z')$ instead of z in (6.63) with polynomial approximation of z' relatively to x' and y' coordinates, as $z' = (C_1 x'^2 + C_2 y'^2)$, and derive plane surface integral from the use of a projection and its Jacobian. For that, we can let:

$$\begin{aligned}\sqrt{\mathcal{R}^2} &= -ik\sqrt{(x'-a_x)^2+(y'-a_y)^2+((C_1x'^2+C_2y'^2)-z)^2}\\ &= -ik\sqrt{h_1'(x'-a_x')^2+h_2'(y'-a_y')^2+z''^2+\epsilon(x',y')}\\ &= \sqrt{\mathcal{R}_0^2-k^2\epsilon(x',y')},\end{aligned} \tag{6.142}$$

with $\epsilon(x',y') = (C_1x'^2+C_2y'^2)^2$ and:

$$\begin{aligned}&h_j' = 1-2zC_j,\ a_x' = \frac{a_x}{h_1'},\ a_y' = \frac{a_y}{h_2'},\ z''^2 = z^2 - (\frac{1-h_1'}{h_1'}a_x^2 + \frac{1-h_2'}{h_2'}a_y^2),\\ &\sqrt{\mathcal{R}_0^2} = -ik\sqrt{h_1'(x'-a_x')^2+h_2'(y'-a_y')^2+z''^2},\end{aligned} \tag{6.143}$$

which is particularly useful when $\epsilon(x',y') \ll z^2$. Similarly to (6.105), we can then use one of complex scaling $\mathcal{P}$ verifying:

$$\begin{aligned}&p^2 = h_2'/h_1', \mathcal{P}(\boldsymbol{r'}) = (\widehat{\boldsymbol{x}}.\boldsymbol{r'})\widehat{\boldsymbol{x}} + p\,(\widehat{\boldsymbol{y}}.\boldsymbol{r'})\widehat{\boldsymbol{y}}, \xi = h_1',\\ &p^2 = h_1'/h_2', \mathcal{P}(\boldsymbol{r'}) = p(\widehat{\boldsymbol{x}}.\boldsymbol{r'})\widehat{\boldsymbol{x}} + (\widehat{\boldsymbol{y}}.\boldsymbol{r'})\widehat{\boldsymbol{y}}, \xi = h_2',\end{aligned} \tag{6.144}$$

such that

$$\begin{aligned}&v'(\boldsymbol{r_p'}-\boldsymbol{a}_p)^2 = h_1'(x'-a_x')^2 + h_2'(y'-a_y')^2,\ d_0+v_1 = z''^2,\\ &\text{with } \boldsymbol{r_p'} = \mathcal{P}(\boldsymbol{r'}),\ \mathcal{P}^{-1}(\boldsymbol{a}_p) = a_x'\widehat{\boldsymbol{x}} + a_y'\widehat{\boldsymbol{y}},\ v' = \xi,\end{aligned} \tag{6.145}$$

and then reduce asymptotically, for large k and z'' when $\epsilon(x',y') \ll z^2$, the surface integral $\int_{S'} g(r')f(\mathcal{R}^2)dS'$ with $f(\mathcal{R}^2) = \frac{e^{2\sqrt{\mathcal{R}^2}}}{\sqrt{\mathcal{R}^2}}$ from theorems 6.1 and 6.12 and lemma 6.2.

THEOREM 6.14.– *Letting $f(\mathcal{R}^2) = \frac{e^{2\sqrt{\mathcal{R}^2}}}{\sqrt{\mathcal{R}^2}}$, we have for its primitive $F_v(\mathcal{R}^2) = e^{2\sqrt{\mathcal{R}^2}}$. We can use theorems 6.1 and 6.12, with lemma 6.2 and complex scaling $\mathcal{P}$, and then write, for large k:*

$$\begin{aligned}&\int_{S'} g(\boldsymbol{r'})f(\mathcal{R}_0^2)dS'\\ &\sim \int_{\mathcal{L}} \frac{\frac{1}{2}(g(\boldsymbol{r'})F_v(\mathcal{R}_0^2(\boldsymbol{r'})) - g(\boldsymbol{a'})F_v(\mathcal{R}_0^2(\boldsymbol{a'})))(\widehat{\boldsymbol{z}}\times(\boldsymbol{r'}-\mathcal{P}^{-1}(\boldsymbol{a}_p))).d\boldsymbol{r'}}{-k^2(h_1'(x'-a_x')^2+h_2'(y'-a_y')^2)}\\ &\sim \int_{\mathcal{L}} \frac{\frac{1}{2}(g(\boldsymbol{r'})F_v(\mathcal{R}_0^2(\boldsymbol{r'})) - g(\boldsymbol{a'})F_v(\mathcal{R}_0^2(\boldsymbol{a'})))(\widehat{\boldsymbol{z}}\times(\boldsymbol{r'}-\boldsymbol{a'})).d\boldsymbol{r'}}{-k^2(h_1'(x'-a_x')^2+h_2'(y'-a_y')^2)}.\end{aligned} \tag{6.146}$$

For small $\epsilon(x', y')$*, we can then choose to take* $\frac{g(\boldsymbol{r'})f(\mathcal{R}^2)}{f(\mathcal{R}_0^2)} = \frac{g(\boldsymbol{r'})\sqrt{\mathcal{R}_0^2}}{\sqrt{\mathcal{R}^2}}$ $e^{2(\sqrt{\mathcal{R}^2}-\sqrt{\mathcal{R}_0^2})}$ *in place of* $g(r')$ *and obtain:*

$$\int_{S'} g(\boldsymbol{r'})f(\mathcal{R}^2)dS',$$

$$\sim \int_{\mathcal{L}} \frac{\frac{1}{2}(g_1(\boldsymbol{r'})F_v(\mathcal{R}^2(\boldsymbol{r'})) - g_1(\boldsymbol{a'})F_v(\mathcal{R}^2(\boldsymbol{a'}))(\widehat{\boldsymbol{z}} \times (\boldsymbol{r'} - \boldsymbol{a'})).d\boldsymbol{r'}}{-k^2(h_1'(x'-a_x')^2 + h_2'(y'-a_y')^2)}, \tag{6.147}$$

with $g_1(\boldsymbol{r'}) = \frac{g(\boldsymbol{r'})\sqrt{\mathcal{R}_0^2}}{\sqrt{\mathcal{R}^2}}$*, when the* h_j' *are both real and of the same sign. Besides, in order to consider* $\frac{e^{2\sqrt{\mathcal{R}^2}}}{(\sqrt{\mathcal{R}^2})^n}$ *in place of* $\frac{e^{2\sqrt{\mathcal{R}^2}}}{\sqrt{\mathcal{R}^2}}$*, we can let* $\frac{g(\boldsymbol{r'})}{(\sqrt{\mathcal{R}^2})^{n-1}}$ *in place of* $g(\boldsymbol{r'})$ *in (6.147).*

For more generality, in particular to cover the case of real h_j' *of different signs, we will consider, in the contour integral:*

$$h_j' - i\delta \text{ instead of } h_j', \text{ with } \delta sign(\arg(ik)) \to 0^+, \tag{6.148}$$

as $0 < |\arg(ik)| \leq \pi/2$*, to avoid any line of singularities of denominator in (6.147).*

PROOF.– These expressions are derived from theorems 6.1 and 6.12, used for primitives defined asymptotically with lemma 6.2 (see appendix D). For that, we take $g(\boldsymbol{r'})f(\mathcal{R}^2)$ in place of $f(\mathcal{R}^2)$ in theorem 6.12, considering that its primitive is asymptotically $\sim g(\boldsymbol{r'})F_v(\mathcal{R}^2)$ from lemma 6.2. The small real parameter δ permits us to avoid any line of singularities. Concerning the sign of δ, we choose it so that $e^{2\sqrt{\mathcal{R}^2}} < \infty$ when $\boldsymbol{r'} \to \infty$ and $0 < |\arg(ik)| \leq \pi/2$, which implies $0 < \delta\text{sign}(\arg(ik)) \to 0^+$.

This theorem will be used in the next section and is validated in numerical results, letting $g_1(\boldsymbol{r'}) \simeq g(\boldsymbol{r'})$.

NOTE 6.24.– Considering (6.92), we can let:

$$v'(\boldsymbol{r_p'} - \boldsymbol{a_p'})^2 = (h_1' - i\delta)(x' - a_x')^2 + (h_2' - i\delta)(y' - a_y')^2 =$$

$$= \prod_{\pm}(\sqrt{(h_1' - i\delta)}(x' - a_x') \pm i\sqrt{(h_2' - i\delta)}(y' - a_y')), \tag{6.149}$$

with small real $\delta \neq 0$ such that $(h_1' - i\delta) \neq -(h_2' - i\delta)$, and $v' = \xi$. The sign chosen for δ has a crucial influence. For example, we note, from propositions 6.9 and 6.1, that:

$$\int_{\mathcal{L}} \frac{\frac{1}{2}(\widehat{\boldsymbol{z}} \times (\boldsymbol{r'} - \boldsymbol{a'})).d\boldsymbol{r'}}{(h_1' - i\delta)(x' - a_x')^2 + (h_2' - i\delta)(y' - a_y')^2} = \frac{1}{2p\xi}\int_{\mathcal{L}_p} (\widehat{\boldsymbol{z}} \times \frac{\boldsymbol{r}_p' - \boldsymbol{a}_p'}{(\boldsymbol{r_p'} - \boldsymbol{a}_p')^2}).d\boldsymbol{r_p'}$$

$$= \frac{\pi(\epsilon^+ 1_{S_+'}(a_x' + ia_y') + \epsilon^- 1_{S_-'}(a_x' - ia_y'))}{2p\xi}, \tag{6.150}$$

where $\xi = (h'_1 - i\delta)$ and $p = \sqrt{\frac{(h'_2 - i\delta)}{\xi}} \sim \sqrt{\frac{h'_2}{h'_1}(1 - i\delta(\frac{1}{h'_2} - \frac{1}{h'_1}))}$ (see (6.105)). For real h'_1 and h'_2 of opposite signs, we note the importance of the sign of δ, since we have:

$$\lim_{\delta \to 0^\pm} (2p\xi) = \pm i h'_1 \mathrm{sign}(h'_2)\sqrt{-(h'_2/h'_1)}. \tag{6.151}$$

NOTE 6.25.– It is worth noting that we have not to take the term $p\xi$ as $\sqrt{(p\xi)^2}$, but as the product of the terms p and ξ evaluated separately. Indeed, for example, we note that, for real δ, the term $(1 - i\delta)\sqrt{\frac{(-1-i\delta)}{(1-i\delta)}}$ is defined, while the term $\sqrt{(1-i\delta)(-1-i\delta)}$ is not, since $(1 - i\delta)(-1 - i\delta) = -1 - \delta^2$ is on the cut of the square root function.

6.9. Reduction in case (c) for surface radiation integrals on curved plates

Let us consider the expressions of the electric $\boldsymbol{E}$ in (6.47), radiated by electric and magnetic currents, $\boldsymbol{J} = 2\widehat{\boldsymbol{z}} \times \boldsymbol{H}_i$ and $\boldsymbol{M} = 0$, excited on a real perfectly conducting curved surface S by weighted dipolar illuminations, as defined with $f_{e,h}$ in (6.62) (note: for flat plate, $\boldsymbol{E}$ was given by (6.63) of section 6.5).

We take, this time, a curved plate S instead of a flat one, with current point $\boldsymbol{R}'$ defined by:

$$\boldsymbol{R}' = \boldsymbol{r}' + \zeta_s(\boldsymbol{r}')\widehat{\boldsymbol{z}},\ \zeta_s(\boldsymbol{r}') = C_1 x'^2 + C_2 y'^2,\ \zeta_s(\boldsymbol{0}) = 0, \tag{6.152}$$

with $C_j < 0$ (respectively, $C_j > 0$) for convex (respectively, concave) surface inflections. We note that, for a purely convex (respectively, purely concave) surface with $C_1C_2 > 0$, S is below (respectively, above) the plane $z = 0$, while, for saddle surfaces with $C_1C_2 < 0$, S is partly below and partly above the plane.

In what follows, the surface S' is the projection of S on the plane $z = 0$, $\boldsymbol{R}'$ is the current point of integration on S, while $\boldsymbol{r}'$ is the one on S', $\mathcal{L}$ is the positively oriented contour of S', $\boldsymbol{R}_1 = (\boldsymbol{a} + z\,\widehat{\boldsymbol{z}})$ is the observation point and $\boldsymbol{R}$ is the vector between the observation point and the current point of S:

$$\boldsymbol{R} = \boldsymbol{R}_1 - \boldsymbol{R}' = (\boldsymbol{a} + z\,\widehat{\boldsymbol{z}}) - (\boldsymbol{r}' + \zeta_s(\boldsymbol{r}')\widehat{\boldsymbol{z}}),$$
$$\boldsymbol{R}^2(r') = (x' - a_x)^2 + (y' - a_y)^2 + ((C_1 x'^2 + C_2 y'^2) - z)^2, \tag{6.153}$$

for which we can consider the equalities in (6.142)–(6.145) and theorem 6.14, when we are in the region Ω_C defined with $z \gg (C_1 x'^2 + C_2 y'^2)$ and large $k\sqrt{\boldsymbol{R}^2}$.

In PO, the radii of curvature $\frac{1}{2|C_j|}$ are assumed to be large relatively to λ, and it is worth noticing that, with this assumption, the domain of validity Ω_C that we consider is generally much wider than the far field region (note: in far field, $|\boldsymbol{R}_1| > d_F = 8\max(\frac{(\boldsymbol{r}')^2|_S}{\lambda})$, $k|\boldsymbol{R}_1|$ is large and $|\boldsymbol{R}'|/|\boldsymbol{R}_1|$ is small, while $d_F = 2D^2/\lambda$ is named Fraunhofer distance [1], $D = 2\max(|\boldsymbol{r}'|_S)$).

NOTE 6.26.– Considering $R_c = \min(\frac{1}{2|C_j|})$, we notice that $|C_1x^2 + C_2y^2|_S \leq z_\lambda \frac{\lambda}{2R_c}$ with $z_\lambda = \max(\frac{(x^2+y^2)|_S}{\lambda}) = \frac{z_F}{8}$, and thus, note that we have $|C_1x^2 + C_2y^2|_S \leq \frac{z}{\alpha}$ when $z \geq z_F(\frac{\alpha\lambda}{16R_c})$ is verified.

6.9.1. *Reduction to a contour integral for curved plates when* $f_h \equiv 0$

6.9.1.1. *A contour integral for radiation at finite distance, large relatively to wavelength*

The PO diffracted field at $\boldsymbol{r} = \boldsymbol{a} + z\widehat{\boldsymbol{z}} \notin S$, when $\boldsymbol{r} = \boldsymbol{r}_0$ (monostatic case) and $f_h \equiv 0$, satisfies:

$$\boldsymbol{E} \sim -\frac{Z_0}{8(\pi k)^2}\frac{ik}{2}\int_{S'}[(\widehat{\boldsymbol{n}}.\boldsymbol{R})J^s](\boldsymbol{r}')\widetilde{R}((\frac{\psi'}{\widetilde{R}})^2)' f_e(-\boldsymbol{u})((\boldsymbol{u}.\widehat{\boldsymbol{d}})\boldsymbol{u} - \widehat{\boldsymbol{d}})dS' \quad (6.154)$$

where $\widehat{n}J^{as} = \partial_x\boldsymbol{R}'|_S \times \partial_y\boldsymbol{R}'|_S$, S' of current points $\boldsymbol{r}'$ is the projection on the plane $z = 0$ of the curved surface S with points $\boldsymbol{R}'$ and outward normal vector $\widehat{n}$, $\psi(\widetilde{R}) = \frac{e^{-ik\widetilde{R}}}{\widetilde{R}}$, $(.)' \equiv \partial_{\widetilde{R}}(.)$. We then use:

$$\boldsymbol{R}^2(r') = (1-2zC_1)(x'-a'_x)^2 + (1-2zC_2)(y'-a'_y)^2 + z''^2 + \epsilon(\boldsymbol{r}'),$$

$$a'_x = \frac{a_x}{1-2zC_1},\ a'_y = \frac{a_y}{1-2zC_2},\ z''^2 = z^2 - (\frac{2zC_1}{1-2zC_1}a_x^2 + \frac{2zC_2}{1-2zC_2}a_y^2),$$

$$J^{as}(\boldsymbol{r}') = \left|\frac{\partial\boldsymbol{R}'}{\partial x'} \times \frac{\partial\boldsymbol{R}'}{\partial y'}\right|,\ \widehat{\boldsymbol{n}} = \left(\frac{\partial\boldsymbol{R}'}{\partial x'} \times \frac{\partial\boldsymbol{R}'}{\partial y'}\right)/\left|\frac{\partial\boldsymbol{R}'}{\partial x'} \times \frac{\partial\boldsymbol{R}'}{\partial y'}\right|,$$

$$[(\widehat{\boldsymbol{n}}.\boldsymbol{R})J^{as}](\boldsymbol{r}') = -2(C_1x'(a_x - x') + C_2y'(a_y - y')) + z - \zeta_s(\boldsymbol{r}'), \quad (6.155)$$

where $\epsilon(\boldsymbol{r}') = (\zeta_s(\boldsymbol{r}'))^2$, $\zeta_s(\boldsymbol{r}') = C_1x'^2 + C_2y'^2$ and $\boldsymbol{R}^2 = v(\boldsymbol{r}'_{\boldsymbol{p}} - \boldsymbol{a}_{\boldsymbol{p}})^2 + z''^2 + \epsilon(\boldsymbol{r}')$ with $\boldsymbol{a}_{\boldsymbol{p}} = \mathcal{P}(\boldsymbol{a}')$ as in (6.142)–(6.145). In what follows, we consider that the term $\epsilon(x, y)$ has moderate influence on the reduction. For that, we consider z sufficiently large so that $|(C_1x^2 + C_2y^2)|_S \ll z$.

THEOREM 6.15.– *Considering theorem 6.14, we can write the surface integral (6.154), when k is large and* $|C_1x^2 + C_2y^2|_S \ll z$, *as the contour integral:*

$$\boldsymbol{E} \sim \frac{-Z_0ik}{(4\pi)^2}\int_{\mathcal{L}} \frac{(\frac{e^{-2ik\widetilde{R}(\boldsymbol{r}')}}{R^2(\boldsymbol{r}')}\mathcal{W}_e(\boldsymbol{r}') - \frac{e^{-2ik\widetilde{R}(\boldsymbol{a}')}}{R^2(\boldsymbol{a}')}\mathcal{W}_e(\boldsymbol{a}'))[(\widehat{z} \times (\boldsymbol{r}' - \boldsymbol{a}')).d\boldsymbol{r}']}{(1-2zC_1-i\delta)(x'-a'_x)^2 + (1-2zC_2-i\delta)(y'-a'_y)^2}, \quad (6.156)$$

where

$$\begin{aligned}\widetilde{R}(\boldsymbol{r}') &= \sqrt{(\boldsymbol{r}'-\boldsymbol{a})^2 + (\zeta_s(\boldsymbol{r}') - z)^2},\\ \mathcal{W}_e(\boldsymbol{r}') &= -f_e(-\boldsymbol{u})((\boldsymbol{u}.\widehat{\boldsymbol{d}})\widehat{\boldsymbol{u}} - \widehat{\boldsymbol{d}})[(\widehat{\boldsymbol{n}}.\boldsymbol{R})J^{as}](\boldsymbol{r}'),\end{aligned} \tag{6.157}$$

with $\mathcal{L}$ the positively oriented contour of S' on the plane $z = 0$. The term δ is a small real constant tending to 0, with $\delta sign(\arg(ik)) \to 0^+$, necessary for the cases where $(1 - 2zC_1)(1 - 2zC_2) < 0$.

When $f_e(-u) = \frac{1}{\sqrt{1-(u.\widehat{d})^2}}$, we then have the VV polarization term $\boldsymbol{E}.\widehat{\boldsymbol{d}}$, with $\widehat{\boldsymbol{d}} = \frac{\boldsymbol{r}_0 \times \widehat{\boldsymbol{z}}}{\sqrt{(\boldsymbol{r}_0\times\widehat{\boldsymbol{z}})^2}}$ as $\boldsymbol{r}_0 \times \widehat{\boldsymbol{z}} \neq 0$ (already considered for a flat plate in (6.69)) following:

$$\begin{aligned}\boldsymbol{E}.\widehat{\boldsymbol{d}} &= -\frac{Z_0 ik}{(4\pi)^2}\int_{\mathcal{L}} \frac{(\mathcal{V}(\boldsymbol{r}') - \mathcal{V}(\boldsymbol{a}'))[(\widehat{\boldsymbol{z}} \times (\boldsymbol{r}' - \boldsymbol{a}')).d\boldsymbol{r}']}{(1 - 2zC_1 - i\delta)(x' - a'_x)^2 + (1 - 2zC_2 - i\delta)(y' - a'_y)^2},\\ \mathcal{V}(\boldsymbol{r}') &= \frac{[(\widehat{\boldsymbol{n}}.\boldsymbol{R})J^s](\boldsymbol{r}')e^{-2ik\widetilde{R}(\boldsymbol{r}')}\sqrt{1 - (\frac{(r'-a)}{\widetilde{R}(\boldsymbol{r}')}.\widehat{d})^2}}{R^2(\boldsymbol{r}')}.\end{aligned} \tag{6.158}$$

6.9.1.2. *Particular asymptotic cases: At very large distance or at focal points*

At very large distance, we can take in surface and contour integral:

$$\sqrt{\boldsymbol{R}^2} \sim \widetilde{r_0} - \frac{\boldsymbol{r_0}}{\widetilde{r_0}}.(\boldsymbol{r}' + \widehat{z}\zeta_s(\boldsymbol{r}')),\ z'' \sim \widetilde{r_0} - \frac{\boldsymbol{r_0}}{\widetilde{r_0}}.(\boldsymbol{a}' + \widehat{z}\zeta_s(\boldsymbol{a}')), \tag{6.159}$$

while $[(\widehat{\boldsymbol{n}}.\boldsymbol{R})J^{as}](\boldsymbol{r}') \to -2(C_1x'a_x + C_2y'a_y) + z$, which implies that the expression (6.156) becomes:

$$\begin{aligned}\boldsymbol{E} \sim &-\frac{Z_0 ik}{(4\pi)^2}\frac{e^{-2ik\widetilde{r_0}}}{\boldsymbol{r}_0^2} f_e(-\frac{\boldsymbol{r_0}}{\widetilde{r_0}})(\widehat{\boldsymbol{d}} - (\frac{\boldsymbol{r_0}}{\widetilde{r_0}}.\widehat{\boldsymbol{d}})\frac{\boldsymbol{r_0}}{\widetilde{r_0}})\\ &\int_{\mathcal{L}} \frac{\mathcal{H}(\boldsymbol{r}')e^{2ik(\frac{\boldsymbol{r_0}}{\widetilde{r_0}}.(\boldsymbol{r}'+\widehat{z}\zeta_s(\boldsymbol{r}'))} - \mathcal{H}(\boldsymbol{a}')e^{2ik(\frac{\boldsymbol{r_0}}{\widetilde{r_0}}.(\boldsymbol{a}'+\widehat{z}\zeta_s(\boldsymbol{a}'))}}{(\frac{2zC_1-1}{2z} + i\delta)(x' - a'_x)^2 + (\frac{2zC_2-1}{2z} + i\delta)(y' - a'_y)^2}[(\widehat{\boldsymbol{z}} \times (\boldsymbol{r}'-\boldsymbol{a}')).d\boldsymbol{r}'],\end{aligned} \tag{6.160}$$

where $\mathcal{H}(\boldsymbol{r}') = (C_1x'(\frac{a_x}{z}) + C_2y'(\frac{a_y}{z}) - \frac{1}{2})$.

At focal points, as $\mathcal{K}_m = \min(1 - 2zC_j)|_{j=1,2} \to 0$, we have $\boldsymbol{a}' \to \infty$ and $z''^2 \to \infty$, but a'/z''^2 and $\mathcal{K}_m a'$ tend to finite constant vectors, and we can write for (6.156):

$$\boldsymbol{E} \sim \frac{Z_0 ik}{(4\pi)^2}\Big(\int_{\mathcal{L}} \frac{\big(\frac{e^{-2ik\widetilde{R}(\boldsymbol{r}')}}{R^2(\boldsymbol{r}')}\mathcal{W}_e(\boldsymbol{r}') - \frac{e^{-2ik\widetilde{R}(\boldsymbol{a}')}}{R^2(\boldsymbol{a}')}\mathcal{W}_e(\boldsymbol{a}')\big)[(\widehat{\boldsymbol{z}} \times \mathcal{K}_m \boldsymbol{a}').d\boldsymbol{r}']}{\mathcal{K}_m(\boldsymbol{R}_\delta^2 - z^2 - \epsilon(\boldsymbol{r}')) + (\frac{2zC_1\mathcal{K}_m}{(1-2zC_1)}a_x^2 + \frac{2zC_2\mathcal{K}_m}{(1-2zC_2)}a_y^2)}, \tag{6.161}$$

with $\boldsymbol{R}_\delta^2 = (1 - 2zC_1 - i\delta)(x' - a_x')^2 + (1 - 2zC_2 - i\delta)(y' - a_y')^2 + z''^2 + \epsilon(\boldsymbol{r}')$.

6.9.2. *Reduction to a contour integral for curved plates when* $f_e \equiv 0$

6.9.2.1. *A contour integral for radiation at finite distance, large relatively to wavelength*

The PO diffracted field at $\boldsymbol{r} = \boldsymbol{a} + z\widehat{\boldsymbol{z}} \notin S'$, when $\boldsymbol{r} = \boldsymbol{r}_0$ (monostatic case) and $f_e \equiv 0$, satisfies:

$$\boldsymbol{E}.\boldsymbol{v}_0 \sim \frac{-1}{8(\pi ik)^2}\int_{S'} [(\widehat{\boldsymbol{n}}.\boldsymbol{R})J^{as}](\boldsymbol{r}')((\frac{\psi'}{\underset{\sim}{R}})')^2 \boldsymbol{w}_0.\boldsymbol{R}\, f_h(-\boldsymbol{u})dS', \tag{6.162}$$

where $\boldsymbol{v}_0 = (\widehat{\boldsymbol{d}} \times \frac{\boldsymbol{r}_0}{\widetilde{r}_0})$ with $\widehat{\boldsymbol{d}} = \frac{\boldsymbol{r}_0 \times \widehat{\boldsymbol{z}}}{\sqrt{(\boldsymbol{r}_0 \times \widehat{\boldsymbol{z}})^2}}$ as $\boldsymbol{r}_0 \times \widehat{\boldsymbol{z}} \neq 0$, $\boldsymbol{w}_0 = \frac{\boldsymbol{r}_0}{\widetilde{r}_0} = \frac{\boldsymbol{a}+z\widehat{\boldsymbol{z}}}{\widetilde{r}_0}$, and $(\widehat{\boldsymbol{n}}.\boldsymbol{R})J^{as}$ is defined as previously.

Considering that $\boldsymbol{w}_0.\boldsymbol{R} = \boldsymbol{w}_0.(\boldsymbol{R}_1 + \boldsymbol{R}_2)$ with $\boldsymbol{R}_1(\boldsymbol{r}') = (\boldsymbol{a}' - \boldsymbol{r}')$ and $\boldsymbol{R}_2(\boldsymbol{r}') = (\boldsymbol{a} - \boldsymbol{a}' + (z - \zeta_s(\boldsymbol{r}'))\widehat{\boldsymbol{z}})$, we could use (6.141) for the part with $\boldsymbol{R}_1$ term, as this was in theorems 6.8–6.10 for a flat plate; however, to simplify, we choose here the asymptotic expression directly derived from theorem 6.14.

THEOREM 6.16.– *Considering theorem 6.14, we can write the surface integral (6.162), when k is large and* $|(C_1x^2 + C_2y^2)||_S \ll z$, *as the contour integral:*

$$\boldsymbol{E}.\boldsymbol{v}_0 \sim \frac{ik}{(4\pi)^2}\int_{\mathcal{L}} \frac{\big(\frac{\mathcal{W}_h(\boldsymbol{r}')e^{-2ik\widetilde{R}(\boldsymbol{r}')}}{(\widetilde{R}(\boldsymbol{r}'))^3} - \frac{\mathcal{W}_h(\boldsymbol{a}')e^{-2ik\widetilde{R}(\boldsymbol{a}')}}{(\widetilde{R}(\boldsymbol{a}'))^3}\big)(\widehat{\boldsymbol{z}} \times (\boldsymbol{r}' - \boldsymbol{a}')).d\boldsymbol{r}'}{(1 - 2zC_1 - i\delta)(x' - a_x')^2 + (1 - 2zC_2 - i\delta)(y' - a_y')^2}, \tag{6.163}$$

with $\mathcal{W}_h(\boldsymbol{r}') = f_h(-\boldsymbol{u})[(\widehat{\boldsymbol{n}}.\boldsymbol{R})J^{as}](\boldsymbol{r}')[\boldsymbol{w}_0.\boldsymbol{R}](\boldsymbol{r}')$. *The term* δ *is a small real constant tending to 0, with* $\delta \text{sign}(\arg(ik)) \to 0^+$, *necessary for the cases where we have* $(1 - 2zC_1)(1 - 2zC_2) < 0$. *The HH polarization is obtained when we let* $\boldsymbol{w}_0 = \frac{\boldsymbol{r}_0}{\widetilde{r}_0} = \frac{\boldsymbol{a}+z\widehat{\boldsymbol{z}}}{\widetilde{r}_0}$. *We can then consider (as in (6.78) for a flat plate) an isotropic illumination with* $f_h(\boldsymbol{u}) = 1/\sqrt{1 - (\frac{(\boldsymbol{r}'-\boldsymbol{a})}{\widetilde{R}}.\widehat{\boldsymbol{d}})^2}$, $\widehat{\boldsymbol{d}} = \frac{\boldsymbol{r}_0 \times \widehat{\boldsymbol{z}}}{\sqrt{(\boldsymbol{r}_0 \times \widehat{\boldsymbol{z}})^2}}$ *as* $\boldsymbol{r}_0 \times \widehat{\boldsymbol{z}} \neq 0$, *and* $\boldsymbol{w}_0 = \frac{\boldsymbol{r}_0}{\widetilde{r}_0} = \frac{\boldsymbol{a}+z\widehat{\boldsymbol{z}}}{\widetilde{r}_0}$.

6.9.2.2. *Particular asymptotic cases: at very large distance or at focal points*

At very large distance, we use (6.159) in the previous expression, and we can write:

$$\boldsymbol{E}.\boldsymbol{v}_0 \sim \frac{ik}{(4\pi)^2}\frac{e^{-2ik\widetilde{R}}}{\boldsymbol{r}_0^2} f_h(-\frac{\boldsymbol{r}_0}{\widetilde{r}_0})(\boldsymbol{w}_0.\boldsymbol{r}_0)$$

$$\int_{\mathcal{L}} \frac{\mathcal{H}(\boldsymbol{r}')e^{2ik(\frac{\boldsymbol{r}_0}{\widetilde{r}_0}.(\boldsymbol{r}'+\widehat{z}\zeta_s(\boldsymbol{r}'))} - \mathcal{H}(\boldsymbol{a}')e^{2ik(\frac{\boldsymbol{r}_0}{\widetilde{r}_0}.(\boldsymbol{a}'+\widehat{z}\zeta_s(\boldsymbol{a}'))}}{(\frac{2zC_1-1}{2z}+i\delta)(x'-a_x')^2+(\frac{2zC_2-1}{2z}+i\delta)(x'-a_y')^2}(\widehat{\boldsymbol{z}}\times(\boldsymbol{r}'-\boldsymbol{a}')).d\boldsymbol{r}', \tag{6.164}$$

where $\mathcal{H}(\boldsymbol{r}') = (C_1 x'(\frac{a_x}{z}) + C_2 y'(\frac{a_y}{z}) - \frac{1}{2})$.

At focal points, as $\mathcal{K}_m = \min(|1-2zC_j|_{j=1,2}) \to 0$, we have $a' \to \infty$ and $z''^2 \to \infty$, but $\boldsymbol{a}'/z''^2$ tends to a finite constant vector, and we can then write:

$$\boldsymbol{E}.\boldsymbol{v}_0 \sim \frac{-ik}{(4\pi)^2}\int_{\mathcal{L}} \frac{(\frac{\mathcal{W}_h(\boldsymbol{r}')e^{-2ik\widetilde{R}(\boldsymbol{r}')}}{\widetilde{R}\boldsymbol{R}^2} - \frac{\mathcal{W}_h(\boldsymbol{a}')e^{-2ik\widetilde{R}(\boldsymbol{a}')}}{(\widetilde{R}(\boldsymbol{a}'))^3})(\widehat{\boldsymbol{z}}\times\mathcal{K}_m\boldsymbol{a}').d\boldsymbol{r}'}{\mathcal{K}_m(\boldsymbol{R}_\delta^2 - z^2 - \epsilon(\boldsymbol{r}')) + (\frac{2zC_1\mathcal{K}_m}{(1-2zC_1)}a_x^2 + \frac{2zC_2\mathcal{K}_m}{(1-2zC_2)}a_y^2)}. \tag{6.165}$$

with $\boldsymbol{R}_\delta^2 = (1-2zC_1-i\delta)(x'-a_x')^2 + (1-2zC_2-i\delta)(y'-a_y')^2 + z''^2 + \epsilon(\boldsymbol{r}')$.

NOTE 6.27.– Considering $x \in [x'-\Delta/2, x'+\Delta/2]$ with $0 < |(cx)^2+y^2| \ll 1$, $0 < \Delta \ll 1$, and smooth function $v(x)$, we can generally write $\int_{x'-\Delta/2}^{x'+\Delta/2}\frac{v(x)dx}{(cx)^2+y^2} \simeq \frac{iv(x')}{2cy}[\ln(\frac{iy+cx}{iy-cx})]_{x'-\Delta/2}^{x'+\Delta/2}$.

6.10. Miscellaneous results

PROPOSITION 6.11.– *Considering:*

$$\begin{aligned}\mathcal{R}^2 &= (\alpha_1 x' + \alpha_2 y') - ik(h_1 x'^2 + h_2 y'^2 + h_3 x'y') \\ &= \boldsymbol{w}'.\boldsymbol{r}' - ik(h_1 x'^2 + h_2 y'^2 + h_3 x'y'),\end{aligned} \tag{6.166}$$

we have, for arbitrary constants $\mathbf{d}$ *and* A_0*:*

$$\int_{S'}(\boldsymbol{d}.\boldsymbol{r}')f(\mathcal{R}^2)dS' = A_1.(\int_{\mathcal{L}}[F_v(\mathcal{R}^2) - A_0]d\boldsymbol{r}' - (\widehat{z}\times\boldsymbol{w}')\int_{S'}f(\mathcal{R}^2)dS'), \tag{6.167}$$

where F_v *is a primitive of* f *and* $A_1 = \frac{\mathcal{P}(\widehat{z}\times\mathcal{P}^{-1}(\mathbf{d}))}{-2ikp\xi}$. *In the particular cases:*

$$\begin{aligned}p^2\xi &= \frac{(h_1+h_2) - \epsilon\sqrt{(h_1-h_2)^2+4(h_3/2)^2}}{2} \to 0, \\ \text{or } \xi &= \frac{(h_1+h_2) + \epsilon\sqrt{(h_1-h_2)^2+4(h_3/2)^2}}{2} \to 0,\end{aligned} \tag{6.168}$$

we have

$$h_3^2/(4h_1h_2) \to 1, \textit{ i.e. } 2\epsilon'\sqrt{h_1h_2} \to h_3 \textit{ with } (\epsilon')^2 = 1,$$
$$(h_1x'^2 + h_2y'^2 + h_3x'y') \to (\sqrt{h_1}x' + \epsilon'\sqrt{h_2}y')^2. \qquad (6.169)$$

In these circumstances, the equality (6.167) has for consequence that:

$$\lim_{\xi\to 0} \int_{S'} f(\mathcal{R}^2)dS' = \lim_{\xi\to 0} \frac{\widehat{\boldsymbol{c}}_p}{\widehat{\boldsymbol{c}}_p.(\widehat{\boldsymbol{z}}\times \boldsymbol{w}')} \cdot \int_{\mathcal{L}} [F(\mathcal{R}^2) - A_0]d\boldsymbol{r}',$$
$$\lim_{p^2\xi\to 0} \int_{S'} f(\mathcal{R}^2)dS' = \lim_{p^2\xi\to 0} \frac{(\widehat{\boldsymbol{z}}\times\widehat{\boldsymbol{c}}_p)}{(\widehat{\boldsymbol{z}}\times\widehat{\boldsymbol{c}}_p).(\widehat{\boldsymbol{z}}\times \boldsymbol{w}')} \cdot \int_{\mathcal{L}} [F(\mathcal{R}^2) - A_0]d\boldsymbol{r}'. \qquad (6.170)$$

Letting

$$\boldsymbol{v}' = ((\xi - h_1)\widehat{x} + \frac{-h_3}{2}\widehat{y}) \textit{ for } \xi \neq h_1,$$
$$\textit{or } \boldsymbol{v}' = (\frac{-h_3}{2}\widehat{x} + (\xi - h_2)\widehat{y}) \textit{ for } \xi \neq h_2. \qquad (6.171)$$

We can then write, when $\xi \neq h_1$ or $\xi \neq h_2$:

$$\lim_{\xi\to 0} \int_{S'} f(\mathcal{R}^2)dS' = \lim_{\xi\to 0} \frac{\boldsymbol{v}'}{\boldsymbol{v}'.(\widehat{\boldsymbol{z}}\times \boldsymbol{w}')} \cdot \int_{\mathcal{L}} [F(\mathcal{R}^2) - A_0]d\boldsymbol{r}',$$
$$\lim_{p^2\xi\to 0} \int_{S'} f(\mathcal{R}^2)dS' = \lim_{p^2\xi\to 0} \frac{(\widehat{\boldsymbol{z}}\times\boldsymbol{v}')}{(\widehat{\boldsymbol{z}}\times\boldsymbol{v}').(\widehat{\boldsymbol{z}}\times \boldsymbol{w}')} \cdot \int_{\mathcal{L}} [F(\mathcal{R}^2) - A_0]d\boldsymbol{r}', \qquad (6.172)$$

while, for $\xi = h_1 = h_2$, which implies $h_3 = 0$ and $p^2\xi = \xi$, we have:

$$(\widehat{\boldsymbol{z}}\times \boldsymbol{w}') \lim_{\xi\to 0} \int_{S'} f(\mathcal{R}^2)dS' = \lim_{\xi\to 0} \int_{\mathcal{L}} [F(\mathcal{R}^2) - A_0]d\boldsymbol{r}'. \qquad (6.173)$$

PROOF.– Using that $\boldsymbol{d}.\boldsymbol{r}' = \mathcal{P}^{-1}(\boldsymbol{d}).(\boldsymbol{r}'_p - \boldsymbol{a}_p) + \mathcal{P}^{-1}(\boldsymbol{d}).\boldsymbol{a}_p$, then proposition 6.3, we derive:

$$\int_{S'} \boldsymbol{d}.\boldsymbol{r}' f(\mathcal{R}^2)dS' = \mathcal{P}^{-1}(\boldsymbol{d}).\boldsymbol{a}_p \int_{S'} f(\mathcal{R}^2)dS'$$
$$+ \frac{1}{p}\int_{S'_p} \mathcal{P}^{-1}(\boldsymbol{d}).(\boldsymbol{r}'_p - \boldsymbol{a}_p) f(\mathcal{R}^2)dS'_p$$
$$= \mathcal{P}^{-1}(\boldsymbol{d}).\boldsymbol{a}_p \int_{S'} f(\mathcal{R}^2)dS'$$
$$+ \frac{1}{2p}\int_{\mathcal{L}} \frac{[F_v(\mathcal{R}^2) - A_0]}{v} [\mathcal{P}\,(\widehat{\boldsymbol{z}}\times\mathcal{P}^{-1}(\boldsymbol{d})).d\boldsymbol{r}'], \qquad (6.174)$$

and thus (6.167), if we notice that:

$$\begin{aligned}\mathcal{P}^{-1}(\boldsymbol{d}).\boldsymbol{a}_p &= (\widehat{\boldsymbol{z}} \times \mathcal{P}^{-1}(\boldsymbol{d})).(\widehat{\boldsymbol{z}} \times \boldsymbol{a}_p) = \frac{\widehat{\boldsymbol{z}} \times \mathcal{P}^{-1}(\boldsymbol{d}).\widehat{\boldsymbol{z}} \times \mathcal{P}^{-1}(\boldsymbol{w}')}{2ik\xi} \\ &= \frac{\widehat{\boldsymbol{z}} \times \mathcal{P}^{-1}(\boldsymbol{d}).\mathcal{P}(\widehat{\boldsymbol{z}} \times \boldsymbol{w}')}{2ikp\xi} = \frac{\mathcal{P}(\widehat{\boldsymbol{z}} \times \mathcal{P}^{-1}(\boldsymbol{d})).(\widehat{\boldsymbol{z}} \times \boldsymbol{w}')}{2ikp\xi},\end{aligned} \tag{6.175}$$

where we have used (6.113)–(6.114) with $\boldsymbol{a}_p = \frac{\mathcal{P}^{-1}(\boldsymbol{w}')}{2ik\xi}$. Considering that:

$$\begin{aligned}\frac{\mathcal{P}(\widehat{\boldsymbol{z}} \times \mathcal{P}^{-1}(\boldsymbol{d}))}{p\xi} &= \frac{(\widehat{\boldsymbol{z}} \times \mathcal{P}^{-2}(\boldsymbol{d}))}{\xi}, \\ \frac{\mathcal{P}^{-2}(\boldsymbol{d})}{\xi} &= \frac{\boldsymbol{d}.\widehat{\boldsymbol{c}}_p}{p^2\xi}\widehat{\boldsymbol{c}}_p + \frac{\boldsymbol{d}.(\widehat{\boldsymbol{z}} \times \widehat{\boldsymbol{c}}_p)}{\xi}\widehat{\boldsymbol{z}} \times \widehat{\boldsymbol{c}}_p,\end{aligned} \tag{6.176}$$

we note that, in right side of (6.167), $\frac{\mathcal{P}(\widehat{\boldsymbol{z}} \times \mathcal{P}^{-1}(\boldsymbol{d}))}{2ikp\xi}$ tends to ∞ for $\xi \to 0$ or $p^2\xi \to 0$, while left side of (6.167) is finite for arbitrary bounded $\boldsymbol{d}$. We can then write:

$$\begin{aligned}&\lim_{\xi \to 0} \widehat{\boldsymbol{c}}_p.[(\widehat{\boldsymbol{z}} \times \boldsymbol{w}')\int_{S'} f(\mathcal{R}^2)dS' - \int_{\mathcal{L}}[F(\mathcal{R}^2) - A_0]d\boldsymbol{r}'] = 0, \\ &\lim_{p^2\xi \to 0} (\widehat{\boldsymbol{z}} \times \widehat{\boldsymbol{c}}_p).[(\widehat{\boldsymbol{z}} \times \boldsymbol{w}')\int_{S'} f(\mathcal{R}^2)dS' - \int_{\mathcal{L}}[F(\mathcal{R}^2) - A_0]d\boldsymbol{r}'] = 0,\end{aligned} \tag{6.177}$$

and thus (6.170). We then obtain (6.172), if we use (6.82)–(6.84) and:

$$\begin{aligned}&c_{p,x} = \epsilon_x\sqrt{\xi - h_1}/\sqrt{\xi},\ c_{p,y} = \epsilon_y\sqrt{\xi - h_2}/\sqrt{\xi}, \\ &\epsilon_x^2 = \epsilon_y^2 = 1, \epsilon_y\epsilon_x = \frac{\epsilon_y}{\epsilon_x} = \frac{\epsilon_x}{\epsilon_y} = \frac{-h_3}{2\sqrt{\xi - h_1}\sqrt{\xi - h_2}}, \\ &\xi = \frac{(h_1 + h_2) + \epsilon\sqrt{(h_1 - h_2)^2 + h_3^2}}{2}, \\ &(\boldsymbol{c}_p)_\epsilon^2 = 1 - \frac{(h_1 + h_2) - \epsilon\sqrt{(h_1 - h_2)^2 + h_3^2}}{(h_1 + h_2) + \epsilon\sqrt{(h_1 - h_2)^2 + h_3^2}}, \\ &p^2 = 1 - (\boldsymbol{c}_p)^2,\ p^2\xi = \frac{(h_1 + h_2) - \epsilon\sqrt{(h_1 - h_2)^2 + h_3^2}}{2},\end{aligned} \tag{6.178}$$

and (6.173) from (6.177), noticing that $p^2\xi \to \xi$ when $\xi \to h_1 = h_2$.

NOTE 6.28.– Using (6.116) of proposition 6.8, we have, as $\epsilon = \pm 1$:

$$\begin{aligned}&(\widehat{\boldsymbol{c}}_p)_\pm = \pm\epsilon_c(\widehat{\boldsymbol{z}} \times (\widehat{\boldsymbol{c}}_p)_\mp), (p^2\xi)_\pm = (\xi)_\mp, \\ &(\frac{\boldsymbol{d}.(\widehat{\boldsymbol{c}}_p)}{p^2\xi}(\widehat{\boldsymbol{c}}_p))_\pm = (\frac{\boldsymbol{d}.(\widehat{\boldsymbol{z}} \times \widehat{\boldsymbol{c}}_p)}{\xi}\widehat{\boldsymbol{z}} \times \widehat{\boldsymbol{c}}_p)_\mp,\end{aligned} \tag{6.179}$$

with $(\epsilon_c)^2 = 1$, and we then note that:

$$\frac{\mathcal{P}^{-2}(\boldsymbol{d})}{\xi} = \frac{\boldsymbol{d}.\widehat{\boldsymbol{c}}_p}{p^2\xi}\widehat{\boldsymbol{c}}_p + \frac{\boldsymbol{d}.(\widehat{\boldsymbol{z}}\times\widehat{\boldsymbol{c}}_p)}{\xi}\widehat{\boldsymbol{z}}\times\widehat{\boldsymbol{c}}_p \text{ is independent of } \epsilon. \tag{6.180}$$

PROPOSITION 6.12.– *Let F_v be a primitive of f, and $\xi = h_1 = h_2$, we then have $h_3 = 0$, and $\mathcal{R}^2 \to \boldsymbol{w}'.\boldsymbol{r}'$ for $\xi \to 0$. Therefore, we can write, from proposition 6.11:*

$$\int_{S'} f(\boldsymbol{w}'.\boldsymbol{r}')dS' = \frac{(\widehat{\boldsymbol{z}}\times\boldsymbol{w}')}{(\boldsymbol{w}')^2}.\int_{\mathcal{L}}(F_v(\boldsymbol{w}'.\boldsymbol{r}') - F_v(0))d\boldsymbol{r}'. \tag{6.181}$$

PROOF.– We can obtain this equality from (6.173) or by using $\text{grad}(F_v(\boldsymbol{w}'.\boldsymbol{r}')) = \boldsymbol{w}'f(\boldsymbol{w}'.\boldsymbol{r}')$.

NOTE 6.29.– Letting $\boldsymbol{w}' = a\boldsymbol{w}'_1$ with $a \to 0$, we note from (6.181) that:

$$\int_{S'} dS' = \frac{(\widehat{\boldsymbol{z}}\times\boldsymbol{w}'_1)}{(\boldsymbol{w}'_1)^2}.\int_{\mathcal{L}}(\boldsymbol{w}'_1.\boldsymbol{r}')d\boldsymbol{r}'. \tag{6.182}$$

6.11. Numerical results

6.11.1. *Exact contour reduction of a surface integral whose integrand has an exponential term with a quadratic argument*

6.11.1.1. *For $e^{\boldsymbol{w}'.\boldsymbol{r}'-ik\xi(\boldsymbol{r}'^2-(\boldsymbol{c}_p.\boldsymbol{r}')^2)}$ integrand in surface integral*

We choose to validate our reduction given in theorem 6.11 by verifying:

$$\begin{aligned}
&\frac{k}{\sqrt{\pi}}\int_{S'} e^{\boldsymbol{w}'.\boldsymbol{r}'-ik\xi((\boldsymbol{r}')^2-(\boldsymbol{c}_p.\boldsymbol{r}')^2)}dS' = \\
&= \frac{k}{2\sqrt{\pi}}\int_{\mathcal{L}}\Big[\frac{(e^{\boldsymbol{w}'.\boldsymbol{r}'-ik\xi((\boldsymbol{r}')^2-(\boldsymbol{c}_p.\boldsymbol{r}')^2)} - e^{\frac{\boldsymbol{w}'^2_m}{4ik\xi}})}{(\boldsymbol{w}'.\boldsymbol{r}' - ik\xi(|\boldsymbol{r}'|^2 - (\boldsymbol{c}_p.\boldsymbol{r}')^2)) - \frac{\boldsymbol{w}'^2_m}{4ik\xi}}\Big] \\
&\Big[(\widehat{\boldsymbol{z}}\times(\boldsymbol{r}' - \frac{\mathcal{M}(\boldsymbol{w}')}{2ik\xi(1-\boldsymbol{c}_p^2)})).d\boldsymbol{r}'\Big],
\end{aligned} \tag{6.183}$$

when $\boldsymbol{w}'$ and $\boldsymbol{c}_p$ vary (Figure 6.3). We take S', a square of 1 m side length, whose center is at $\boldsymbol{r}_c = (0.5\,\widehat{\boldsymbol{x}} + 0.5\,\widehat{\boldsymbol{y}})$, $k = 2\pi/\lambda$, $\lambda = 2\,\text{cm}$, $\xi = e^{i\Phi}/|\boldsymbol{r}|$, $|\boldsymbol{r}| = 1\,\text{m}$. The term $\boldsymbol{c}_p = \sin\theta\cos\varphi\widehat{\boldsymbol{x}} + \sin\theta\sin\varphi\widehat{\boldsymbol{y}}$ is considered for $\varphi = \pi/4$ as θ varies and $\boldsymbol{w}' = 2ik\boldsymbol{c}_p$. The modulus of each side in (6.183) are given in dB, for $\xi = e^{i\Phi}/|\boldsymbol{r}|$ with $\Phi = 0$, $\Phi = -\frac{\pi}{18}$ and $\Phi = -\frac{\pi}{2}$. The respective complexity of integrands in surface and contour integrals are equal, and the save in computation time grows linearly with the ratio of surface to its contour. We notice the perfect agreement of the respective results.

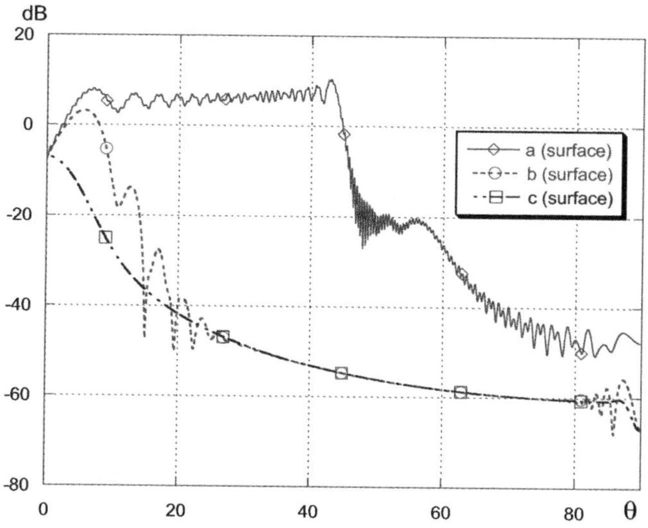

(a)

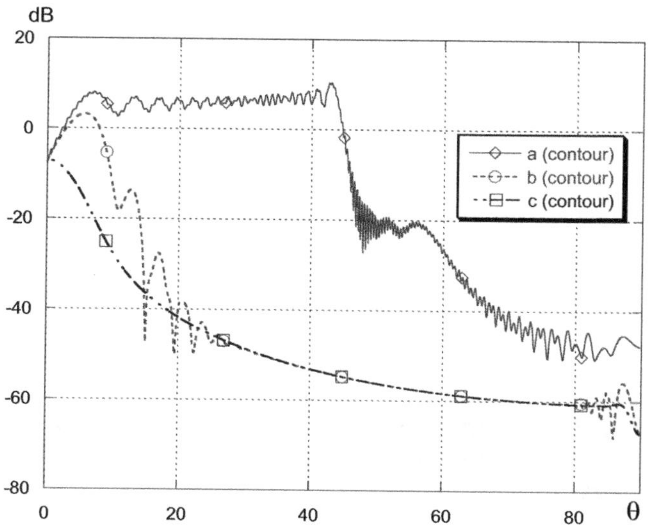

(b)

Figure 6.3. *Comparison of (a) surface and (b) contour integral terms in (6.183), for $\xi = e^{i\Phi}/|\boldsymbol{r}|$ with $\Phi = 0$ (case a), $\Phi = -\frac{\pi}{18}$ (case b), and $\Phi = -\frac{\pi}{2}$ (case c). For a color version of this figure, see www.iste.co.uk/bernard/theoryofscattering.zip*

6.11.1.2. *For* $\sqrt{a'+b'x'+cx'^2}e^{\boldsymbol{w}'.\boldsymbol{r}'-ik\xi(\boldsymbol{r}'^2-(\boldsymbol{c}_p.\boldsymbol{r}')^2)}$ *integrand in surface integral*

We here choose the example with $\sqrt{a'+b'x'+cx'^2}e^{w'.r'-ik\xi(|r'|^2-(c_p.r')^2)}$ integrand. The exponential argument is similar to the ones considered in the previous section, with $\Phi = -\frac{\pi}{4}$. The constants a', b' and c' considered in Figure 6.4 are, respectively, $a' = 1$ with $b' = c' = 0$, and $a' = 1$ with $b' = c' = 100$. In both cases, we then apply theorem 6.19 of appendix E with $b_s = 4$. The agreement of results given by contour and surface integrals are excellent.

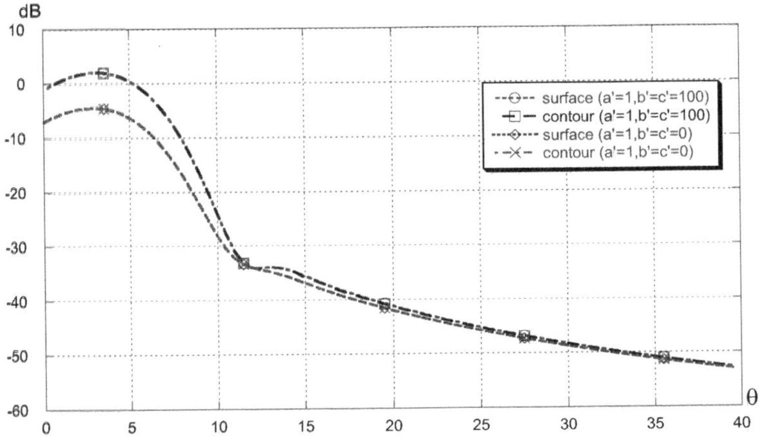

Figure 6.4. *Comparison of the surface and contour integrals with* $\sqrt{a'+b'x'+cx'^2}e^{w'.r'-ik\xi(|r'|^2-(c_p.r')^2)}$ *integrand (with its contour reduction given by theorem 6.19 in appendix E): for* $\xi = e^{i\Phi}/|\boldsymbol{r}|$ *with* $\Phi = -\pi/4$, $a' = 1$ *with* $b' = c' = 0$*, and* $a' = 1$ *with* $b' = c' = 100$*. For a color version of this figure, see www.iste.co.uk/bernard/theoryofscattering.zip*

6.11.2. *Asymptotic contour reduction of near-field radiation surface integral in physical optics for a large plate illuminated by a point source with spherical pattern*

6.11.2.1. *Asymptotic reduction for large flat plates*

We consider the monostatic PO radiation term $U = \boldsymbol{E}.\widehat{\boldsymbol{d}}$ in VV polarization (respectively, $U = \boldsymbol{E}.\boldsymbol{v}_0$ in HH) at constant φ when θ varies (Figure 6.5), for a point source at $\boldsymbol{r}_0$ with isotropic pattern, where $\widehat{\boldsymbol{d}} = \frac{\boldsymbol{r}_0\times\widehat{\boldsymbol{z}}}{\sqrt{(\boldsymbol{r}_0\times\widehat{\boldsymbol{z}})^2}}$ and $\boldsymbol{v}_0 = \widehat{\boldsymbol{d}} \times \frac{\boldsymbol{r}_0}{\tilde{r}_0}$ as $\boldsymbol{r}_0 \times \widehat{\boldsymbol{z}} \neq 0$, and its reduction (6.69) (respectively, (6.78)) from theorem 6.7 (respectively, 6.9) for a flat plate S' of large dimension.

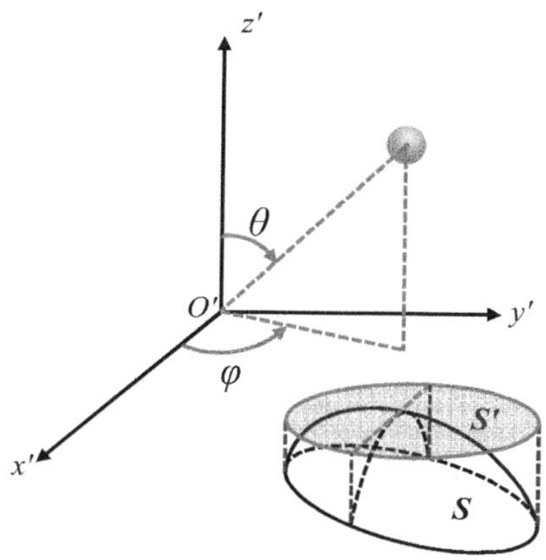

Figure 6.5. *Geometry for the near-field RCS for source illuminating scatterers with a spherical pattern*

We then compare the backscattering pattern $\frac{\sqrt{4\pi}\tilde{r}U}{|E_i|_{O'}|}$ (RCS) given by contour and surface integral expressions in Figure 6.6 when we consider a square plate with 1 m side length, shifted from the origin O' and centered at $(0.0.5\,\mathrm{m})$, observed in the plane $\varphi = 20$ deg, at the distance $\tilde{r} = \tilde{r_0} = 1\,\mathrm{m}$, for $\lambda = 2$ cm.

In Figure 6.7, we consider a larger shift of the plate center, disposed at $(0.3\mathrm{m})$, and we observe the field at $\tilde{r} = \tilde{r_0} = 3.5\,\mathrm{m}$ in the plane $\varphi = 80°$. We then distinguish more clearly differences between the polarizations VV and HH (VV is clearly larger than HH), and notice that the agreement between our contour integral and the initial surface integral continues to be excellent.

6.11.2.2. *Asymptotic reduction for large curved plates: convex, concave and saddle surfaces*

The exact reduction of surface integrals to contour integrals, described in particular in section 6.7, necessitates the evaluations of some primitives, which we need to define in closed form or by asymptotics. Among the different results derived from our method, we give, in section 6.9, a simple asymptotic reduction for the case of PO scattering by curved plates in near field, that we now test, comparing the respective results given by surface integrals and its contour reduction.

In what follows, we consider, in different cases of geometries, the monostatic PO surface radiation and its reduction, for $U = \boldsymbol{E}.\widehat{\boldsymbol{d}}$ in VV polarization (respectively, $U = \boldsymbol{E}.\boldsymbol{v}_0$ in HH), at constant φ when θ varies, for a point source at $\boldsymbol{r}_0$ with isotropic pattern (Figure 6.5), where $\widehat{\boldsymbol{d}} = \frac{\boldsymbol{r}_0 \times \widehat{\boldsymbol{z}}}{\sqrt{(\boldsymbol{r}_0 \times \widehat{\boldsymbol{z}})^2}}$ and $\boldsymbol{v}_0 = \widehat{\boldsymbol{d}} \times \frac{\boldsymbol{r}_0}{\tilde{r_0}}$ as $\boldsymbol{r}_0 \times \widehat{\boldsymbol{z}} \neq 0$.

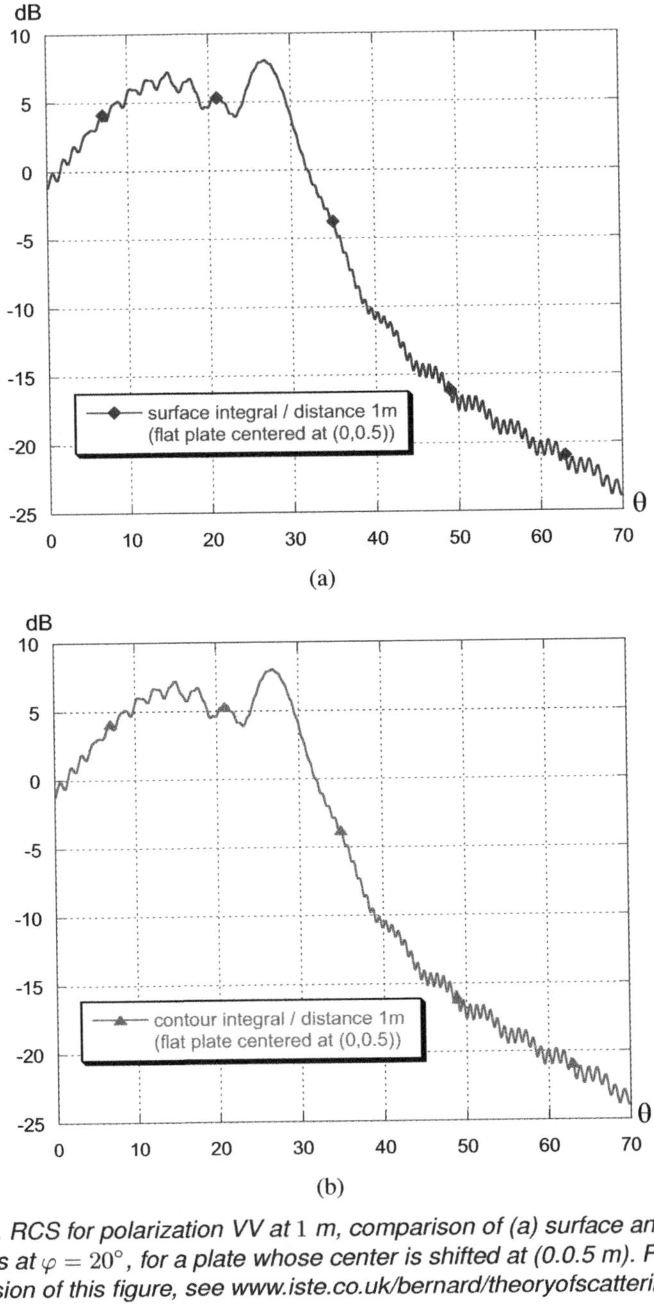

Figure 6.6. *RCS for polarization VV at* 1 *m, comparison of (a) surface and (b) contour integrals at* $\varphi = 20°$*, for a plate whose center is shifted at (0.0.5 m). For a color version of this figure, see www.iste.co.uk/bernard/theoryofscattering.zip*

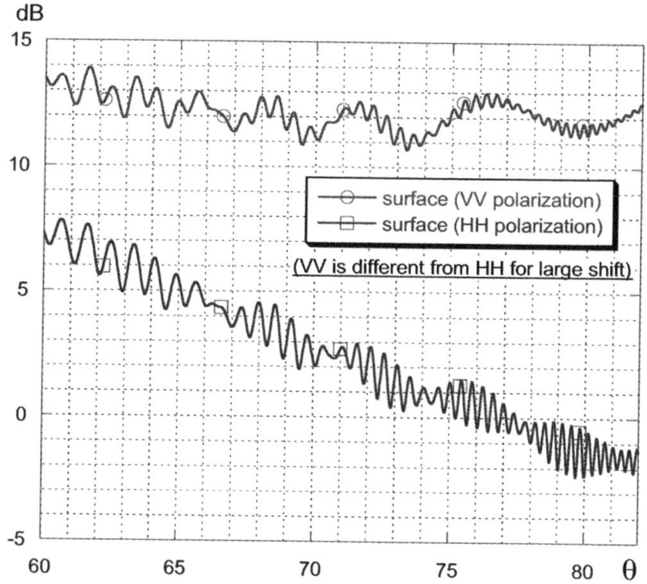

(a)

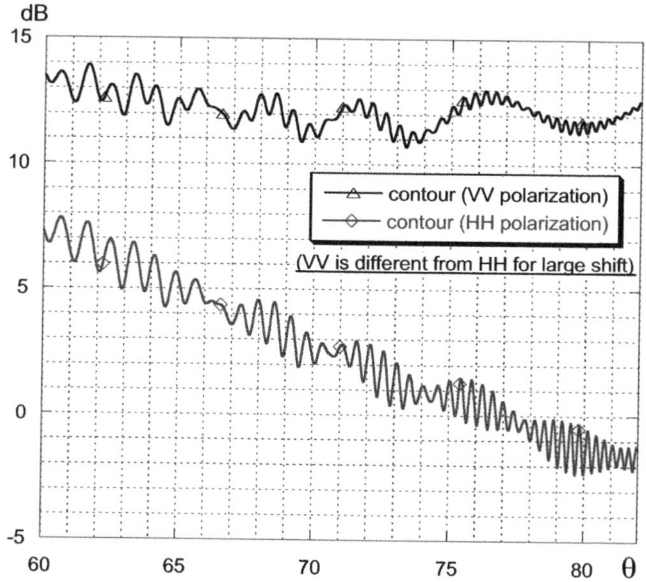

(b)

Figure 6.7. *RCS for polarization VV and HH at* 3.5 *m for a large shift, comparison of (a) surface and (b) contour integrals at* $\varphi = 80°$. *For a color version of this figure, see www.iste.co.uk/bernard/theoryofscattering.zip*

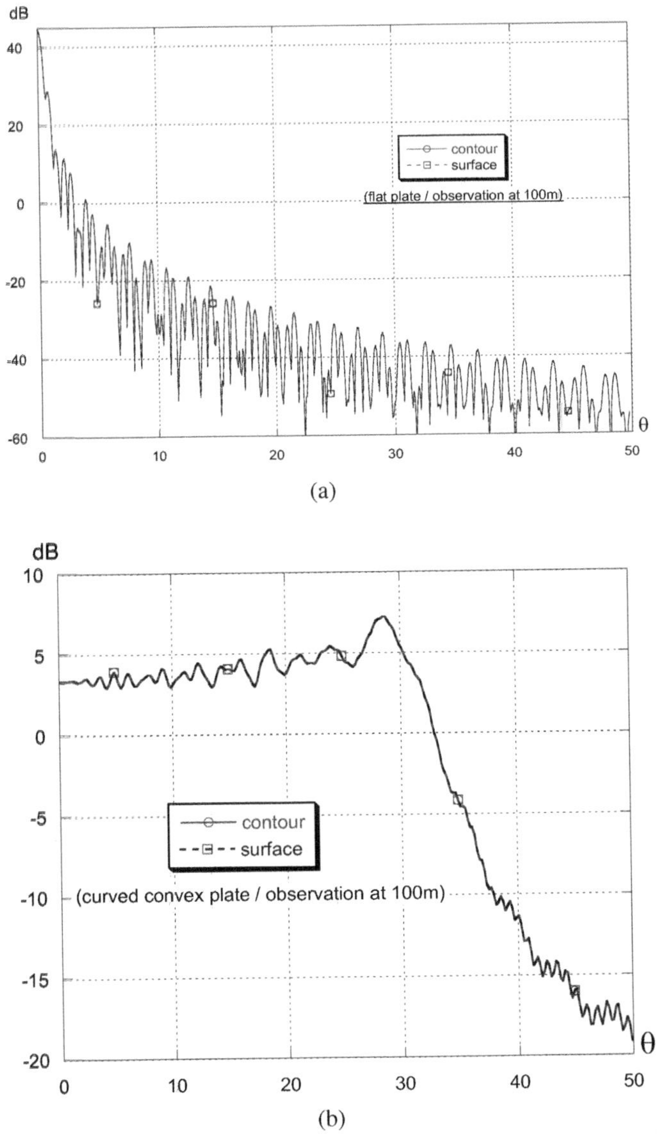

Figure 6.8. *RCS VV at* 100 *m for (a) flat plate* $C_{1,2} = 0$*, (b) convex plate with* $C_{1,2} = -0.6$ *or (c) concave with* $C_{1,2} = +0.6$*, with surface and contour integrals at* $\varphi = 70°$*. For a color version of this figure, see www.iste.co.uk/bernard/theoryofscattering.zip*

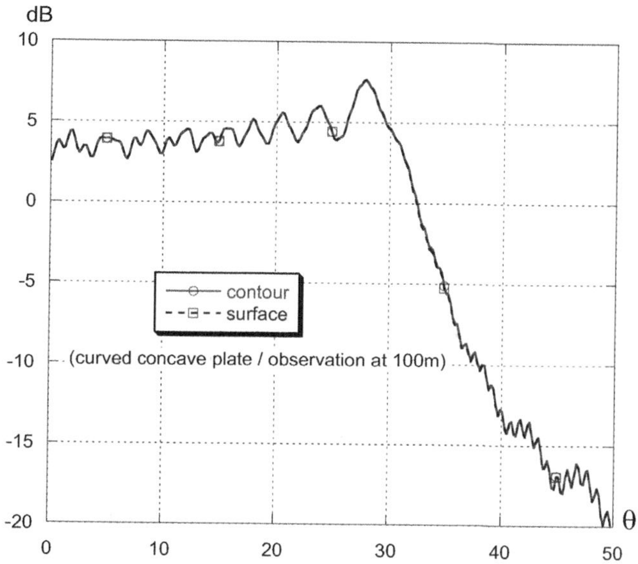

(c)

Figure 6.8. *Continued*

The contour reductions of the surface radiation integrals are then calculated by (6.158) (respectively, (6.163)) from theorem 6.15 (respectively, 6.16). The curved plate S, whose projection is a square of side length $D = 1$ m, is from now centered at the origin, and the field is observed in planes $\varphi = 70°$ and $20°$, for $k = 100\pi$ ($\lambda = 2$ cm), with $|C_{1,2}|D^2 \ll 4z$. We then compare the backscattering pattern $\frac{\sqrt{4\pi}\tilde{r}U}{|E_i|_{O'}|}$ given by contour and surface integral expressions, and analyze the modifications, as distance, angles and curvatures vary for convex, concave and saddle plates.

In Figure 6.8, at large distance $z = 100$ m, we begin by noting the influence of curvature on the width of the plateau. It is large, and similar in convex and concave cases, as curvature sign changes, while it almost vanishes when the plate is flat.

In Figures 6.9 and 6.10, we let vary the curvature in the near field, at a distance of 6 m, and give the results from surface and contour integrals, in convex and concave cases.

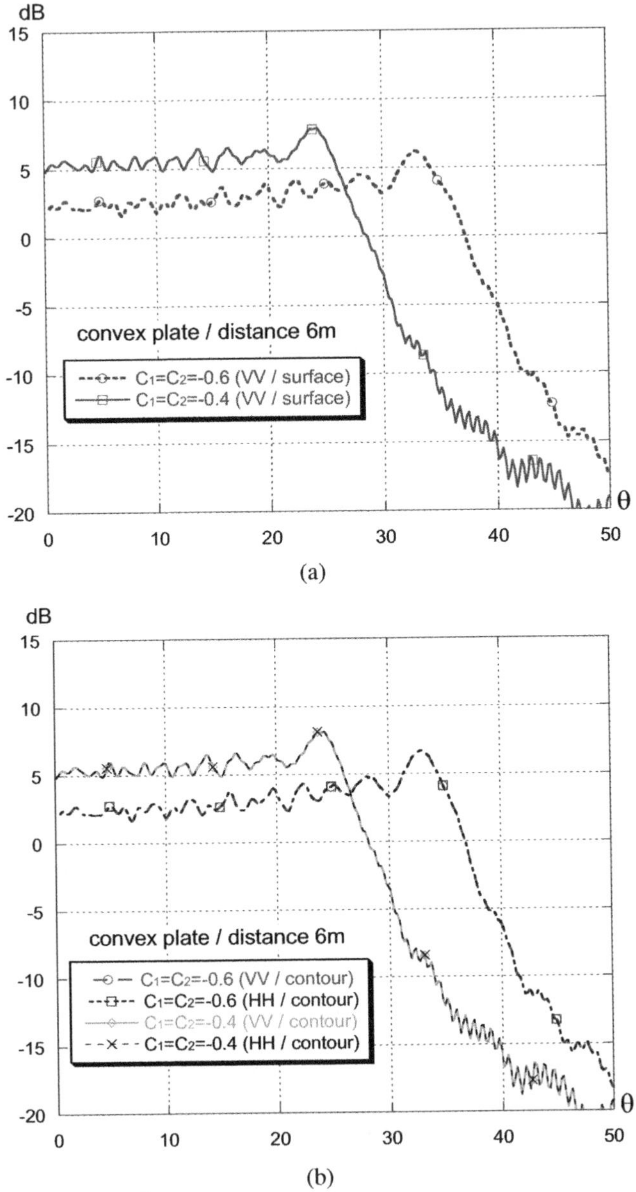

Figure 6.9. *RCS VV and HH at 6 m, comparison of (a) surface and (b) contour integrals, for* $C_1 = C_2 = -0.4$ *and* $C_1 = C_2 = -0.6$ *(convex), at* $\varphi = 70°$*. For a color version of this figure, see www.iste.co.uk/bernard/theoryofscattering.zip*

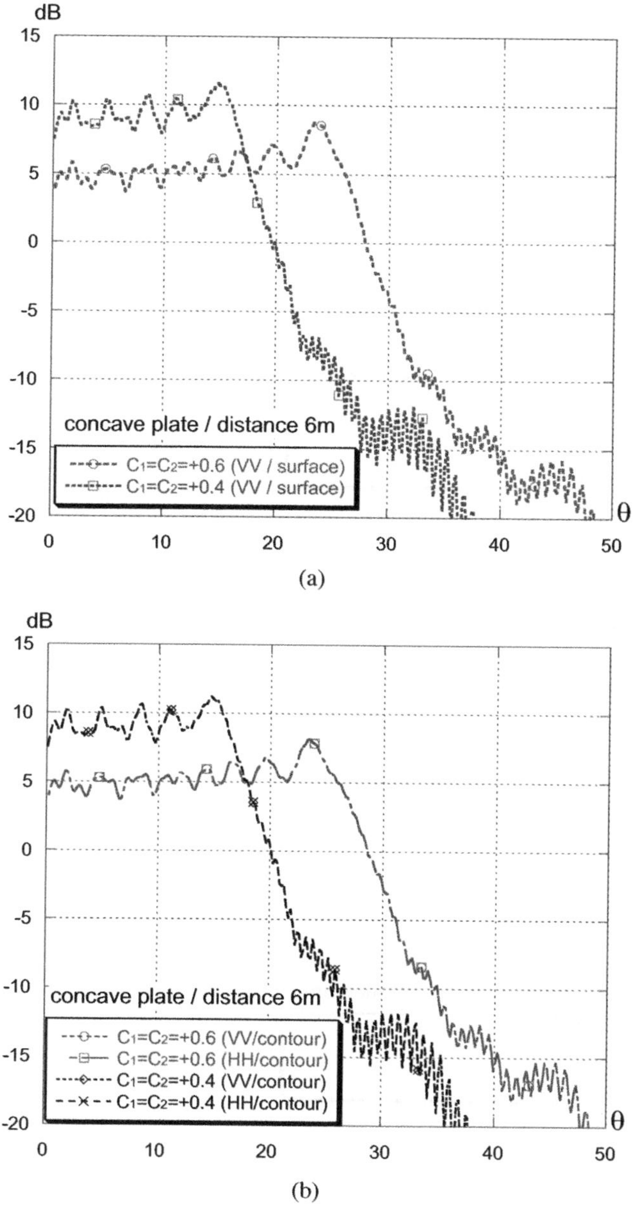

Figure 6.10. *RCS VV and HH at 6m, comparison of (a) surface and (b) contour integrals, for $C_1 = C_2 = 0.4$ and $C_1 = C_2 = 0.6$ (concave), at $\varphi = 70°$. For a color version of this figure, see www.iste.co.uk/bernard/theoryofscattering.zip*

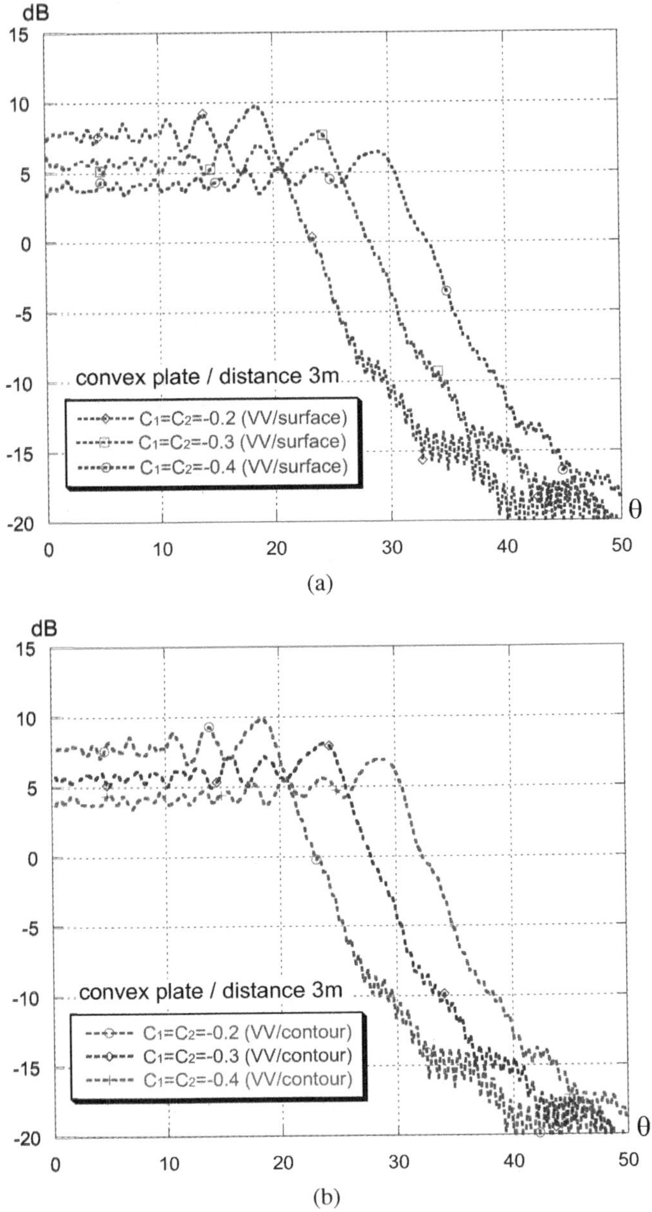

Figure 6.11. *RCS VV and HH at* 3 *m, comparison of (a) surface integrals and (b) contour integrals, for* $C_1 = C_2 = -0.2, = -0.3, = -0.4$ *at* $\varphi = 70°$. *For a color version of this figure, see www.iste.co.uk/bernard/theoryofscattering.zip*

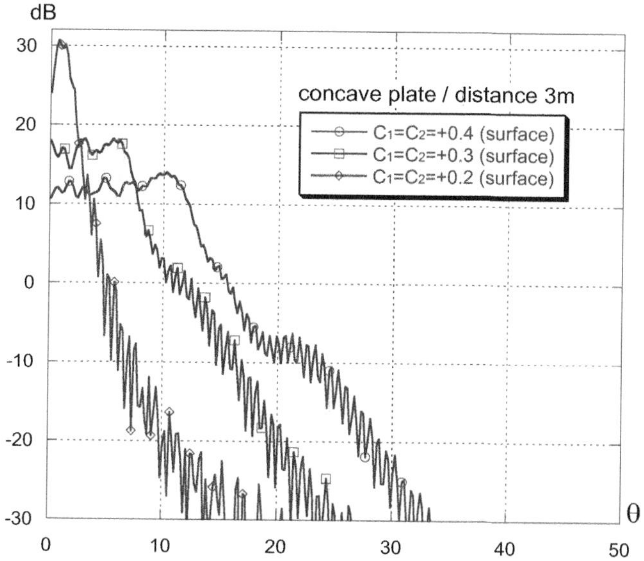

(a)

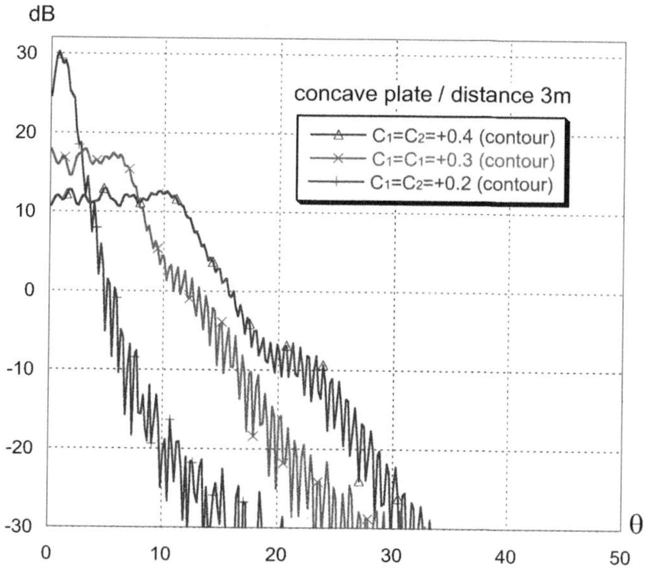

(b)

Figure 6.12. *RCS VV and HH at* 3 *m, comparison of (a) surface and (b) contour integrals, for* $C_1 = C_2 = 0.2, = 0.3, = 0.4$ *at* $\varphi = 70°$. *For a color version of this figure, see www.iste.co.uk/bernard/theoryofscattering.zip*

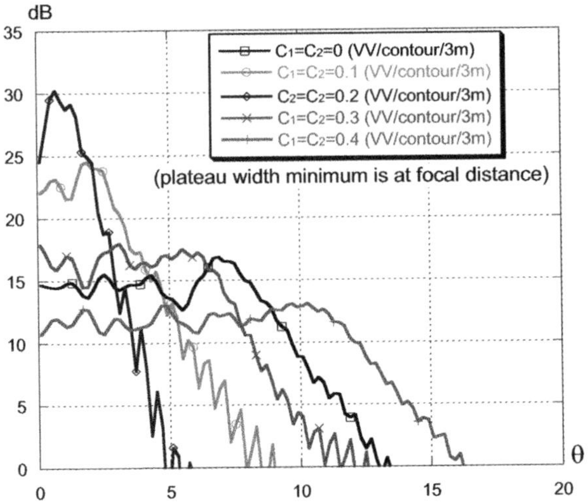

Figure 6.13. *Variation in purely concave cases with $C_j = C$, in vicinity of $1 - 2zC = 0$ (focal case) at 3 m, for $C = 0.4$, $C = 0.3$, $C = 0.2$, $C = 0.1$, $C = 0$, at $\varphi = 70°$. For a color version of this figure, see www.iste.co.uk/bernard/theoryofscattering.zip*

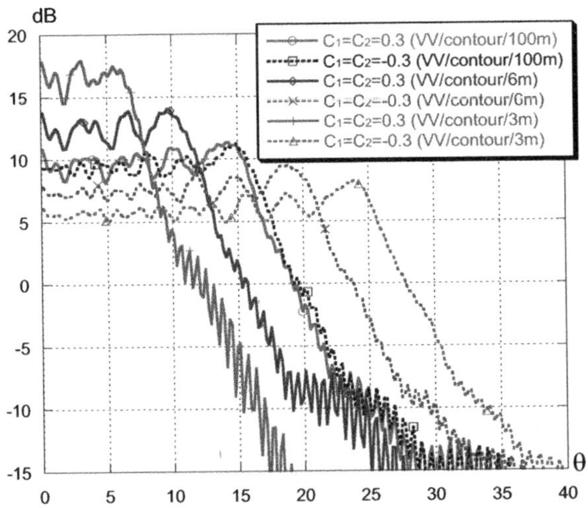

Figure 6.14. *Purely convex cases with negative C_j and concave cases with positive C_j, in VV, for $|C_1| = |C_2| = 0.3$, at $3, 6$ and 100 m at $\varphi = 70°$. For a color version of this figure, see www.iste.co.uk/bernard/theoryofscattering.zip*

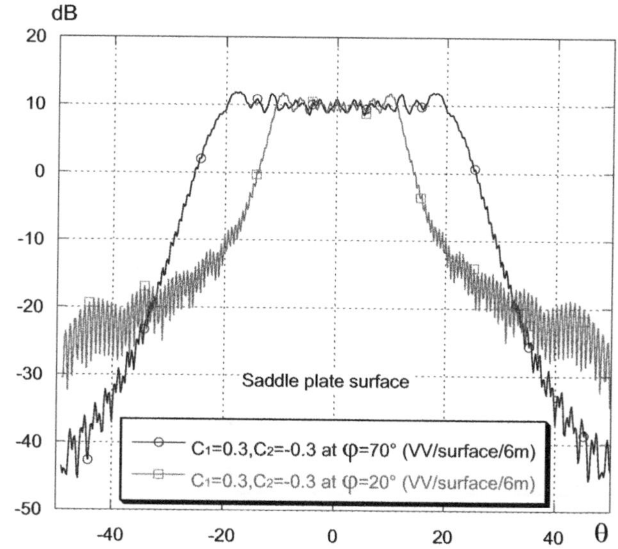

(a)

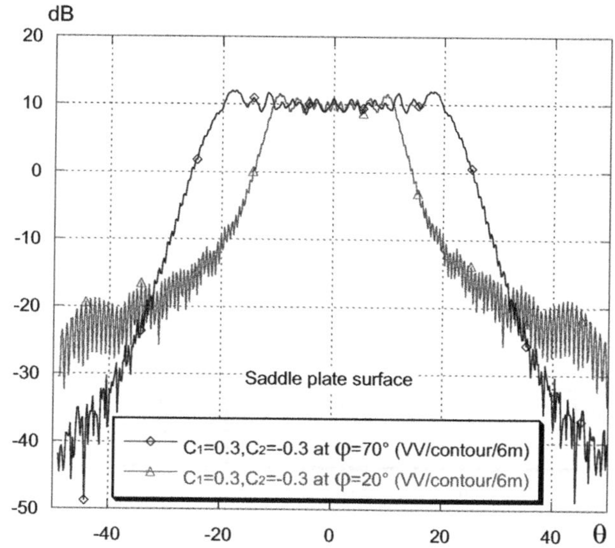

(b)

Figure 6.15. *RCS VV, comparison of (a) surface and (b) contour integrals, at* 6 *m, for a saddle curved plate with* $C_1 = 0.3$ *and* $C_2 = -0.3$ *at* $\varphi = 70°$ *and at* $\varphi = 20°$*. For a color version of this figure, see www.iste.co.uk/bernard/theoryofscattering.zip*

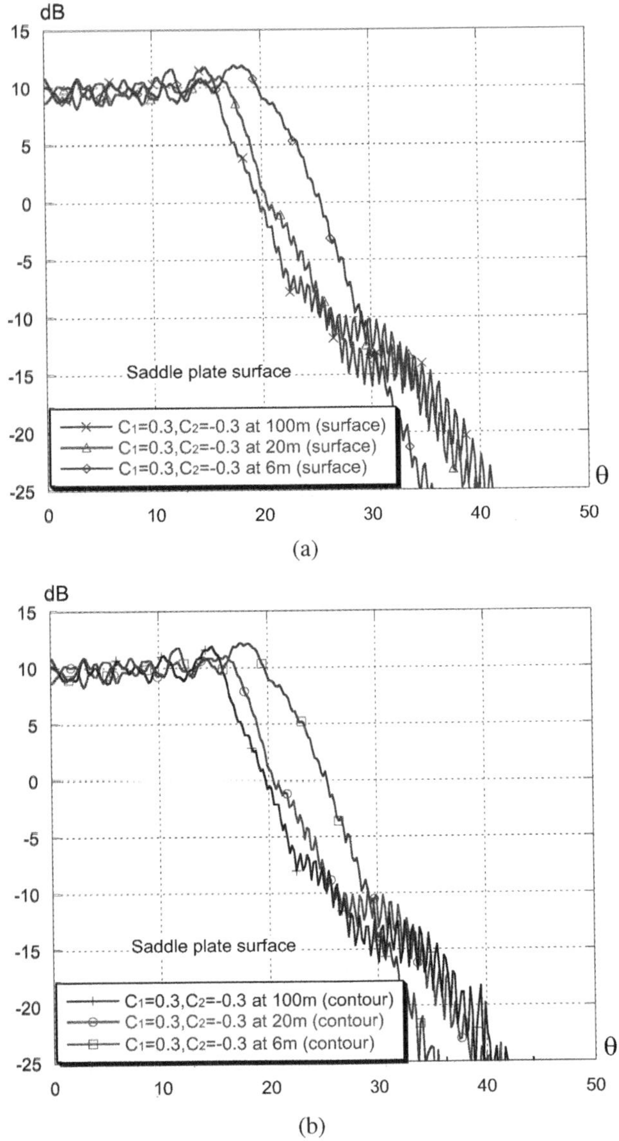

Figure 6.16. *RCS VV, comparison of (a) surface and (b) contour integrals, at* 6*,* 20 *and* 100 *m, for a saddle curved plate with* $C_1 = 0.3$ *and* $C_2 = -0.3$ *(convex in the plane of incidence) at* $\varphi = 70°$*. For a color version of this figure, see www.iste.co.uk/bernard/theoryofscattering.zip*

At 3 m, Figures 6.11–6.13 show the variation of patterns as we let vary the curvature. In Figure 6.13, we detail the concave case and show a minimum of the width of the plateau near the focal point where $1 - 2zC = 0$, which varies monotically else.

In Figure 6.14, we let vary the distance and observe that the width of plateaus, very different in near field for convex and concave plates, becomes similar at large distance when curvatures have the same absolute values.

In Figures 6.15 and 6.16, we give results for a saddle surface. The asymmetry of the plate is illustrated by the change in patterns in the cases $\varphi = 70°$ and $\varphi = 20°$ in Figure 6.15, while we observe the changes as we let vary the distance in Figure 6.16.

In all these cases, we notice the very good agreement of the results given by surface and contour integrals. The contour integrals then used are derived from exact or asymptotic applications of general exact results of section 6.7. However, it should be noted that the reduction accuracy, concerning in particular the near field PO case for curved plates, could be yet improved by applications of higher order asymptotics.

6.12. References

[1] Jones, D.S. (1964). *The Theory of Electromagnetism*. Pergamon Press, Oxford.

[2] Pouliguen, P. and Hemon, R. (2008). Analytical formulae for radar cross section of flat plates in near-field and normal incidence. *PIERS, B*, 9, 263–279.

[3] Bourlier, C. and Pouliguen, P. (2009). Useful analytical formulae for near-field monostatic radar cross section under the physical optics: Far-field criterion. *IEEE Transactions on AP*, 57, 205–214.

[4] Pouliguen, P. and Desclos, L. (1996). A physical optics approach to near-field RCS computations. *Ann. Telecom.*, 51(5–6), 219–226.

[5] Legault, S.R. (2004). Refining physical optics for near-field computations. *Electronics Letters*, 40, 1.

[6] Albani, M. and Maci, S. (2002). An exact line integral representation of the PO radiation integral from a flat perfectly conducting surfaces illuminated by elementary electric and magnetic dipoles. *Turk J. Elec. Engin.*, 10(2), 291–305.

[7] Albani, M. (2011). Boundary diffracted wave and incremental geometrical optics. *IEEE Transactions on AP*, 59(2), 586–594.

[8] Rubinowicz, A. (1962). Simple derivation of the Miyamoto-Wolf formula for the vector potential associated with a solution of the Helmholtz equation. *J. Opt. Soc. Am.*, 52(6), 717–718.

[9] Asvestas, J.S. (1985). Line integrals and physical optics. Part I. *J.Opt. Soc. Am.*, 2(6), 891–895.

[10] Asvestas, J.S. (1985). Line integrals and physical optics. Part II. *J.Opt. Soc. Am.*, 2(6), 896–902.

[11] Asvestas, J.S. (1986). The physical optics fields of an aperture on a perfectly conducting screen in terms of line integrals. *IEEE Transactions on AP*, 34(9), 1155–1159.

[12] Johansen, P.M. and Breinbjerg, O. (1995). An exact line integral representation of the physical optics scattered field: The case of a perfectly conducting polyhedral structure illuminated by electric hertzian dipoles. *IEEE Transactions on AP*, 43(7), 689–696.

[13] Delgado, C., Gomez, J.M., Catedra, M.F. (2007). Analytical field calculation involving current modes and quadratic phase expressions. *IEEE Transactions on AP*, 55(1), 233–240.

[14] Wu, Y.M., Jiang, L.J., Chew, W.C. (2012). An efficient method for computing highly oscillatory physical optics integral. *PIER*, 127, 211–257.

[15] Hillairet, J., Sokoloff, J., Bolioli, S. (2009). Uniform analytic scattered fields of a PEC plate illuminated by a vector paraxial Gaussian beam. *PIER B*, 14, 203–217.

[16] Gordon, W.B. (1975). Far-field to the Kirchhoff-Helmholtz representations of scattered fields. *IEEE Transactions on AP*, 23, 590–592.

[17] Carluccio, G., Albani, M., Pathak, P.H. (2010). Uniform asymptotic evaluation of surface integrals with polygonal integration domains in terms of UTD transition functions. *IEEE Transactions on AP*, 58(4), 1155–1163.

[18] Gordon, W.B. (2003). Calculating scatter from surfaces with zero curvature. *IEEE Transactions on AP*, 51(9), 2506–2508.

[19] Gordon, W.B. (2012). Contour integral representation for near-field backscatter from a flat plate. *IEEE Transactions on AP*, 60(5), 2587–2589.

[20] Asvestas, J.S. and Bilow, H.J. (2007). Line integral approach to computing impedance matrix elements. *IEEE Transactions on AP*, 55(10), 2767–2772.

[21] Knockaert, L. (1991). A general Gauss theorem for evaluating singular integrals over polyhedral domains. *Elelectromagnetics*, 11, 269–280.

[22] Gordon, W.B. and Bilow, H.J. (2002). Reduction of surface integrals to contour integrals. *IEEE Transactions on AP*, 50(3), 308–311.

[23] Bernard, J.M.L. (2016). On novel non-singular contour integrals for radiation surface integrals and near-field physical optics. *Proceedings of ICEAA Conference*, Turin (erratum: in section 4, replace $F(0)$ by $F(v_1 + d_o)$).

[24] Garabedian, P.R. (1998). *Partial Differential Equations*. AMS Chelsea Publishing, New York.

[25] Chen, S.-C. and Shaw, M.-C. (2001). *Partial Differential Equations in Several Complex Variables*. American Mathematical Society, Providence.

[26] Arnold, V.I. (1989). *Mathematical Methods of Classical Mechanics*. Springer-Verlag, New York.

[27] Arapura, D. (2016). Introduction to differential forms. Purdue University.

[28] Vladimirov, V.S. (2007). *Methods of the Theory of Function of Many Complex Variables*. Dover Publications Inc., New York.

[29] Bernard, J.M.L. (2014). *Advanced Theory of Diffraction by a Semi-infinite Impedance Cone*. Alpha Sciences International Ltd., Oxford.

[30] Bernard, J.M.L. (2011). On a novel expression of the field scattered by an arbitrary constant impedance plane. *Wave Motion*, 48, 634–645.

[31] Gradshteyn, I.S. and Ryzhik, I.M. (1994). *Table of Integrals, Series, and Products*. Academic Press, London.

6.A. Appendix A: primitives for theorems 6.5 and 6.6, and properties

6.A.1. *Primitives for the theorem 6.5*

Let us show that:

$$\int \frac{1}{\widetilde{R}}((\frac{\psi'}{\widetilde{R}})^2)' d\boldsymbol{R}^2 = 2(\frac{\psi'}{\widetilde{R}})^2 = 2\frac{(ik+\frac{1}{\widetilde{R}})^2}{\boldsymbol{R}^4}e^{-i2k\widetilde{R}},$$

$$\int (\frac{\psi'}{\widetilde{R}})^2 d\boldsymbol{R}^2 = -2\frac{(1+2ik\widetilde{R})}{4\boldsymbol{R}^4}e^{-i2k\widetilde{R}},$$

$$\int \widetilde{R}((\frac{\psi'}{\widetilde{R}})^2)' d\boldsymbol{R}^2 = 2(-k^2+\frac{3ik}{\widetilde{R}}+\frac{3}{2\boldsymbol{R}^2})\frac{e^{-i2k\widetilde{R}}}{\boldsymbol{R}^2}, \tag{6.184}$$

with $\widetilde{R} = \sqrt{\boldsymbol{R}.\boldsymbol{R}}$, $(.)' \equiv \partial_{\widetilde{R}}(.)$, $\psi(\widetilde{R}) = \frac{e^{-ik\widetilde{R}}}{\widetilde{R}}$, $\boldsymbol{R}^2 = (\boldsymbol{r'}-\boldsymbol{a})^2 + z^2 \neq 0$. The first line is trivial since $d\boldsymbol{R}^2 = 2\widetilde{R}d\widetilde{R}$, while the second and third line need some manipulations.

We remark first that ψ verifies $\Delta\psi + k^2\psi = 0$ as $\boldsymbol{R}^2 \neq 0$, and thus:

$$\frac{1}{\boldsymbol{R}^2}((\widetilde{R})^3(\frac{\psi'}{\widetilde{R}}))' = \widetilde{R}(\frac{\psi'}{\widetilde{R}})' + 3\frac{\psi'}{\widetilde{R}} = -k^2\psi, \tag{6.185}$$

which multiplied by $\frac{\psi'}{\widetilde{R}}$, leads us to:

$$\widetilde{R}((\frac{\psi'}{\widetilde{R}})^2)' = -k^2\frac{((\psi)^2)'}{\widetilde{R}} - 6(\frac{\psi'}{\widetilde{R}})^2. \tag{6.186}$$

Considering that we have, by integration by parts:

$$\int \widetilde{R}((\frac{\psi'}{\widetilde{R}})^2)'\widetilde{R}d\widetilde{R} = (\psi')^2 - 2\int (\frac{\psi'}{\widetilde{R}})^2\widetilde{R}d\widetilde{R}, \tag{6.187}$$

we can use (6.186), and write:

$$\int -k^2((\psi)^2)'d\widetilde{R} - \int 6(\frac{\psi'}{\widetilde{R}})^2\widetilde{R}d\widetilde{R} = (\psi')^2 - 2\int (\frac{\psi'}{\widetilde{R}})^2\widetilde{R}d\widetilde{R}, \tag{6.188}$$

so that we obtain the second line of (6.184), following:

$$\int (\frac{\psi'}{\widetilde{R}})^2 \widetilde{R} d\widetilde{R} = \frac{1}{4}((ik\psi)^2 - (\psi')^2) = -(2ik + \frac{1}{\widetilde{R}})\frac{\psi^2}{4\widetilde{R}}(= \frac{(\widetilde{R}\psi^2)'}{4\boldsymbol{R}^2}). \tag{6.189}$$

We can then rewrite (6.187) and derive the third line of (6.184):

$$\int \widetilde{R}((\frac{\psi'}{\widetilde{R}})^2)' \widetilde{R} d\widetilde{R} = \frac{3(\psi')^2 - (ik\psi)^2}{2} = (-k^2 + \frac{3ik}{\widetilde{R}} + \frac{3}{2\boldsymbol{R}^2})\psi^2. \tag{6.190}$$

6.A.2. *Primitives for the theorem 6.6*

For theorem 6.6, we have to complete (6.184). For this, we write:

$$\begin{aligned} &\int (\frac{\psi'}{\widetilde{R}})'(\frac{\psi'}{\widetilde{R}})' d\boldsymbol{R}^2 = -2\int (\frac{\psi'}{\widetilde{R}})'(k^2\psi + 3\frac{\psi'}{\widetilde{R}})d\widetilde{R} = \\ &= (-2k^2\frac{\psi'\psi}{\widetilde{R}} + k^2\int (\frac{\psi'}{\widetilde{R}})^2 d\boldsymbol{R}^2) - 3(\frac{\psi'}{\widetilde{R}})^2 \\ &= 2k^2(ik + \frac{1}{\widetilde{R}})\frac{\psi^2}{\widetilde{R}} - k^2(ik + \frac{1}{2\widetilde{R}})\frac{\psi^2}{\widetilde{R}} - 3(\frac{\psi'}{\widetilde{R}})^2 \\ &= k^2(ik + \frac{3}{2\widetilde{R}})\frac{\psi^2}{\widetilde{R}} - 3(ik + \frac{1}{\widetilde{R}})^2(\frac{\psi}{\widetilde{R}})^2, \end{aligned} \tag{6.191}$$

where we have successively used (6.185), an integration by parts, and (6.189).

6.A.3. *Expressions and properties of successive primitives of* $\frac{e^{-v\widetilde{R}}}{(\widetilde{R})^n}$

Letting $\widetilde{R} = \sqrt{\boldsymbol{R}.\boldsymbol{R}}$, we have, for $n \geq 1$:

$$\begin{aligned} &\int_{\infty}^{\boldsymbol{R}^2} \frac{e^{-v\widetilde{R}}}{(\widetilde{R})^n} d\boldsymbol{R}^2 = 2v^{n-2}V_{n-1}(v\widetilde{R}) \sim -\frac{2}{v}\frac{e^{-v\widetilde{R}}}{(\widetilde{R})^{n-1}}(1 - \frac{n-1}{v\widetilde{R}} + \ldots) \text{ if } v\widetilde{R} \to \infty, \\ &V_n(u) = \int_{\infty}^{u} \frac{e^{-t}}{t^n} dt = -\frac{E_n(u)}{u^{n-1}},\ V_n(u) \sim -\frac{e^{-u}}{u^n}(1 - \frac{n}{u} + \frac{n(n+1)}{u^2} - \ldots), \\ &E_{n+1}(u) = \frac{1}{n}(e^{-u} - uE_n(u)),\ V_n(u) + nV_{n+1}(u) = -\frac{e^{-u}}{(u)^n}, \end{aligned} \tag{6.192}$$

while, for p successive primitives, we can write:

$$(p-1)! \int_{\infty}^{\boldsymbol{R}'^2} d\boldsymbol{R}^2 ..... \int_{\infty}^{\boldsymbol{R}'^2} d\boldsymbol{R}^2 \frac{e^{-v\widetilde{R}}}{\widetilde{R}^n} = \int_{\infty}^{\boldsymbol{R}'^2} (\boldsymbol{R}'^2 - \boldsymbol{R}^2)^{p-1} \frac{e^{-v\widetilde{R}}}{\widetilde{R}^n} d\boldsymbol{R}^2$$

$$= \sum_{0 \le m \le p-1} (-1)^{p-1-m} C_{p-1}^m (\widetilde{R}')^{2m} \int_{\infty}^{\boldsymbol{R}'^2} \frac{e^{-v\widetilde{R}}}{(\widetilde{R})^{n-2(p-1-m)}} d\boldsymbol{R}^2$$

$$= 2 \sum_{0 \le m \le p-1} (-1)^{p-1-m} C_{p-1}^m (\widetilde{R}')^{2m} (v^{n-2(p-1-m)-2} V_{n-2(p-1-m)-1}(v\widetilde{R}'))$$

$$= -2 \sum_{0 \le m \le p-1} (-1)^{p-1-m} C_{p-1}^m (\widetilde{R}')^{2m} (\frac{E_{n-2(p-1-m)-1}(v\widetilde{R}')}{(\widetilde{R}')^{n-2(p-1-m)-2}})$$

$$= \frac{-2}{(\widetilde{R}')^{n-2p}} \sum_{0 \le m \le p-1} (-1)^{p-1-m} C_{p-1}^m E_{n-2(p-1-m)-1}(v\widetilde{R}'), \tag{6.193}$$

We can remark that the first asymptotic orders of the sum, as $v\widetilde{R} \gg 1$, vanish. Indeed, for example, using $\sum_{0 \le m \le p-1} (-1)^{p-1-m} C_{p-1}^m = 0$ in this expression for $p = 2$, we have:

$$\int_{\infty}^{\boldsymbol{R}'^2} d\boldsymbol{R}^2 \int_{\infty}^{\boldsymbol{R}'^2} d\boldsymbol{R}^2 \frac{e^{-v\widetilde{R}}}{(\widetilde{R})^n} \sim$$

$$\sim \frac{-2}{v} \frac{e^{-v\widetilde{R}'}}{(\widetilde{R}')^{n-(2p-1)}} \sum_{0 \le m \le 1} (-1)^{p-1-m} C_{p-1}^m (1 - \frac{n - 2(p-m) + 1}{v\widetilde{R}'})|_{p=2}$$

$$\sim \frac{4}{v^2} \frac{e^{-v\widetilde{R}'}}{(\widetilde{R}')^{n-(2p-2)}} \sum_{0 \le m \le 1} (-1)^{p-1-m} m C_{p-1}^m |_{p=2} = \frac{4}{v^2} \frac{e^{-v|R'|}}{(\widetilde{R}')^{n-2}}. \tag{6.194}$$

NOTE 6.30.– We note that from:

$$\int_{\infty}^{\boldsymbol{R}'^2} \frac{-ve^{-v\widetilde{R}}(1 + \frac{n}{v\widetilde{R}})}{(\widetilde{R})^{n+1}} d\boldsymbol{R}^2 = \frac{2e^{-v\widetilde{R}'}}{(\widetilde{R}')^n}, \tag{6.195}$$

we recover

$$(v\widetilde{R}')^n e^{v\widetilde{R}'} (V_n(v\widetilde{R}') + nV_{n+1}(v\widetilde{R}')) = -1. \tag{6.196}$$

6.B. Appendix B: reduction of physical optics surface integrals to contour integrals in bistatic case and quadratic approximation of exponential arguments

6.B.1. *Physical optics surface integrals and asymptotics*

Let us consider (6.47) for the fields, diffracted by the surface S, when it is illuminated by incident field $\boldsymbol{E}_i$ and $\boldsymbol{H}_i$, with tangential fields on scatterer approximated, as in PO. At the observation point $\boldsymbol{R}_1 = \boldsymbol{a} + z\widehat{z} \notin S$, we can then write:

$$\boldsymbol{E} = \frac{Z_0}{4\pi ik}\int_S (\tilde{R}(\frac{\psi'}{\tilde{R}})'\boldsymbol{u}\times(\boldsymbol{u}\times\boldsymbol{J}) - \frac{2\psi'}{\tilde{R}}\boldsymbol{J})dS - \frac{1}{4\pi}\int_S \frac{\psi'}{\tilde{R}}\boldsymbol{R}\times\boldsymbol{M}dS,$$

$$Z_0\boldsymbol{H} = \frac{Z_0}{4\pi}\int_S \frac{\psi'}{\tilde{R}}\boldsymbol{R}\times\boldsymbol{J}dS + \frac{1}{4\pi ik}\int_S (\tilde{R}(\frac{\psi'}{\tilde{R}})'\boldsymbol{u}\times(\boldsymbol{u}\times\boldsymbol{M}) - \frac{2\psi'}{\tilde{R}}\boldsymbol{M})dS,$$

$$\boldsymbol{J} = \widehat{\boldsymbol{n}}\times\boldsymbol{H}\sim\widehat{\boldsymbol{n}}\times\boldsymbol{H}_{GO}, \boldsymbol{M} = -\widehat{\boldsymbol{n}}\times\boldsymbol{E}\sim-\widehat{\boldsymbol{n}}\times\boldsymbol{E}_{GO}, \tag{6.197}$$

where

$$\psi(\tilde{R}) = \frac{e^{-ik\tilde{R}}}{\tilde{R}}, (.)' \equiv \partial_{\tilde{R}}(.), \tilde{R} = \sqrt{\boldsymbol{R}.\boldsymbol{R}},$$

$$\boldsymbol{R} = \boldsymbol{R}_1 - \boldsymbol{R}' = \tilde{R}\boldsymbol{u}, \boldsymbol{R}_1 = \sqrt{\boldsymbol{R}_1.\boldsymbol{R}_1}\boldsymbol{u}_1, \tag{6.198}$$

with $\boldsymbol{R}' = (\boldsymbol{r}' + z'\widehat{z})$ the point of integration on S. The fields $\boldsymbol{E}_{GO}$ and $\boldsymbol{H}_{GO}$ derive from the geometrical optics fields on S, and we can write:

$$\boldsymbol{H}_{GO}\sim\mathcal{D}_H(\boldsymbol{u}_0,\mathcal{F}_i(\boldsymbol{u}_0);\boldsymbol{R}')\psi(\tilde{R}_0),\ \boldsymbol{E}_{GO}\sim\mathcal{D}_E(\boldsymbol{u}_0,\mathcal{F}_i(\boldsymbol{u}_0);\boldsymbol{R}')\psi(\tilde{R}_0),$$

$$\boldsymbol{R}_0 = \boldsymbol{R}' - \boldsymbol{R}_i = \tilde{R}_0\boldsymbol{u}_0,\ \tilde{R}_0 = \sqrt{\boldsymbol{R}_0.\boldsymbol{R}_0},\ \boldsymbol{R}_i = \sqrt{\boldsymbol{R}_i.\boldsymbol{R}_i}\boldsymbol{u}_i,$$

$$\boldsymbol{E}_i = \boldsymbol{u}_0\times\mathcal{F}_i(\boldsymbol{u}_0)\psi(\tilde{R}_0),\ Z_0\boldsymbol{H}_i = \boldsymbol{u}_0\times\boldsymbol{E}_i, \tag{6.199}$$

where $\boldsymbol{R}_i = (\boldsymbol{a}_i + z_i\widehat{z}) \notin S$ is the origin of phase for $\boldsymbol{E}_i$ and $\boldsymbol{H}_i$ (see (6.62) for an example of $\boldsymbol{E}_i$ and $\boldsymbol{H}_i$). We can take the example of a curved plate S defined by:

$$\boldsymbol{R}' = \boldsymbol{r}' + \zeta_s(\boldsymbol{r}')\widehat{z},\ \zeta_s(\boldsymbol{r}') = C_1x'^2 + C_2y'^2,\ \zeta_s(\boldsymbol{0}) = 0, \tag{6.200}$$

with $C_j < 0$ (respectively, $C_j > 0$) for convex (respectively, concave) surface inflection, where $\boldsymbol{R}'$ is the current point on S', the projection of S on the plane $z = 0$. We then have:

$$\boldsymbol{R} = (\boldsymbol{a} + z\widehat{z}) - (\boldsymbol{r}' + \zeta_s(\boldsymbol{r}')\widehat{z}),$$

$$\boldsymbol{R}^2 = (x' - a_x)^2 + (y' - a_y)^2 + ((C_1x'^2 + C_2y'^2) - z)^2, \tag{6.201}$$

and we can express the fields as integrals on S' following:

$$\begin{aligned}
\boldsymbol{E} &= \frac{Z_0}{4\pi ik}\int_{S'}(\widetilde{R}(\frac{\psi'}{\widetilde{R}})'\boldsymbol{u}\times(\boldsymbol{u}\times\boldsymbol{J_p}) - \frac{2\psi'}{\widetilde{R}}\boldsymbol{J_p})dS_{r'} \\
&- \frac{1}{4\pi}\int_{S'}\frac{\psi'}{\widetilde{R}}\boldsymbol{R}\times\boldsymbol{M_p}dS_{r'}, \\
Z_0\boldsymbol{H} &= \frac{1}{4\pi ik}\int_{S'}(\widetilde{R}(\frac{\psi'}{\widetilde{R}})'\boldsymbol{u}\times(\boldsymbol{u}\times\boldsymbol{M_p}) - \frac{2\psi'}{\widetilde{R}}\boldsymbol{M_p})dS_{r'} \\
&+ \frac{Z_0}{4\pi}\int_{S'}\frac{\psi'}{\widetilde{R}}\boldsymbol{R}\times\boldsymbol{J_p}dS_{r'}, \\
\boldsymbol{J_p} &\sim \widehat{\boldsymbol{n}}J^s(\boldsymbol{r'})\times\boldsymbol{H_{GO}},\ \boldsymbol{M_p}\sim -\,\widehat{\boldsymbol{n}}J^s(\boldsymbol{r'})\times\boldsymbol{E_{GO}},
\end{aligned} \tag{6.202}$$

where

$$\begin{aligned}
J^s(\boldsymbol{r'}) &= \left|\frac{\partial\boldsymbol{R'}}{\partial x'}\wedge\frac{\partial\boldsymbol{R'}}{\partial y'}\right|,\ \widehat{\boldsymbol{n}} = \left(\frac{\partial\boldsymbol{R'}}{\partial x'}\wedge\frac{\partial\boldsymbol{R'}}{\partial y'}\right)/\left|\frac{\partial\boldsymbol{R'}}{\partial x'}\wedge\frac{\partial\boldsymbol{R'}}{\partial y'}\right|, \\
\widehat{\boldsymbol{n}}J^s(\boldsymbol{r'}) &= -2C_1x'\widehat{\boldsymbol{x}} - 2C_2y'\widehat{\boldsymbol{y}} + \widehat{\boldsymbol{z}}.
\end{aligned} \tag{6.203}$$

Considering large $k\boldsymbol{R}$, we can neglect high order of $1/\widetilde{R}$ in the integral following:

$$\begin{aligned}
\boldsymbol{E} &= \frac{ikZ_0}{4\pi(\widetilde{R})^3}\int_{S'}e^{-ik\widetilde{R}}\,\boldsymbol{R}\times(\boldsymbol{R}\times\boldsymbol{J_p})dS + \frac{ik}{4\pi(\widetilde{R})^2}\int_{S'}e^{-ik\widetilde{R}}\,\boldsymbol{R}\times\boldsymbol{M_p}dS, \\
Z_0\boldsymbol{H} &= \frac{-ikZ_0}{4\pi(\widetilde{R})^2}\int_{S'}e^{-ik\widetilde{R}}\boldsymbol{R}\times\boldsymbol{J_p}dS + \frac{ik}{4\pi(\widetilde{R})^3}\int_{S'}e^{-ik\widetilde{R}}\,\boldsymbol{R}\times(\boldsymbol{R}\times\boldsymbol{M_p})dS,
\end{aligned} \tag{6.204}$$

where

$$\begin{aligned}
\boldsymbol{J_p} &\sim \widehat{\boldsymbol{n}}J^s\times\boldsymbol{H}_{GO}\sim\widehat{\boldsymbol{n}}J^s\times\frac{e^{-ik\widetilde{R}_0}}{\widetilde{R}_0}\mathcal{D}_H(\boldsymbol{u_0},\mathcal{F}_i(\boldsymbol{u_0});\boldsymbol{R'}), \\
\boldsymbol{M_p} &\sim -\widehat{\boldsymbol{n}}J^s(\boldsymbol{r'})\times\boldsymbol{E}_{GO}\sim -\widehat{\boldsymbol{n}}J^s\times\frac{e^{-ik\widetilde{R}_0}}{\widetilde{R}_0}\mathcal{D}_E(\boldsymbol{u_0},\mathcal{F}_i(\boldsymbol{u_0});\boldsymbol{R'}),
\end{aligned} \tag{6.205}$$

with $\widehat{\boldsymbol{n}}J^s(\boldsymbol{r'}) = -2C_1x'\widehat{\boldsymbol{x}} - 2C_2y'\widehat{\boldsymbol{y}} + \widehat{\boldsymbol{z}}$.

6.B.2. *An asymptotic development of* $\widetilde{R}$

Let us consider the asymptotics of $\widetilde{R} = \sqrt{\boldsymbol{R}^2}$ when $\boldsymbol{R} = \boldsymbol{R_1} - \boldsymbol{R_2}$ as $|\boldsymbol{R_1}| \gg |\boldsymbol{R_2}|$. For that, we write:

$$\begin{aligned} \boldsymbol{R}^2 &= (\boldsymbol{R_1} - \boldsymbol{R_2})^2 = (\sqrt{\boldsymbol{R}_1^2}\boldsymbol{u}_1 - \boldsymbol{R_2})^2 \\ &= (\sqrt{\boldsymbol{R}_1^2} - \boldsymbol{R_2}.\boldsymbol{u}_1)^2 + \boldsymbol{R}_2^2 - (\boldsymbol{R_2}.\boldsymbol{u}_1)^2, \end{aligned} \tag{6.206}$$

which leads us to:

$$\begin{aligned} \sqrt{\boldsymbol{R}^2} &= (\sqrt{\boldsymbol{R}_1^2} - \boldsymbol{R_2}.\boldsymbol{u}_1) + \frac{\boldsymbol{R}_2^2 - (\boldsymbol{R_2}.\boldsymbol{u}_1)^2}{\sqrt{\boldsymbol{R}^2} + (\sqrt{\boldsymbol{R}_1^2} - \boldsymbol{R_2}.\boldsymbol{u}_1)}, \\ \frac{\sqrt{\boldsymbol{R}^2}}{\sqrt{\boldsymbol{R}_1^2}} &= 1 - \frac{\boldsymbol{R_2}.\boldsymbol{u}_1}{\sqrt{\boldsymbol{R}_1^2}} + \frac{\boldsymbol{R}_2^2 - (\boldsymbol{R_2}.\boldsymbol{u}_1)^2}{2\boldsymbol{R}_1^2(1 + \frac{1}{2}((\frac{\sqrt{\boldsymbol{R}^2}}{\sqrt{\boldsymbol{R}_1^2}} - 1) - \frac{\boldsymbol{R_2}.\boldsymbol{u}_1}{\sqrt{\boldsymbol{R}_1^2}}))}. \end{aligned} \tag{6.207}$$

Letting $\delta_{as} = \frac{1}{2}((\frac{\sqrt{\boldsymbol{R}^2}}{\sqrt{\boldsymbol{R}_1^2}} - 1) - \frac{\boldsymbol{R_2}.\boldsymbol{u}_1}{\sqrt{\boldsymbol{R}_1^2}})$, we can then write:

$$\begin{aligned} \frac{\sqrt{\boldsymbol{R}^2}}{\sqrt{\boldsymbol{R}_1^2}} &= 1 - \frac{\boldsymbol{R_2}.\boldsymbol{u}_1}{\sqrt{\boldsymbol{R}_1^2}} + \frac{\boldsymbol{R}_2^2 - (\boldsymbol{R_2}.\boldsymbol{u}_1)^2}{2\boldsymbol{R}_1^2} \\ &+ \frac{\boldsymbol{R}_2^2 - (\boldsymbol{R_2}.\boldsymbol{u}_1)^2}{2\boldsymbol{R}_1^2}(\sum_{n=1}^{N}(-\delta_{as})^n + \frac{(-\delta_{as})^{N+1}}{1 + \delta_{as}}), \end{aligned} \tag{6.208}$$

which gives us a complete asymptotics of $\sqrt{\boldsymbol{R}^2}$ for $|\boldsymbol{R_1}| \gg |\boldsymbol{R_2}|$ at any order N with explicit remaining term.

6.B.3. *Quadratic approximation of exponential arguments and reduction of surface integrals to contour ones*

Let us consider quadratic exponential arguments approximations in (6.204) from (6.208) following:

$$\begin{aligned} \widetilde{R}_0 &\sim \sqrt{\boldsymbol{R}_i^2} - \boldsymbol{R}'.\boldsymbol{u}_i + \frac{\boldsymbol{R}'^2 - (\boldsymbol{R}'.\boldsymbol{u}_i)^2}{2\sqrt{\boldsymbol{R}_i^2}}, \\ \widetilde{R} &\sim \sqrt{\boldsymbol{R}_1^2} - \boldsymbol{R}'.\boldsymbol{u}_1 + \frac{\boldsymbol{R}'^2 - (\boldsymbol{R}'.\boldsymbol{u}_1)^2}{2\sqrt{\boldsymbol{R}_1^2}}, \end{aligned} \tag{6.209}$$

as $\sqrt{\boldsymbol{R}_{i}^{2}}$ and $\sqrt{\boldsymbol{R}_{1}^{2}}$ are large relatively to $|\boldsymbol{R}'|\sqrt{k|\boldsymbol{R}'|}$, so that higher order terms can be neglected in the calculus of $k\widetilde{R}_0$ and $k\widetilde{R}$. We then obtain (6.204) as combinations of elementary integrals of the form:

$$\int_{S'} g(\boldsymbol{r}')f(\mathcal{R}^2)dS', \tag{6.210}$$

where f is the exponential function, while $\mathcal{R}^2$ and $g(\boldsymbol{r}')$ are the polynomial functions:

$$\begin{aligned}\mathcal{R}^2 &= (\alpha_1 x' + \alpha_2 y') - ik(h_1 x'^2 + h_2 y'^2 + h_3 x'y')\\ &= v(r'_p - a_p)^2 + v_1,\\ g(r') &= \sum_{j=0}^{p}\frac{1}{j!}\sum_{k=0}^{j}\binom{j}{k}\frac{\partial^j g|_{r'=\mathcal{P}^{-1}(a_p)}}{\partial x'^k \partial y'^{j-k}}[(\mathcal{P}^{-1}(\widehat{x}).(r'_p - a_p))^k\\ &(\mathcal{P}^{-1}(\widehat{y}).(r'_p - a_p))^{j-k}],\end{aligned} \tag{6.211}$$

with $r'_p = \mathcal{P}(r')$ the complex scaling considered in (6.135) and (6.136). We can then reduce the surface integrals to contour integrals, using theorems 6.12 and 6.13.

NOTE 6.31.– We can also write:

$$\begin{aligned}\boldsymbol{R}^2(r') &= (x' - a_x)^2 + (y' - a_y)^2 + ((C_1 x'^2 + C_2 y'^2) - z)^2\\ &= (1 - 2zC_1)(x' - a'_x)^2 + (1 - 2zC_2)(y' - a'_y)^2 + z''^2 + \epsilon(\boldsymbol{r}'),\\ a'_x &= \frac{a_x}{1 - 2zC_1},\ a'_y = \frac{a_y}{1 - 2zC_2},\ z''^2 = z^2 - (\frac{2zC_1}{1 - 2zC_1}a_x^2 + \frac{2zC_2}{1 - 2zC_2}a_y^2),\end{aligned} \tag{6.212}$$

with $\epsilon(\boldsymbol{r}') = (\zeta_s(\boldsymbol{r}'))^2$, $\zeta_s(\boldsymbol{r}') = C_1 x'^2 + C_2 y'^2$. Neglecting $\epsilon(\boldsymbol{r}')$, we then note that, as in (6.145), $\boldsymbol{R}^2 = v'(\boldsymbol{r}'_p - \boldsymbol{a}_p)^2 + z''^2$, with $\boldsymbol{a}_p = \mathcal{P}(\boldsymbol{a}')$.

6.C. Appendix C: reduction of physical optics radiation integrals in monostatic case for an imperfectly conducting flat plate

We consider a point source illuminating at large distance (relatively to the wavelength) an isotropic impedance plate S' with normal $\widehat{z}$. We consider the incident field as locally plane, combination of TE ($\boldsymbol{E}$//surface) and TM ($\boldsymbol{H}$// surface) waves, respectively, denoted $(\boldsymbol{E}^i_e, \boldsymbol{H}^i_e)$ and $(\boldsymbol{E}^i_h, \boldsymbol{H}^i_h)$. The currents $\boldsymbol{J}$ and $\boldsymbol{M}$, respectively, denoted $\boldsymbol{J}_{e(h)}|_S = \widehat{z} \times \boldsymbol{H}_{e(h)}$ and $\boldsymbol{M}_{e(h)}|_S = -\widehat{z} \times \boldsymbol{E}_{e(h)}$ then satisfy:

$$\boldsymbol{J}_{e(h)} = 2b'_{e(h)}(-\boldsymbol{u}_0.\widehat{z})[\widehat{z} \times \boldsymbol{H}^i_{e(h)}],\ \boldsymbol{M}_{e(h)} = -c'(-\boldsymbol{u}_0.\widehat{z})[\widehat{z} \times \boldsymbol{J}_{e(h)}], \tag{6.213}$$

where c' and $b'_{e,h}$ are regular functions, depending on the angle of the incidence vector $\boldsymbol{u}_0$ with $\widehat{z}$. In the case of constant impedance, the terms $b'_{e(h)}$ and c' are constants [29]–[30], while $b'_{e(h)} \equiv 1$ and $c' \equiv 0$ in perfectly conducting case.

6.C.1. *Physical optics fields and reduction for an electric dipolar source*

THEOREM 6.17.– *The PO field, diffracted by the surface S' with an impedance condition, illuminated by the electric dipole $\boldsymbol{J}_0 = \widehat{\mathbf{d}}\delta(\boldsymbol{r'} - \boldsymbol{r}_0)$ is given by (6.47) with (6.213). It can be written, in monostatic case, when $\widehat{\mathbf{d}}.\widehat{z} = 0$ following:*

$$\boldsymbol{E} \sim \frac{Z_0}{8\pi^2 ik}\int_{S'} \frac{z}{2}\tilde{R}((\frac{\psi'}{\tilde{R}})^2)'((\boldsymbol{u}.\boldsymbol{h})\boldsymbol{u} - \boldsymbol{h})dS' +$$
$$+\frac{1}{8\pi^2}\int_{S'} c'(\frac{z}{\tilde{R}})(\psi')^2(\frac{\boldsymbol{R}^2 - z^2}{\boldsymbol{R}^2}(\frac{(\boldsymbol{r'} - \boldsymbol{a})}{(\boldsymbol{r'} - \boldsymbol{a})^2} \times \widehat{z})(((\boldsymbol{r'} - \boldsymbol{a}) \times \widehat{z}).\boldsymbol{h})$$
$$+ ((\boldsymbol{u}.\boldsymbol{h})\boldsymbol{u} - \boldsymbol{h}))dS', \tag{6.214}$$

for large $k\tilde{R}$, where

$$\boldsymbol{h} = b_h'(\frac{z}{\tilde{R}})(\widehat{\mathbf{d}}.\widehat{\mathbf{d}}_1)\widehat{\mathbf{d}}_1 + b_e'(\frac{z}{\tilde{R}})(\widehat{\mathbf{d}}.\widehat{\mathbf{d}}_2)\widehat{\mathbf{d}}_2,$$
$$\widehat{\mathbf{d}}_1 = \frac{(\boldsymbol{a} - \boldsymbol{r'})}{\sqrt{(\boldsymbol{a} - \boldsymbol{r'})^2}},\ \widehat{\mathbf{d}}_2 = \widehat{z} \times \widehat{\mathbf{d}}_1.$$
$$(\boldsymbol{u}.\boldsymbol{h})\boldsymbol{u} = \frac{z\widehat{z}(\boldsymbol{h}.(\boldsymbol{a} - \boldsymbol{r'}))}{\boldsymbol{R}^2} + \frac{(\boldsymbol{R}^2 - z^2)}{\boldsymbol{R}^2}\frac{(\boldsymbol{h}.(\boldsymbol{a} - \boldsymbol{r'}))(\boldsymbol{a} - \boldsymbol{r'})}{(\boldsymbol{a} - \boldsymbol{r'})^2}. \tag{6.215}$$

Considering the propositions 6.13 and 6.14, we can use:

$$\int (\psi')^2 d\boldsymbol{R}^2 \sim -\frac{ike^{-i2k\tilde{R}}}{\tilde{R}},\ \int z\tilde{R}((\frac{\psi'}{\tilde{R}})^2)' d\boldsymbol{R}^2 \sim -\frac{2k^2 ze^{-i2k\tilde{R}}}{\boldsymbol{R}^2}, \tag{6.216}$$

and asymptotically reduce each integral to a non-singular contour integral.

PROOF.– We consider, using (6.213), that S' is illuminated by the radiation of the electrical dipole $\boldsymbol{J}_0 = \widehat{\mathbf{d}}\delta(\boldsymbol{r'} - \boldsymbol{r}_0)$ with $\widehat{\mathbf{d}}.\widehat{z} = 0$ ($\widehat{d}$//plane). We distinguish locally TE and TM waves in incident waves and apply $b_{e,h}$ coefficients. We can then write:

$$2\pi\boldsymbol{J}|_S \sim \widehat{z} \times (\boldsymbol{R}_0 \times \frac{\psi_0'}{\tilde{R}_0}\boldsymbol{h}),$$
$$2\pi\boldsymbol{M}|_S \sim -c'(-\boldsymbol{u}_0.\widehat{z})(\widehat{z} \times (\widehat{z} \times (\boldsymbol{R}_0 \times \frac{\psi_0'}{\tilde{R}_0}\boldsymbol{h})))$$
$$= -c'(-\boldsymbol{u}_0.\widehat{z})((\widehat{z}.(\boldsymbol{R}_0 \times \frac{\psi_0'}{\tilde{R}_0}\boldsymbol{h}))\widehat{z} - (\boldsymbol{R}_0 \times \frac{\psi_0'}{\tilde{R}_0}\boldsymbol{h})),$$
$$\boldsymbol{h} = b_h'(-\boldsymbol{u}_0.\widehat{z})(\widehat{\mathbf{d}}.\widehat{\mathbf{d}}_1)\widehat{\mathbf{d}}_1 + b_e'(-\boldsymbol{u}_0.\widehat{z})(\widehat{\mathbf{d}}.\widehat{\mathbf{d}}_2)\widehat{\mathbf{d}}_2,\ \widehat{\mathbf{d}}_2 = \widehat{z} \times \widehat{\mathbf{d}}_1, \tag{6.217}$$

with $\boldsymbol{R}_0 = (\boldsymbol{r}' - \boldsymbol{r}_0)$ and $\boldsymbol{u}_0 = \frac{\boldsymbol{R}_0}{\widetilde{R}_0}$. In the case where $\boldsymbol{R} = -\boldsymbol{R}_0 = z\widehat{\boldsymbol{z}} + \boldsymbol{a} - \boldsymbol{r}'$ (monostatic case), $\widehat{\boldsymbol{d}}_1 = \frac{(\boldsymbol{a}-\boldsymbol{r}')}{\sqrt{(\boldsymbol{a}-\boldsymbol{r}')^2}}$ and we then obtain:

$$\boldsymbol{E} \sim \frac{Z_0}{8\pi^2 ik}\int_{S'}\frac{z}{2}\widetilde{R}((\frac{\psi'}{\widetilde{R}})^2)'(\boldsymbol{u}\times(\boldsymbol{u}\times\boldsymbol{h}))dS'$$

$$+\frac{1}{8\pi^2}\int_{S'}c'(\frac{z}{\widetilde{R}})(\frac{\psi'}{\widetilde{R}})^2(-\boldsymbol{R}\times\widehat{\boldsymbol{z}}\,(\widehat{\boldsymbol{z}}.(\boldsymbol{R}\times\boldsymbol{h}))+\boldsymbol{R}\times(\boldsymbol{R}\times\boldsymbol{h}))dS', \qquad (6.218)$$

and thus (6.214). We can then, following theorems 6.1 and 6.2, and propositions 6.13 and 6.14, reduce the field $\boldsymbol{E}$ to a contour integral. For this, we can use (6.184), and also:

$$\int(\psi')^2 d\boldsymbol{R}^2 = \int \boldsymbol{R}^2(\frac{\psi'}{\widetilde{R}})^2 d\boldsymbol{R}^2 = \frac{(\widetilde{R}\psi^2)'}{2} - \int\frac{(\widetilde{R}\psi^2)'}{2R^2}d\boldsymbol{R}^2$$

$$= -\frac{(3+2ik\widetilde{R})}{2\boldsymbol{R}^2}e^{-i2k\widetilde{R}} - \int\frac{e^{-i2k\widetilde{R}}}{2\boldsymbol{R}^4}d\boldsymbol{R}^2 = -\frac{ik}{\widetilde{R}}e^{-i2k\widetilde{R}}(1+O(\frac{1}{k\widetilde{R}})). \qquad (6.219)$$

6.C.2. Physical optics fields and reduction for a magnetic dipolar source

THEOREM 6.18.– *The PO field, diffracted by the surface S' with an impedance condition, illuminated by the magnetic dipole $\boldsymbol{M}_0 = \widehat{\mathbf{d}}\delta(\boldsymbol{r}' - \boldsymbol{r}_0)$, is given by (6.47) with (6.213). It can be written, in monostatic case, when $\widehat{\mathbf{d}}.\widehat{z} = 0$ following:*

$$\boldsymbol{E} \sim \frac{-z}{8(\pi ik)^2}\int_{S'}\boldsymbol{h}\times\mathbf{R}((\frac{\psi'}{\widetilde{R}})')^2 dS' - \frac{1}{16\pi^2 ikZ_0}\int_{S'}\boldsymbol{h}\times\mathbf{R}c'(\frac{z}{\widetilde{R}})\widetilde{R}((\frac{\psi'}{\widetilde{R}})^2)'dS'$$

$$+\frac{1}{16\pi^2 ikZ_0}\int_{S'}(\boldsymbol{R}\times\widehat{z})c'(\frac{z}{\widetilde{R}})\widetilde{R}((\frac{\psi'}{\widetilde{R}})^2)'(\frac{z}{\boldsymbol{R}^2}(\boldsymbol{R}.\boldsymbol{h}))dS', \qquad (6.220)$$

for large $k\widetilde{R}$, with

$$\boldsymbol{h} = b_e'(\frac{z}{\widetilde{R}})(\widehat{\boldsymbol{d}}.\widehat{\boldsymbol{d}}_1)\widehat{\boldsymbol{d}}_1 + b_h'(\frac{z}{\widetilde{R}})(\widehat{\boldsymbol{d}}.\widehat{\boldsymbol{d}}_2)\widehat{\boldsymbol{d}}_2,$$

$$\widehat{\boldsymbol{d}}_1 = \frac{(\boldsymbol{a}-\boldsymbol{r}')}{\sqrt{(\boldsymbol{a}-\boldsymbol{r}')^2}},\ \widehat{\boldsymbol{d}}_2 = \widehat{\boldsymbol{z}}\times\widehat{\boldsymbol{d}}_1,$$

$$(\boldsymbol{R}\times\widehat{z})(\boldsymbol{R}.\boldsymbol{h}) = (\boldsymbol{R}^2 - z^2)\frac{(\boldsymbol{h}.(\boldsymbol{a}-\boldsymbol{r}'))(\boldsymbol{a}-\boldsymbol{r}')\times\widehat{z}}{(\boldsymbol{a}-\boldsymbol{r}')^2}. \qquad (6.221)$$

Considering propositions 6.13 and 6.14, we can use:

$$\int (\frac{\psi'}{\widetilde{R}})'(\frac{\psi'}{\widetilde{R}})' d\boldsymbol{R}^2 \sim k^2 ik \frac{e^{-i2k\widetilde{R}}}{(\widetilde{R})^3}, \int \widetilde{R}((\frac{\psi'}{\widetilde{R}})^2)' d\boldsymbol{R}^2 \sim -\frac{2k^2 e^{-i2k\widetilde{R}}}{\boldsymbol{R}^2},$$

$$\int \frac{1}{\widetilde{R}}((\frac{\psi'}{\widetilde{R}})^2)' d\boldsymbol{R}^2 \sim -\frac{2k^2 e^{-i2k\widetilde{R}}}{\boldsymbol{R}^4}, \tag{6.222}$$

and asymptotically reduce each surface integral to a non-singular contour integral.

PROOF.– We consider, using (6.213), that S' is illuminated by the radiation of the magnetic dipole $\boldsymbol{M}_0 = \widehat{\boldsymbol{d}}\delta(\boldsymbol{r}' - \boldsymbol{r}_0)$ with $\widehat{\boldsymbol{d}}.\widehat{z} = 0$ ($\widehat{\boldsymbol{d}}$//plane), and thus

$$\boldsymbol{J}|_S = \frac{1}{2\pi ik Z_0}\widehat{z} \times (\widetilde{R}_0(\frac{\psi'}{\widetilde{R}_0})'\boldsymbol{u}_0 \times (\boldsymbol{u}_0 \times \boldsymbol{h}) - \frac{2\psi'}{\widetilde{R}}\boldsymbol{h}), \ \boldsymbol{R}_0 = (\boldsymbol{r}' - \boldsymbol{r}_0),$$

$$\boldsymbol{h} = b_e'(-\boldsymbol{u}_0.\widehat{z})(\widehat{\boldsymbol{d}}.\widehat{\boldsymbol{d}}_1)\widehat{\boldsymbol{d}}_1 + b_h'(-\boldsymbol{u}_0.\widehat{z})(\widehat{\boldsymbol{d}}.\widehat{\boldsymbol{d}}_2)\widehat{\boldsymbol{d}}_2, \ \widehat{\boldsymbol{d}}_2 = \widehat{z} \times \widehat{\boldsymbol{d}}_1,$$

$$\boldsymbol{M}|_S = \frac{-c'(\frac{z}{\widetilde{R}})}{2\pi ik Z_0}\widehat{z} \times (\widehat{z} \times (\widetilde{R}_0(\frac{\psi'}{\widetilde{R}_0})'\boldsymbol{u}_0 \times (\boldsymbol{u}_0 \times \boldsymbol{h}) - \frac{2\psi'}{\widetilde{R}}\boldsymbol{h})), \tag{6.223}$$

where $\boldsymbol{R}_0 = (\boldsymbol{r}' - \boldsymbol{r}_0)$, $\boldsymbol{u}_0 = \frac{\boldsymbol{R}_0}{\widetilde{R}_0}$. In the case where $\boldsymbol{R} = -\boldsymbol{R}_0 = z\widehat{z} + \boldsymbol{a} - \boldsymbol{r}'$ (monostatic case), $\widehat{\boldsymbol{d}}_1 = \frac{(\boldsymbol{a}-\boldsymbol{r}')}{\sqrt{(\boldsymbol{a}-\boldsymbol{r}')^2}}$ and we then obtain:

$$\begin{aligned}
\boldsymbol{E} = & \frac{1}{8(\pi ik)^2}\int_{S'} z(\frac{\psi'}{\widetilde{R}})'(\frac{\psi'}{\widetilde{R}})'(R \times \boldsymbol{h})dS' \\
& - \frac{1}{4(\pi ik)^2}\int_{S'} \widehat{z} \times (\widetilde{R}((\frac{\psi'}{\widetilde{R}})^2)'(\boldsymbol{u}(\boldsymbol{u}.\boldsymbol{h}) - \boldsymbol{h}) - 2(\frac{\psi'}{\widetilde{R}})^2\boldsymbol{h})dS' \\
& + \frac{1}{8\pi^2 ik Z_0}\int_{S'} c'(\frac{z}{\widetilde{R}})\frac{\psi'}{\widetilde{R}}\boldsymbol{R} \times (\widehat{z} \times (\widehat{z} \times (\widetilde{R}(\frac{\psi'}{\widetilde{R}})'\boldsymbol{u} \times (\boldsymbol{u} \times \boldsymbol{h}) \\
& - \frac{2\psi'}{\widetilde{R}}\boldsymbol{h})))dS' \sim \frac{-z}{8(\pi ik)^2}\int_{S'} ((\frac{\psi'}{\widetilde{R}})')^2\boldsymbol{h} \times \boldsymbol{R}dS' \\
& + \frac{1}{8\pi^2 ik Z_0}\int c'(\frac{z}{\widetilde{R}})\widetilde{R}\frac{\psi'}{\widetilde{R}}(\boldsymbol{R} \times \widehat{z})(\widehat{z}.(\boldsymbol{u} \times (\boldsymbol{u} \times \boldsymbol{h}))(\frac{\psi'}{\widetilde{R}})'dS' \\
& - \frac{1}{8\pi^2 ik Z_0}\int c'(\frac{z}{\widetilde{R}})\widetilde{R}\frac{\psi'}{\widetilde{R}}\boldsymbol{R} \times ((\frac{\psi'}{\widetilde{R}})'\boldsymbol{u} \times (\boldsymbol{u} \times \boldsymbol{h}))dS',
\end{aligned} \tag{6.224}$$

and thus (6.220). We can then, following theorems 6.1 and 6.2, and propositions 6.13 and 6.14, reduce the field $\boldsymbol{E}$ to a contour integral. For this, we use (6.184) and (6.191) and:

$$(\boldsymbol{R}\times\widehat{\boldsymbol{z}})\widetilde{R}\frac{(\boldsymbol{R}.\boldsymbol{h})}{\boldsymbol{R}^2}=\frac{\boldsymbol{R}^2-z^2}{\widetilde{R}}\frac{(\boldsymbol{h}.(\boldsymbol{a}-\boldsymbol{r}'))((\boldsymbol{a}-\boldsymbol{r}')\times\widehat{\boldsymbol{z}})}{(\boldsymbol{a}-\boldsymbol{r}')^2}. \tag{6.225}$$

6.D. Appendix D: particular uses of asymptotics with theorems 6.1 and 6.2

The reduction to a contour integral generally supposes the knowledge of a primitive, as in theorems 6.1 and 6.2. When it cannot be expressed in closed form, its expression can often be derived asymptotically.

LEMMA 6.2.– *Let* $F_1((\boldsymbol{r}'-\boldsymbol{a})^2,\varphi)=\frac{e^{-v\widetilde{R}}}{\widetilde{R}^n}$ *be analytic on the surface* S' *of complex points* $\boldsymbol{r}'=x'\widehat{\boldsymbol{x}}+y'\widehat{\boldsymbol{y}}$ *of the plane* $z=0$, *with* $\widetilde{R}=\sqrt{\boldsymbol{R}.\boldsymbol{R}}=\sqrt{(\boldsymbol{r}'-\boldsymbol{a})^2+z^2}$, *constants* n *and* v *with* $Re(v)>0$, $n\geq 1$. *We define functions so that:*

$$\frac{\partial(F((\boldsymbol{r}'-\boldsymbol{a})^2,\varphi))}{\partial((\boldsymbol{r}'-\boldsymbol{a})^2)}\equiv g_0((\boldsymbol{r}'-\boldsymbol{a})^2,\varphi)\frac{\partial(F_1((\boldsymbol{r}'-\boldsymbol{a})^2,\varphi))}{\partial((\boldsymbol{r}'-\boldsymbol{a})^2)}, \tag{6.226}$$

remains $O(e^{-v\widetilde{R}})$ *as we let, from any definite point on* S', *the parameter* $w=(\boldsymbol{r}'-\boldsymbol{a})^2$ *continuously vary to* ∞ *at* φ *constant. Considering* $g_0(w,\varphi)$ *and* $\frac{\partial g_0(w,\varphi)}{\partial(w)}$ *as analytic functions, relatively to* $(x'-a_x)$ *and* $(y'-a_y)$ *variables, smooth and* $O(1+|w|^m)$ *in the domain where we let* w *and* v *vary, we can write, for large* $v\widetilde{R}$:

$$\int_{\infty}^{(\boldsymbol{r}'-\boldsymbol{a})^2}\frac{\partial F(w,\varphi)}{\partial w}dw=$$

$$=g_0((\boldsymbol{r}'-\boldsymbol{a})^2,\varphi)F_1((\boldsymbol{r}'-\boldsymbol{a})^2,\varphi)-\int_{\infty}^{(\boldsymbol{r}'-\boldsymbol{a})^2}\frac{\partial g_0(w,\varphi)}{\partial w}F_1(w,\varphi)dw$$

$$=F_1((\boldsymbol{r}'-\boldsymbol{a})^2,\varphi)(g_0((\boldsymbol{r}'-\boldsymbol{a})^2,\varphi)+\frac{2\widetilde{R}\partial g_0((\boldsymbol{r}'-\boldsymbol{a})^2,\varphi)}{v\partial((\boldsymbol{r}'-\boldsymbol{a})^2)}+o(1/v)). \tag{6.227}$$

More generally, with the weaker assumption that $\frac{\partial g_0(w,\varphi)}{\partial(\sqrt{w})}$ *remains smooth when* w *varies, we can write for large* vz:

$$\int_{\infty}^{(\boldsymbol{r}'-\boldsymbol{a})^2}\frac{\partial(F(w,\varphi)}{\partial w}dw=F_1((\boldsymbol{r}'-\boldsymbol{a})^2,\varphi)(g_0((\boldsymbol{r}'-\boldsymbol{a})^2,\varphi)+O(\frac{1}{z^{n-1}\sqrt{vz}})). \tag{6.228}$$

These asymptotic results can be used in combination with theorems 6.1 and 6.2.

PROOF.– The function $F_1((\boldsymbol{r'}-\boldsymbol{a})^2,\varphi)=\frac{e^{-v\tilde{R}}}{\tilde{R}^n}$ is analytic on S'. We note that:

$$\int_{\infty}^{(\boldsymbol{r'}-\boldsymbol{a})^2}\frac{2e^{-v\tilde{R}}}{(\tilde{R})^{n-1}}(\tilde{R})'d(\boldsymbol{r'}-\boldsymbol{a})^2=\int_{\infty}^{(\boldsymbol{r'}-\boldsymbol{a})^2}\frac{2e^{-v\tilde{R}}}{(\tilde{R})^{n-1}}d\tilde{R}\sim-\frac{2\tilde{R}}{v}F_1((\boldsymbol{r'}-\boldsymbol{a})^2),\tag{6.229}$$

with $(\tilde{R})'=\partial_{(\boldsymbol{r'}-\boldsymbol{a})^2}(\tilde{R})=1/2\tilde{R}$, and we can derive (6.227) from the use of integration by parts. More generally, if $g_1(\sqrt{w})=\frac{\partial g_0(w,\varphi)}{\partial(\sqrt{w})}$ remains smooth when w varies, we note that, for large vz:

$$\int_{\infty}^{(\boldsymbol{r'}-\boldsymbol{a})^2}\frac{\partial g_0(w,\varphi)}{\partial w}F_1(w,\varphi)dw=\int_{\infty}^{(\boldsymbol{r'}-\boldsymbol{a})^2}\frac{2\sqrt{w}\partial g_0(w,\varphi)}{\partial w}\frac{e^{-v\tilde{R}}}{(\tilde{R})^n}d\sqrt{w}$$

$$=g_1(\sqrt{(\boldsymbol{r'}-\boldsymbol{a})^2})\int_{\infty}^{\sqrt{(\boldsymbol{r'}-\boldsymbol{a})^2}}\frac{e^{-v\sqrt{u^2+z^2}}}{\sqrt{u^2+z^2}^n}du=O(\frac{1/z^{n-1}}{\sqrt{vz}}),\tag{6.230}$$

which implies (6.228).

6.D.1. *A particular use of lemma 6.2 with theorem 6.1*

PROPOSITION 6.13.– *Let $F_1((\boldsymbol{r'}-\boldsymbol{a})^2,\varphi)=\frac{e^{-v\tilde{R}}}{\tilde{R}^n}$ on a flat real plate S', with $\tilde{R}=\sqrt{\boldsymbol{R}.\boldsymbol{R}}=\sqrt{(\boldsymbol{r'}-\boldsymbol{a})^2+z^2}$. For large $v\tilde{R}$ on S', and constants n, v, $z_1\neq 0$, we can write:*

$$\int_{S'}h_0(\frac{z_1}{\tilde{R}},\varphi)\frac{\partial(F_1((\boldsymbol{r'}-\boldsymbol{a})^2))}{\partial((\boldsymbol{r'}-\boldsymbol{a})^2)}dS'$$

$$\sim\frac{1}{2}\int_{\mathcal{L}}\Big(h_0(\frac{z_1}{\tilde{R}},\varphi)F_1((\boldsymbol{r'}-\boldsymbol{a})^2)(1-\frac{1}{v\tilde{R}}(\frac{z_1}{\tilde{R}}\frac{h_0'}{h_0}(\frac{z_1}{\tilde{R}},\varphi)))$$

$$-F_1(0)(h_0(\frac{z_1}{\tilde{R}},\varphi)-\frac{z_1}{z}\frac{h_0'(\frac{z_1}{\tilde{R}},\varphi)}{v\tilde{R}})|_{(\boldsymbol{r'}-\boldsymbol{a})^2=0}\Big)(\frac{\hat{\boldsymbol{z}}\times(\boldsymbol{r'}-\boldsymbol{a})}{(\boldsymbol{r'}-\boldsymbol{a})^2}).d\boldsymbol{r'},\tag{6.231}$$

when the functions $\frac{\partial(F((\boldsymbol{r'}-\boldsymbol{a})^2,\varphi))}{\partial((\boldsymbol{r'}-\boldsymbol{a})^2)}\equiv h_0(\frac{z_1}{\tilde{R}},\varphi)\frac{\partial(F_1((\boldsymbol{r'}-\boldsymbol{a})^2,\varphi))}{\partial((\boldsymbol{r'}-\boldsymbol{a})^2)}$, and $h_0'(\tau,\varphi)=\partial_\tau h_0(\tau,\varphi)$ with $\tau=\frac{z_1}{\tilde{R}}$, are smooth functions on S' permitting the application of *lemma 6.2. Using* $\partial_{(\boldsymbol{r'}-\boldsymbol{a})^2}(\boldsymbol{R^2}) = 2\widetilde{R}(\widetilde{R})' = 1$, *we then have:*

$$\int_{S'} h_0(\frac{z_1}{\widetilde{R}},\varphi)\frac{e^{-v\widetilde{R}}}{\widetilde{R}^{n+1}}dS'$$

$$\sim \frac{-1}{v}\int_{\mathcal{L}}(h_0(\frac{z_1}{\widetilde{R}},\varphi)\frac{e^{-v\widetilde{R}}}{\widetilde{R}^{n}} - h_0(\frac{z_1}{z},\varphi)\frac{e^{-vz}}{z^n})(\frac{\widehat{\boldsymbol{z}}\times(\boldsymbol{r'}-\boldsymbol{a})}{(\boldsymbol{r'}-\boldsymbol{a})^2}).d\boldsymbol{r'}. \tag{6.232}$$

This asymptotic expression also remains valid when $\sqrt{(\boldsymbol{r'}-\boldsymbol{a})^2}h_0'(\frac{z_1}{\widetilde{R}},\varphi)$ *has a smooth behavior on* S'.

PROOF.– We consider theorem 6.1 with

$$\frac{\partial(F((\boldsymbol{r'}-\boldsymbol{a})^2,\varphi))}{\partial((\boldsymbol{r'}-\boldsymbol{a})^2)} = h_0(\frac{z_1}{\widetilde{R}},\varphi)\frac{\partial(F_1((\boldsymbol{r'}-\boldsymbol{a})^2),\varphi)}{\partial((\boldsymbol{r'}-\boldsymbol{a})^2)}, \tag{6.233}$$

when $h_0(\frac{z_1}{\widetilde{R}},\varphi)$ is a smooth function allowing the application of lemma. In this case, we can generally use an asymptotic expression of F if F_1 has steep variation for large $\widetilde{R}$, following, as in lemma 6.2:

$$\int_{\infty}^{(\boldsymbol{r'}-\boldsymbol{a})^2}\frac{\partial F(w,\varphi)}{\partial w}dw =$$

$$= h_0(\frac{z_1}{\widetilde{R}},\varphi)F_1((\boldsymbol{r'}-\boldsymbol{a})^2,\varphi) + \int_{\infty}^{(\boldsymbol{r'}-\boldsymbol{a})^2}\frac{F_1(w,\varphi)(\widetilde{R})'}{\widetilde{R}}\frac{z_1h_0'(\frac{z_1}{\widetilde{R}},\varphi)}{\widetilde{R}}dw$$

$$\sim h_0(\frac{z_1}{\widetilde{R}},\varphi)F_1((\boldsymbol{r'}-\boldsymbol{a})^2,\varphi) + \frac{z_1h_0'(\frac{z_1}{\widetilde{R}},\varphi)}{(\widetilde{R})^2}\int_{\infty}^{(\boldsymbol{r'}-\boldsymbol{a})^2}F_1(w,\varphi)(\widetilde{R})'dw, \tag{6.234}$$

with $h_0'(\tau,\varphi) = \partial_\tau h_0(\tau,\varphi)$, $(\widetilde{R})' = \partial_w(\widetilde{R})$, $\widetilde{R} = \sqrt{w+z^2}$. We then derive (6.231) from theorem 6.1 and then (6.232). More generally, from lemma 6.2, (6.232) can also be applied when $\sqrt{(\boldsymbol{r'}-\boldsymbol{a})^2}h_0'(\frac{z_1}{\widetilde{R}},\varphi)$ has a smooth behavior.

6.D.2. *A particular use of lemma 6.2 with theorem 6.2*

PROPOSITION 6.14.– *Let* $w = (\boldsymbol{r'}-\boldsymbol{a})^2$, $\boldsymbol{R}^2 = w + z^2$ *and* $F(w,\varphi) = F_0(w)h_1(\frac{w^q}{\boldsymbol{R}^{2q}}G(\varphi))$ *be analytic relatively to* $(x'-a_x)$ *and* $(y'-a_y)$ *on a flat* *real plate S'. Considering $F_0(w) = \frac{e^{-v\widetilde{R}}}{\widetilde{R}^{n-1}}$ with $\widetilde{R} = \sqrt{\boldsymbol{R}^2}$, we have:*

$$\int_{S'} \mathbf{b}.(\boldsymbol{r'} - \boldsymbol{a})h_1(\frac{(\boldsymbol{r'} - \boldsymbol{a})^{2q}}{\boldsymbol{R}^{2q}}G(\varphi))\frac{\partial F_0((\boldsymbol{r'} - \boldsymbol{a})^2)}{\partial((\boldsymbol{r'} - \boldsymbol{a})^2)}dS'$$

$$\sim \frac{1}{2}\int_{\mathcal{L}} F((\boldsymbol{r'} - \boldsymbol{a})^2, \varphi)(\widehat{z} \times \mathbf{b}).d\boldsymbol{r'}, \qquad (6.235)$$

for large $v\widetilde{R}$ on S', with v, $\mathbf{b}$, $z \neq 0$ some constants, and n, $q \geq 1$. Generalizing this result to a vector function $\boldsymbol{h_v}$ with $F_v(w, \varphi) = F_0(w)\boldsymbol{h_v}(\frac{w^q}{\boldsymbol{R}^{2q}}G(\varphi))$, we have:

$$\int_{S'} (\boldsymbol{r'} - \boldsymbol{a}).\boldsymbol{h_v}(\frac{(\boldsymbol{r'} - \boldsymbol{a})^{2q}}{\boldsymbol{R}^{2q}}G(\varphi))\frac{\partial F_0((\boldsymbol{r'} - \boldsymbol{a})^2)}{\partial((\boldsymbol{r'} - \boldsymbol{a})^2)}dS'$$

$$\sim \frac{1}{2}\int_{\mathcal{L}} (\widehat{z} \times F_v((\boldsymbol{r'} - \boldsymbol{a})^2, \varphi)).d\boldsymbol{r'}. \qquad (6.236)$$

Let us note that $\frac{\partial F_0((\boldsymbol{r'}-\boldsymbol{a})^2)}{\partial((\boldsymbol{r'}-\boldsymbol{a})^2)} \sim (\frac{-v\widetilde{R}F_0((\boldsymbol{r'}-\boldsymbol{a})^2)}{2\boldsymbol{R}^2})$, and that these results remain valid for not constant z as $grad_{\boldsymbol{r'}}(v\boldsymbol{R}^2) = 2v(\boldsymbol{r'} - \boldsymbol{a}) + o(1)$, for large $v\widetilde{R}$ on S'.

PROOF.– Using $\frac{\partial}{\partial((\boldsymbol{r'}-\boldsymbol{a})^2)}(F_0((\boldsymbol{r'} - \boldsymbol{a})^2)) = (\frac{-v\widetilde{R}F_0((\boldsymbol{r'}-\boldsymbol{a})^2)}{2\boldsymbol{R}^2})(1 + \frac{n}{v\widetilde{R}})$ and:

$$\frac{\partial(h_1(\frac{(\boldsymbol{r'}-\boldsymbol{a})^{2q}}{\boldsymbol{R}^{2q}}G(\varphi)))}{\partial((\boldsymbol{r'} - \boldsymbol{a})^2)} = \frac{q}{\boldsymbol{R}^2}(\frac{z^2}{\boldsymbol{R}^2}\frac{(\boldsymbol{r'} - \boldsymbol{a})^{2q-2}}{\boldsymbol{R}^{2q-2}}G(\varphi)h_1'(\frac{(\boldsymbol{r'} - \boldsymbol{a})^{2q}}{\boldsymbol{R}^{2q}}G(\varphi))),$$

$$\frac{\partial(h_1(\frac{(\boldsymbol{r'}-\boldsymbol{a})^{2q}}{\boldsymbol{R}^{2q}}G(\varphi)))}{(\boldsymbol{r'} - \boldsymbol{a})^2\partial\varphi} = \frac{1}{\boldsymbol{R}^2}(\frac{(\boldsymbol{r'} - \boldsymbol{a})^{2q-2}}{\boldsymbol{R}^{2q-2}}G'(\varphi)h_1'(\frac{(\boldsymbol{r'} - \boldsymbol{a})^{2q}}{\boldsymbol{R}^{2q}}G(\varphi))), \qquad (6.237)$$

we have, for large $v\widetilde{R}$,

$$h(\frac{(\boldsymbol{r'} - \boldsymbol{a})^{2q}}{\boldsymbol{R}^{2q}}G(\varphi))\text{grad}_{\boldsymbol{r'}}(F_0((\boldsymbol{r'} - \boldsymbol{a})^2)) \sim \text{grad}_{\boldsymbol{r'}}(F((\boldsymbol{r'} - \boldsymbol{a})^2, \varphi)), \qquad (6.238)$$

which implies, following theorem 6.2, our asymptotic reduction.

We can then derive (6.235) from theorem 6.2 and use it for each component of $\boldsymbol{h_v}$ to derive (6.236).

6.E. Appendix E: complements on the use of proposition 6.2 for $s(r') = g(r')f(\mathcal{R}^2)$, when we have $f(\mathcal{R}^2) = e^{\mathcal{R}^2}$ and $g(r')$ is a smooth analytical function on S'

Let us note $\mathcal{R}^2$ a polynomial of order 2 as in (6.129) following:

$$\mathcal{R}^2 = d_0 + (\alpha_1 x' + \alpha_2 y') - ik(h_1 x'^2 + h_2 y'^2 + h_3 x'y') = d_0 + vh'^2_p + v_1, \qquad (6.239)$$

with $h'^2_p = (\boldsymbol{r'_p} - \boldsymbol{a}_p)^2$, $(\boldsymbol{r'_p} - \boldsymbol{a}_p) = \mathcal{P}(\boldsymbol{r'} - \boldsymbol{a})$. The reduction of $\int_{S'} g(r')f(\mathcal{R}^2)dS'$, when $f(\mathcal{R}^2) = e^{\mathcal{R}^2}$ has large variations on S', can be obtained from theorems 6.1 and 6.2 in combination with proposition 6.2 and complex scaling $\mathcal{P}$.

For this purpose, we present here some complements on evaluation of primitives using in parts asymptotic approximations.

PROPOSITION 6.15.– *We consider the proposition 6.2 with:*

$$\frac{\partial F((\boldsymbol{r'} - \boldsymbol{a})^2, \varphi)}{\partial(\boldsymbol{r'_p} - \boldsymbol{a_p})^2} = s(\boldsymbol{r'}) = g(\boldsymbol{r'})e^{\mathcal{R}^2}, \tag{6.240}$$

as $\boldsymbol{a} = \mathcal{P}^{-1}(\boldsymbol{a}_p)$. We then assume that, for any point $\boldsymbol{r'} \in S'$, there exists a path $\mathcal{T_L}$, in the t-complex plane from $t = 0$ to $t = 1$, defined so that $t(\boldsymbol{r'} - \boldsymbol{a}) + \boldsymbol{a}$ remains in an open simply connected subset $\mathcal{W}_s$ of the domain of analyticity of s. A primitive $F_{vs} \equiv vF$ of s relatively to $vh'^2_p = v(\boldsymbol{r'_p} - \boldsymbol{a_p})^2$ at φ_p constant, which satisfies:

$$\frac{\partial F_{vs}((\boldsymbol{r'} - \boldsymbol{a})^2, \varphi)}{v\partial(\boldsymbol{r'_p} - \boldsymbol{a_p})^2} = s(\boldsymbol{r'}) = g(\boldsymbol{r'})e^{\mathcal{R}^2}, \tag{6.241}$$

then verifies

$$F_{vs}(\boldsymbol{r'}) - F_{vs}(\boldsymbol{a}) = 2v(\boldsymbol{r'_p} - \boldsymbol{a_p})^2 e^{d_0+v_1} \int_{\mathcal{T_L}} g(t(\boldsymbol{r'} - \boldsymbol{a'}) + \boldsymbol{a'})e^{t^2 vh'^2_p} tdt. \tag{6.242}$$

PROOF.– From its definition and proposition 6.2, the function F_{vs} verifies:

$$F_{vs}(\boldsymbol{r'} = \mathcal{P}^{-1}(\boldsymbol{a}_p) + h'_p\, \boldsymbol{u}_{\varphi_p}) - F_{vs}(\boldsymbol{r'} = \mathcal{P}^{-1}(\boldsymbol{a}_p))$$

$$= \int_0^{vh'^2_p} s(\mathcal{P}^{-1}(\boldsymbol{a}_p) + h'\,\boldsymbol{u}_{\varphi_p})vdh'^2$$

$$= 2v(\boldsymbol{r'_p} - \boldsymbol{a}_p)^2 \int_{\mathcal{T_L}} g(t\mathcal{P}^{-1}(\boldsymbol{r'_p} - \boldsymbol{a}_p) + \mathcal{P}^{-1}(\boldsymbol{a}_p))\, f(d_0 + v_1 + t^2 vh'^2_p)tdt, \tag{6.243}$$

where $\boldsymbol{r'_p} - \boldsymbol{a}_p = h'_p\mathcal{P}(\boldsymbol{u}_{\varphi_p})$, $h'^2_p = (\boldsymbol{r'_p} - \boldsymbol{a}_p)^2$, $\boldsymbol{u}_{\varphi_p} = \mathcal{P}^{-1}(\cos\varphi_p\widehat{\boldsymbol{x}} + \sin\varphi_p\widehat{\boldsymbol{y}})$, and $h' = th'_p$, which finally implies (6.242).

THEOREM 6.19.– *When the proposition 6.15 applies, the surface integral $\int_{S'} g(\boldsymbol{r'})e^{\mathcal{R}^2}dS'$ can be reduced semi-asymptotically for large v, by changing in theorem 6.12, $f(\mathcal{R}^2)$ for $g(\boldsymbol{r'})e^{\mathcal{R}^2}$, $F_v(\mathcal{R}^2)$ for $F_a(\boldsymbol{r'})$ and $F_v(\frac{\boldsymbol{w}'^2_m}{4ik\xi} + d_0)$ for 0. We then obtain:*

$$F_a(\boldsymbol{r'}) = \mathcal{K}_s(\boldsymbol{r'}) + U_s(|vh'^2_p| - b_s)(\mathcal{K}_l(\boldsymbol{r'})), \tag{6.244}$$

with $vh'^2_p = v(\boldsymbol{r'_p} - \boldsymbol{a}_p)^2$, *while, letting* $u(t) = g(t\mathcal{P}^{-1}(\boldsymbol{r'_p} - \boldsymbol{a}_p) + \mathcal{P}^{-1}(\boldsymbol{a}_p))$, *we define:*

$$\mathcal{K}_l(\boldsymbol{r'}) = (u(1) - \frac{\partial_t u(1)}{2vh'^2_p})e^{(vh'^2_p + d_0 + v_1)} - (u(t_s) - \frac{\partial_t u(t_s)}{2t_s vh'^2_p})e^{(t_s^2 vh'^2_p + d_0 + v_1)},$$

$$\mathcal{K}_s(\boldsymbol{r'}) = 2vh'^2_p e^{d_0 + v_1} \int_0^{t_s} u(t)\, e^{t^2 vh'^2_p} t dt, \tag{6.245}$$

with $t_s = 1 + U_s(|vh'^2_p| - b_s)(\sqrt{\frac{b_s}{|vh'^2_p|}} - 1) \leq 1$, $U_s(x) = |^{1\, if\, x \geq 0}_{0\, if\, x < 0}$. *The number* b_s *is some chosen arbitrary large constant, and* $\mathcal{K}_l$ *is the asymptotic expression of the part of integral with* $t^2 vh'^2_p \geq b_s$ *used in (6.244) for* $|vh'^2_p| \geq b_s$.

PROOF.– For large $t^2 vh'^2_p > b_s$, we can use an integration by parts in (6.242), and remark that:

$$\begin{aligned}
&2vh'^2_p e^{d_0 + v_1} \int_{t_s}^{1} g(t(\boldsymbol{r'} - \mathcal{P}^{-1}(\boldsymbol{a}_p)) + \mathcal{P}^{-1}(\boldsymbol{a}_p))\, (e^{t^2 vh'^2_p} t) dt \\
&= e^{d_0 + v_1}(u(1)e^{vh'^2_p} - u(t_s)e^{t_s^2 vh'^2_p} \\
&- \int_{t_s}^{1} \partial_t g(t(\boldsymbol{r'} - \mathcal{P}^{-1}(\boldsymbol{a}_p)) + \mathcal{P}^{-1}(\boldsymbol{a}_p)))e^{t^2 vh'^2_p} dt) \\
&\sim e^{d_0 + v_1}((u(1) - \frac{\partial_t u(1)}{2vh'^2_p})e^{vh'^2_p} - (u(t_s) - \frac{\partial_t u(t_s)}{2vh'^2_p t_s})e^{t_s^2 vh'^2_p}).
\end{aligned} \tag{6.246}$$

Dividing the path of integration in (6.242) into two parts, we then derive (6.244).

NOTE 6.32.– From proposition 6.10, we have, when $\frac{\mp i\sqrt{h_1 h_2 - (\frac{h_3}{2})^2} + \frac{h_3}{2}}{h_2}$ are not real, two roots $\boldsymbol{r'}_\pm$ with $h'^2_p = 0$ on the real plane, and if $g(r_\pm) = g(\mathcal{P}^{-1}(a_p))$, we note that $\frac{(r'_p - a_p)(g(r') - g(\mathcal{P}^{-1}(a_p)))}{(r'_p - a_p)^2} < \infty$ since we have, as $r' \to r_\pm$:

$$(r'_p - a_p)\frac{g(r') - g(r_\pm)}{z'_{a+} z'_{a-}} = \frac{(r'_p - a_p)}{z'_{a\mp}}[\frac{(r' - r_\pm)}{z'_{a\pm}}.\text{grad}_{r'}(g(\boldsymbol{r'}))] < \infty. \tag{6.247}$$

6.F. Appendix F: complements on theorem 6.19 for some complicated case

Usual methods (as Filon's) can be used for $\mathcal{K}_s$ in (6.244) with some precaution concerning analyticity of u. For example, we can let $\int_0^{t_s} g(s_1 t + s_0)e^{t^2 vh'^2_p} t dt \simeq \sum_{0 \leq t_{s,n} \leq t_s} e^{t_{s,n}^2 vh'^2_p} \int_{t_{s,n-1}}^{t_{s,n}} g(s_1 t + s_0) t dt$, and generally, we can assume that $g(s_1 t +$

s_0) is such that, when the path from 0 to t_s is real, it does not cross any branch cut of g. However, even when cuts are present on S', we have the choice to split S' in two parts, so that we suppress the integration through the cut.

Sometimes, a closed form primitive of $tg(s_1t + s_0)$ can be used. So, for $g(\boldsymbol{r}') = \sqrt{a' + bx' + c'x'^2}$, that is, $g(s_1t + s_0) = \sqrt{a + bt + ct^2}$, the primitive $H(t) = \int^t t\sqrt{R}dt$ with $R = a+bt+ct^2 = \frac{(2ct+b)^2+\Delta'}{4c} = c\prod_{\pm}(t-t_{\pm}), t_{\pm} = \frac{b(-1\pm\sqrt{1-\frac{4ac}{b^2}})}{2c}$, $\Delta' = 4ac - b^2$, is given by (see [31, 2.262.2]):

$$H|_{c\neq 0} = \frac{1}{c^2\sqrt{c}}(\sqrt{c}\sqrt{R}(\frac{cR}{3} - \frac{(2ct+b)b}{8}) - \frac{b\Delta'}{16}\ln(\frac{2\sqrt{c}\sqrt{R} + 2ct + b}{b\sqrt{1 - \frac{4ac}{b^2}}})),$$

$$H|_{c=0,b\neq 0} = \frac{2t\sqrt{R^3}}{3b} - \frac{4\sqrt{R^5}}{15b^2}, \; H|_{c=b=0} = \frac{t^2\sqrt{a}}{2}. \tag{6.248}$$

Considering a Taylor expansion for $H|_{|c|\ll|a|,|c|\ll|b|}$ when $0 \leq t \leq 1$, we have:

$$H|_{|c|\ll|a|,|c|\ll|b|\neq 0} \sim H|_{c=0,b\neq 0} + c\frac{\sqrt{R_{c=0}}((\frac{R_{c=0}}{7} - \frac{3a}{5})R_{c=0} + a^2)R_{c=0} - a^3)}{b^4},$$

$$H|_{|c|\ll|b|\ll|a|} \sim H|_{c=0,|b|\ll|a|} + c\int\frac{t^3}{2\sqrt{a+bt}}dt - \frac{c^2}{2}\int\frac{t^5}{4(\sqrt{a+bt})^3}dt$$

$$\sim H|_{c=b=0} + b\frac{t^3}{2\sqrt{a}}(\frac{1}{3} - b\frac{t}{16a}) + c\frac{t^4}{4\sqrt{a}}(\frac{1}{2} - \frac{t}{a}(\frac{b}{5} + \frac{ct}{12})). \tag{6.249}$$

In practice, we can consider for (6.249) that $u \ll v$ if $u \leq v/20$.

NOTE 6.33.– Let $P_{2n}(z) = z^{2n} + \sum_{k=0}^{2n-1} a_k z^k = z^{2n}(1 + z^{-1}\sum_{k=0}^{2n-1} a_k z^{k+1-2n})$. For $g \equiv (P_{2n})^{1/2}$, it exists a polynomial $g_0(z) = z^n(1 + \sum_{k=1}^{n} b_k z^{-k})$ such that $g(z) = g_0(z) + O(z^{-1})$.

6.G. Appendix G: some equalities in relation with proposition 6.11

The expression of $\frac{\mathcal{P}^{-2}(\boldsymbol{d})}{\xi}$, used in proof of proposition 6.11 and given by:

$$\frac{\mathcal{P}^{-2}(\boldsymbol{d})}{\xi} = \frac{\boldsymbol{d}.\widehat{\boldsymbol{c}}_p}{p^2\xi}\widehat{\boldsymbol{c}}_p + \frac{\boldsymbol{d}.(\widehat{\boldsymbol{z}} \times \widehat{\boldsymbol{c}}_p)}{\xi}\widehat{\boldsymbol{z}} \times \widehat{\boldsymbol{c}}_p, \tag{6.250}$$

is independent of the choice of the term ϵ used in the expression of ξ. In this expression, we have:

$$
\begin{aligned}
(\mathbf{d}.\widehat{\mathbf{c}}_p)\widehat{\mathbf{c}}_p &= \frac{\mathbf{d}.(\sqrt{\xi - h_1}\widehat{x} + \frac{\epsilon_y}{\epsilon_x}\sqrt{\xi - h_2}\widehat{y})}{\epsilon\sqrt{(h_1-h_2)^2 + h_3^2}}(\sqrt{\xi - h_1}\widehat{x} + \frac{\epsilon_y}{\epsilon_x}\sqrt{\xi - h_2}\widehat{y}) \\
&= \frac{\mathbf{d}.((\xi - h_1)\widehat{x} + \frac{-h_3}{2}\widehat{y})}{(\xi - h_1)\epsilon\sqrt{(h_1-h_2)^2 + h_3^2}}((\xi - h_1)\widehat{x} + \frac{-h_3}{2}\widehat{y}) \\
&= \frac{\mathbf{d}.(\frac{-h_3}{2}\widehat{x} + (\xi - h_2)\widehat{y})}{(\xi - h_2)\epsilon\sqrt{(h_1-h_2)^2 + h_3^2}}(\frac{-h_3}{2}\widehat{x} + (\xi - h_2)\widehat{y}),
\end{aligned} \tag{6.251}
$$

and

$$
\begin{aligned}
(\mathbf{d}.(\widehat{\mathbf{z}} \times \widehat{\mathbf{c}}_p))(\widehat{\mathbf{z}} \times \widehat{\mathbf{c}}_p) &= \frac{(\xi - h_1)\mathbf{d}.(\widehat{y} - \frac{\epsilon_y}{\epsilon_x}\frac{\sqrt{\xi - h_2}}{\sqrt{\xi - h_1}}\widehat{x})}{\epsilon\sqrt{(h_1-h_2)^2 + h_3^2}}(\widehat{y} - \frac{\epsilon_y}{\epsilon_x}\frac{\sqrt{\xi - h_2}}{\sqrt{\xi - h_1}}\widehat{x}) \\
&= \frac{\mathbf{d}.((\xi - h_1)\widehat{y} - \frac{-h_3}{2}\widehat{x})}{(\xi - h_1)\epsilon\sqrt{(h_1-h_2)^2 + h_3^2}}((\xi - h_1)\widehat{y} - \frac{-h_3}{2}\widehat{x}) \\
&= \frac{\mathbf{d}.(\frac{-h_3}{2}\widehat{y} - (\xi - h_2)\widehat{x})}{(\xi - h_2)\epsilon\sqrt{(h_1-h_2)^2 + h_3^2}}(\frac{-h_3}{2}\widehat{y} - (\xi - h_2)\widehat{x}),
\end{aligned} \tag{6.252}
$$

where $\epsilon\sqrt{(h_1-h_2)^2 + h_3^2} = \xi(\mathbf{c}_p)^2_\epsilon$. So, we note, for example, that we have:

$$
\begin{aligned}
\frac{\widehat{x}\mathcal{P}^{-2}(\widehat{x})}{\xi} &= \frac{(\widehat{x}.\widehat{\mathbf{c}}_p)^2}{p^2\xi} + \frac{(\widehat{x}.(\widehat{\mathbf{z}} \times \widehat{\mathbf{c}}_p))^2}{\xi} = \frac{1}{\xi(\mathbf{c}_p)^2_\epsilon}(\frac{\xi - h_1}{p^2\xi} + \frac{\xi - h_2}{\xi}) \\
&= \frac{1}{\xi(\mathbf{c}_p)^2_\epsilon}\sum_{\pm}\frac{\pm(h_2 - h_1) + \epsilon\sqrt{(h_1-h_2)^2 + h_3^2}}{(h_1 + h_2) \mp \epsilon\sqrt{(h_1-h_2)^2 + h_3^2}} \\
&= \frac{4h_2\epsilon\sqrt{(h_1-h_2)^2 + h_3^2}}{\xi(\mathbf{c}_p)^2_\epsilon}\prod_{\pm}\frac{1}{(h_1 + h_2) \pm \epsilon\sqrt{(h_1-h_2)^2 + h_3^2}} \\
&= \frac{h_2}{(h_1h_2 - h_3^2/4)}.
\end{aligned} \tag{6.253}
$$

Index

Other titles from

iSTE

in

Waves

2024

RÉVEILLAC Jean-Michel
Synthesizers and Subtractive Synthesis 1: Theory and Overview
Synthesizers and Subtractive Synthesis 2: Application and Practice

SAKHO Ibrahima
Nuclear Physics 2: Radiochronometers and Radiopharmaceuticals
Nuclear Physics 3: Radiopharmaceuticals used in Nuclear Medicine

2022

GONTRAND Christian
Electromagnetism: Links to Special Relativity

LACAZE Pierre Camille, LACROIX Jean-Christophe
Nanotechnology and Nanomaterials for Energy

RÉVEILLAC Jean-Michel
Recording and Voice Processing 1: History and Generalities
Recording and Voice Processing 2: Working in the Studio

ROYER Daniel, VALIER-BRASIER Tony
Elastic Waves in Solids 1: Propagation

SAKHO Ibrahima
Nuclear Physics 1: Nuclear Deexcitations, Spontaneous Nuclear Reactions

VALIER-BRASIER Tony, ROYER Daniel
Elastic Waves in Solids 2: Radiation, Scattering, Generation

2021

DAHOO Pierre-Richard, LAKHLIFI Azzedine
Infrared Spectroscopy of Symmetric and Spherical Top Molecules for Space Observation 1
(Infrared Spectroscopy Set – Volume 3)
Infrared Spectroscopy of Symmetric and Spherical Top Molecules for Space Observation 2
(Infrared Spectroscopy Set – Volume 4)

2020

DANIELE Vito G., LOMBARDI Guido
Scattering and Diffraction by Wedges 1: The Wiener-Hopf Solution - Advanced Applications
(Waves and Scattering Set – Volume 1)
Scattering and Diffraction by Wedges 2: The Wiener-Hopf Solution - Advanced Applications
(Waves and Scattering Set – Volume 2)

SAKHO Ibrahima
Introduction to Quantum Mechanics 2: Wave-Corpuscle, Quantization & Schrödinger's Equation

2019

BERTRAND Pierre, DEL SARTO Daniele, GHIZZO Alain
The Vlasov Equation 1: History and General Properties

DAHOO Pierre-Richard, LAKHLIFI Azzedine
Infrared Spectroscopy of Triatomics for Space Observation
(Infrared Spectroscopy Set – Volume 2)

RÉVEILLAC Jean-Michel
Electronic Music Machines: The New Musical Instruments

ROMERO-GARCIA Vicente, HLADKY-HENNION Anne-Christine
Fundamentals and Applications of Acoustic Metamaterials: From Seismic to Radio Frequency
(Metamaterials Applied to Waves Set – Volume 1)

SAKHO Ibrahima
Introduction to Quantum Mechanics 1: Thermal Radiation and Experimental Facts Regarding the Quantization of Matter

2018

SAKHO Ibrahima
Screening Constant by Unit Nuclear Charge Method: Description and Application to the Photoionization of Atomic Systems

2017

DAHOO Pierre-Richard, LAKHLIFI Azzedine
Infrared Spectroscopy of Diatomics for Space Observation
(Infrared Spectroscopy Set – Volume 1)

PARET Dominique, HUON Jean-Paul
Secure Connected Objects

PARET Dominque, SIBONY Serge
Musical Techniques: Frequencies and Harmony

RÉVEILLAC Jean-Michel
Analog and Digital Sound Processing

STAEBLER Patrick
Human Exposure to Electromagnetic Fields

2016

ANSELMET Fabien, MATTEI Pierre-Olivier
Acoustics, Aeroacoustics and Vibrations

BAUDRAND Henri, TITAOUINE Mohammed, RAVEU Nathalie
The Wave Concept in Electromagnetism and Circuits: Theory and Applications

PARET Dominique
Antennas Designs for NFC Devices

PARET Dominique
Design Constraints for NFC Devices

WIART Joe
Radio-Frequency Human Exposure Assessment

2015

PICART Pascal
New Techniques in Digital Holography

2014

APPRIOU Alain
Uncertainty Theories and Multisensor Data Fusion

JARRY Pierre, BENEAT Jacques N.
RF and Microwave Electromagnetism

LAHEURTE Jean-Marc
UHF RFID Technologies for Identification and Traceability

SAVAUX Vincent, LOUËT Yves
MMSE-based Algorithm for Joint Signal Detection, Channel and Noise Variance Estimation for OFDM Systems

THOMAS Jean-Hugh, YAAKOUBI Nourdin
New Sensors and Processing Chain

TING Michael
Molecular Imaging in Nano MRI

VALIÈRE Jean-Christophe
Acoustic Particle Velocity Measurements using Laser: Principles, Signal Processing and Applications

VANBÉSIEN Olivier, CENTENO Emmanuel
Dispersion Engineering for Integrated Nanophotonics

2013

BENMAMMAR Badr, AMRAOUI Asma
Radio Resource Allocation and Dynamic Spectrum Access

BOURLIER Christophe, PINEL Nicolas, KUBICKÉ Gildas
Method of Moments for 2D Scattering Problems: Basic Concepts and Applications

GOURE Jean-Pierre
Optics in Instruments: Applications in Biology and Medicine

LAZAROV Andon, KOSTADINOV Todor Pavlov
Bistatic SAR/GISAR/FISAR Theory Algorithms and Program Implementation

LHEURETTE Eric
Metamaterials and Wave Control

PINEL Nicolas, BOURLIER Christophe
Electromagnetic Wave Scattering from Random Rough Surfaces: Asymptotic Models

SHINOHARA Naoki
Wireless Power Transfer via Radiowaves

TERRE Michel, PISCHELLA Mylène, VIVIER Emmanuelle
Wireless Telecommunication Systems

2012

LALAUZE René
Chemical Sensors and Biosensors

LE MENN Marc
Instrumentation and Metrology in Oceanography

LI Jun-chang, PICART Pascal
Digital Holography

2011

BECHERRAWY Tamer
Mechanical and Electromagnetic Vibrations and Waves

BESNIER Philippe, DÉMOULIN Bernard
Electromagnetic Reverberation Chambers

GOURE Jean-Pierre
Optics in Instruments

GROUS Ammar
Applied Metrology for Manufacturing Engineering

LE CHEVALIER François, LESSELIER Dominique, STARAJ Robert
Non-standard Antennas

2010

BEGAUD Xavier
Ultra Wide Band Antennas

MARAGE Jean-Paul, MORI Yvon
Sonar and Underwater Acoustics

2009

BOUDRIOUA Azzedine
Photonic Waveguides

BRUNEAU Michel, POTEL Catherine
Materials and Acoustics Handbook

DE FORNEL Frédérique, FAVENNEC Pierre-Noël
Measurements using Optic and RF Waves

FRENCH COLLEGE OF METROLOGY
Transverse Disciplines in Metrology

2008

FILIPPI Paul J.T.
Vibrations and Acoustic Radiation of Thin Structures

LALAUZE René
Physical Chemistry of Solid-Gas Interfaces

2007

KUNDU Tribikram
Advanced Ultrasonic Methods for Material and Structure Inspection

PLACKO Dominique
Fundamentals of Instrumentation and Measurement

RIPKA Pavel, TIPEK Alois
Modern Sensors Handbook

2006

BALAGEAS Daniel *et al.*
Structural Health Monitoring

BOUCHET Olivier *et al.*
Free-Space Optics

BRUNEAU Michel, SCELO Thomas
Fundamentals of Acoustics

FRENCH COLLEGE OF METROLOGY
Metrology in Industry

GUILLAUME Philippe
Music and Acoustics

GUYADER Jean-Louis
Vibration in Continuous Media

Printed and bound by CPI Group (UK) Ltd, Croydon, CR0 4YY
13/09/2026
14961042-0008